D1090827

HVAC PRINCIPLES AND APPLICATIONS MANUAL

Other HVAC Titles of Interest

HVAC PRINCIPLES AND APPLICATIONS MANUAL

Thomas E. Mull, P.E.

McGRAW-HILL

New York St. Louis San Francisco Auckland Bogotá
Caracas Lisbon London Madrid Mexico Milan
Montreal New Delhi Paris San Juan São Paulo
Singapore Sydney Tokyo Toronto

Library of Congress Cataloging-in-Publication Data

Mull, Thomas E.
    HVAC principles and applications manual / Thomas E. Mull.
        p.   cm.
    Includes index.
    ISBN 0-07-044451-X (alk. paper)
    1. Heating--Equipment and supplies--Design and construction.
2. Air conditioning--Equipment and supplies--Design and
construction.   3. Ventilation--Equipment and supplies--Design and
construction.   I. Title.
    TH7345.M85   1997
    697--dc21   97-20717
                                                                                                              CIP

McGraw-Hill

A Division of The McGraw·Hill Companies

1 2 3 4 5 6 7 8 9 0   DOC/DOC   9 0 2 1 0 9 8 7

ISBN 0-07-044451-X

*The sponsoring editor for this book was Harold B. Crawford and the production
supervisor was Sherri Souffrance.
It was set in Times Roman by Douglas & Gayle, Limited.*

Printed and bound by R. R. Donnelley & Sons Company.

McGraw-Hill books are available at special quantity discounts to use as premiums
and sales promotions, or for use in corporate training programs. For more
information, please write to the Director of Special Sales, McGraw-Hill,
11 West 19th Street, New York, NY 10011. Or contact your local bookstore.

The book is printed on recycled, acid-free paper containing a minimum of 50%
recycled, de-inked fiber.

DEDICATION

This book is dedicated to my wife Norma, my late father Elmer, and my late mother Marie Mull. Without their inspiration and support, this book would not have been possible.

PREFACE

This manual is intended as an introduction to heating, ventilating, and air conditioning (HVAC) design. The major emphasis is on the basics of designing HVAC systems. The methods and procedures presented here are based upon those recommended by the American Society of Heating, Refrigerating, and Air-Conditioning Engineers, Inc. (ASHRAE) as well as other methods that are recognized industry wide.

Many books and manuals on HVAC design that are available today are intended, primarily, as handbooks or reference books for degreed engineers practicing HVAC design. This manual is intended for engineers not already familiar with HVAC design, non-degreed people who are currently working in the HVAC design field, people who are considering entering the field, or people in related fields who have an interest in HVAC design. It is intended for readers who have at least completed high school and, perhaps, are at the community college level. While every attempt has been made to keep the mathematics as simple as possible, readers should have at least completed high school algebra and trigonometry. Readers should also have some background and interest in science and technology.

Chapter 1, *Introduction to HVAC Design*, is a broad brush look at HVAC design and some of the major considerations involved in HVAC design. The reader is introduced to some of the major issues involved when selecting and designing systems.

Chapter 2, *Basic Scientific Principles*, includes definitions and an introduction to the basic scientific principles involved in heating and air conditioning. The reader is introduced to basic thermodynamics, heat transfer, and fluid mechanics.

Chapter 3, *Climatic Conditions*, begins the actual HVAC design process by discussing climatic conditions necessary for sizing HVAC equipment and some considerations regarding climatic conditions.

Chapter 4, *Building Heat Transmission Surfaces*, instructs the reader on information regarding the building itself, such as heat transmission areas for various surfaces in a building. This chapter also instructs the reader on the procedure for calculating heat transmission factors.

Chapter 5, *Infiltration and Ventilation*, includes discussion of infiltration, ventilation, infiltration estimation methods, and applicable ventilation codes.

Chapter 6, *Heating Loads*, provides the reader with the step by step procedure for calculating the heat loss rate for a building, including transmission heat losses, infiltration heat losses, and ventilation heat losses.

Chapter 7, *External Heat Gains and Cooling Loads*, introduces the reader to external heat gains and cooling loads for buildings. It begins with a discussion of non-steady state heat transfer and latent heat gains. The chapter goes on to provide step by step procedures for calculating cooling loads for walls, roofs, and glass.

Chapter 8, *Internal Heat Gains and Cooling Loads*, introduces the reader to internal heat gains and cooling loads for a building. The chapter provides a step by step procedure for calculating cooling loads for lights, occupants, equipment, etc. using the cooling load factor method.

Chapter 9, *HVAC Psychrometrics*, introduces the reader to psychrometrics. It instructs the reader on the use of a psychrometric chart, how to perform psychrometric calculations using the chart, and how to calculate the necessary air quantities required for given cooling loads.

Chapter 10, *Overview of HVAC Systems*, is a broad brush look at major components and equipment used in building heating and cooling systems. The reader is introduced to equipment such as boilers, chillers, rooftop air conditioners, etc.

Chapter 11, *Codes and Standards for HVAC Systems Design*, introduces the reader to the codes and standards with which a designer must be familiar to successfully design a safe HVAC system.

Chapter 12, *Acoustics and Vibration*, introduces the reader to acoustics and vibrations which are important considerations in the design of HVAC systems.

Chapter 13, *Human Comfort*, is a discussion of human comfort which usually is the major objective of the HVAC system.

Chapter 14, *Air Distribution*, discusses the requirements of room air distribution and the selection of air distribution devices to provide adequate air distribution in the conditioned space.

Chapter 15, *Duct System Design*, introduces the reader to various types of air distribution systems and the various methods used to size and design duct systems.

Chapter 16, *Fans and Central Air Systems*, is a discussion of the various types of fans and their application to HVAC systems.

Chapter 17, *Air System Heating and Cooling*, includes a discussion of various types of heating and cooling coils and the selection of coils for HVAC systems.

Chapter 18, *Air Cleaning and Filtration*, informs the reader about the various types of air filtration and air cleaning devices and how they may be used to control indoor air quality.

Chapter 19, *Introduction to Electrical Systems*, is an overview of electrical systems and electrical equipment, such as motors, used in HVAC systems.

Chapter 20, *Controls for Air Distribution Systems*, is a discussion of controls for HVAC systems. Various control components are discussed as well as basic control systems for air systems.

CONTENTS

HVAC PRINCIPLES AND APPLICATIONS MANUAL

Chapter 1

INTRODUCTION TO HVAC DESIGN

When designing and selecting a heating, ventilating, and air-conditioning (HVAC) system for a building, whether for residential, office, commercial, industrial, or whatever type, there are a number of general considerations that the designer should bear in mind. The system should provide a comfortable environment for the occupants, in addition to meeting any special process requirements, such as special humidity control. The system should operate efficiently, not waste energy, or be unnecessarily expensive to operate. The designer must be aware of indoor air quality and attempt to reduce the possibility of a buildup in indoor air pollutants.

An air conditioner may be defined as any apparatus controlling and, especially, lowering the temperature and humidity of an enclosure. In most cases, the enclosure is a building or a space within a building. By controlling the air temperature and humidity, an HVAC system is attempting to provide a comfortable indoor environment for the building's occupants, or a controlled temperature and humidity for a process in a space. The HVAC system attempts to maintain the desired indoor conditions, despite changes in outdoor weather conditions, indoor usage, or activity within the space. As an example, a system is typically required to maintain an indoor temperature between 70°F and 78°F when the outdoor temperature is 0°F and when it is 100°F. The system must also maintain the desired conditions when the space is empty, or when it is occupied with many people and there are many interior heat gains, such as lights and equipment. It may also be necessary for the system to maintain the indoor relative humidity between a minimum of 30% and a maximum of 50%. The system must accomplish all these tasks without being unreasonably costly to install and operate.

In addition to maintaining comfortable temperature and humidity conditions in a space, an HVAC system must also provide a healthy indoor environment. It has been estimated that people in the United States spend 65% to 90% of their time indoors. With so much time spent indoors, the quality of the air we breathe has a significant effect on our health. Unfortunately, the indoor air we breathe is frequently much more polluted than the outdoor air. Construction materials such as paints, plastics, carpeting, wall coverings, etc., are major sources of indoor air pollution. When designing an HVAC system, the designer should always be aware of any pollution causing materials, systems, equipment or processes. The HVAC system design should attempt to minimize indoor air pollution. Most building codes provide guidelines for outdoor ventilation air which must be brought into a building in order to dilute indoor air pollutants. In addition to bringing in outdoor air, which may contain some pollutants, steps must be taken in the design of the HVAC system to minimize air pollutants from all known sources.

Another important consideration when selecting and designing HVAC systems is the need for energy conservation and efficiency. The availability, cost, and probable consumption of various fuel types are important considerations in the selection and design of HVAC systems. HVAC systems frequently use over 60% of the energy consumed in commercial and institutional buildings. It is important that systems be selected and designed with energy efficiency in mind.

Finally, HVAC systems must be designed and installed with safety in mind. All HVAC systems must provide for the safety of the building's occupants, as well as others. Generally, safety guidelines for HVAC systems, as well as buildings in general, are provided in various building codes and standards. Codes impose requirements on new construction and renovation of existing facilities for the protection of public health, safety, and welfare. Usually, the applicable codes are enforced by local governments, such as city or county departments of public works.

With the basic requirements of the HVAC system in mind, the designer must select the type of system that best suits the needs of the building and its occupants. The designer should consider the relative advantages and disadvantages of each type of system.

1.1 PROVIDING A COMFORTABLE INDOOR ENVIRONMENT

Air conditioning is the simultaneous control of temperature, humidity, air movement, and quality of air within a space. The HVAC system attempts to maintain desired indoor conditions despite changes in outdoor weather, and indoor activity, or usage of the space. The use of the space, as well as the heating and cooling loads, determines the requirements of the HVAC system. As an example, a computer room may require special temperature and humidity conditions, where an office may only require comfortable conditions for the occupants.

Comfort may be defined as any condition, which, when changed, will make a person uncomfortable. While many things may make a person uncomfortable, such as an uncomfortable chair to sit in, we are primarily concerned with air temperature, humidity, air movement, and indoor air quality within a space. People must be kept comfortable with varying types of clothing and with various types of activity, such as sitting or exercising. Individual comfort depends on how fast the body is losing heat. The objective of the HVAC system is to assist the body in controlling its cooling rate in both summer and winter outdoor conditions. In the summer, it is necessary to increase the body's cooling rate while in winter, it is necessary to decrease the body's cooling rate.

The human body loses or gains heat from the surroundings by three different methods —sensible heat loss or gain, latent heat loss, and radiant heat loss or gain. Sensible heat loss or gain occurs as a result of direct contact of the skin with the surrounding air. The amount of sensible heat transferred depends upon the difference in temperature between the skin and the surrounding air. As the *difference* in temperature between the skin and the air decreases, the amount of sensible heat lost or gained by the body decreases. When the surrounding air temperature is around 70°F, most people lose sensible heat at a rate such that they are comfortable. If the air temperature rises to 80°F, the sensible heat loss from the body approaches zero and people begin to feel hot. If the air temperature rises above 80°F, the human body will begin absorbing or gaining heat.

The human body also loses heat through latent heat loss or perspiration. Latent heat is lost by the body as moisture evaporates. As the moisture evaporates, or changes from a liquid to a vapor, heat is absorbed from the body and carried away by the vapor. At 80°F, sensible heat loss by the body is negligible; however, the latent heat loss from the body can be around 400 Btu/hr. As the surrounding air temperature increases, the body is able to continue losing heat through increased perspiration. If the body did not give off latent heat, it would overheat whenever the ambient air temperature rose above 80°F.

The third way that the human body may gain or lose heat is by radiation. Radiant heat transfer does not require direct contact between the body and the object with which the body is exchanging heat. An example of radiant heat gain is the feeling of warmth when the sun is shining on your face on a cold sunny day. There is no direct contact between the sun and the skin; however, heat is still received from the sun. In order for radiant heat exchange to occur, the body must "see" the object with which it is exchanging heat. It must not be blocked or shaded by another object.

1.2   SYSTEM HEATING AND COOLING LOADS

When designing and selecting HVAC systems, the first and most basic question that must be answered is that of the required capacities of the systems. It is important that the heating loads, cooling loads, and air flow requirements be determined before a final system type is selected. A brief overview of system heating loads and cooling loads is presented here. Detailed methods for determining the heating, cooling, and ventilation requirements of HVAC systems are presented later in this book.

Heat is a form of energy that, when added to a substance, causes the substance to rise in temperature, fuse, evaporate, or expand. Heat occurs in two forms—sensible heat and latent heat. Sensible heat is heat that changes the temperature of the substance. Latent heat is heat that causes a change in state or phase change of the substance. The standard unit of measurement for heat is the British thermal unit (Btu) and the standard for the rate at which heat is transferred is Btu per hour (Btu/hr). Heat is discussed in more detail in Chapter 2 and Chapters 4 through 9.

The heating and cooling loads are actually the rate at which energy, in the form of heat, is transferred to or from the conditioned space. If heat is lost from the conditioned space, there is a heating load. If heat is added to the space from sources other than the HVAC system, the load is a cooling load for the system. Ideally, the HVAC system should supply heat to a space at the same rate at which it is lost. It should also remove heat from the conditioned space at the same rate that heat is entering the space or being released within the space. If heat is added or removed from the conditioned space at the same rate as the space heat loss or cooling load, then the space will be in equilibrium and the space temperature will remain constant. However, this situation rarely occurs in actual applications.

1.3   ZONING

A zone is a space with uniform thermal loads (heating or cooling loads) or thermal characteristics. A zone may be defined as an area of a building (actually a volume) that is considered as one thermal unit. It is a space that is to be maintained at a uniform space temperature. Heating and cooling loads are usually determined on a per zone basis. When a building is zoned, it is divided into separately controlled spaces in which different conditions may be simultaneously maintained.

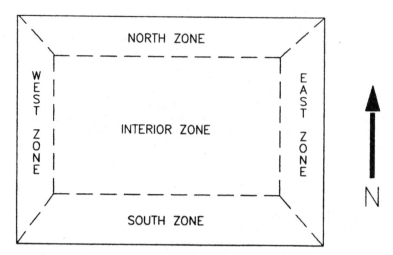

FIGURE 1.1   Zoned Rectangular Building

Each zone usually has a thermostat located within the zone to control the temperature of that zone. Ideally, the areas served by each thermostat in a building would constitute a zone. Realistically, however, it is necessary to combine those areas in a building with similar load profiles into a zone. Most buildings, larger than single family residences, have more than one zone. Even large single family residences have multiple zones. As an example, the heating and cooling loads for the top floor of a two story house usually has a different thermal load from the first floor. The zoning of a typical rectangular one story building is shown in Figure 1.1.

There is not one fixed, or correct, way to zone a building. It must be done on the basis of experience and judgment. The zones of a building must reflect both architectural and HVAC system configurations. It is typical to have one core, or interior, zone and one or more exterior zones. Perimeter zones have exterior walls and windows. The loads in perimeter zones are strongly driven by outdoor temperatures and solar gains. Exterior areas are often divided into zones according to their exposure (i.e., north, south, east, and west). This is due to the fact that the amount of solar heat gain varies widely from face-to-face throughout the day and throughout the year.

An interior zone, however, typically has relatively constant interior gains—occupants, lights, etc.—and little, or no, transmission losses or gains. Interior zones typically have only one exterior surface, if any, which is a roof. Interior zones tend to have relatively constant loads since they are isolated from exterior influences. They usually require cooling throughout the year, with the possible exception of an interior zone with a roof.

Buildings may also be divided, or zoned, according to their usage, occupancy, and activity level of the occupants. For example, a south-facing office may have a different load profile than a south-facing computer room, with much heat generating equipment, or conference room with many occupants.

The zoning of a building has a significant effect on system selection and design. As will be seen in later chapters, it is sometimes difficult to have one system serve a building with multiple zones. In order to account for variable zone loads, the HVAC system designer must choose between a number of system types.

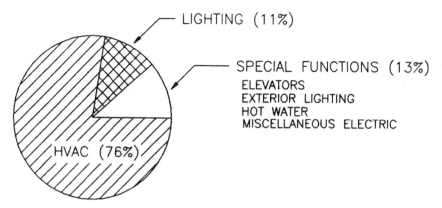

FIGURE 1.2   Energy Consumption in Office Type Buildings

1.4   THE NEED FOR ENERGY CONSERVATION

The availability, cost, and probable consumption of various fuel types are important considerations in the selection of the type of HVAC system to use in a building. HVAC systems frequently use over 60% of the energy consumed in buildings today; thus it is important that systems be selected and designed with energy conservation in mind. As can be observed in Figure 1.2, HVAC systems typically consume over 75% of the energy used in office type buildings.

Of course, this percentage will vary with the location or climate. The consumption percentages may also vary slightly with building usage; however, it can easily be observed that space conditioning consumes the largest amount of energy.

Figure 1.3 indicates the relative energy consumption for educational buildings.

Again, it can easily be observed that the HVAC systems are the major consumers of energy in the buildings. Although the percentages may vary slightly for different climatic zones, HVAC systems account for about 65% of the energy consumed in educational buildings.

As can be observed from Figures 1.4 and 1.5, HVAC systems also are large consumers of energy in long-term care facilities, such as nursing homes, and hospitals.

Again, it is evident that HVAC systems are the major consumers of energy at around 60% in both long-term care and hospital type buildings. While these are a few typical examples, it may readily be seen that HVAC systems are the major consumers of energy in most buildings. Unfortunately, most HVAC systems are designed and selected to produce the desired space conditions at minimum first cost or installed cost. Operating costs are frequently secondary.

Frequently, HVAC systems are designed using excessively conservative methods, resulting in over-sized systems. Most systems operate at maximum efficiency at, or near, their maximum load or capacity. Over-sized systems can be as great a problem as under-sized systems. HVAC systems are frequently sized based on peak heating and cooling loads which occur only a few hours each year. During most of the service life of an HVAC system, the system is operating at considerably less than peak design and less than peak efficiency.

The question of energy conservation in HVAC systems is not really a technical question, but more an economic question. When considering the economic costs of HVAC systems,

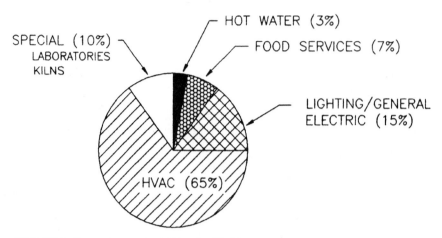

FIGURE 1.3   Energy Consumption in Educational Buildings

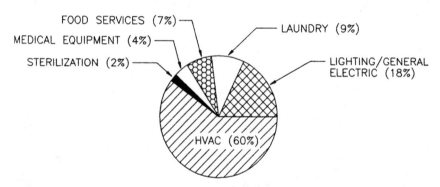

FIGURE 1.4   Energy Consumption in Long-Term Care Facilities

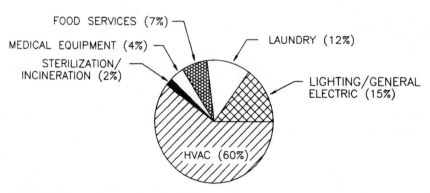

FIGURE 1.5   Energy Consumption in Hospitals

it is best to look at systems on the basis of life cycle cost, or the total cost to own and operate for the entire life of the system. First cost, or installation cost, is only one factor in the economic equation.

In addition to economic considerations, there are code considerations affecting HVAC systems. Many codes and standards set maximum energy consumption budgets for buildings in Btu per square foot (Btu/ft^2) of floor space per year. It is possible to deviate from these energy budget guidelines, in the design of a building, if it can be shown that the deviation will result in no additional energy usage. Any combinations of building design, construction materials, and HVAC system types may be used as long as the maximum budget figure is not exceeded. The most widely used energy conservation standards are published by the American Society of Heating, Refrigerating, and Air Conditioning Engineers (ASHRAE). ASHRAE standards provide minimum guidelines for energy conservation design and operation. There are two types of energy standards:

1. Prescriptive standards which specify the materials and methods for the design and construction of buildings.

2. Component performance standards which set requirements for each component, system or subsystem within a building.

When designing HVAC systems for buildings, the reader should consult ASHRAE Standard 90, "Energy Conservation on New Building Design" or ASHRAE Standard 100-P, "Energy Conservation in Existing Buildings."

1.5   INDOOR AIR QUALITY

It is estimated that people in the United States spend 65% to 90% of their time indoors. With so much time spent indoors, the quality of the indoor air we breathe has a significant affect on our health. Unfortunately, the indoor air we breathe is frequently much more polluted than outdoor air. Generally speaking, if more than 20% of a building's occupants complain of symptoms, such as headaches, breathing problems, excessive colds, etc., and no specific cause can be found, the building is probably suffering from the "sick building syndrome" (poor indoor air quality). The National Institute of Occupational Safety and Health (NIOSH) investigated several hundred buildings and determined that 50% of occupants' complaints were due to building material contaminants and 11% were due to outdoor contaminants. It is estimated that the number of "sick" buildings in the United States is between 20% and 30% of all existing buildings. Health complaints include itchy eyes, headaches, rashes, nausea, chronic fatigue, and respiratory problems. When a building is suffering from the sick building syndrome, productivity of its occupants can drop off as much as 5%.

Construction materials, such as paints, plastics, carpeting, and wall coverings, are major sources of indoor air pollution. Approximately 20% of all commercial type buildings contain asbestos. Formaldehyde is used in over 3000 building products. Accumulated dust and moisture can become "environmental niches" that promote the growth of microorganisms or molds. These can be drawn into the air stream and circulated throughout the building by the air distribution system. Environmental niches combine organisms with near optimum temperature, humidity, and moisture to cause concentrations to grow very rapidly. An example is Legionnaires' Disease, which is associated with standing water. Another common and obvious source of indoor air pollution is cigarette smoke. Since most air distribution systems recirculate air within the building, isolation of smokers and other sources of pollution is not very effective.

Although indoor air quality is an important consideration when selecting and designing HVAC systems, designers often neglect to give proper attention to IAQ. System designers are frequently so concerned about temperature control and installed cost, they tend to forget about indoor air quality. In some variable air volume systems, the reduced need for cooling can reduce the amount of air flow to a space to such an extent that a stuffy, and possibly unhealthy, environment can result. Air filtration is not always effective in reducing indoor air pollution, particularly gaseous air pollution.

When designing an HVAC system, the designer should always be aware of any pollution causing materials, systems, equipment, or processes within a building. The design should attempt to minimize indoor air pollution. Many building mechanical codes, such as the Building Officials and Code Administrators (BOCA) Code, prescribe minimum outdoor air requirements for buildings to control indoor air pollution. The ASHRAE Standard 62 provides guidelines for minimum outdoor ventilation brought into a building to dilute indoor air pollutants. Codes and standards for HVAC systems are discussed in Chapter 11.

In addition to bringing outdoor air into a building, which may contain some pollutants, steps can be taken in the design of the HVAC system to minimize indoor pollutants. Electronic filters and other types of high efficiency filters can remove more than 90% of airborne particles; however, they cannot remove gaseous contaminants. In order to reduce indoor air pollution, the designer should consider the following:

- Consider differential pressure control within the building to reduce contamination from one area to other spaces (e.g., a smoking room).
- Design the air distribution system to exclude materials that promote contaminants such as non-metallic duct linings.
- Avoid standing water in condensate drains, equipment, ducts, etc.
- Install efficient upstream filtration.
- Provide adequate space for maintenance of all equipment.
- Fully balance air systems to provide designated air flow rates.
- Avoid locating ceiling supply and return grilles in close proximity to each other, causing the air to "short circuit."
- Locate outdoor air intakes so there will be no re-entry of exhaust fumes from toilet exhausts, plumbing, vent stacks, cooling towers, vehicle exhaust, etc.
- Air systems should be designed so they are easily balanced.
- Pay particular attention to reduced air flow rates in variable air volume systems.

1.6 BASIC HVAC SYSTEM TYPES

There are several major heating, ventilating, and air conditioning system types in widespread use today. These are air systems, hydronic and steam systems, and unitary type systems. Most systems in use today fall into one of these categories, or are a combination or variation of them. Each type of system has advantages and disadvantages.

1.6.1 Air Systems

Air type HVAC systems are systems that provide heating, ventilation, and cooling by circulating air through spaces, or rooms, to maintain the desired conditions in the space. A typical air system generally consists of a central air handling unit, with heating and/or cooling capabilities, and an air distribution system. The air distribution system generally consists of supply air ductwork to the space or room, return air ductwork from the space back to the air handling unit, and air distribution devices in the room, such as grilles and registers that provide desirable air patterns within the room. Heating of the air in the air handling unit (AHU) is usually accomplished with hot water coils, steam coils, gas furnaces, or electric heating coils. Cooling and dehumidification is generally accomplished by passing the air through direct expansion refrigerant coils, chilled water cooling coils, or by the introduction of outdoor air when conditions are favorable.

Air systems offer a number of advantages over other types of systems. These advantages include:

- Central location of major equipment components.
- Provide ventilation directly to the space.
- Easy to use outdoor air for cooling when conditions are favorable (economizer).
- Widely used for HVAC systems with many variations to choose from, i.e., multizone, variable air volume, dual duct, etc.

Air type systems also have a number of disadvantages when compared to other system types. These disadvantages include:

- Central air handling unit (AHU) can require a large amount of space and ductwork can require large amounts of space above ceilings.
- It is difficult to efficiently provide heating to one space while providing cooling to another space served by the same system.
- Air flows into spaces need to be properly balanced to obtain the design air flow rates.
- Air systems may tend to over cool and cause draftiness in the space.
- Air systems can be quite noisy.

1.6.2 Hydronic and Steam Systems

Hydronic systems are systems that circulate water as a circulating medium for heating and/or cooling. Hydronic systems circulate hot water for heating and chilled water for cooling. Steam heating systems are similar to heating water systems, except steam is used as the heating medium instead of water.

Hydronic heating systems generally consist of a boiler, hot water circulating pumps, distribution piping, and a fan coil unit or a radiator located in the room or space. Steam heating systems are similar except no circulating pumps are required.

Hydronic cooling systems consist of a water chiller, circulating pumps, distribution piping, and a fan coil unit in the space or room. Chilled water systems usually require a cooling tower or some other device to reject heat to the outdoors.

Hydronic systems have several advantages over other types of systems. These include:

- Individual room temperature control is easily achieved.
- The same distribution piping system can sometimes be used for heating water and chilled water if simultaneous heating and cooling is not required.
- The distribution piping requires considerably less space than ductwork.
- System components have a long service life, between 15 to 20 years.
- Heating is usually supplied at the perimeter of the building, where it is most needed.

Hydronic systems also have several disadvantages, which include:

- It is difficult to provide adequate ventilation to rooms.
- It is difficult to take advantage of outdoor air cooling when conditions are favorable (economizer).
- Fan coil units in the occupied space can be noisy.
- An elaborate condensate drainage system may be required for individual fan coil cooling units.
- Hydronic cooling is usually not economical for smaller buildings.
- Fan coil units take up space in the room.
- It may be difficult to control humidity levels in the rooms.
- Hydronic systems may be subject to freezing when used in air handling units with outdoor air.

1.6.3   Unitary Systems

Unitary systems consist of self-contained units that heat and/or cool a single space. Each space may have its own unitary system. The units are self contained and produce heating and/or cooling with the application of the proper energy source, usually electricity.

Cooling units are usually completely self contained, including cooling coils, compressors, refrigerant piping, condenser coils, and controls. Unitary cooling units need to be located near an outside wall or roof so the refrigeration system can reject heat to the outdoors.

Unitary heating is usually accomplished with electric heating coils or by a gas furnace within the unit. Occasionally, heating is accomplished with a heat pump cycle.

Examples of unitary equipment include window air conditioners, through-the-wall air conditioners and heat pumps, package rooftop air conditioners, electric baseboard radiators, and water source heat pumps.

Unitary systems have several advantages over other types of systems. These include:

- Lower installation costs.
- Individual control in each room.
- Units installed on perimeter of building can easily bring ventilation air into building.
- Simple, easy to use temperature controls.
- Easy to service and maintain.

Unitary systems have several disadvantages which include:

- Units with refrigeration compressors are usually quite noisy.
- Precise temperature and humidity control is difficult, if not impossible.

- Cooling units usually must be located near an outside wall or on the roof in order to reject heat to the atmosphere.
- Units can be unsightly and aesthetically unpleasant.
- Condensate drainage can be a problem with some cooling units.
- Many units can be very time consuming to maintain.
- Compressors have a relatively short service life, between 5 to 10 years.

1.7   SYSTEM SELECTION CRITERIA

The decision of which type of system is best suited for a particular application is based upon several selection criteria. Some of the major considerations are discussed herein. No one type of system will be best suited for every application. Each application must be considered on an individual basis. Some of the major considerations are as follows:

- The heating, cooling and ventilation loads, and requirements for the conditioned space.
- Installed Cost: Some building owners are primarily concerned with initial installed costs. Some may have specialized operating requirements, such as a computer room, while others are more concerned about operating costs and are willing to pay for a higher performance system.
- Operating Cost: Some systems are inherently more costly to operate than others.
- Individual Temperature Control: Some systems can more readily provide individual room or zone temperature control than others.
- Adequacy of Ventilation: Some systems, by their very nature, provide better ventilation than others.
- Specialized Temperature and Humidity Control: Areas, such as computer rooms and special manufacturing processes, frequently require precise temperature and humidity control.
- Maintenance Requirements: Consideration should be given to the capabilities of the owner's maintenance staff.
- Aesthetics: Some owners are very conscious of the appearance of the building, both inside and outside.
- Space requirements: Consideration must be given to space requirements of the HVAC system, including space for maintenance, and how much space is likely to be available for the system.
- Availability of Equipment: Some equipment is "off-the-shelf," whereas others may have long lead times to manufacture or may even have to be custom built.
- Climate: Some systems, such as air source heat pumps, are better suited for one climate than another.
- Availability of Energy Sources: Some fuel sources, such as natural gas, are not always available at the building site.
- Acoustics: It is always important to consider the amount of noise and vibration generated by various HVAC system types. Some applications are more sensitive than others.
- Complexity: A system should be selected that can adequately handle the requirements of the building, yet not be so overly complex that it is difficult to have the desired performance.

Chapter 2

BASIC SCIENTIFIC PRINCIPLES

Before a heating and air conditioning system can actually be designed, the heating and cooling loads, or requirements, must first be determined. The design of HVAC systems is actually the application of science and scientific principles, along with technology, to achieve a desired result. In this case, the desired result is the control of the indoor environment. In order to understand how HVAC systems are designed and how they actually operate, it is first necessary to understand the basic underlying scientific principles. Among the physical sciences that are essential to HVAC design are the fundamental sciences of thermodynamics, heat transfer, and fluid mechanics.

This chapter will provide an overview of the scientific principles used in HVAC design. It is assumed that you already have a basic understanding of the underlying scientific principles such as physics and chemistry.

2.1 THE CONCEPT OF HEAT AND WORK

Energy can be defined as the capacity to produce an effect or the capacity for doing work. It can also be defined as the potential for providing useful work or heat. Energy can neither be created nor destroyed, and can only change from one form to another. The lowest form of energy is heat.

There are two general classes of energy: energy in transition and stored (or internal) energy. Stored energy is energy that resides in the matter itself, either in a state of aggregation or motion. An example of stored energy is a coiled spring with potential energy stored in its coils. Examples of energy in transition are work and heat.

There are essentially six basic forms of energy: heat (thermal energy), work, kinetic energy, potential energy, chemical energy and electromagnetic energy.

2.1.1 Heat

Heat is a form of energy that, when added to a substance, causes the substance to rise in temperature, fuse, evaporate, or expand. Heat itself cannot be seen or observed. It is only possible to observe the effects of heat. The primary effect of adding heat to a substance is an increase in the vibration of the molecules of the body. Heat always flows from the high state (high temperature) to a lower state (lower temperature). In other words, heat always flows from a hotter area to a colder area. The cold does not flow to the warmer area; rather, heat flows to the colder area. An analogy would be a bucket full of water with a leak. The water flows out of the bucket, the emptiness does not flow in.

The standard unit of measurement for heat is the British thermal unit or Btu. A Btu is the amount of heat required to raise the temperature of 1 lb of water from 59.5°F to 60.5°F. Another commonly used measurement of heat in S.I. units is the kilowatt-hour, which is equal to 3412 Btu. Conversion factors for units of measurement commonly used in the HVAC industry can be found in Appendix A.

Heat occurs in two forms: sensible heat and latent heat. Sensible heat is heat that changes the temperature of a substance. An example of sensible heat is a substance, such as water, air, or a solid that increases or decreases in temperature. Latent heat is heat that produces a change in state or phase change of a material. Examples of latent heating of a substance are evaporation, condensation, melting, and boiling. In each case, heat is added to or removed from the substance, while the temperature of the material remains constant as the phase change occurs.

2.1.2 Work

Work is another important concept in HVAC design. Work is a transient form of mechanical energy. Work is done on an object whenever a force is exerted on the object and the object is displaced or accelerated in the direction of the force. The commonly used units for work are foot-pounds per minute. Power is the time rate at which work is done and is measured in horsepower. Examples of work commonly found in HVAC systems include fans circulating air, pumps circulating liquids, and compressors compressing gases.

$$\text{Work} = \text{Force} \times \text{Distance} \tag{2.1}$$

The force acting may be constant or variable. The work required to lift a 200 lb weight a distance of 10 ft would be equal to 2000 ft•lb (neglecting any friction).

Heat may be transformed into work. Such an example would be an internal combustion engine where fuel is burned and converted to work at a drive shaft or flywheel. Work may also be transformed into heat directly, as in the case of a heat pump, or indirectly through friction. In all cases, some energy will be lost due to inefficiency of the process.

Power is the rate at which work is done and is usually measured in horsepower (hp).

Examples of work found in HVAC systems include fans circulating air, pumps circulating fluids, and compressors compressing gases.

2.1.3 Kinetic Energy

The energy a quantity of matter has, or possesses, by virtue of its motion is called kinetic energy. Every body in motion that has mass possesses kinetic energy. Whenever a body is in motion, it is able to exert a force and do work in coming to rest.

$$\text{Kinetic Energy} = \frac{mV^2}{2g} \tag{2.2}$$

where

m = the mass of the object

V = the velocity of the object

g = the acceleration due to gravity (32.2 ft/sec^2)

2.1.4 Potential Energy

The energy which matter has by virtue of its position or configuration is called potential energy. A body that has been moved or distorted to a new position is able to do work because of its position or configuration.

When a body undergoes a displacement (change in elevation or configuration), a force does work on the body. An example of vertical displacement would be the lifting of a weight. An example of a change in configuration would be the compression of a spring. In each case, a force does work on a body, resulting in stored potential energy.

2.1.5 Chemical Energy

Chemical energy is stored energy by virtue of the chemical composition of the material. An example of chemical energy is gasoline which, when burned, releases thermal energy. Another example of chemical energy is atomic energy which exists due to the atomic make-up of certain materials.

2.1.6 Electromagnetic Energy

Electromagnetic energy is potential energy resulting from magnetism. Electrical energy can be generated by rotating coils in a magnetic field.

2.2 THERMODYNAMICS

Central to the study of heating and air conditioning is the **First Law of Thermodynamics** and the concept of a control volume. A control volume can be any part of the system in which we want to examine the flow of energy into and out. An example might be a room, which is part of the system as a whole, that we wish to know the net heat flowing into and out. The room is the control volume and, the walls, ceiling, and floor are boundaries of the control volume. We may wish to examine the heat flowing out through walls, so we can determine the amount of heat that must be *added* to maintain the room temperature. Another control volume may be the air handling system itself. Heat is added to the control volume by a heating device, such as a coil. Work, or mechanical energy, is added by a fan and energy flows out of the control volume through air grilles and registers. Another control volume that we may wish to consider is the heating coil. If the coil is a steam heating coil, heat flows into the control volume in the form of steam and out as heat is transferred to the air stream. Control volumes are used frequently when studying HVAC systems. A control volume is used whenever it is necessary to isolate a particular component in order to analyze the flow of energy into and out of the control volume or component.

The science of thermodynamics is based upon two laws. The first law deals with the quantities involved in energy transformations, and the second law with the direction in which transformations take place. The reader should consult Reference 8 for a more detailed discussion of thermodynamics.

2.2.1   The First Law of Thermodynamics

Basically, the First Law of Thermodynamics (the conservation of energy law) states that the energy flowing into a control volume is equal to the energy flowing out of a control volume less the energy stored within the control volume. A simple mathematical expression for the First Law of Thermodynamics is

$$\text{Energy In} = \text{Energy Out} - \text{Energy Stored} \tag{2.3}$$

and is shown schematically in Figure 2.1 where the dashed box indicates the boundaries of the control volume.

Energy can cross the system boundaries in any form, i.e., heat, work, chemical energy, or potential energy.

Using the example of a room as a control volume, let's look at a typical HVAC application. Suppose a room had an outside wall exposed to a cold outside temperature and it also had a radiator to supply heat. Energy in the form of heat would flow out of the room, or control volume, and heat energy would flow into the room from the radiator. If the heat flowing out of the room through the wall (heat loss) was equal to the heat flowing into the room (heat gain), no energy would be stored and the room temperature would remain constant. If however, the heat supplied by the radiator was greater than the heat loss, energy would be stored in the room in the form of heat. The result of this addition of stored heat would result in an increase in room air temperature. Conversely, if the heat loss was greater than the heat gain from the radiator, a net heat loss would occur, resulting in negative heat storage and a resultant decrease in the room air temperature.

The First Law of Thermodynamics, and the concept of the control volume, are applied over and over again in HVAC design. A control volume can be as large as an entire building, it can be an entire system, such as an air conditioning system, or it can be as small as an individual component in the system. In each case, the First Law of Thermodynamics is used to perform an energy balance on the control volume in question. The usefulness of the First Law will become obvious in later chapters when the calculation process for heating and cooling loads is studied and when system components are studied.

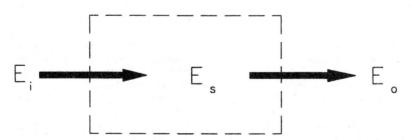

FIGURE 2.1   Control Volume for the First Law of Thermodynamics

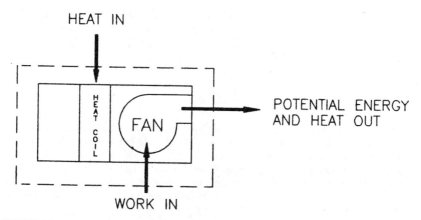

FIGURE 2.2   Energy Balance for an Air Handling Unit

As an example, consider an air handling unit with a hot water heating coil and fan shown in Figure 2.2.

Heat energy enters the controlled volume as heat in the form of heated water to the heating coil. Mechanical energy (work) enters the controlled volume via a shaft driving the fan. The fan converts the mechanical energy into potential energy (pressure). Heat and potential energy leave the controlled volume as heated pressurized air. Heat energy may be lost through the casing of the air handling unit if it is not well insulated. Some of the mechanical energy will be converted to heat energy due to friction losses in the fan drives. Energy in the form of heat will also be stored within the air handling unit as the components inside the unit heat up. All the energy flowing into the unit is either stored or it flows out of the unit, even though it may change form.

2.2.2   Forms of Heat

Heat may be stored in substances in two forms, sensible heat and latent heat. Sensible heat is the type of heat that is most recognizable. A change in the sensible heat content of a substance causes a change in temperature of the substances. As an example, heat applied to water will cause an increase in the temperature of water. The change in temperature caused by sensible heat is measured by an ordinary thermometer. When measuring the temperature change of air due to a change in sensible heat, the temperatures measured are called "dry bulb" temperatures.

Latent heat is a heat associated with a phase change of a substance. A phase of a substance may be defined as a quantity of the substance that is homogeneous throughout. As an example, water may be in the liquid phase, the vapor phase (steam), or the solid phase (ice). If water were in the vapor phase, it would be completely steam with no liquid or solid water present. A change of phase can be from a liquid to a vapor or gas (evaporation) from a vapor or gas to a liquid (condensation), from a liquid to a solid (solidification), or from a solid to a liquid (melting). A substance may also change directly from a solid to a gas or vapor (sublimation). In each case, heat is either added to, or removed from, the substance as it changes from one phase to another. When heat is added to, or removed from, a substance during a change of phase, it is called latent heat. During the change of phase, the temperature of the substance remains constant as heat is added, or removed, until all of the substance has completely changed phase. As an example, when water is boiling, the temperature of the water will remain constant at 212°F as heat is added until all of the water has boiled away and become steam.

2.2.3   Enthalpy

Enthalpy is a thermal property of a substance which consists of the internal energy of the substance plus energy in the form of work due to flow into or out of a control volume. The energy due to flow is the product of the pressure times the volume. Enthalpy is a measure of the total energy content of the substance measured from an arbitrary datum level. Internal energy, which is the major component of enthalpy, is energy associated with motions and configurations of a substance's interior particles primarily due to temperature as the result of the addition of heat.

Enthalpy is used frequently in HVAC calculations, although it is actually the total heat content of a substance. The flow work component in HVAC applications is usually ignored since it is relatively small. HVAC calculations are primarily concerned with the total heat of air, including any suspended water vapor in the form of latent heat. When the term enthalpy of the air is used, it is really referring to the total heat content of the air/vapor mixture.

2.2.4   Specific Heat

Specific heat, or heat capacity as it is sometimes called, is the quantity of heat required to change one unit mass of a substance one degree. Specific heat may be expressed mathematically as

$$q = mc \, (T_2 - T_1) \tag{2.4}$$

where

q = quantity of heat added or removed from the substance in Btu

m = mass of the substance in pounds

c = specific heat of the substances in Btu/lb•°F

$T_2 - T_1$ = temperature change in °F

Since water is a common substance, it has arbitrarily been chosen as the standard measure of specific heat. In the English units system, water has a specific heat of 1.0. The heat required to raise 1 lb mass of water from 59.5°F to 60.5°F is one British thermal unit (Btu). Therefore the specific heat of water is 1.0 Btu/lb•°F.

As a comparison, the specific heat of air is 0.24 Btu/lb mass °F. It can be observed that it takes more than 4 times as much heat to raise 1 lb mass of water 1°F as it does to raise 1 lb of air 1°F.

In general, the specific heat of a substance remains constant. Specific heat does vary slightly with temperature, but for HVAC calculations, it may be assumed a constant.

Example 2.1   Dry air has a specific heat of 0.24 Btu/lb•°F and pure water has a specific heat of 1.0 Btu/lb•°F. How much heat is required to raise 1 lb of each substance, 1°F?

Using Equation 2.4,

For the air: $q = (1 \text{ lb})(0.24 \text{ Btu/lb•°F})(1°F) = 0.24 \text{ Btu}$

For the water: $q = (1)(1)(1) = 1.0 \text{ Btu}$

2.2.5 The Second Law of Thermodynamics

There is a universal law of nature that says differences of energy levels will always tend to disappear. All energy levels tend to level out. Energy will always flow from a region with a higher energy state to a region with a lower energy state. Heat will always flow from an area of higher temperature to an area of lower temperature. Heat cannot naturally flow in the other direction. Potential energy, such as pressure, will flow naturally from an area of high pressure to an area of low pressure.

The Second Law of Thermodynamics is also known as the Law of Degradation of Energy. The basic underlying principle of the second law is that heat cannot be extracted from a body at a low temperature and be delivered to another at a higher temperature unless work is done to accomplish this result. If energy were to flow from a region of lower energy to a higher energy region, the First Law of Thermodynamics would still have been obeyed; however, such a flow does not occur in nature and would be a violation of the second law. The second law is the basic principle that makes mechanical refrigeration necessary.

A reversible process is one which can be made to traverse completely, in the reverse order, the same steps or states it passed through during the direct process and, in the end, the condition of the working substance is restored to its initial state. In order for a process to be reversible, the following conditions must be completely and precisely met.

1. The process may proceed in either direction.

2. The process may be reversed by infinitely small changes in the values of the conditions controlling it.

3. The working substance is always in a state of equilibrium throughout the process.

4. No energy transformations occur as the result of friction, i.e., no losses due to friction.

5. All energy transformations which occur as the process proceeds in one direction must equal, in both form and amount, those which take place in the reverse order.

6. The working substance must be returned to the initial condition by retracing every step of the original process in reverse order.

It is obvious that it is impossible to build a process or machine that meets the requirements for a reversible process. It is not possible to construct a completely reversible process, nor is it possible to construct a process that is 100% efficient. There will always be energy losses, such as friction losses and heat losses, in every process.

In HVAC system design, it is the second law that makes it necessary to have refrigeration systems to cool buildings. As an example, consider a space that is to be maintained at 75°F with an exterior wall to an outside temperature of 95°F. The natural flow of heat is from the region of higher energy to the region of lower energy, i.e., from outside to the inside through the exterior wall. In order to maintain the space at 75°F, it is necessary to transfer heat from the cooler inside to the warmer outside. To accomplish this process, it is necessary to use a refrigeration process (an air conditioner) to transfer the heat. The refrigeration process cannot transfer the heat from the cooler inside to the warmer outside without the addition of mechanical energy in the form of work. A compressor is used to add mechanical energy to the refrigeration system. A simple refrigeration system schematic is shown in Figure 2.3.

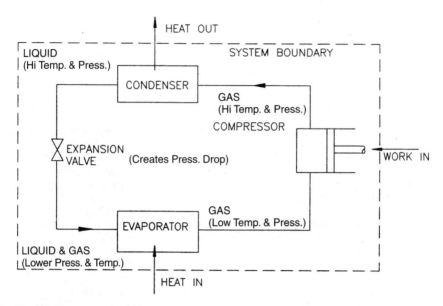

FIGURE 2.3 Refrigeration System Schematic

During the refrigeration process, some of the energy is lost due to friction in the compressor and the refrigerant flow in the piping, etc. Energy may also be lost due to uncontrolled heat transfer from the system. These losses make the refrigeration process irreversible. It is, therefore, necessary to add more energy to the system than would ideally be necessary due to the inherent inefficiencies of the system.

2.2.6 Psychrometrics

Psychrometrics can be defined as the science involving the thermodynamic properties of moist air and the effect of atmospheric moisture on materials and human comfort. All HVAC heating, cooling, humidification, and dehumidification processes involve the application of psychrometrics. It is especially important in air cooling and dehumidification processes. Psychrometric processes for HVAC systems are discussed in more detail in Chapter 9.

2.3 HEAT TRANSFER

Heat transfer involves the transfer, or flow of energy, in the form of heat. Whenever a temperature difference (or gradient) exists within a substance, from one side of a substance to another, or whenever a temperature difference exists between two bodies, heat will transfer from the high temperature region to the low temperature region. Heat transfer is energy transfer in the form of heat, that takes place between two bodies, or through a single body, solely due to a temperature difference.

The study of heat transfer attempts to predict the rate at which heat is transferred through a substance, or from one body to another. Unlike thermodynamics, heat transfer does not always involve the transfer, or flow of mass, in order for heat to flow from the high temperature region to the low temperature region. There are three methods by which heat is transferred through a substance or between bodies. They are conduction, convection, and radiation.

2.3.1 Conduction Heat Transfer

Conduction heat transfer occurs through, and by means of, intervening matter. It does not involve any obvious motion of the matter. It is a mechanism of internal energy exchange, from one part of a substance to another part, by the exchange of energy between molecules of the substance by direct communication. The molecules physically touch each other and transfer heat as they touch.

Consider a long metal rod with one end placed in a flame and the other out of the flame. In time, the end of the rod that is out of the flame will get hotter, even though it is not in direct contact with the flame. Heat reaches the cooler end by conduction through the material of the rod. Microscopically, the molecules at the hot end of the rod increase in energy and increase the energy of their vibrations as the temperature increases. The heated molecules then interact with their more slowly moving neighbors farther from the flame. As they interact, they share some of their energy with these neighbors, who in turn pass energy along to those still farther from the flame. In this manner, energy associated with thermal motion is passed from molecule to molecule, while each individual molecule remains in its original position.

As another example, suppose a thick homogenous concrete wall had a hot plate applied to one side and a cold plate applied to the other side as shown in Figure 2.4. Assume also that the hot and cold plates had been applied for a long period of time. Heat would flow through the concrete wall from the hot plate to the cold plate at a constant rate. The rate of heat flow would be a function of the wall material, in this case concrete, and the thickness of the material.

The concrete offers some resistance to the heat as it flows. Some materials transmit or conduct heat more readily than others. As an example, glass is frequently used for cookware since it conducts heat rather quickly and easily. Metals are also excellent conductors of heat. On the other hand, other materials are poor conductors of heat. Wood and polystyrene (Styrofoam) are poor conductors of heat and, therefore, are frequently used as insulators. Thermal conductance is a measure of the ease with which heat travels through material of a specified thickness. Knowing the thermal conductance, or just conductance, of a given material, the rate at which heat is transferred can be determined. Equation 2.5 may be used to predict the rate at which heat is conducted through a given material of a given thickness.

$$q = CA\,(T_2 - T_1) \tag{2.5}$$

where

q = rate of heat transfer in Btu/hr

C = thermal conductance of material of specified thickness in Btu/hr•ft²•°F

A = surface area of the material in square feet

$T_2 - T_1$ = difference in temperature in °F

Equation 2.5 is used to predict the steady state conduction of heat. In other words, it is assumed that the temperature difference $T_2 - T_1$ has existed for a long period of time and the rate of heat transfer is a constant (not changing with time).

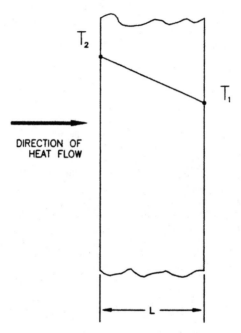

FIGURE 2.4   Heat Conduction Through a Wall

Example 2.2   Assume that 100 ft² of ¹/₂ in thick gypsum board ($C = 2.2$ Btu/hr•ft²•°F), has a temperature of 70° on one side and 100° on the other side. At what rate is heat conducted through the gypsum board?

By using Equation 2.5

$$q = (2.2 \text{ Btu/hr•ft}^2\text{•°F})(100 \text{ ft}^2)(70 - 10)°F = 13{,}230 \text{ Btu/hr}$$

Thermal conductance is a measure of the ease with which a material of a given thickness transmits heat. Conversely, the reciprocal of the conductance gives the resistance to heat flow (R).

$$R = \frac{1}{C} \tag{2.6}$$

The higher the R value, the greater the resistance to the flow of heat.

Thermal conductance is given for a material of a given thickness. Frequently though, tables give the conductance per unit thickness which is known as thermal conductivity of the material.

$$C = \frac{k}{x} \tag{2.7}$$

where

k = thermal conductivity in Btu/hr•ft²•°F•in or Btu/hr•ft²•°F•ft

x = material thickness in inches or feet

Note: Care must be taken to ensure that all units are consistent.

Example 2.3 What is the thermal conductance of $1/2$ in gypsum board that has a thermal conductivity of $k = 0.0926$ Btu/hr•ft²•°F•in.

$$x = \left(\frac{1}{2} \text{ in}\right)\left(\frac{1 \text{ ft}}{12 \text{ in}}\right) = 0.0417 \text{ ft}$$

from Equation 2.7

$$C = \frac{k}{x} = (0.0926)\left(\frac{1}{0.0417}\right) = 2.22 \text{ Btu/hr•ft}^2\text{•°F}$$

Thermal conductances and other thermal properties are given in Appendix B for common building materials.

For a given material of a given thickness, Equations 2.5 and 2.7 may be combined to yield Equation 2.8. Equation 2.8 is used frequently for calculating the rate of conduction heat transfer through building exterior surfaces, such as walls and roofs.

$$q = \frac{kA(T_2 - T_1)}{L} \tag{2.8}$$

where

q = rate of conduction heat transfer in Btu/hr

k = the thermal conductivity of the material in Btu/hr•ft•°F

A = the area of heat transfer in ft²

L = the thickness of the material in feet

T_2 = the high temperature side of the material in °F

T_1 = the low temperature side of the material in °F

The thermal conductivity of most common materials may be obtained from Reference 1, Reference 10 or any heat transfer text book.

Example 2.4 A 75 ft² concrete wall that is 8 in thick, has a temperature of 95°F on one side and a temperature of 25°F on the other side. The thermal conductivity of concrete is 0.17 Btu/hr•ft•°F. Assuming the temperatures have been maintained for a long period of time, what is the rate of heat transferred through the wall?

Using Equation 2.5,

$$q = (0.17 \text{ Btu/hr•ft•°F})(75 \text{ ft}^2)(95 - 25°F)\left(\frac{12}{8}\right) = 1339 \text{ Btu/hr}$$

Note: All units must be consistent, i.e., all feet or all inches.

Example 2.5 Assume the wall in Example 2.4 had 2 in of polystyrene insulation with a thermal conductivity of 0.18 Btu/hr•in•°F applied to one side. What is the rate of heat transfer through the composite wall?

For the polystyrene,

$$k = (0.18 \text{ Btu/hr•in•°F})(12 \text{ in/ft}) = 2.16 \text{ Btu/hr•ft•°F}$$

Modifying Equation 2.8,

$$q = \frac{75(95 - 25)}{\left(\dfrac{1}{0.17}\right)\left(\dfrac{8}{12}\right) + \left(\dfrac{1}{2.16}\right)\left(\dfrac{2}{12}\right)} = 1313 \text{ Btu/hr}$$

In the above examples, it was assumed that the conditions (the temperature difference) had existed for a long period of time. This assumption implies that the heat flow had reached a steady state condition. In other words, the temperatures were not varying and there was no thermal storage within the wall itself. Non-steady state heat conduction was briefly discussed in Section 1.2; refer to References 1 and 10 for more information about non-steady heat flow and thermal storage.

2.3.2 Convection Heat Transfer

Convection is a heat transfer mechanism which occurs in a fluid (liquid or gas) by the motion of the fluid. Heat is transferred within the fluid as it flows by the mixing of the molecules within the fluid. An example of convection heat transfer would be cold air flowing over a warm surface. As the air flows, the air close to the surface is heated as it comes in contact with the surface. As the air flows, it also mixes and the air heated at the surface mixes with the free stream air.

Convection heat transfer, or just convection, can be due to either forced flow or natural flow. In the above example, the flow of the air over the surface could be forced flow from a fan.

An example of natural convection would be a cold fluid next to a warm vertical surface. As the fluid next to the surface is heated, the natural buoyancy forces, due to the change in density of the fluid, would cause the warmed fluid to rise. This would cause a natural flow of fluid past the surface. Conversely, if the fluid was cooled by the surface, the fluid next to the surface would drop.

Whenever a fluid moves past a solid surface, it may be observed that the fluid velocity varies from zero immediately adjacent to the surface, to a "free stream" velocity some distance away. The flow of the fluid can be forced flow from a fan or pump, or it can be natural flow. Natural convection occurs due to changes in density of the fluid at different temperatures.

The resistance to heat transfer at the solid-fluid boundary is a function of the convection, or film, coefficient, h. The units for the convection are Btu per hour•square foot of surface area per degree F temperature difference. The equation for convection heat transfer is known as Newton's Law of Cooling and may be expressed as

$$q = hA \ (T_s - T_f) \tag{2.9}$$

where

q = rate of heat transfer in Btu/hr

h = convection heat transfer coefficient in Btu/hr•ft^2•°F

A = surface area in square feet

T_s = surface temperature in °F

T_f = free stream fluid temperature in °F

The convection coefficient is a function of the fluid velocity (if any)—the difference in temperature between the fluid and the surface, the orientation of the surface (vertical, horizontal, etc.), the roughness of the surface, and the viscosity of the fluid. Fluid viscosity is discussed later in this chapter. Convection coefficients are frequently determined experimentally, although they may also be calculated mathematically from empirical formulas.

Example 2.6 Air at 50°F is flowing over a flat surface that is maintained at 130°F. The surface is 8 ft × 10 ft. The convection coefficient has been determined to be 5 Btu/hr•ft²•°F. What is the rate of heat transfer between the surface and the air?

Using Equation 2.6,

$$q = (8 \text{ ft} \times 10 \text{ ft})(5 \text{ Btu/hr•ft}^2\text{•°F})(130 - 50\text{°F}) = 32,000 \text{ Btu/hr}$$

Values far the convection heat transfer coefficient h are determined experimentally. In a manner similar to the resistance for thermal conductance, the resistance for the convection coefficient is given as

$$R = \frac{1}{h} \tag{2.10}$$

Since air is used so frequently in HVAC applications, resistances for air are tabulated in Table 4.1.

2.3.3 Radiation Heat Transfer

Radiation heat transfer is heat that is transferred via electromagnetic waves. In order for radiation heat transfer, or just radiation, to occur between two bodies, it is not necessary that the two bodies be in direct physical contact with each other. It also is not necessary that there be a medium through which heat is transferred because radiation can pass through a complete vacuum. An example of radiation is the heating of an object by another object, such as the sun heating a surface on earth.

All objects radiate energy, regardless of their temperature. The greater the temperature of the object, the greater the energy radiated. All objects also receive radiant energy from all other objects. The radiant exchange takes place continuously. When an object reaches a constant temperature, it does not stop radiating energy, but it is receiving energy at the same rate that it is giving off energy.

The rate at which a body emits radiant energy increases very rapidly as the temperature increases. The amount of radiant energy emitted is proportional to the fourth power of the body's absolute temperature. Examples of radiant energy include visible light, infrared radiation, and ultra-violet radiation. The type of radiation is a function of the frequency and wavelength of the electro-magnetic radiation, which is a function of the absolute temperature of the radiating object. As the temperature increases, the wavelength of the radiation shortens. At 800°C, a body would emit enough visible radiation to be self-luminous and appear to be "red hot." At 3000°C, the temperature would be hot enough for the object to be "white hot" and most of the radiant energy would be in the visible light spectrum. The thermal radiation spectrum is in the range of 0.1 to 100 m and the visible is from 0.39 to 0.78 m (1 m = 10^{-6} meters).

The ability of a surface to emit and absorb radiant energy is referred to as the emittance or emissivity of a surface. Emittance is a dimensionless quantity that varies from 0 to 1.0, with 1.0 being a perfect emitter. A body that is a good emitter of energy is also a good receiver of radiant energy. An object with a black rough surface is an excellent emitter (as well as an excellent absorber) of radiant energy. A highly polished surface is a poor emitter and absorber of radiant energy since it has a low emittance.

The radiant energy exchange between two objects is also strongly dependent on the geometry of the two objects. In order for objects to exchange radiant energy, they must "see each other." If objects are shielded by another object, they cannot exchange radiant energy unless the shield is transparent to the radiation. If the objects are partially shielded, the radiant energy will be reduced proportionally. This is accounted for by a shape factor (configuration factor) between the objects. A shape factor of 0 means that the objects cannot see each other at all. A shape factor of 1.0 indicates that the objects only "see each other" and nothing else (all radiant energy leaving one object falls on the other and nothing else). The distance between the objects also affects the shape factor.

All the radiation falling on a surface will be absorbed by the surface, reflected by the surface, or transmitted through the surface. Figure 2.5 shows the distribution of radiant energy falling on a surface. The relative amounts of absorbed, reflected, or transmitted energy are a function of the surface absorptance, reflectance, and transmittance.

For any given surface, the relative amounts are also a function of the angle of incidence of the radiation. The more directly the radiation falls on a surface, the greater the intensity of the radiation and the greater the absorptance.

The amount of radiant exchange between two objects may be predicted by Equation 2.11 which is known as the Stefan-Boltzman Law.

$$q = \sigma A F_e F_s (T_1^4 - T_2^4) \tag{2.11}$$

where

q = the rate of heat transfer in Btu/hr

σ = a constant, 0.1714×10^{-8}, Btu/hr•ft²•°R⁴

F_e = absorption factor

F_s = shape or configuration factor

A = surface area

T_1 and T_2 = the absolute temperature of the bodies in °R

Note: The relative temperature of 0°F is equal to the absolute temperature of 460°R.

Example 2.7 Two flat black plates that are infinitely large in each direction are separated by a space that is occupied by a vacuum. One plate is maintained at 0°F and the other at 150°F. What is the radiant heat transfer, per square foot, between the plates?

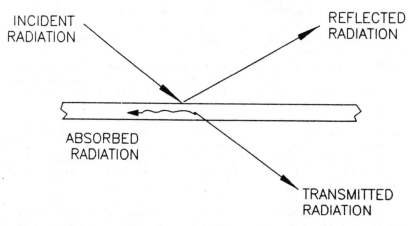

FIGURE 2.5 Distribution of Radiant Energy on a Surface

Since the plates are infinitely long in each direction, they can only see each other and therefore their shape factor is equal to 1.0. Since they are flat black, their emissivity of each can be assumed to be 1.0.

To convert the temperatures of the two plates to absolute temperatures, their temperatures must be converted from degrees Fahrenheit to degrees Rankine. To make the conversion, it is necessary to add 460°F to their relative temperatures. The temperature of the cold plate would be 460°R and the hotter plate would be 610°R.

Using Equation 2.11

$$\frac{q}{A} = (0.1714 \times 10^{-8} \text{Btu/hr•ft}^2 \text{•°R}^4)(1)(1)[(610°R)^4 - (460°R)^4] = 160.6 \text{ Btu/hr•ft}^2$$

2.3.4 Transient Heat Transfer

Previously, all of the heat transfer examples were steady state, one directional. In other words, the rate of heat transfer was assumed to be constant over a long period of time and that the heat flowed in only one direction. In steady state heat transfer, heat entering an object, or control volume, is equal to the heat leaving and this has been the case for a long period of time. No heat is stored. Frequently, in actual HVAC applications, this is not the case.

Consider a situation where a concrete wall is shaded from the sun's rays by some external object. Assume that the object is suddenly moved and the sun's rays shine on the outside surface of the wall. The exterior surface immediately begins to increase in temperature, however, the interior surface does not. There will be a considerable time lag between the time that the sun's rays initially strike the exterior surface and when the inside surface begins to increase in temperature. The rate of heat transfer through the wall, then, is not constant but is varying with time.

The time lag between the application of heat to one surface and when the heat begins to appear on the other side is due to thermal or heat storage of the material. The difference between the instantaneous heat gain on a building, and the actual cooling load the air condi-

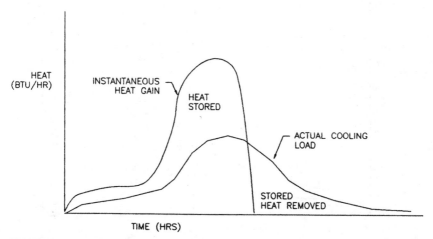

FIGURE 2.6   The Effect of Thermal Storage on Cooling Loads

tioning system must absorb, is illustrated in Figure 2.6.

Note in Figure 2.6 that the instantaneous heat gain peaks sooner and is greater than the actual cooling load. Only where the instantaneous heat and the cooling load curves cross, is the heat gain equal to the cooling load. Prior to the time when the curves cross, is the heat gain is greater than the cooling load and heat is stored by the building mass. After the curves cross, the building is no longer storing heat. The building mass is rejecting heat to the cooling system.

The thermal mass of a building directly affects the time lag between the peak heat gain and peak cooling load, and it inversely affects the intensity of the peak cooling load. The greater the thermal mass of the building, the greater the time lag, and the lower the peak intensity of the cooling load. This is illustrated in Figure 2.7.

In some cases, the time lag between the heat gain and the cooling load is so great, that the peak cooling load for the air conditioning system occurs at night when the building is empty. This will be discussed again in Chapters 4, 7 and 8.

2.3.5   Log Mean Temperature Difference

Heat exchange devices (heat exchangers) are used frequently in HVAC systems. Examples of heat exchange devices include heating and cooling coils, furnace heat exchangers that exchange heat between flue gases and the supply air, and steam to heating water heat exchangers. Whenever heat is transferred between two fluid streams that do not intermix, a heat exchanger is used.

The amount of heat that is transferred between two fluids, passing through a heat exchanger, is a direct function of the temperature difference between the two fluid streams. The log mean temperature difference is the effective temperature difference between the two fluid streams. As an example, consider a hot water heating coil in an air handling unit. Assume the air enters the coil at 40°F and leaves the coil at 100°F. Also assume the heating water enters the coil at 200°F and leaves at 170°F. What is the effective temperature difference between the air and the water in the coil? It is necessary to determine the log mean temperature difference to answer this question and similar questions involving heat ex-

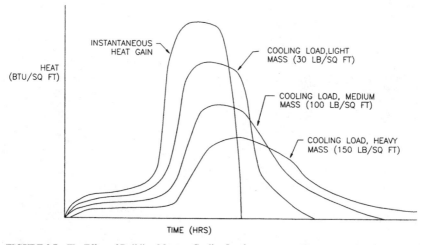

FIGURE 2.7   The Effect of Building Mass on Cooling Loads

change devices.

The log mean temperature difference (LMTD) is the temperature difference at one end of the heat exchanger, less the temperature difference at the other end of the heat exchanger, divided by the natural logarithm of the ratio of these two temperature differences, and expressed mathematically by Equation 2.12.

$$\text{LMTD} = \frac{(T_{b2} - T_{c2}) - (T_{b1} - T_{c1})}{\ln\left[\frac{(T_{b2} - T_{c2})}{(T_{b1} - T_{c1})}\right]} \tag{2.12}$$

where

T_{h1} = entering temperature of the hotter fluid

T_{h2} = leaving temperature of the hotter fluid

T_{c1} = entering temperature of the cooler fluid

T_{c2} = leaving temperature of the cooler fluid

A flow schematic of a heat exchanger is shown in Figure 2.8.

Equation 2.12 is based on three important assumptions, which are:

1. the specific heat of the two fluids remains constant.

2. the convection heat transfer coefficients of the two fluids are constant.

3. the flow rates of the two fluid streams remain constant.

From Figure 2.7, it may be observed that the schematic diagram shows a counter flow heat exchanger. That is, the flow of the two fluid streams inside the heat exchanger run

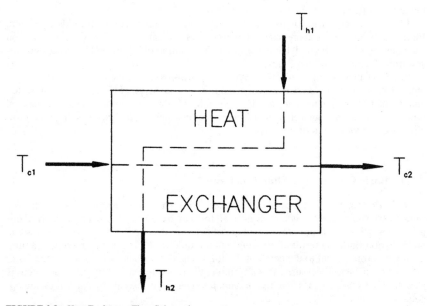

FIGURE 2.8   Heat Exchanger Flow Schematic

opposite, or counter to, each other. In this case, the entering cold fluid, which is the coldest, meets the leaving hot fluid, which is also the coolest. Had the hot fluid entered at the top left of the heat exchanger schematic and exited at the bottom right, the heat exchanger would have been a parallel flow heat exchanger since the flow streams would be parallel inside the heat exchanger.

Example 2.8   Air enters a heating coil at 40°F and is heated to 100°F. The coil has heating water entering at 200°F and leaving at 170°F. Assuming constant flow and fluid properties for both streams, what is the effective temperature difference between the air and water?

Using Equation 2.12

$$\text{LMTD} = \frac{(170 - 100) - (200 - 40)}{\ln\left[\dfrac{170 - 100}{200 - 40}\right]} = \frac{-90}{\ln(0.4375)} = 108.9°\text{F}$$

2.4 CONCEPTS OF FLUID MECHANICS

A fluid may be defined as a substance that deforms continuously when subjected to a shear stress, no matter how small that stress may be. A fluid is a substance that is capable of flowing, having particles that easily move and change their relative position without separation of the mass, and that yields easily to pressure. A fluid may be a liquid, such as water, or it may be a gas, such as air.

Gases are very readily compressible, whereas liquids are only slightly compressible. For most HVAC design applications, the pressures are low enough that compressibility of air may be ignored. In other words, the density of air may be considered to be a constant without any significant error in the calculations.

Fluid statics deal with fluids at rest, while fluid dynamics deal with fluids in motion. Fluid statics involve the forces exerted by a body of fluid at rest, such as fluid pressure, on a submerged object. Fluid dynamics deal with the motion of a fluid and the forces that result from the motion of that fluid.

Fluid mechanics, especially fluid dynamics, is important in the design of HVAC systems. Almost all HVAC systems involve the flow of some type of fluid, such as air, water, and steam. Calculation methods for fluid dynamics are the basis for air duct system design, piping system design, as well as many other systems. This section reviews the basic principles used in the design of these systems.

2.4.1 Basis Concepts in Fluid Mechanics

2.4.1.1 Pressure   Fluid pressure is a concept with which most people are familiar. Any fluid at rest exerts a perpendicular, or normal, force on its boundary and any submerged objects. This is known as pressure. This type of pressure is called static pressure. The static pressure of a fluid is a result of the weight of the fluid itself. The static pressure in a fluid is directly proportional to the depth of the fluid. The deeper the location within the fluid, the greater the static pressure. As an example, the pressure exerted by the atmosphere is greater at sea level than it is on high mountains. The forces exerted by static pressure are equal in all directions.

When performing calculations, or making measurements involving static pressure in a fluid, the fluid pressure is usually stated with respect to a reference pressure. Normally, the reference pressure is atmospheric pressure and the resulting measured pressure is excess pressure above atmospheric pressure, or gage pressure. Pressure measured relative to a perfect vacuum (absolute 0 pressure) is called absolute pressure. The relationship between pressure measurements is shown in Figure 2.9.

The relationship between gage pressure and absolute pressure is given by Equation 2.13.

$$P_{abs} = P_{gage} + P_{atm} \tag{2.13}$$

where

P_{abs} = absolute pressure

P_{gage} = gage pressure

P_{atm} = atmospheric pressure

Vacuum is the depression of pressure below atmospheric pressure. Standard atmospheric pressure is assumed to be 14.7 psia (29.92 in mercury). Measurements of gage pres-

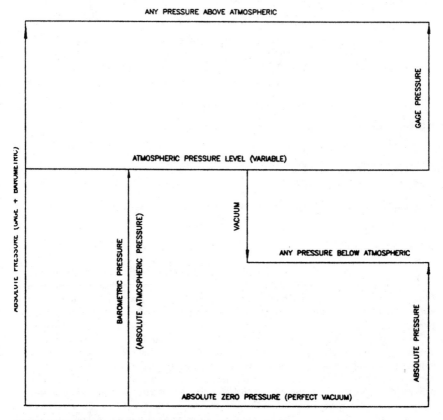

FIGURE 2.9 Relationship Between Pressure Measurements

sure are normally in pounds per square inch (psi) or feet of water column. Low pressures, less than 1 psi are usually expressed in inches of mercury or inches of water gage (wg) or inches of water column (wc).

A fluid in motion exerts an additional pressure due to its momentum. This pressure is exerted in the direction of flow. The pressure, due to fluid motion, is known as velocity pressure. For any given fluid, the magnitude of the velocity pressure is a direct function of the velocity of the fluid. The greater the velocity of the fluid, the greater the velocity pressure.

2.4.1.2 Density   The density of a fluid is its mass per unit volume. A homogenous material has the same density throughout. The density of a liquid is given by Equation 2.14.

$$\sigma = \frac{m}{V} \tag{2.14}$$

where

   m = mass in pounds

   V = volume in ft^3

For liquids, which are incompressible for the most part, the density may be assumed as a constant. The density of most liquids does, however, vary to some extent with temperature. Unless liquid temperatures are quite high, the density of a liquid may be assumed a constant without any significant error.

The density of gases and vapors cannot be assumed to be constant. The density of any given gas varies with the absolute pressure and absolute temperature. The density of a gas may be calculated using Equation 2.15.

$$\sigma = \frac{P}{RT} \tag{2.15}$$

where

   P = the absolute pressure in psia

   R = a gas constant for the particular gas in ft•lb f/lbm•°R

   T = the absolute temperature in °R

Equation 2.15 is applicable only for ideal or perfect gases, such as air.

2.4.2   Viscosity

Viscosity is a measure of the ease with which a fluid flows. It is that property of a fluid which causes resistance to relative motion within the fluid and is observed only when a fluid is in motion. Viscosity enables a fluid to develop and maintain an amount of shearing stress, dependent upon the velocity of the flow, and then to offer continued resistance to flow. Energy is lost, due to friction in the flowing fluid, due to the presence of its viscosity.

Viscosity is caused by the molecular structure of the fluid. In the liquid state, the molecules are packed closely together and the viscosity is due to the cohesiveness of the molecules. As the temperature of a liquid increases, the cohesion decreases and the viscosity decreases. For a gas, the molecules are spaced much farther apart and the viscosity is due to the activity of the molecules which strike against each other. As the temperature increases, the activity of the molecules increases, the collisions increase, and the viscosity of the gas increases.

The absolute viscosity of a fluid is a measure of its resistance to internal deformation of shear. Kinematic viscosity is the ratio of the absolute viscosity to the mass density.

The effect of pressure on the viscosity of liquids and gases is so small that it may be ignored without causing any significant error. With an increase in temperature, the viscosity of most liquids decreases, while the viscosity of most gases increases.

When performing calculations involving fluid flow, it is important to determine the character of the flow. Under some conditions, the fluid will appear to flow in layers in a smooth and regular manner. This type of flow is referred to as laminar flow. In laminar flow, the layers of fluid slide over one another. If the flow is not regular and smooth, the flow is referred to as turbulent flow. In turbulent flow there is flow in random directions within the general flow stream.

Energy is lost, due to friction in a flowing fluid, as the result of fluid viscosity. The magnitude of the energy loss for any given fluid is directly dependent on whether the flow is laminar or turbulent.

2.4.3   The Continuity Equation

According to the continuity equation, which is known as the conservation of mass for fluid flow, no fluid is created, or destroyed, as fluid flows in a closed conduit. In more basic terms, it states that the fluid flow out of a system is equal to the fluid flow into the system, plus any storage within the system. For incompressible fluids, the mass flow, past one point in a closed conduit, must be equal to the flow past another point in the conduit if there is no storage within. If the cross-sectional area of the conduit changes, the fluid velocity must change accordingly. The continuity equation is important in the study of the flow of fluids in closed conduits such as pipes and ducts. Equation 2.16 is a mathematical expression of the continuity equation.

$$Q = A_1 V_1 = A_2 V_2 = \text{constant} \tag{2.16}$$

where

Q = the rate of fluid flow in ft³/min, gpm, ft³/sec, etc.

A = the cross-sectional area of the conduit in feet

V = the velocity of the fluid in ft/min or ft/sec

Equation 2.16 assumes steady state flow, that is, a constant flow rate and a non-compressible fluid.

Example 2.9   Figure 2.10 shows a pipe that increases from an inside diameter of 8 in at point 1 to an inside diameter of 12 in at point 2. Water at 60°F is flowing through the pipe. If the water velocity is 5.72 ft/sec at point 1, how much water is flowing through the pipe and what is the velocity at point 2?

$$A_1 = \left(\frac{1}{4}\right)(\pi)(8^2)\left(\frac{1}{144}\right) = 0.35 \text{ ft}^2$$

$$A_2 = \left(\frac{1}{4}\right)(\pi)(12^2)\left(\frac{1}{144}\right) = 0.79 \text{ ft}^2$$

Using Equation 2.16, the flow rate is

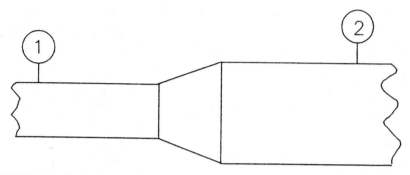

FIGURE 2.10   Pipe for Example 2.9

$$Q = A_1 V_1 = (0.35 \text{ ft}^2)(5.72 \text{ ft/sec}) = 2.0 \text{ ft}^3/\text{sec}$$

Again, using Equation 2.16, the velocity at point 2 is

$$V_2 = \frac{Q}{A_2} = \frac{(2.0 \text{ ft}^3/\text{sec})}{(0.79 \text{ ft}^2)} = 2.53 \text{ ft/sec}$$

Equation 2.16 is only applicable to non-compressible fluids and, technically, it is not applicable to calculations involving a compressible fluid, such as air. However, for most calculations involving HVAC systems, the pressures and velocities are so low that air may be assumed to be incompressible and Equation 2.16 can be used without any significant error. For applications that do involve higher pressures and velocities, Equation 2.17 can be used for the continuity equation; it is applicable to any fluid flow at velocities less than sonic velocity.

$$Q = \sigma_1 A_1 V_1 = \sigma_2 A_2 V_2 = \text{constant} \tag{2.17}$$

where

σ = the fluid density in lbm/ft^3

2.5.3   Bernoulli's Equation

Fluid flow also follows the law of conservation of energy, which states that energy within a system is neither created nor destroyed. It can, however, be transformed from one form to another. Bernoulli's Equation is a means of expressing and applying the law of conservation of energy to the flow of fluids in a closed system such as a conduit. It relates the pressure, flow velocity and height for flow of a fluid.

In order for fluid to flow through a conduit, energy must be applied to the fluid in the form of work (pressure). Bernoulli's Equation states mathematically that as an incompressible fluid flows, the total head (pressure) remains constant. In other words, the energy (work) remains constant as the fluid flows.

There are three forms of energy which should be considered in any fluid flow analysis. They are potential energy, kinetic energy, and flow energy or work. They may be expressed as follows:

1. Potential Energy. Energy due to the elevation of the fluid relative to some reference elevation.

$$PE = wZ \tag{2.18}$$

where

w = the weight of the fluid in pounds

Z = the elevation of the fluid above a reference elevation in feet.

2. Kinetic Energy. Energy due to the velocity and momentum of the fluid

$$KE = \frac{1}{2}\left(\frac{wV^2}{g}\right) \tag{2.19}$$

where

V = the velocity of the fluid in ft/sec

g = the acceleration due to gravity (32.2 ft/sec²)

3. Flow Energy or Work. Also called pressure energy or flow work, it represents the amount of work necessary to move the fluid against a pressure.

$$FE = \frac{wP}{\gamma} \tag{2.20}$$

where

P = the pressure in lb/ft²

γ = the specific weight of the fluid in lb/ft³

The total amount of energy of these three forms of energy, possessed by the fluid as it flows, is expressed in Equation 2.21.

$$E = FE + PE + KE = \frac{wP}{\gamma} + wZ + \frac{wV^2}{2g} \tag{2.21}$$

Since energy is conserved within the system, the energy between two points in the system must remain constant. Assuming the mass flow remains constant, Equation 2.21 may be expressed as

$$\frac{P_1}{\gamma} + Z_1 + \frac{V_1^2}{2g} = \frac{P_2}{\gamma} + Z_2 + \frac{V_2^2}{2g} \tag{2.22}$$

Equation 2.22 is referred to as Bernoulli's Equation.

In the design of air distribution systems for HVAC systems, the potential energy due to elevation changes can usually be neglected since elevation changes are relatively small. In the design of HVAC systems, potential energy and kinetic energy cannot be ignored. Since potential energy can usually be neglected in the design of air systems, Equation 2.22 may be rewritten as Equation 2.23.

$$\left(\frac{1}{\gamma}\right)(P_2 - P_1) + \left(\frac{1}{2g}\right)(V_2^2 - V_1^2) = 0 \tag{2.23}$$

Equation 2.23 states mathematically that the only transformation of energy in an air sys-

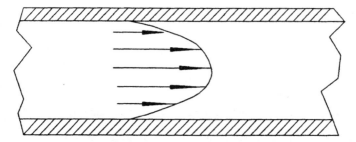

FIGURE 2.11   Fluid Velocity Profile

tem is due to changes in pressure. The changes in pressure in air distribution are due to friction losses and changes in velocity. Energy may be lost from a system because the friction converts pressure energy to heat.

2.4.4   Fluid Flow Friction Losses

In Section 2.3.2 Convection Heat Transfer, it was noted that whenever a fluid flows in a closed conduit or along a solid surface, the fluid next to the surface is actually not moving and has zero velocity. This is due to friction between the flowing fluid molecules and the solid surface. Figure 2.11 shows a typical profile for a fluid in a conduit, such as a duct.

In Figure 2.11, the arrows represent relative velocities of the fluid, with the longer arrows being higher velocities and the shorter arrows indicating lower velocities. It may be observed from Figure 2.11 that, while the fluid near the walls of the conduit has a velocity of zero, the velocity near the center of the conduit is moving at a much greater velocity. The velocity near the center is known as the free stream velocity.

The velocity gradient exists between the fluid at the walls and the center due to friction between the individual molecules of the fluid. Friction loss also results from the friction between the fluid and the walls of the conduit. The energy loss due to friction may be expressed as

$$h_L = \frac{C_L V^2}{2g} \tag{2.24}$$

where

h_L = the energy loss due to friction

C_L = a loss coefficient or friction factor

V = the velocity of the fluid

g = the acceleration due to gravity

The loss coefficient, or friction factor, is a function of the fluid velocity, fluid viscosity, and the roughness of the conduit walls. Equation 2.24 is a simplified version of the Darcy Equation. The application of fluid flow to HVAC design will be discussed in more detail in a later chapter on the design of duct systems.

Chapter 3

CLIMATIC CONDITIONS

When selecting a suitable heating and air conditioning system for a building, it is first necessary to determine the heating and cooling loads that are attributable to the outdoor climatic conditions for the building location. This chapter presents detailed climatic data and considerations used when calculating design heating and cooling loads.

3.1 DESIGN TEMPERATURE CRITERIA

Design temperatures, along with other weather data for various locations, are presented in Table 3.1. The weather data were measured at first-order weather stations, airports, and Air Force bases by trained observers. The recorded values were statistically analyzed for a 15-year period and tabulated.

In Table 3.1, winter and summer design temperatures are given for various percentages. The percentages represent frequency risk levels for the dry bulb temperature listed. It is assumed that the frequency level for any given dry bulb temperature will repeat itself in the future. The percentages can be used to ascertain the level of risk that the design temperature will not be exceeded. The winter percentages are for the months of December, January, and February in the northern hemisphere. The three winter months equal a total of 2160 hr. It may be assumed that the design dry bulb temperature, given in the 99% column, will not be exceeded (the temperature will not drop below) 99% of the time or, conversely, will be exceeded only 1% of the time during the three winter months. This would be equal to a total of 22 hr. For the $97^1/_2\%$ frequency level, the design temperature will be exceeded only $2^1/_2\%$ of the time or 54 hr. Similarly, the summer percentages and dry bulb temperatures are for the four summer months of June, July, August, and September (2928 hr) in the northern hemisphere. The wet bulb temperatures are mean coincident temperatures at the given dry bulb temperature.

Mean daily range temperatures are also tabulated. These are the difference between the average daily maximum and the average daily minimum dry bulb temperatures for the warmest month. When selecting a percentage frequency and the corresponding temperature, it is necessary to consider the effect on the HVAC system and the usage of the building. The lower the frequency risk level, the greater the difference will be in the indoor and outdoor design temperatures. This will result in an HVAC system of greater capacity and greater cost. The intended use of the building is usually the determining factor in the selection of the frequency risk level. In a building where it is critical to maintain the indoor design conditions, such as a hospital, nursing home, or computer room, the lower frequency risk level may be chosen, e.g., 99% for winter and 1% for summer. On the other hand, a building that is not so critical, such as a retail building or office building, may only require a higher frequency risk level (e.g., $97^1/_2\%$ for winter and 5% for summer).

Based on the above considerations, the design outdoor temperatures, and other weather related data, for a given location may be chosen from Table 3.

3.2 SOLAR DATA

The intensity of solar radiation falling on a surface is a function of the location of the surface, the direction the surface is facing, the clearness of the atmosphere, and the time of year. The intensity of solar radiation reaching a surface outside the earth's atmosphere, perpendicular to the sun's rays, is equal to approximately 430 Btu/hr•ft². The intensity reaching an object on the earth's surface is considerably less.

Solar intensity for a given location and direction is a function of the distance that the surface is from the equator. In the northern hemisphere, the greater the north latitude of the surface, the lower the solar intensity, with locations on the equator being the greatest intensity. Solar radiation at locations away from the equator is lower, due to the fact that the radiation is striking the surface at a sharper angle and due to the fact that the radiation must travel a greater distance through the earth's atmosphere.

Solar radiation entering the earth's atmosphere is either reflected (diffused), absorbed, or transmitted through the atmosphere. Approximately 35% of the solar radiation is reflected back into space. Since the solar radiation is diffused and absorbed by the earth's atmosphere, radiation falls on surfaces from all parts of the skydrome. This partially accounts for the fact that north facing surfaces, which never receive direct solar radiation, still receive indirect solar radiation. Ground reflectance also contributes to the intensity of indirect radiation falling on a surface. The total solar intensity falling on surfaces for various north latitudes is given in Table 7.9.

The solar intensity is also a function of the clearness of the local atmosphere. In industrial and urban areas, the solar intensity is reduced due to airborne dust particles; water vapor; carbon dioxide; ozone; and, of course, clouds.

TABLE 3.1 Climatic Design Criteria (Copyright 1993 by the American Society of Heating, Refrigerating, and Air Conditioning Engineers, Inc. from the ASHRAE Handbook-Fundamentals. Used by permission.)

Col. 1	Col. 2	Col. 3	Col. 4	Col. 5		Col. 6			Col. 7	Col. 8			Col. 9		Col. 10	
	Lat.	Long.	Elev.	Winter, °F Design Dry-Bulb		Summer, °F Design Dry-Bulb and Mean Coincident Wet-Bulb			Mean Daily Range	Design Wet-Bulb			Prevailing Wind Knots[d]		Temp. °F Median of Annual Extr.	
State and Station[a]	°N ° '	°W ° '	Feet	99%	97.5%	1%	2.5%	5%		1%	2.5%	5%	Winter	Summer	Max.	Min.
ALABAMA																
Alexander City	32 57	85 57	660	18	22	96/77	93/76	91/76	21	79	78	78			98.4	12.4
Anniston AP	33 35	85 51	599	18	22	97/77	94/76	92/76	21	79	78	78	SW 5	SW	99.8	14.6
Auburn	32 36	85 30	652	18	22	96/77	93/76	91/76	21	79	78	78			98.5	12.9
Birmingham AP	33 34	86 45	620	17	21	96/74	94/75	91/74	21	78	77	76	NNW 8	WNW		
Decatur	34 37	86 59	580	11	16	95/75	93/74	91/74	22	78	77	76				
Dothan AP	31 19	85 27	374	23	27	94/76	92/76	91/76	20	80	79	78				
Florence AP	34 48	87 40	581	17	21	97/74	94/74	92/74	22	78	77	76	NW 7	NW		
Gadsden	34 01	86 00	554	16	20	96/75	94/75	92/74	22	78	77	76	NNW 8	WNW		
Huntsville AP	34 42	86 35	606	11	16	95/75	93/74	91/74	23	78	77	76	N 9	SW		
Mobile AP	30 41	88 15	211	25	29	95/77	93/77	91/76	18	80	79	78	N 10	N		
Mobile Co	30 40	88 15	211	25	29	95/77	93/77	91/76	16	80	79	78				
Montgomery AP	32 23	86 22	169	22	25	96/76	95/76	93/76	21	79	79	78	NW 7	W	97.9	22.3
Selma, Craig AFB	32 20	87 59	166	22	26	97/78	95/77	93/77	21	81	80	79	N 9	SW	98.9	18.2
Talladega	33 27	86 06	565	18	22	97/77	94/76	92/76	21	79	78	78			100.1	17.6
Tuscaloosa AP	33 13	87 37	169	20	23	98/75	96/76	94/76	22	79	78	77	N 5	WNW	99.6	11.2
ALASKA																
Anchorage AP	61 10	150 01	114	-23	-18	71/59	68/58	66/56	15	60	59	57	SE 3	WNW		
Barrow	71 18	156 47	31	-45	-41	57/53	53/50	49/47	12	54	50	47	SW 8	SE		
Fairbanks AP	64 49	147 52	436	-51	-47	82/62	78/60	75/59	24	64	62	60	N 5	S		
Juneau AP	58 22	134 35	12	-4	1	74/60	70/58	67/57	15	61	59	58	N 7	W		
Kodiak	57 45	152 29	73	10	13	69/58	65/56	62/55	10	60	59	58	NW 14	NW		
Nome AP	64 30	165 26	13	-31	-27	66/57	62/55	59/54	10	58	56	55	N 4	W		
ARIZONA																
Douglas AP	31 27	109 36	4098	27	31	98/63	95/63	93/63	31	70	69	68	NE 5	SW	104.4	14.0
Flagstaff AP	35 08	111 40	7006	-2	4	84/55	82/55	80/54	31	61	60	59	SW 5	W	90.0	-11.6
Fort Huachuca AP	31 35	110 20	4664	24	28	95/62	92/62	90/62	27	69	68	67	N 7	S		
Kingman AP	35 12	114 01	3539	18	25	100/64	97/64	94/64	30	70	69	69	N 7	W		
Nogales	31 21	110 55	3800	28	32	99/64	96/64	94/64	31	71	70	69	NW 4	NW		
Phoenix AP	33 26	112 01	1112	31	34	109/71	107/71	105/71	27	76	75	75	SW 5	W	112.8	26.7
Prescott AP	34 39	112 26	5010	4	9	96/61	94/60	92/60	30	66	65	64	E 4	W		

[a] AP, AFB, following the station name designates airport or military airbase temperature observations. Co designates office locations within an urban area that are affected by the surrounding area. Undersigned stations are semirural and may be compared to airport data.

[b] Winter design data are based on the 3-month period, December through February.
[c] Summer design data are based on the 4-month period, June through September.
[d] Mean wind speeds occurring coincidentally with the 99.5% dry-bulb winter design temperature.

TABLE 3.1 Climatic Design Criteria (Continued)

Col. 1	Col. 2	Col. 3	Col. 4	Winter,[b] °F Col. 5		Summer,[c] °F Col. 6			Col. 7	Col. 8			Prevailing Wind Col. 9			Temp., °F Col. 10	
	Lat.	Long.	Elev.	Design Dry-Bulb		Design Dry-Bulb and Mean Coincident Wet-Bulb			Mean Daily Range	Design Wet-Bulb			Winter	Summer		Median of Annual Extr.	
State and Station[a]	°'N	°'W	Feet	99%	97.5%	1%	2.5%	5%		1%	2.5%	5%	___ Knots[d] ___			Max.	Min.
Tucson AP	32 07	110 56	2558	28	32	104/66	102/66	100/66	26	72	71	71	SE	WNW	6	108.9	0.3
Winslow AP	35 01	110 44	4895	5	10	97/61	95/60	93/60	32	66	65	64	SW	WSW	6	102.7	-0.4
Yuma AP	32 39	114 37	213	36	39	111/72	109/72	107/71	27	79	78	77	NNE	WSW	6	114.8	30.8
ARKANSAS																	
Blytheville AFB	35 57	89 57	264	10	15	96/78	94/77	91/76	21	81	80	78	N	SSW	8		
Camden	33 36	92 49	116	18	23	98/76	96/76	94/76	21	80	79	78					
El Dorado AP	33 13	92 49	277	18	23	98/76	96/76	94/76	21	80	79	78	S	SE	6	101.0	13.9
Fayetteville AP	36 00	94 10	1251	7	12	97/72	94/73	92/73	23	77	76	75	NE	SSW	9	99.4	-0.4
Fort Smith AP	35 20	94 22	463	12	17	101/75	98/76	95/76	24	80	79	78	NW	SW	8	101.9	7.0
Hot Springs	34 29	93 06	535	17	23	101/77	97/77	94/77	22	80	79	78	N	SW	8	103.0	10.6
Jonesboro	35 50	90 42	345	10	15	96/78	94/77	91/76	21	81	80	78				101.7	7.3
Little Rock AP	34 44	92 14	257	15	20	99/76	96/77	94/77	22	80	79	78	N	SW	9	99.0	11.2
Pine Bluff AP	34 18	92 05	241	16	22	100/78	97/77	95/78	22	81	80	80	N	SSW	7	102.2	13.1
Texarkana AP	33 27	93 59	389	18	23	98/76	96/77	93/76	21	80	79	78	WNW	SSW	9	104.8	14.0
CALIFORNIA																	
Bakersfield AP	35 25	119 03	475	30	32	104/70	101/69	98/68	32	73	71	70	ENE	WNW	5	109.8	25.3
Barstow AP	34 51	116 47	1927	26	29	106/68	104/68	102/67	37	73	71	70	WNW	W	7	110.4	17.4
Blythe AP	33 37	114 43	395	30	33	112/71	110/71	108/70	28	75	75	74				116.8	24.1
Burbank AP	34 12	118 21	775	37	39	95/68	91/68	88/67	25	71	70	69					
Chico	39 48	121 51	238	28	30	103/69	101/68	98/67	36	71	70	68	NW	S	3	109.0	22.6
Concord	37 58	121 59	200	28	27	100/69	97/68	94/67	32	71	71	70	NW	SSE	5		
Covina	34 05	117 52	575	32	35	98/69	95/68	92/67	31	73	71	70	WNW	NW	5		
Crescent City AP	41 46	124 12	40	31	33	68/60	65/59	63/58	18	62	60	59					
Downey	33 56	118 08	116	37	40	93/70	89/70	86/69	22	72	71	70					
El Cajon	32 49	116 58	367	42	44	83/69	80/69	78/68	30	71	70	68	W	SE	6		
El Centro AP	32 49	115 40	-43	35	38	112/74	110/74	108/74	34	81	80	78					
Escondido	33 07	117 05	660	39	41	89/68	85/68	82/68	30	71	70	69					
Eureka, Arcata AP	40 59	124 06	218	31	33	68/60	65/59	63/58	11	62	60	59	E	NW	5	75.8	29.7
Fairfield, Travis AFB	38 16	121 56	62	29	32	99/68	95/67	91/66	34	70	68	67	N	WSW	5		
Fresno AP	36 46	119 43	328	28	30	102/70	100/69	97/68	34	72	71	70	E	WNW	4	108.7	25.8
Hamilton AFB	38 04	122 30	3	30	32	89/68	84/66	80/65	28	72	69	67	N	SE	4		
Laguna Beach	33 33	117 47	35	41	43	83/68	80/68	77/67	18	70	69	68					
Livermore	37 42	121 57	545	24	27	100/69	97/68	93/67	24	71	70	68	WNW	NW	4		
Lompoc, Vandenberg AFB	34 43	120 34	368	35	38	75/61	70/61	67/60	20	63	61	60	ESE	NW	5		
Long Beach AP	33 49	118 09	30	41	43	83/68	80/68	77/67	22	70	69	68	NW	WNW	4		

Col. 1	Col. 2 Lat. °′N	Col. 3 Long. °′W	Col. 4 Elev. Feet	Col. 5 Winter Design Dry-Bulb 99%	97.5%	Col. 6 Summer Design Dry-Bulb and Mean Coincident Wet-Bulb 1%	2.5%	5%	Col. 7 Mean Daily Range	Col. 8 Design Wet-Bulb 1%	2.5%	5%	Col. 9 Prevailing Wind Winter	Knots[d]	Summer	Col. 10 Median of Annual Extr. Max.	Min.
Los Angeles AP	33 56	118 24	97	41	43	83/68	80/68	77/67	15	70	69	68	E	4	WSW	98.1	35.9
Los Angeles Co	34 03	118 14	270	37	40	93/70	89/70	86/69	20	72	71	70	NW	4	NW		
Merced, Castle AFB	37 23	120 34	188	29	31	102/70	99/69	95/68	36	72	71	70	ESE	4	NW	105.8	26.2
Modesto	37 39	121 00	91	28	30	101/69	98/68	95/67	36	71	70	69				103.1	25.8
Monterey	36 36	121 54	39	35	38	75/63	71/61	68/61	20	64	62	61	SE	4	NW		
Napa	38 13	122 17	56	30	32	100/69	96/68	92/67	30	71	69	68					
Needles AP	34 46	114 37	913	30	33	112/71	110/71	108/70	27	75	75	74				116.4	26.7
Oakland AP	37 49	122 19	5	34	36	85/64	80/63	75/62	19	66	64	63	E	5	WNW	93.0	31.8
Oceanside	33 14	117 25	26	41	43	83/68	80/68	77/67	13	70	69	68					
Ontario	34 03	117 36	952	31	33	102/70	99/69	96/67	36	74	72	71	E	4	WSW		
Oxnard	34 12	119 11	49	34	36	83/66	80/64	77/63	19	70	68	67					
Palmdale AP	34 38	118 06	2542	18	22	103/65	101/65	98/64	35	69	67	66	SW	5	WSW		
Palm Springs	33 49	116 32	411	33	35	112/71	110/70	108/70	35	76	74	73					
Pasadena	34 09	118 09	864	32	35	98/69	95/68	92/67	29	73	71	70				102.8	30.4
Petaluma	38 14	122 38	16	26	29	94/68	90/66	87/65	31	72	70	68				102.0	24.2
Pomona Co	34 03	117 45	934	28	30	102/70	99/69	95/68	36	74	72	71	E	4	W	105.7	26.2
Redding AP	40 31	122 18	495	29	31	105/68	102/67	100/66	32	71	69	68				109.2	26.0
Redlands	34 03	117 11	1318	31	33	102/70	99/69	96/68	33	74	72	71				106.7	27.1
Richmond	37 56	122 21	55	34	36	85/64	80/63	75/62	17	66	64	63					
Riverside, March AFB	33 54	117 15	1532	29	32	100/68	98/68	95/67	37	72	71	70	N	4	NW	107.6	26.6
Sacramento AP	38 31	121 30	17	30	32	101/70	98/70	94/69	36	72	71	70	NNW	6	SW	105.1	27.6
Salinas AP	36 40	121 36	75	30	32	74/61	70/60	67/59	24	62	61	59					
San Bernardino, Norton AFB	34 08	117 16	1125	31	33	102/70	99/69	96/68	38	74	72	71	E	3	W	109.3	25.3
San Diego AP	32 44	117 10	13	42	44	83/69	80/69	78/68	12	71	70	68	NE	3	WNW	91.2	37.4
San Fernando	34 17	118 28	965	37	39	95/68	91/68	88/67	38	71	70	69					
San Francisco AP	37 37	122 23	8	35	38	82/64	77/63	73/62	20	65	64	62	S	5	NW	91.3	35.9
San Francisco Co	37 46	122 26	72	38	40	74/63	71/62	69/61	14	64	62	61	W	5	NNW		
San Jose AP	37 22	121 56	56	34	36	85/66	81/65	77/64	26	68	67	65	SE	4	W	98.6	28.2
San Luis Obispo	35 20	120 43	250	33	35	92/69	88/70	84/69	26	73	71	70	E	4	NNW	99.8	29.3
Santa Ana AP	33 45	117 52	115	37	39	89/69	85/68	82/68	28	71	70	69	SE	4	W	101.0	29.9
Santa Barbara AP	34 26	119 50	10	34	36	81/67	77/66	75/65	24	68	67	66	E	3	SW	97.1	31.7
Santa Cruz	36 59	122 01	125	35	38	75/63	71/61	68/61	28	64	62	61	NE	3	SW		
Santa Maria AP	34 54	120 27	236	31	33	81/64	76/63	73/62	23	65	64	63	E	4	WNW	97.5	26.8
Santa Monica Co	34 01	118 29	64	41	43	83/68	80/68	77/67	16	70	69	68					
Santa Paula	34 21	119 05	263	33	35	90/68	86/67	84/66	36	70	69	68					
Santa Rosa	38 31	122 49	125	27	29	99/68	95/67	91/66	34	70	68	67	N	5	SE	102.5	23.4
Stockton AP	37 54	121 15	22	28	30	100/69	97/68	94/67	37	71	70	68	WNW	4	NW	104.1	24.5
Ukiah	39 09	123 12	623	27	29	99/69	95/68	91/67	40	70	68	67				108.1	21.6

TABLE 3.1 Climatic Design Criteria (Continued)

Col. 1	Col. 2		Col. 3		Col. 4	Winter,[b] °F Col. 5		Summer,[c] °F Col. 6			Col. 7	Col. 8			Prevailing Wind Col. 9			Temp., °F Col. 10	
State and Station[a]	Lat. °N	'	Long. °W	'	Elev. Feet	Design Dry-Bulb 99%	97.5%	Design Dry-Bulb and Mean Coincident Wet-Bulb 1%	2.5%	5%	Mean Daily Range	Design Wet-Bulb 1%	2.5%	5%	Winter	Summer	Knots[d]	Median of Annual Extr. Max.	Min.
Visalia	36	20	119	18	325	28	30	102/70	100/69	97/68	38	72	71	70				108.4	25.1
Yreka	41	43	122	38	2625	13	17	95/65	92/64	89/63	38	67	65	64				102.8	7.1
Yuba City	39	08	121	36	80	29	31	104/68	101/67	99/66	36	71	69	68					
COLORADO																			
Alamosa AP	37	27	105	52	7537	-21	-16	84/57	82/57	80/57	35	62	61	60				96.0	-8.4
Boulder	40	00	105	16	5445	-2	8	93/59	91/59	89/59	27	64	63	62				92.3	-12.1
Colorado Springs AP	38	49	104	43	6145	-3	2	91/58	88/57	86/57	30	63	62	61	N	S	9	96.8	-10.4
Denver AP	39	45	104	52	5283	-5	1	93/59	91/59	89/59	28	64	63	62	S	SE	8	92.4	-11.2
Durango	37	17	107	53	6550	-1	4	89/59	87/59	85/59	30	64	63	62				95.2	-18.1
Fort Collins	40	35	105	05	4999	-10	-4	93/59	91/59	89/59	28	64	63	62					
Grand Junction AP	39	07	108	32	4843	2	7	96/59	94/59	92/59	29	64	63	62	ESE	WNW	5	99.9	-3.4
Greeley	40	26	104	38	4648	-11	-5	96/60	94/60	92/60	29	65	64	63					
Lajunta AP	38	03	103	30	4160	-3	3	100/68	98/68	95/67	31	72	70	69	W	S	8		
Leadville	39	15	106	18	10155	-8	-4	84/52	81/51	78/50	30	56	55	54				79.7	-17.8
Pueblo AP	38	18	104	29	4641	-7	0	97/61	95/61	92/61	31	67	66	65	W	SE	5	100.5	-12.2
Sterling	40	37	103	12	3939	-7	2	95/62	93/62	90/62	30	67	66	65				100.3	-15.4
Trinidad AP	37	15	104	20	5740	-2	3	93/61	91/61	89/61	32	66	65	64	W	WSW	7	96.8	-10.5
CONNECTICUT																			
Bridgeport AP	41	11	73	11	25	6	9	86/73	84/71	81/70	18	75	74	73	NNW	WSW	13	95.7	-4.4
Hartford, Brainard Field	41	44	72	39	19	3	7	91/74	88/73	85/72	22	77	75	74	N	SSW	5	93.0	-0.2
New Haven AP	41	19	73	55	6	3	7	88/75	84/73	82/72	17	76	75	74	NNE	SW	7		
New London	41	21	72	06	59	5	9	88/73	85/72	83/71	16	76	75	74					
Norwalk	41	07	73	25	37	6	9	86/73	84/71	81/70	19	75	74	73					
Norwich	41	32	72	04	20	3	7	89/75	86/73	83/72	18	76	75	74					
Waterbury	41	35	73	04	843	-4	2	88/83	85/71	82/70	21	75	74	72	N	SW	8		
Windsor Locks, Bradley Fld	41	56	72	41	169	0	4	91/74	88/72	85/71	22	76	75	73	N	SW	8		
DELAWARE																			
Dover AFB	39	08	75	28	28	11	15	92/75	90/75	87/74	18	79	77	76	W	SW	9	97.0	7.0
Wilmington AP	39	40	75	36	74	10	14	92/74	89/74	87/73	20	77	76	75	WNW	WSW	9	95.4	4.9
DISTRICT OF COLUMBIA																			
Andrews AFB	38	05	76	05	279	10	14	92/75	90/74	87/73	18	78	76	75					
Washington. National AP	38	51	77	02	14	14	17	93/75	91/74	89/74	18	78	77	76	WNW	S	11	97.6	7.4

3.6

Climatic conditions — design values

Col. 1 State and Station[a]	Col. 2 Lat. °N	Col. 3 Long. °W	Col. 4 Elev. Feet	Winter °F — Col. 5 Design Dry-Bulb 99%	97.5%	Summer °F — Col. 6 Design Dry-Bulb and Mean Coincident Wet-Bulb 1%	2.5%	5%	Col. 7 Mean Daily Range	Col. 8 Design Wet-Bulb 1%	2.5%	5%	Col. 9 Prevailing Wind Winter Knots[d]	Summer	Temp. °F — Col. 10 Median of Annual Extr. Max.	Min.
FLORIDA																
Belle Glade	26 39	80 39	16	41	44	92/76	91/76	89/76	16	79	78	78			94.7	30.9
Cape Kennedy AP	28 29	80 34	16	35	38	90/78	88/78	87/78	15	80	79	79				
Daytona Beach AP	29 11	81 03	31	32	35	92/78	90/77	88/77	15	80	79	78	NW 8			
E Fort Lauderdale	26 04	80 09	10	42	46	91/78	91/78	90/78	15	80	79	79	NW 9	ESE		
Fort Myers AP	26 35	81 52	15	41	44	93/78	92/78	91/77	18	80	79	79	NNE 7	W	94.9	34.9
Fort Pierce	27 28	80 21	25	38	42	91/78	90/78	89/78	15	80	79	79			96.1	34.0
Gainesville AP	29 41	82 16	152	28	31	95/77	93/77	92/77	18	80	79	78	W 6	W	97.8	23.3
Jacksonville AP	30 30	81 42	26	29	32	96/77	94/77	92/76	19	79	79	78	NW 7	SW	97.5	25.4
Key West AP	24 33	81 45	4	55	57	90/78	90/78	89/78	09	80	79	79	NNE 12	SE	92.0	51.5
Lakeland Co	28 02	81 57	214	39	41	93/76	91/76	89/76	17	79	78	78	NNW 9	SSW		
Miami AP	25 48	80 16	7	44	47	91/77	90/77	89/77	15	79	79	78	NNW 8	SE	92.5	39.0
Miami Beach Co	25 47	80 17	10	45	48	90/77	89/77	88/77	10	79	79	78	NNW 8	SE		
Ocala	29 11	82 08	89	31	34	95/77	93/77	92/76	18	80	79	78			98.6	24.8
Orlando AP	28 33	81 23	100	35	38	94/76	93/76	91/76	17	79	78	78	NNW 9	SSW		
Panama City, Tyndall AFB	30 04	85 35	18	29	33	92/78	90/77	89/77	14	81	80	79	N 8	WSW	96.3	23.3
Pensacola Co	30 25	87 13	56	25	29	94/77	93/77	91/77	14	80	79	79	NNE 7	SW	97.6	25.8
St. Augustine	29 58	81 20	10	31	35	92/78	89/78	87/78	16	80	79	79	N 7	W	94.8	35.6
St. Petersburg	27 46	82 80	35	36	40	92/77	91/77	90/76	16	79	79	78	N 8	W		
Sanford	28 46	81 17	89	35	38	94/76	93/76	91/76	17	79	78	78				
Sarasota	27 23	82 33	26	39	42	93/77	92/77	90/76	17	79	78	78				
Tallahassee AP	30 23	84 22	55	27	30	94/77	92/76	90/76	19	79	78	78	NW 6	NW	97.6	20.9
Tampa AP	27 58	82 32	19	36	40	92/77	91/77	90/76	17	79	78	78	N 8	W	95.0	31.5
West Palm Beach AP	26 41	80 06	15	41	45	92/78	91/78	90/78	16	80	79	79	NW 9	ESE		
GEORGIA																
Albany, Turner AFB	31 36	84 05	223	25	29	97/77	95/76	93/76	20	80	79	78	N 7	W	100.6	19.9
Americus	32 03	84 14	456	21	25	97/77	94/76	92/75	20	79	78	77			100.4	16.5
Athens	33 57	83 19	802	18	22	94/74	92/74	90/74	21	78	77	76	NW 9	WNW	98.7	13.5
Atlanta AP	33 39	84 26	1010	17	22	94/74	92/74	90/73	19	77	76	75	NW 11	NW	95.7	11.9
Augusta AP	33 22	81 58	145	20	23	97/77	95/76	93/76	19	80	79	78	W 4	WSW	99.0	17.5
Brunswick	31 15	81 29	25	29	32	92/78	89/78	87/78	18	80	79	79			99.3	24.7
Columbus, Lawson AFB	32 31	84 56	242	21	24	95/76	93/76	91/75	21	80	79	78	NW 8	W		
Dalton	34 34	84 57	720	17	22	94/76	93/76	91/76	22	79	78	77				
Dublin	32 20	82 54	215	21	25	96/77	93/76	91/75	20	79	78	77				
Gainesville	34 11	83 41	50	21	27	96/77	93/77	91/77	21	79	78	77			101.0	16.7
Griffin	33 13	84 16	981	24	27	93/76	90/75	88/74	21	80	79	78	WNW 7	SW	98.7	21.9
LaGrange	33 01	85 04	709	19	22	94/76	91/75	89/74	21	78	77	76			97.7	11.9
Macon AP	32 42	83 39	354	21	25	96/77	93/75	91/75	22	79	78	77	NW 8	WNW	99.6	16.9

3.7

TABLE 3.1 Climatic Design Criteria (Continued)

Col. 1	Col. 2	Col. 3	Col. 4	Winter,[b] °F Col. 5		Summer,[c] °F Col. 6			Col. 7	Col. 8			Prevailing Wind Col. 9			Temp., °F Col. 10	
State and Station[a]	Lat. °N	Long. °W	Elev. Feet	Design Dry-Bulb 99%	97.5%	Design Dry-Bulb and Mean Coincident Wet-Bulb 1%	2.5%	5%	Mean Daily Range	Design Wet-Bulb 1%	2.5%	5%	Winter	Summer	Knots[d]	Median of Annual Extr. Max.	Min.
Marietta, Dobbins AFB	33 55	84 31	1068	17	21	94/74	92/74	90/74	21	78	77	76	NNW	NW	12		
Savannah	32 08	81 12	50	24	27	96/77	93/77	91/77	20	80	79	78	WNW	SW	7	98.7	21.9
Valdosta-Moody AFB	30 58	83 12	233	28	31	96/77	94/77	92/76	20	80	79	78	WNW	W	6		
Waycross	31 15	82 24	148	26	29	96/77	94/77	91/76	20	80	79	78				100.0	19.5
HAWAII																	
Hilo AP	19 43	155 05	36	61	62	84/73	83/72	82/72	15	75	74	74	SW	NE	6		
Honolulu AP	21 20	157 55	13	62	63	87/73	86/73	85/72	12	76	75	74	ENE	ENE	12		
Kaneohe Bay MCAS	21 27	157 46	18	65	66	85/75	84/74	83/74	12	76	76	75	NNE	NE	9		
Wahiawa	21 03	158 02	900	58	59	86/73	85/72	84/72	14	75	74	73	WNW	E	5		
IDAHO																	
Boise AP	43 34	116 13	2838	3	10	96/65	94/64	91/64	31	68	66	65	SE	NW	6	103.2	0.6
Burley	42 32	113 46	4156	-3	2	99/62	95/61	92/66	35	64	63	61				98.6	-8.3
Coeur D'Alene AP	47 46	116 49	2972	-8	-1	89/62	86/61	83/60	31	64	63	61				99.9	-4.5
Idaho Falls AP	43 31	112 04	4741	-11	-6	89/61	87/61	84/59	38	65	63	61	N	S	9	96.2	-16.0
Lewiston AP	46 23	117 01	1413	-1	6	96/65	93/64	90/63	32	67	66	64	W	WNW	3	105.9	2.7
Moscow	46 44	116 58	2660	-7	0	90/63	87/62	84/61	32	65	64	62				98.0	-5.9
Mountain Home AFB	43 02	115 54	2996	6	12	97/63	94/62	94/62	36	66	65	63	ESE	NW	7	103.2	-6.5
Pocatello AP	42 55	112 36	4454	-8	-1	94/61	91/60	89/59	35	64	63	61	NE	W	5	97.9	-11.4
Twin Falls AP	42 29	114 29	4150	-3	2	99/62	95/61	92/60	34	64	63	61	SE	NW	6	100.9	-5.1
ILLINOIS																	
Aurora	41 45	88 20	744	-6	-1	93/76	91/76	88/75	20	79	78	76				96.7	-13.0
Belleville, Scott AFB	38 33	89 51	453	1	6	94/76	92/76	89/75	21	79	78	76	WNW	S	8		
Bloomington	40 29	88 57	876	-6	-2	92/75	90/74	88/73	21	78	76	75				98.4	-9.6
Carbondale	37 47	89 15	417	2	7	95/77	93/77	90/76	21	80	79	77				100.9	-0.8
Champaign/Urbana	40 02	88 17	777	-3	2	95/75	92/74	90/73	21	78	77	75					
Chicago, Midway AP	41 47	87 45	607	-5	0	94/74	91/73	88/72	20	78	76	74	NW	SW	11		
Chicago, O'Hare AP	41 59	87 54	658	-8	-4	91/74	89/74	86/72	20	77	75	74	WNW	SW	9		
Chicago Co	41 53	87 38	590	-3	2	94/75	91/74	88/73	15	79	77	75				96.1	-8.3
Danville	40 12	87 36	695	-4	1	93/75	90/74	88/73	21	78	77	75				98.2	-8.4
Decatur	39 50	88 52	679	-3	2	94/75	91/74	88/73	21	78	77	75	W	SSW	10	99.0	-8.1
Dixon	41 50	89 29	696	-7	-2	93/75	90/74	88/73	23	78	77	75	NW	SW	10	97.5	-13.5
Elgin	42 02	88 16	758	-7	-2	91/75	88/74	86/73	21	78	77	75					
Freeport	42 18	89 37	780	-9	-4	91/74	89/73	87/72	24	77	76	74					
Galesburg	40 56	90 26	764	-7	-2	93/75	91/75	88/74	22	78	77	75	WNW	SW	8		

Col. 1 State and Station[a]	Col. 2 Lat. °'N	Col. 3 Long. °'W	Col. 4 Elev. Feet	Winter[b] Col. 5 Design Dry-Bulb 99%	97.5%	Summer[c] Col. 6 Design Dry-Bulb and Mean Coincident Wet-Bulb 1%	2.5%	5%	Col. 7 Mean Daily Range	Col. 8 Design Wet-Bulb 1%	2.5%	5%	Prevailing Wind Col. 9 Winter Knots[d]	Summer	Temp. °F Col. 10 Median of Annual Extr. Max.	Min.
Greenville	38 53	89 24	563	-1	4	94/76	92/75	89/74	21	79	78	76				
Joliet	41 31	88 10	582	-5	0	93/75	90/74	88/73	20	78	77	75	NW 11	SW		
Kankakee	41 05	87 55	625	-4	1	93/75	90/74	88/73	21	78	77	75				
La Salle/Peru	41 19	89 06	520	-7	-2	93/75	91/75	88/74	22	78	77	75				
Macomb	40 28	90 40	702	-5	0	95/76	92/76	89/75	22	79	78	75				
Moline AP	41 27	90 31	582	-9	-4	93/75	91/75	88/74	23	78	77	75	WNW 8	SW	96.8	-12.7
Mt Vernon	38 19	88 52	479	0	5	95/76	92/75	89/74	21	79	78	76			100.5	-2.9
Peoria AP	40 40	89 41	652	-8	-4	91/75	89/74	87/73	22	78	76	75	WNW 8	SW	98.0	-10.9
Quincy AP	39 57	91 12	769	-2	3	96/76	93/76	90/76	21	80	78	77	NW 11	SSW	101.1	-6.7
Rantoul, Chanute AFB	40 18	88 08	753	-4	4	94/75	91/74	89/73	21	78	77	75	W 10	SSW		
Rockford	42 21	89 03	741	-9	-4	91/74	89/73	87/73	24	77	76	74			97.4	-13.8
Springfield AP	39 50	89 40	588	-3	2	94/75	92/74	89/74	21	78	77	76	NW 10	SW	98.1	-7.2
Waukegan	42 21	87 53	700	-6	-3	92/76	89/74	87/73	21	78	76	75			96.5	-10.6
INDIANA																
Anderson	40 06	85 37	919	0	6	95/76	92/75	89/74	22	79	78	76	W 9	SW	95.1	-6.0
Bedford	38 51	86 30	670	0	5	95/76	92/75	89/74	22	79	78	76			97.5	-4.4
Bloomington	39 08	86 37	847	0	5	95/76	92/75	89/74	22	79	78	76	W 9	SW	97.8	-4.6
Columbus, Bakalar AFB	39 16	85 54	651	3	7	94/75	92/75	90/74	22	79	78	76	W 9	SW	98.3	-6.4
Crawfordsville	40 03	86 54	679	-2	3	94/75	91/74	88/73	22	79	77	76			98.4	-7.6
Evansville AP	38 03	87 32	381	4	9	95/76	93/75	91/75	22	79	78	77			98.2	0.2
Fort Wayne AP	41 00	85 12	791	-4	1	92/73	89/72	87/72	24	78	75	74	NW 9	SW	96.8	-10.5
Goshen AP	41 32	85 48	827	-3	2	91/73	89/73	87/72	23	77	75	74	WSW 10	SW	98.5	-8.5
Hobart	41 32	87 15	600	-4	2	91/73	88/73	85/72	21	77	75	74			96.9	-8.1
Huntington	40 53	85 30	802	-4	2	92/73	89/72	87/72	23	77	75	74			96	-.7
Indianapolis AP	39 44	85 17	792	-2	2	92/74	90/74	87/73	22	78	76	75	WNW 10	SW	98	2
Jeffersonville	38 17	85 45	455	5	10	95/74	93/74	90/74	23	78	77	76			98.2	-7.5
Kokomo	40 25	86 03	855	-4	0	91/74	90/73	88/73	22	77	76	75				
Lafayette	40 2	86 5	600	-3	3	94/74	91/73	88/73	22	78	76	75				
La Porte	41 36	86 43	810	-3	3	93/74	90/73	87/73	22	78	76	75			98.1	-10.5
Marion	40 29	85 41	859	-4	0	91/74	90/73	88/73	23	77	75	74			97.0	-8.6
Muncie	40 11	85 21	957	-3	3	92/74	90/73	87/73	22	78	76	75				
Peru, Grissom AFB	40 39	86 09	813	-6	-1	90/74	88/73	86/73	22	76	75	74	W 10	SW		
Richmond AP	39 46	84 50	1141	-2	2	92/74	90/74	87/73	22	77	76	75			94.8	-8.5
Shelbyville	39 31	85 47	750	-1	3	93/74	91/74	88/73	22	78	76	75			97.7	-6.0
South Bend AP	41 42	86 19	773	-3	1	91/73	89/73	86/72	22	77	75	74	SW 11	SSW	96.2	-9.2
Terre Haute AP	39 27	87 18	585	-2	4	95/75	92/74	89/73	22	79	77	76	NNW 7	SSW	98.3	-4.9
Valparaiso	41 31	87 02	801	-3	3	93/74	90/74	87/73	22	78	76	75			95.5	-11.0
Vincennes	38 41	87 32	420	1	6	95/75	92/74	90/73	22	79	77	76			100.3	-2.8

TABLE 3.1 Climatic Design Criteria (Continued)

Col. 1	Col. 2	Col. 3	Col. 4	Col. 5 Winter,[b] °F Design Dry-Bulb		Col. 6 Summer,[c] °F Design Dry-Bulb and Coincident Wet-Bulb			Col. 7 Mean Daily Range	Col. 8 Design Wet-Bulb			Col. 9 Prevailing Wind Knots[d]		Col. 10 Temp., °F Median of Annual Extr.	
State and Station[a]	Lat. °N	Long. °W	Elev. Feet	99%	97.5%	1%	2.5%	5%	Range	1%	2.5%	5%	Winter	Summer	Max.	Min.
IOWA																
Ames	42 02	93 48	1099	−11	−6	93/75	90/74	87/73	23	78	76	75	NW 9	SSW	97.4	−17.8
Burlington AP	40 47	91 07	692	−7	−3	94/74	91/75	88/73	22	78	77	75	NW 9	S	98.6	−11.0
Cedar Rapids AP	41 53	91 42	863	−10	−5	91/76	88/75	86/74	23	78	77	75			97.7	−15.6
Clinton	41 50	90 13	595	−8	−3	92/75	90/75	87/74	23	78	77	75			97.5	−13.8
Council Bluffs	41 20	95 49	1210	−8	−3	94/76	91/75	88/74	22	78	77	75	NW 11	S		
Des Moines AP	41 32	93 39	938	−10	−5	94/75	91/74	88/74	23	78	75	74	N 10	SSW	98.2	−14.2
Dubuque	42 24	90 42	1056	−12	−7	90/74	88/73	86/72	22	77	75	74	NW 11	S	95.2	−15.0
Fort Dodge	42 33	94 11	1162	−12	−7	91/74	88/74	86/72	23	77	75	74	NW 9	SSW	98.5	−19.1
Iowa City	41 38	91 33	661	−11	−6	92/76	89/76	87/74	22	80	78	76			97.4	−15.2
Keokuk	40 24	91 24	574	−5	0	95/75	92/75	89/74	22	79	77	76			98.4	−8.8
Marshalltown	42 04	92 56	898	−12	−7	92/76	90/75	88/74	23	78	77	75	NW 11	S	98.5	−13.4
Mason City AP	43 09	93 20	1213	−15	−11	90/74	88/74	85/72	24	77	75	74			96.5	−21.7
Newton	41 41	93 02	936	−10	−5	94/75	91/74	88/73	23	78	77	75			98.2	−14.7
Ottumwa AP	41 06	92 27	840	−8	−4	94/75	91/74	88/73	22	78	77	75	NNW 9	S	99.1	−12.0
Sioux City AP	42 24	96 23	1095	−11	−7	95/74	92/74	89/73	24	78	77	75	NW 9	S	99.9	−17.7
Waterloo	42 33	92 24	868	−15	−10	91/76	89/75	86/74	23	78	77	75			97.7	−19.8
KANSAS																
Atchison	39 34	95 07	945	−2	2	96/77	93/76	91/76	23	81	79	77	NNW 11	SSW	100.5	−8.8
Chanute AP	37 40	95 29	981	3	7	100/74	97/74	94/74	23	78	77	76	N 12	SSW	102.8	−2.8
Dodge City AP	37 46	99 58	2582	0	5	100/69	97/69	95/69	25	74	73	71			102.9	−7.0
El Dorado	37 49	96 50	1282	3	7	101/72	98/73	96/73	24	77	76	75			103.5	−5.0
Emporia	38 20	96 12	1210	1	5	100/74	97/74	94/73	25	78	77	76			102.4	−6.4
Garden City AP	37 56	100 44	2880	−1	4	99/69	96/69	94/69	28	74	73	71				
Goodland AP	39 22	101 42	3654	−5	0	99/66	96/65	93/66	31	71	70	68	WSW 10	S	103.2	−10.4
Great Bend	38 21	98 52	1889	0	4	101/73	98/73	95/73	28	78	76	75				
Hutchinson AP	38 04	97 52	1542	4	8	102/72	99/72	97/72	28	77	75	74	N 14	S	105.3	−6.1
Liberal	37 03	100 58	2870	2	7	99/68	96/68	94/68	28	73	72	71	NNE 8	S	105.8	−3.8
Manhattan, Ft Riley	39 03	96 46	1065	−1	3	99/75	96/75	92/74	24	78	77	76	NNW 11	SSW	104.5	−8.6
Parsons	37 20	95 31	899	5	9	100/74	97/74	94/74	23	79	77	76				
Russell AP	38 52	98 49	1866	0	4	101/73	98/73	95/73	29	78	76	75				
Salina	38 48	97 39	1272	0	5	103/74	100/74	97/73	26	78	77	75	N 8	SSW		
Topeka AP	39 04	97 38	877	0	4	99/75	96/75	93/74	24	79	78	77	NNW 10	S	101.8	−6.4
Wichita AP	37 39	97 25	1321	3	7	101/72	98/73	96/73	23	77	76	75	NNW 12	SSW	102.5	−2.8

3.10

Table Col. 1 – Col. 10: Climatic Design Conditions

State and Station[a]	Lat. °'N	Long. °'W	Elev. Feet	Winter[b] Design Dry-Bulb 99%	97.5%	Summer[c] Design Dry-Bulb and Mean Coincident Wet-Bulb 1%	2.5%	5%	Mean Daily Range	Design Wet-Bulb 1%	2.5%	5%	Prevailing Wind Winter	Summer	Knots[d]	Median of Annual Extr. Max	Min
KENTUCKY																	
Ashland	38 33	82 44	546	5	10	94/76	91/74	89/73	22	78	77	75	W	SW	6	97.4	0.8
Bowling Green AP	35 58	86 28	535	4	10	94/77	92/75	89/74	21	79	77	76				99.9	1.2
Corbin AP	36 57	84 06	1174	4	9	94/73	92/75	89/72	23	77	76	75					
Covington AP	39 03	84 40	869	1	6	92/73	90/72	88/72	22	77	75	74	W	SW	9		
Hopkinsville, Ft Campbell	36 40	87 29	571	4	10	94/77	92/75	89/74	21	79	77	76	N	W	6	100.1	-0.4
Lexington AP	38 02	84 36	966	3	8	93/73	91/73	88/72	22	77	76	75	WNW	SW	9	95.3	-0.5
Louisville AP	38 11	85 44	477	5	10	95/74	93/74	90/74	23	79	77	76	NW	SW	8	97.4	1.2
Madisonville	37 19	87 29	439	5	10	96/76	93/75	90/75	22	79	78	77					
Owensboro	37 45	87 10	407	5	10	97/76	94/75	91/75	23	79	78	77					
Paducah AP	37 04	88 46	413	7	12	98/76	95/75	92/75	20	79	78	77	NW	SW	9	98.0	-0.2
LOUISIANA																	
Alexandria AP	31 24	92 18	92	23	27	95/77	94/77	92/77	20	80	79	78	N	S	7	100.1	25.7
Baton Rouge AP	30 32	91 09	64	25	29	95/77	93/77	92/77	19	80	80	79	ENE	W	8	98.0	21.4
Bogalusa	30 47	89 52	103	24	28	95/77	93/77	92/77	19	80	80	79				99.3	20.2
Houma	29 31	90 40	13	31	35	95/78	93/78	92/77	15	81	80	79				97.2	22.5
Lafayette AP	30 12	92 00	42	26	30	95/78	94/78	92/78	18	81	80	79	N	SW	8	98.2	22.6
Lake Charles AP	30 07	93 13	9	27	31	95/77	93/77	92/77	17	80	80	79	N	SSW	9	99.2	20.5
Minden	32 36	93 18	250	20	25	99/77	96/76	94/76	20	79	79	78				101.7	24.9
Monroe AP	32 31	92 02	79	22	26	99/77	96/76	94/76	20	79	79	78	N	S	9	101.1	25.9
Natchitoches	31 46	93 05	130	29	33	97/77	95/77	93/77	16	80	79	78					
New Orleans AP	29 59	90 15	4	20	25	93/78	92/78	90/77	16	81	80	79	NNE	SSW	9	96.3	27.7
Shreveport AP	32 28	93 49	254	20	25	99/77	96/76	94/76	20	79	79	78	S	S	9		
MAINE																	
Augusta AP	44 19	69 48	353	-7	-3	88/73	85/70	82/68	22	74	72	70	NNE	WNW	10		
Bangor, Dow AFB	44 48	68 50	192	-11	-6	86/70	83/68	80/67	22	73	71	69	WNW	S	7		
Caribou AP	46 52	68 01	624	-18	-13	84/69	81/67	78/66	21	71	69	67	WSW	SW	10		
Lewiston	44 02	70 15	200	-7	-2	88/73	85/70	82/68	22	74	72	70				94.0	-13.7
Millinocket AP	45 39	68 42	413	-13	-9	87/69	83/68	80/66	22	72	70	68	WNW	WNW	11	92.4	-23.0
Portland	43 39	70 19	43	-6	-1	87/72	84/71	81/69	22	74	72	70	W	S	7	93.5	-9.9
Waterville	44 32	69 40	302	-8	-4	87/72	84/69	81/68	22	74	72	70					

TABLE 3.1 Climatic Design Criteria (Continued)

Col. 1	Col. 2	Col. 3	Col. 4	Col. 5		Col. 6			Col. 7	Col. 8			Col. 9		Col. 10	
	Lat.	Long.	Elev.	Winter,[b] °F Design Dry-Bulb		Summer,[c] °F Design Dry-Bulb and Mean Coincident Wet-Bulb			Mean Daily Range	Design Wet-Bulb			Prevailing Wind		Temp., °F Median of Annual Extr.	
State and Station[a]	°N	°W	Feet	99%	97.5%	1%	2.5%	5%		1%	2.5%	5%	Winter	Summer Knots[d]	Max.	Min.
MARYLAND																
Baltimore AP	39 11	76 40	148	10	13	94/75	91/75	89/74	21	78	77	76	W 9	WSW 9		
Baltimore Co	39 20	76 25	20	14	17	92/77	89/76	87/75	17	80	78	76	WNW 9	S 9	97.9	7.2
Cumberland	39 37	78 46	790	6	10	92/75	89/74	87/74	22	77	76	75	WNW 10	W 10		
Frederick AP	39 27	77 25	313	8	12	94/76	91/75	88/74	22	78	77	76	N 9	WNW 9		
Hagerstown	39 42	77 44	704	8	12	94/75	91/74	89/74	22	77	76	75	WNW 10	W 10		
Salisbury	38 20	75 30	59	12	16	93/75	91/75	88/74	18	79	77	76			96.8	7.4
MASSACHUSETTS																
Boston AP	42 22	71 02	15	6	9	91/73	88/71	85/70	16	75	74	72	WNW 16	SW	95.7	-1.2
Clinton	42 24	71 41	398	-2	9	90/72	87/71	84/69	17	75	73	72			91.7	-8.5
Fall River	41 43	71 08	190	5	9	87/72	84/71	81/69	18	74	73	71	NW 10	SW	92.1	-1.0
Framingham	42 17	71 25	170	3	6	89/72	86/71	83/69	17	74	73	72			96.0	-7.7
Gloucester	42 35	70 41	10	-2	5	89/73	86/71	83/70	15	75	74	72				
Greenfield	42 3	72 4	205	-7	2	88/72	85/71	82/69	23	74	73	71				
Lawrence	42 42	71 10	57	-6	0	90/73	87/72	84/70	22	76	74	73	NW 8	WSW	95.2	-9.0
Lowell	42 39	71 19	90	-4	1	91/73	88/72	85/70	21	76	74	73	NW 10	SW	95.1	-8.5
New Bedford	41 41	70 58	79	5	9	85/72	82/71	80/69	19	74	73	72	NW 12	SW	91.4	2.2
Pittsfield AP	42 26	73 18	1194	-8	-3	87/71	84/70	81/68	23	74	73	70	N 8	SSW		
Springfield, Westover AFB	42 12	72 32	245	-5	0	90/72	87/71	84/69	19	75	73	73	W 14	W	95.7	-4.7
Taunton	41 54	71 04	20	5	9	89/73	86/72	83/70	18	75	74	73			92.9	-9.8
Worcester AP	42 16	71 52	986	0	4	87/71	84/70	81/68	18	73	72	70				
MICHIGAN																
Adrian	41 55	84 01	754	-1	3	91/73	88/72	85/71	23	76	75	73			97.2	-7.0
Alpena AP	45 04	83 26	610	-11	-6	89/70	85/70	83/69	27	73	72	70	W 5	SW	93.9	-14.8
Battle Creek AP	42 19	85 15	941	1	5	92/74	88/72	85/70	23	76	74	73	SW 8	SW		
Benton Harbor AP	42 08	86 26	643	1	5	91/72	88/72	85/70	20	75	74	72	SSW 8	WSW		
Detroit	42 25	83 01	619	3	6	91/73	88/72	86/71	20	76	74	73	W 11	SW	95.1	-2.6
Escanaba	45 44	87 05	607	-11	-7	83/69	83/69	80/68	17	73	71	69			88.8	-16.1
Flint AP	42 58	83 44	771	-4	1	90/73	87/72	85/71	25	74	74	72			95.3	-9.9
Grand Rapids AP	42 53	85 31	784	1	5	91/72	88/72	85/70	24	75	73	72	SW 8	SW	95.4	-5.6
Holland	42 42	86 06	678	2		88/72	85/71	83/70	22	75	73	72	WNW 8	WSW	94.1	-6.8
Jackson AP	42 16	84 28	1020	1	5	92/74	88/72	85/70	23	76	74	73			96.5	-7.8
Kalamazoo	42 17	85 36	955	1		92/74	88/72	85/70	23	76	74	73			95.9	-6.7
Lansing AP	42 47	85 36	873	-3	1	90/73	87/72	84/70	24	75	74	72	SW 12	W	94.6	-11.0
Marquette Co	46 34	87 24	735	-12	-8	84/70	81/69	77/66	18	72	70	68			94.5	-11.8
Mt Pleasant	43 35	84 46	796	0	3	91/73	87/72	84/71	24	76	74	72			95.4	-11.1
Muskegon AP	43 10	86 14	625	2	6	86/72	84/70	82/70	21	75	74	73	E 8	SW		
Pontiac	42 40	83 25	981	0	4	90/73	87/72	84/70	21	76	74	73			95.0	-6.8

3.12

Col. 1	Col. 2	Col. 3	Col. 4	Winter,[b] °F Col. 5		Summer,[c] °F Col. 6			Col. 7	Col. 8			Prevailing Wind Col. 9		Temp., °F Col. 10	
	Lat.	Long.	Elev.	Design Dry-Bulb		Design Dry-Bulb and Mean Coincident Wet-Bulb			Mean Daily Range	Design Wet-Bulb			Winter	Summer	Median of Annual Extr.	
State and Station[a]	°N '	°W '	Feet	99%	97.5%	1%	2.5%	5%		1%	2.5%	5%	Knots[d]		Max.	Min.
Port Huron	42 59	82 25	586	0	4	90/73	87/72	83/71	21	76	74	73	W 8	S	96.1	-7.6
Saginaw AP	43 32	84 05	667	0	4	91/73	87/72	84/71	23	76	74	72	WSW 7	SW	89.8	-21.0
Sault Ste. Marie AP	46 28	84 22	721	-12	-8	84/70	81/69	77/66	23	72	70	68	E 7	SW		
Traverse City AP	44 45	85 35	624	-3	1	89/72	86/71	83/69	22	75	73	71	SSW 9	SW	95.4	-10.7
Ypsilanti	42 14	83 32	716	1	5	92/72	89/71	86/70	22	75	74	72	SW 10	SW		
MINNESOTA																
Albert Lea	43 39	93 21	1220	-17	-12	90/74	87/72	84/71	24	77	75	73			95.1	-28.0
Alexandria AP	45 52	95 23	1430	-22	-16	91/72	88/72	85/70	24	76	74	72	N 8	S	94.5	-36.9
Bemidji AP	47 31	94 56	1389	-31	-26	88/69	85/69	81/67	24	73	71	69				
Brainerd	46 24	94 08	1227	-20	-16	90/73	87/71	84/69	24	75	73	71				
Duluth AP	46 50	92 11	1428	-21	-16	85/70	82/68	79/66	22	72	70	68	WNW 12	WSW	90.9	-27.4
Fairbault	44 18	93 16	940	-17	-12	91/74	88/72	85/71	24	77	75	73			95.8	-24.3
Fergus Falls	46 16	96 04	1210	-21	-17	91/72	88/72	85/70	24	76	74	72			96.9	-27.8
International Falls AP	48 34	93 23	1179	-29	-25	85/68	83/68	80/66	26	71	70	68	N 9	S	93.4	-36.5
Mankato	44 09	93 59	1004	-17	-12	91/72	88/72	85/70	24	77	75	73				
Minneapolis/St. Paul AP	44 53	93 13	834	-16	-12	92/75	89/73	86/71	22	77	75	73	NW 8	S	96.5	-22.0
Rochester AP	43 55	92 30	1297	-17	-12	90/74	87/72	84/71	24	77	75	73	NW 9	SSW		
St. Cloud AP	45 35	94 11	1043	-15	-11	91/74	88/72	85/70	24	76	74	72				
Virginia	47 30	92 33	1435	-25	-21	85/69	83/68	80/66	23	71	70	68			92.6	-33.0
Willmar	45 07	95 05	1128	-15	-11	91/74	88/72	85/71	24	76	74	72			96.8	-24.3
Winona	44 03	91 38	652	-14	-10	91/75	88/73	85/72	24	77	75	74				
MISSISSIPPI																
Biloxi, Keesler AFB	30 25	88 55	26	28	31	94/79	92/79	90/78	16	82	81	80	N 8	S	98.0	23.0
Clarksdale	34 12	90 34	178	14	19	96/77	94/77	92/76	21	80	79	78			100.9	13.2
Columbus AFB	33 39	88 27	219	15	20	95/77	93/77	91/76	22	80	79	78	N 7	W	101.6	12.7
Greenville AFB	33 29	90 59	138	15	20	95/77	93/77	91/76	21	80	79	78			99.5	14.9
Greenwood	33 30	90 05	148	15	20	95/77	93/77	91/76	21	80	79	78			100.6	15.3
Hattiesburg	31 16	89 15	148	24	27	96/78	94/77	92/77	21	81	80	79			99.9	18.2
Jackson AP	32 19	90 05	310	21	25	97/76	95/76	93/76	21	79	80	79			99.8	16.0
Laurel	31 40	89 10	236	24	27	96/78	94/77	92/77	21	81	80	79			99.7	17.8
McComb AP	31 15	90 28	469	21	26	96/77	94/76	92/76	18	80	79	78	NNW 6	NW		
Meridian AP	32 20	88 45	290	19	23	97/77	95/76	93/76	22	81	80	79			98.3	15.7
Natchez	31 33	91 23	195	23	27	96/78	94/78	92/77	21	81	80	79			98.4	18.4
Tupelo	34 16	88 46	361	14	19	96/77	94/77	92/77	22	80	79	78	N 6	WSW	100.7	11.8
Vicksburg Co	32 24	90 47	262	22	26	97/78	95/78	93/77	21	81	80	79			96.9	18.0

TABLE 3.1 Climatic Design Criteria (Continued)

Col. 1	Col. 2 Lat. °N		Col. 3 Long. °W		Col. 4 Elev.	Col. 5 Winter,[b] Design Dry-Bulb		Col. 6 Summer,[c] Design Dry-Bulb and Mean Coincident Wet-Bulb			Col. 7 Mean Daily	Col. 8 Design Wet-Bulb			Col. 9 Prevailing Wind		Col. 10 Temp., °F Median of Annual Extr.	
State and Station[a]	°	′	°	′	Feet	99%	97.5%	1%	2.5%	5%	Range	1%	2.5%	5%	Winter (Knots[d])	Summer	Max.	Min.
MISSOURI																		
Cape Girardeau	37	14	89	35	351	8	13	98/76	95/75	92/75	21	79	78	77	WNW 9	WSW	99.5	-6.2
Columbia AP	38	58	92	22	778	-1	4	97/74	94/74	91/73	22	78	77	76			99.9	-2.1
Farmington AP	37	46	90	24	928	3	8	96/76	93/75	90/74	22	78	77	75	NNW 11	SSW	98.4	-7.6
Hannibal	39	42	91	21	489	-2	3	96/76	93/76	90/76	22	80	78	77			101.2	-6.1
Jefferson City	38	34	92	11	640	2	7	98/75	95/74	92/74	23	78	77	76				
Joplin AP	37	09	94	30	980	6	10	100/73	97/73	94/73	24	78	77	76	NNW 12	SSW		
Kansas City AP	39	07	94	35	791	2	6	99/75	96/74	93/74	20	78	77	76	NW 9	S	100.2	-4.3
Kirksville AP	40	06	92	33	964	-5	0	96/74	93/74	90/73	24	78	77	76			98.3	-10.8
Mexico	39	11	91	54	775	-1	4	97/74	94/74	91/73	22	78	77	76			101.2	-8.0
Moberly	39	24	92	26	850	-2	3	97/74	94/74	91/73	23	78	77	76				
Poplar Bluff	36	46	90	25	380	11	16	98/78	95/76	92/76	22	81	79	78				
Rolla	37	59	91	43	1204	-3	2	94/77	91/75	89/74	22	78	77	77			99.4	-3.1
St. Joseph AP	39	46	94	55	825	3	2	96/77	93/76	91/76	23	81	79	77	NNW 9	S	100.6	-8.0
St. Louis AP	38	45	90	23	535	2	6	97/75	94/75	91/74	21	78	77	76	NW 9	WSW		
St. Louis Co	38	39	90	38	462	9	8	98/77	94/75	91/74	18	78	78	76	NW 6	S	99.1	-2.7
Sikeston	36	53	89	36	325	-1	4	98/77	95/76	92/75	21	80	78	77				
Sedalia, Whiteman AFB	38	43	93	33	869	3	15	95/76	92/76	90/75	22	79	78	76	NNW 7	SSW	100.0	-5.1
Sikeston	36	53	89	36	325			98/77	95/76	92/75	21	80	78	77				
Springfield AP	37	14	93	23	1268	3	9	96/73	93/74	91/74	23	78	77	75	NNW 10	S	97.2	-2.4
MONTANA																		
Billings AP	45	48	108	32	3567	-15	-10	94/64	91/64	88/63	31	67	66	64	NE 9	SW	100.5	-19.1
Bozeman	45	47	111	09	4448	-20	-14	90/61	87/60	84/59	32	63	62	60	S 5	NW	92.2	-23.2
Butte AP	45	57	112	30	5553	-24	-17	86/58	83/56	80/56	35	60	58	57			91.8	-26.3
Cut Bank AP	48	37	112	22	3838	-25	-20	88/61	85/61	82/60	35	64	62	61			94.7	-30.9
Glasgow AP	48	25	106	32	2760	-22	-18	92/64	89/63	85/62	29	68	66	64	E 8	S	103.3	-29.8
Glendive	47	08	104	48	2476	-18	-13	95/66	92/64	89/62	29	69	67	65			98.0	-25.1
Great Falls AP	47	29	111	22	3662	-21	-15	91/60	88/60	85/59	28	64	62	60	SW 7	WSW	99.7	-31.3
Havre	48	34	109	40	2492	-18	-11	94/65	90/64	87/63	33	68	66	65			95.6	-23.7
Helena AP	46	36	112	00	3828	-21	-16	91/60	88/60	85/59	32	64	62	61	N 12	WNW	94.4	-16.8
Kalispell AP	48	18	114	16	2974	-14	-7	91/62	87/61	84/60	34	65	63	62			96.2	-27.7
Lewistown AP	47	04	109	27	4122	-22	-16	90/62	87/61	83/60	30	65	63	62	NW 9	NW	97.2	-21.2
Livingston AP	45	42	110	26	4618	-20	-14	90/61	87/60	84/59	32	63	62	60				
Miles City AP	46	26	105	52	2634	-20	-15	98/66	95/66	92/65	30	70	68	67	NW 7	SE	103.6	-27.7
Missoula AP	46	55	114	05	3190	-13	-6	92/62	88/61	85/60	36	65	63	62	ESE 7	NW	98.6	-13.9

3.14

State and Station[a]	Lat. °'N	Long. °'W	Elev. Feet	Winter[b] Design Dry-Bulb 99%	97.5%	Summer[c] Design Dry-Bulb & Mean Coincident Wet-Bulb 1%	2.5%	5%	Mean Daily Range	Design Wet-Bulb 1%	2.5%	5%	Prevailing Wind Winter (Knots[d])	Summer	Median Annual Extr. Max	Min
NEBRASKA																
Beatrice	40 16	96 45	1235	-5	-2	99/75	95/74	92/74	24	78	77	76			103.1	-11.3
Chadron AP	42 50	103 05	3313	-8	-3	97/66	94/65	91/65	30	71	69	68				
Columbus	41 20	97 20	1450	-6	-2	98/74	95/73	92/73	25	77	76	75				
Fremont	41 26	96 29	1200	-6	-2	98/75	95/74	92/74	22	78	77	76				
Grand Island AP	40 59	98 19	1860	-8	-3	97/72	94/71	91/71	28	75	74	73	NNW 10	S	103.3	-14.2
Hastings	40 36	98 26	1954	-7	-3	97/72	94/71	91/71	27	75	74	73	NNW 10	S	103.5	-10.7
Kearney	40 44	99 01	2132	-9	-4	96/71	93/70	90/70	28	74	73	72			102.9	-13.7
Lincoln Co	40 51	96 45	1180	-5	-2	99/75	95/74	92/74	24	78	77	76	N 8	S	102.0	-12.4
McCook	40 12	100 38	2768	-6	-2	98/69	95/69	91/69	28	74	72	71				
Norfolk	41 59	97 26	1551	-8	-4	97/74	93/74	90/73	30	78	77	75				
North Platte AP	41 08	100 41	2779	-8	-4	97/69	94/69	90/69	28	74	72	71	NW 9	SSE	102.0	-20.0
Omaha AP	41 18	95 54	977	-8	-3	94/76	91/75	88/74	22	78	77	75	NW 8	S	100.8	-15.8
Scottsbluff AP	41 52	103 36	3958	-8	-3	95/65	92/65	90/64	31	70	68	67	NW 9	SE	100.2	-13.2
Sidney AP	41 13	103 06	4399	-8	-3	95/65	92/65	90/64	31	70	68	67			101.6	-18.9
NEVADA																
Carson City	39 10	119 46	4675	4	9	94/60	91/59	89/58	42	63	61	60	SSW 3	WNW	99.2	-5.0
Elko AP	40 50	115 47	5050	-8	-2	94/59	92/59	90/58	42	63	62	60	E 4	SW		
Ely AP	39 17	114 51	6253	-10	-4	89/57	87/56	85/55	39	60	59	58	S 9	SSW		
Las Vegas AP	36 05	115 10	2178	25	28	108/66	106/65	104/65	30	71	70	69	ENE 7	SW	103.0	-1.0
Lovelock AP	40 04	118 33	3903	8	12	98/63	96/63	93/62	42	66	65	64				
Reno AP	39 30	119 47	4404	5	10	95/61	92/60	90/59	45	64	62	61	SSW 3	WNW		
Reno Co	39 30	119 47	4408	6	11	96/61	93/60	91/59	45	64	62	61			98.9	0.2
Tonopah AP	38 04	117 05	5426	5	10	94/60	92/59	90/58	40	64	62	61	N 8	S		
Winnemucca AP	40 54	117 48	4301	-1	3	96/60	94/60	92/60	42	64	62	61	SE 10	W	100.1	-8.1
NEW HAMPSHIRE																
Berlin	44 03	71 01	1110	-14	-9	87/71	84/69	81/68	22	73	71	70				
Claremont	43 02	72 02	420	-9	-4	89/72	86/70	83/69	24	74	73	71				
Concord AP	43 12	71 30	342	-10	-3	90/72	87/70	84/69	26	74	73	71	NW 7	SW	93.2	-24.7
Keene	42 55	72 17	490	-12	-7	90/72	87/70	83/69	24	74	73	71				
Laconia	43 03	71 03	505	-10	-5	89/72	86/70	83/69	25	74	73	71			94.8	-16.0
Manchester, Grenier AFB	42 56	71 26	233	-8	-3	91/72	88/71	85/70	24	75	74	72	N 11	SW	94.6	-18.9
Portsmouth, Pease AFB	43 04	70 49	101	-2	2	89/73	85/71	83/70	22	75	74	72	W 8	W	93.7	-12.6

TABLE 3.1 Climatic Design Criteria (Continued)

Col. 1	Col. 2	Col. 3	Col. 4	Winter,[b] °F Col. 5		Summer,[c] °F Col. 6 Design Dry-Bulb and Mean Coincident Wet-Bulb			Col. 7	Col. 8 Design Wet-Bulb			Prevailing Wind Col. 9		Temp., °F Col. 10 Median of Annual Extr.	
State and Station[a]	Lat. °N	Long. °W	Elev. Feet	Design Dry-Bulb 99%	97.5%	1%	2.5%	5%	Mean Daily Range	1%	2.5%	5%	Winter Knots[d]	Summer	Max.	Min.
NEW JERSEY																
Atlantic City Co	39 23	74 26	11	10	13	92/74	89/74	86/72	18	78	77	75	NW 11	WSW	93.0	7.5
Long Branch	40 19	74 01	15	10	13	93/74	90/73	87/72	18	78	77	75	WNW 11	WSW	95.9	4.3
Newark AP	40 42	74 10	7	10	14	94/74	91/73	88/72	20	77	76	75	WNW 11	WSW		
New Brunswick	40 29	74 26	125	6	10	92/74	89/73	86/72	19	77	76	75				
Paterson	40 54	74 09	100	6	10	94/74	91/73	88/72	21	77	76	75				
Phillipsburg	40 41	75 11	180	1	6	92/73	89/72	86/71	21	76	75	74			97.4	-0.7
Trenton Co	40 13	74 46	56	11	14	91/75	88/74	85/73	19	78	76	75	W 9	SW	96.2	4.2
Vineland	39 29	75 00	112	8	11	91/75	89/74	86/73	19	78	76	75				
NEW MEXICO																
Alamagordo, Holloman AFB	32 51	106 06	4093	14	19	98/64	96/64	94/64	30	69	68	67				
Albuquerque AP	35 03	106 37	5311	12	16	96/61	94/61	92/61	27	66	65	64	N 7	W	98.1	5.1
Artesia	32 46	104 23	3320	13	19	103/67	100/67	97/67	30	72	71	70			105.5	3.7
Carlsbad AP	32 20	104 16	3293	13	19	103/67	100/67	97/67	28	72	71	70	N 6	SSE		
Clovis AP	34 23	103 19	4294	8	13	95/65	93/65	91/65	28	69	68	67			102.0	2.5
Farmington AP	36 44	108 14	5503	1	6	95/63	93/62	91/61	30	67	65	64				
Gallup	35 31	108 47	6465	0	5	90/59	90/59	86/58	32	64	62	61	ENE 5	SW		
Grants	35 10	107 54	6524	-1	4	89/59	88/58	85/57	32	64	62	61				
Hobbs AP	32 45	103 13	3690	13	18	101/66	99/66	97/66	29	71	70	69				
Las Cruces	32 18	106 55	4544	15	20	99/64	96/64	94/64	30	69	68	67	SE 5	SE		
Los Alamos	35 52	106 19	7410	5	9	89/60	87/60	85/60	32	62	61	60			89.8	-2.3
Raton AP	36 45	104 30	6373	-4	1	91/60	89/60	87/60	34	65	64	63				
Roswell, Walker AFB	33 18	104 32	3676	13	18	100/66	98/66	96/66	33	71	70	69	N 6	SSE	103.0	2.7
Santa Fe Co	35 37	106 05	6307	6	10	90/61	88/61	86/61	28	63	62	61			90.1	-1.2
Silver City AP	32 38	108 10	5442	5	10	95/61	94/60	91/60	30	66	64	63				
Socorro AP	34 03	106 53	4624	13	17	97/62	95/62	93/62	30	67	66	65				
Tucumcari AP	35 11	103 36	4039	8	13	99/66	97/66	95/65	28	70	69	68	NE 8	SW	102.7	1.1

Col. 1 State and Station[a]	Col. 2 Lat. °N	Col. 3 Long. °W	Col. 4 Elev. Feet	Col. 5 Winter[b] Design Dry-Bulb 99%	97.5%	Col. 6 Summer[c] Design Dry-Bulb and Mean Coincident Wet-Bulb 1%	2.5%	5%	Col. 7 Mean Daily Range	Col. 8 Design Wet-Bulb 1%	2.5%	5%	Col. 9 Prevailing Wind Winter (Knots[d])	Summer	Col. 10 Temp. °F Median of Annual Extr. Max.	Min.
NEW YORK																
Albany AP	42 45	73 48	275	- 6	- 1	91/73	88/72	85/70	23	75	74	72	WNW 8	S	95.2	-11.4
Albany Co	42 39	73 45	19	- 4	- 3	91/73	88/72	85/70	20	75	74	72			92.4	-9.5
Auburn	42 54	76 32	715	- 3	2	90/73	87/71	84/70	22	75	73	72			92.2	-7.5
Batavia	43 00	78 11	922	1	5	90/72	87/71	84/70	22	75	73	72			92.9	-9.3
Binghamton AP	42 13	75 59	1590	- 2	1	86/71	83/69	81/68	20	73	72	70	WSW 10	WSW		
Buffalo AP	42 56	78 44	705	2	6	88/71	85/70	83/69	21	74	73	72	W 10	SW	90.0	-3.2
Cortland	42 36	76 11	1129	- 5	0	88/71	85/71	82/70	23	74	73	71			93.8	-11.2
Dunkirk	42 29	79 16	692	4	9	88/73	85/72	83/71	18	75	74	72	SSW 10	WSW		
Elmira AP	42 10	76 54	955	- 4	1	89/71	86/71	83/70	24	74	73	71			96.2	-6.7
Geneva	42 45	76 54	613	- 3	2	90/73	87/71	84/70	22	75	73	72			96.1	-6.5
Glens Falls	43 20	73 37	328	- 11	- 5	88/72	85/71	82/69	23	74	73	71	NNW 6	S		
Gloversville	43 02	74 21	760	- 8	- 2	89/72	86/71	83/69	23	75	74	72			93.2	-14.6
Hornell	42 21	77 42	1325	- 4	0	88/71	85/70	82/69	24	74	73	71				
Ithaca	42 27	76 29	928	- 5	0	88/71	85/71	82/70	24	74	73	71	W 6	SW		
Jamestown	42 07	79 14	1390	- 1	3	88/70	86/70	83/69	20	74	72	71	WSW 9	WSW		
Kingston	41 56	74 00	279	- 3	2	91/73	88/72	85/70	22	76	74	73				
Lockport	43 09	79 15	638	- 4	7	89/74	86/72	84/71	21	76	74	73	N 9	SW	92.2	-4.8
Massena AP	44 56	74 51	207	- 13	- 8	86/70	83/69	80/68	20	73	72	70				
Newburgh, Stewart AFB	41 30	74 06	471	- 1	4	90/73	88/72	85/70	21	76	74	73	W 10	W		
NYC-Central Park	40 47	73 58	157	11	15	92/74	89/73	87/72	17	76	75	74			94.9	3.8
NYC-Kennedy AP	40 39	3 47	13	12	15	90/73	87/72	84/71	16	76	75	74	WNW 4	SSW		
NYC-La Guardia AP	40 46	73 54	11	11	15	92/74	89/73	87/72	16	76	74	73	WNW 15	SW		
Niagara Falls AP	43 06	79 57	590	4	7	89/74	86/72	84/71	20	76	74	73	W 9	SW		
Olean	42 14	78 22	2119	- 2	2	87/71	84/71	81/70	23	74	73	71				
Oneonta	42 31	75 04	1775	- 7	- 4	86/71	83/69	80/68	24	73	72	70				
Oswego Co	43 28	76 33	300	1	7	86/73	83/71	80/70	20	75	73	72	E 7	WSW	91.3	-7.4
Plattsburg AFB	44 39	73 28	235	- 13	- 8	86/70	83/69	80/68	22	73	73	70	NW 6	SE		
Poughkeepsie	41 38	73 55	165	0	6	92/74	89/74	86/72	21	77	75	74	NNE 6	SSW	98.1	-5.6
Rochester AP	43 07	77 40	547	1	5	91/73	88/71	85/70	22	75	73	72	WSW 11	WSW		
Rome, Griffiss AFB	43 14	75 25	514	- 11	- 5	88/71	85/70	83/69	22	75	73	71	NW 5	W		
Schenectady	42 51	73 57	377	- 4	1	90/73	87/72	84/70	22	75	74	72	WNW 8	S		
Suffolk County AFB	40 51	72 38	67	7	10	86/72	83/71	80/70	16	76	74	73	NW 9	SW		
Syracuse AP	43 07	76 07	410	- 3	2	90/73	87/71	84/70	20	75	73	72	N 7	WNW		
Utica	43 09	75 23	714	- 12	- 6	88/73	85/71	82/70	22	75	73	71	NW 12	W	93.	-10.0
Watertown	43 59	76 01	325	- 11	- 6	86/73	83/71	81/70	20	75	73	72	E 7	WSW	91.7	-19.6

TABLE 3.1 Climatic Design Criteria (Continued)

Col. 1	Col. 2	Col. 3	Col. 4	Col. 5 Winter,[b] °F Design Dry-Bulb		Col. 6 Summer,[c] °F Design Dry-Bulb and Mean Coincident Wet-Bulb			Col. 7 Mean Daily Range	Col. 8 Design Wet-Bulb			Col. 9 Prevailing Wind		Col. 10 Temp., °F Median of Annual Extr.	
State and Station[a]	Lat. ° 'N	Long. ° 'W	Elev. Feet	99%	97.5%	Mean 1%	Coincident 2.5%	5%	Range	1%	2.5%	5%	Winter	Summer Knots[d]	Max.	Min.
NORTH CAROLINA																
Asheville AP	35 26	82 32	2140	10	14	89/73	87/72	85/71	21	75	74	72	NNW 12	NNW	91.9	5.8
Charlotte AP	35 13	80 56	736	18	22	95/74	93/74	91/74	20	77	76	76	NNW 6	SW	97.8	12.6
Durham	35 52	78 47	434	16	20	94/75	92/75	90/75	20	78	77	76			98.9	9.6
Elizabeth City AP	36 16	76 11	12	12	19	93/76	91/77	89/76	18	80	78	78	NW 8	SW		
Fayetteville, Pope AFB	35 10	79 01	218	17	20	95/76	92/76	90/75	20	79	78	77	N 6	SSW	99.1	13.1
Goldsboro	35 20	77 58	109	18	21	94/77	91/76	89/75	18	79	78	77	N 8	SW	99.8	13.0
Greensboro AP	36 05	79 57	897	14	18	93/74	91/73	89/73	21	77	76	75	NE 7	SW	97.7	9.7
Greenville	35 37	77 25	75	18	21	93/77	91/76	89/75	19	79	78	77				
Henderson	36 22	78 25	480	12	15	95/77	92/76	90/76	20	79	78	77				
Hickory	35 45	81 23	1187	14	18	92/73	90/72	88/72	21	75	74	73			96.5	9.6
Jacksonville	34 50	77 37	95	20	24	92/78	90/78	88/77	18	80	79	78				
Lumberton	34 37	79 04	129	18	21	95/76	92/76	90/75	20	79	78	77				
New Bern AP	35 05	77 03	20	20	24	92/78	90/78	88/77	18	80	79	78			98.2	15.1
Raleigh/Durham AP	35 52	78 47	434	16	20	94/75	92/75	90/75	20	78	77	76	N 7	SW	97.7	12.2
Rocky Mount	35 58	77 48	121	18	21	94/77	91/76	89/75	19	79	78	77				
Wilmington AP	34 16	77 55	28	23	26	93/79	91/78	89/77	18	81	80	79	N 8	SW	96.9	18.2
Winston-Salem AP	36 08	80 13	969	16	20	94/74	91/73	89/73	20	76	75	74	NW 8	WSW		
NORTH DAKOTA																
Bismarck AP	46 46	100 45	1647	-23	-19	95/68	91/68	88/67	27	73	71	70	WNW 7	S	100.3	-31.5
Devils Lake	48 07	98 54	1450	-25	-21	91/69	88/68	85/66	25	73	71	69			97.5	-30.4
Dickinson AP	46 48	102 48	2585	-21	-17	94/68	90/66	87/65	25	71	69	68	WNW 12	SSE	101.0	-31.3
Fargo AP	46 54	96 48	896	-22	-18	92/73	89/71	85/69	25	76	74	72	SSE 11	S	97.3	-29.7
Grand Forks AP	47 57	97 24	911	-26	-22	91/70	87/70	84/68	25	74	72	70	N 8	S	97.6	-29.0
Jamestown AP	46 55	98 41	1492	-22	-18	94/70	90/69	87/68	26	74	74	71			101.3	-27.9
Minot AP	48 25	101 21	1668	-24	-20	92/68	89/67	86/65	25	72	70	68	WSW 10	S		
Williston	48 09	103 35	1876	-25	-21	91/68	88/67	85/65	25	72	70	68			99.7	-32.9

Col. 1 State and Station[a]	Col. 2 Lat. °N	Col. 3 Long. °W	Col. 4 Elev. Feet	Col. 5 Design Dry-Bulb 99%	97.5%	Col. 6 Design Dry-Bulb and Mean Coincident Wet-Bulb 1%	2.5%	5%	Col. 7 Mean Daily Range	Col. 8 Design Wet-Bulb 1%	2.5%	5%	Col. 9 Winter (Knots[d])	Summer	Col. 10 Median of Annual Extr. Max.	Min.
OHIO																
Akron, Canton AP	40 55	81 26	1208	1	6	89/72	86/71	84/70	21	75	73	72	SW 9	SW	94.4	-4.6
Ashtabula	41 51	80 48	690	4	9	88/73	85/72	83/71	18	75	74	72				
Athens	39 20	82 06	700	0	6	95/75	92/74	90/73	22	78	76	74				
Bowling Green	41 23	83 38	675	-2	2	92/73	89/73	86/71	23	76	75	73			96.7	-7.3
Cambridge	40 04	81 35	807	1	7	93/75	90/74	87/73	23	78	76	75				
Chillicothe	39 21	83 00	640	0	6	95/75	92/74	90/73	22	78	76	74	W 8	WSW	98.2	-2.1
Cincinnati Co	39 09	84 31	758	1	6	92/73	90/72	88/72	21	77	75	74	W 9	SW	97.2	-0.2
Cleveland AP	41 24	81 51	777	1	5	91/73	88/72	86/71	22	76	74	73	SW 12	N	94.7	-3.1
Columbus AP	40 00	82 53	812	0	5	92/73	90/73	87/72	24	77	75	74	W 8	SSW	96.0	-3.4
Dayton AP	39 54	84 13	1002	-1	4	91/73	89/72	86/71	20	76	75	73	WNW 11	SW	96.6	-4.5
Defiance	41 17	84 23	700	-1	4	94/74	91/73	88/72	24	77	76	74				
Findlay AP	41 01	83 40	804	-2	3	92/74	90/73	87/72	24	77	76	74			97.4	-7.4
Fremont	41 20	83 07	600	-3	1	90/73	88/73	85/71	24	76	75	73				
Hamilton	39 24	84 35	650	0	5	92/73	90/72	87/71	22	76	75	73			98.2	-2.8
Lancaster	39 44	82 38	860	0	5	93/74	91/73	88/72	23	77	75	74				
Lima	40 42	84 02	975	-1	4	94/74	91/73	88/72	24	77	76	74	WNW 11	SW	96.0	-6.5
Mansfield AP	40 49	82 31	1295	0	5	90/73	87/72	85/72	22	76	74	73	W 8	SW	93.8	-10.7
Marion	40 36	83 10	920	0	5	93/74	91/73	88/72	23	77	76	74				
Middletown	39 31	84 25	635	0	5	92/73	90/72	87/71	22	76	75	73				
Newark	40 01	82 28	880	-1	5	94/73	92/73	89/72	23	77	75	74	W 8	SSW	95.8	-6.8
Norwalk	41 16	82 37	670	-3	1	90/73	88/73	85/71	22	76	75	73			97.3	-8.3
Portsmouth	38 45	82 55	540	5	10	95/76	92/74	89/73	22	78	77	75	W 8	SW	97.9	1.0
Sandusky Co	41 27	82 43	606	1	6	93/73	91/72	88/71	21	76	74	73			96.7	-1.9
Springfield	39 50	83 50	1052	-1	3	91/74	89/73	87/72	21	77	76	74	W 7	W		
Steubenville	40 23	80 38	992	-1	5	89/72	86/71	84/70	22	74	73	72				
Toledo AP	41 36	83 48	669	-3	1	90/73	88/73	85/71	25	76	75	73	WSW 8	SW	95.4	-5.2
Warren	41 20	80 51	928	0	5	89/71	87/71	85/70	23	74	73	71				
Wooster	40 47	81 55	1020	-1	6	89/72	86/71	84/70	22	75	73	72			94.0	-7.7
Youngstown AP	41 16	80 40	1178	-1	4	88/71	86/71	84/70	23	74	73	71	SW 10	SW		
Zanesville AP	39 57	81 54	900	1	7	93/75	90/74	87/73	23	78	76	75	W 6	WSW		

TABLE 3.1 Climatic Design Criteria (Continued)

Col. 1	Col. 2	Col. 3	Col. 4	Col. 5		Col. 6			Col. 7	Col. 8			Col. 9		Col. 10	
	Lat.	Long.	Elev.	Winter,[b] °F		Summer,[c] °F							Prevailing Wind		Temp., °F	
				Design Dry-Bulb		Design Dry-Bulb and Coincident Wet-Bulb			Mean Daily Range	Design Wet-Bulb					Median of Annual Extr.	
State and Station[a]	°N	°W	Feet	99%	97.5%	1%	2.5%	5%		1%	2.5%	5%	Winter	Summer	Max.	Min.
													Knots[d]			
OKLAHOMA																
Ada	34 47	96 41	1015	10	14	100/74	97/74	95/74	23	77	76	75				
Altus AFB	34 39	99 16	1378	11	16	102/73	100/73	98/73	25	77	76	75	N 10	S		
Ardmore	34 18	97 01	771	13	17	100/74	98/74	95/74	23	77	77	76				
Bartlesville	36 45	96 00	715	6	10	101/73	98/74	95/74	23	77	77	76				
Chickasha	35 03	97 55	1085	10	14	101/74	98/74	95/74	24	78	77	76				
Enid, Vance AFB	36 21	97 55	1307	9	13	103/74	100/74	97/74	24	79	77	76				
Lawton AP	34 34	98 25	1096	12	16	101/74	99/74	96/74	24	78	77	76				
McAlester	34 50	95 55	776	14	19	99/74	96/74	93/74	23	77	76	75	N 10	S		
Muskogee AP	35 40	95 22	610	10	15	101/74	98/75	95/75	23	79	78	77				
Norman	35 15	97 29	1181	9	13	99/74	96/74	94/74	24	77	76	75	N 10	S		
Oklahoma City AP	35 24	97 36	1285	9	13	100/74	97/74	95/73	23	78	77	76	N 14	SSW		
Ponca City	36 44	97 06	997	5	9	100/74	97/74	94/74	24	77	76	76				
Seminole	35 14	96 40	865	11	15	99/74	96/74	94/73	23	77	76	75				
Stillwater	36 10	97 05	984	8	13	100/74	96/74	93/74	24	77	76	75	N 12	SSW	103.7	1.6
Tulsa AP	36 12	95 54	650	8	13	101/74	98/75	95/75	22	79	78	77	N 11	SSW		
Woodward	36 36	99 31	2165	6	10	100/73	97/73	94/73	26	78	77	75			107.1	-1.3
OREGON																
Albany	44 38	123 07	230	18	22	92/67	89/66	86/65	31	69	67	66				
Astoria AP	46 09	123 53	8	25	29	75/65	71/62	68/61	16	65	63	62	ESE 7	NNW	97.5	-6.8
Baker AP	44 50	117 49	3372	-1	6	92/63	89/61	86/60	30	65	63	61			96.4	-5.8
Bend	44 04	121 19	3595	-3	4	90/62	87/60	84/59	33	64	62	60			98.5	17.1
Corvallis	44 30	123 17	246	18	22	92/67	89/66	86/65	31	69	67	66	N 6	N		
Eugene AP	44 07	123 13	359	17	22	92/67	89/66	86/65	31	69	67	66	N 7	N		
Grants Pass	42 26	123 19	925	20	24	99/69	96/68	93/67	33	71	69	68	N 5	N	103.6	16.4
Klamath Falls AP	42 09	121 44	4092	4	9	90/61	87/60	84/59	36	63	61	60	N 4	W	96.3	0.9
Medford AP	42 22	122 52	1298	19	23	98/68	94/67	91/66	35	70	68	67	S 4	WMW	103.8	15.0
Pendleton AP	45 41	118 51	1482	-2	5	97/65	93/64	90/62	29	66	65	63	NNW 6	WNW		
Portland AP	45 36	122 36	21	17	23	89/68	85/67	81/65	23	69	67	66	ESE 12	NW	96.6	18.3
Portland Co	45 32	122 40	75	18	24	90/68	86/67	82/65	21	69	67	66			97.6	20.5
Roseburg AP	43 14	123 22	525	18	23	93/67	90/66	87/65	30	69	67	66			99.6	19.5
Salem AP	44 55	123 01	196	18	23	92/68	88/66	84/65	31	69	68	66	N 6	N	98.9	15.9
The Dalles	45 36	121 12	100	13	19	93/69	89/68	85/66	28	70	68	67			105.1	7.9

Col. 1	Col. 2	Col. 3	Col. 4	Col. 5		Col. 6			Col. 7	Col. 8			Col. 9		Col. 10	
State and Station	Lat. °'N	Long. °'W	Elev. Feet	Winter Design Dry-Bulb 99%	97.5%	Summer Design Dry-Bulb and Mean Coincident Wet-Bulb 1%	2.5%	5%	Mean Daily Range	Design Wet-Bulb 1%	2.5%	5%	Prevailing Wind Winter (Knots)	Summer	Median of Annual Extr. Max.	Min.
PENNSYLVANIA																
Allentown AP	40 39	75 26	387	4	9	92/73	88/72	86/72	22	76	75	73	W 11	SW	93.7	−5.2
Altoona Co	40 18	78 19	1504	0	5	90/72	87/71	84/70	23	74	73	72	WNW 11	WSW		
Butler	40 52	79 54	1100	1	6	90/73	87/72	85/71	22	75	74	73			97.4	−0.3
Chambersburg	39 56	77 38	640	4	8	93/75	90/74	87/73	23	77	76	75	SSW 0	WSW	91.3	−2.2
Erie AP	42 05	80 11	731	4	9	88/73	85/72	83/71	18	75	74	72	NW 11	WSW	96.5	3.7
Harrisburg AP	40 12	76 46	308	7	11	94/75	91/74	88/73	21	77	76	75				
Johnstown	40 19	78 50	2284	−3	2	86/70	83/70	80/68	23	72	71	70	WNW 8	WSW	96.4	−1.8
Lancaster	40 07	76 18	403	4	8	93/75	90/74	87/73	22	77	76	75	NW 11	WSW		
Meadville	41 38	80 10	1065	0	4	88/71	85/70	83/69	21	73	72	71			93.2	−8.5
New Castle	41 01	80 22	825	2	7	91/73	88/72	86/71	23	75	74	73	WSW 10	WSW	94.7	−6.4
Philadelphia AP	39 53	75 15	5	10	14	93/75	90/74	87/72	21	77	76	75	WNW 10	WSW	96.4	5.9
Pittsburgh AP	40 30	80 13	1137	1	5	89/72	86/71	84/70	22	74	73	72	WSW 10	WSW	94.6	−1.1
Pittsburgh Co	40 27	80 00	1017	3	7	91/72	88/71	86/70	19	74	73	72			97.0	3.6
Reading Co	40 20	75 38	266	9	13	92/73	89/72	86/72	19	76	75	73	W 11	SW	94.8	−2.2
Scranton/Wilkes-Barre	41 20	75 44	930	1	5	90/72	87/71	84/70	19	74	73	72	SW 8	WSW	93.2	−3.6
State College	40 48	77 52	1175	3	7	90/72	87/71	84/70	23	74	73	72	NNW 8	WSW		
Sunbury	40 53	76 46	446	2	7	92/73	89/72	86/70	22	75	74	73				
Uniontown	39 55	79 43	956	5	9	91/74	88/73	85/72	22	76	75	74			93.9	−2.5
Warren	41 51	79 08	1280	−2	4	89/71	86/71	83/70	24	74	73	72			93.3	−10.7
West Chester	39 58	75 38	450	9	13	92/75	89/74	86/72	20	77	76	75	w 9	WSW	95.5	−4.6
Williamsport AP	41 15	76 55	524	2	7	92/73	89/72	86/70	23	75	74	73			97.0	−2.4
York	39 55	76 45	390	8	12	94/75	91/74	88/73	22	77	76	75				
RHODE ISLAND																
Newport	41 30	71 20	10	5	9	88/73	85/72	82/70	16	76	75	73	WNW 10	SW		
Providence AP	41 44	71 26	51	5	9	89/73	86/72	83/70	19	75	74	73	WNW 11	SW	94.6	−0.5
SOUTH CAROLINA																
Anderson	34 30	82 43	774	19	23	94/74	92/74	90/74	21	77	76	75	NNE 8	SW	99.5	13.3
Charleston AFB	32 54	80 02	45	24	27	93/78	91/78	89/77	18	81	80	79			97.8	21.4
Charleston Co	32 54	79 58	3	25	28	94/78	92/78	90/77	13	81	80	79	W 6	SW	100.6	16.2
Columbia AP	33 57	81 07	213	20	24	97/76	95/75	93/75	22	79	78	77	N 7	SW	99.5	16.5
Florence AP	34 11	79 43	147	22	25	94/77	92/77	90/76	21	80	79	78	N 7	SSW	98.2	19.1
Georgetown	33 23	79 17	14	23	26	92/79	90/78	88/77	18	81	80	79				

TABLE 3.1 Climatic Design Criteria (Continued)

Col. 1	Col. 2	Col. 3	Col. 4	Col. 5		Col. 6			Col. 7	Col. 8			Col. 9		Col. 10	
	Lat.	Long.	Elev.	Winter,[b] °F		Summer,[c] °F							Prevailing Wind		Temp., °F	
				Design Dry-Bulb		Design Dry-Bulb and Mean Coincident Wet-Bulb			Mean Daily Range	Design Wet-Bulb			Winter	Summer	Median of Annual Extr.	
State and Station[a]	°N	°W	Feet	99%	97.5%	1%	2.5%	5%		1%	2.5%	5%	Knots[d]		Max.	Min.
Greenville AP	34 54	82 13	957	18	22	93/74	91/74	89/74	21	77	76	75	NW 8	SW	97.3	12.6
Greenwood	34 10	82 07	620	18	22	95/75	93/74	91/74	21	78	77	76			99.5	14.1
Orangeburg	33 30	80 52	260	20	24	97/76	95/75	93/75	20	79	78	77			101.2	18.0
Rock Hill	34 59	80 58	470	19	23	96/75	94/74	92/74	20	78	77	76				
Spartanburg AP	34 58	82 00	823	18	22	93/74	91/74	89/74	20	77	76	75	NNE 6		99.5	13.9
Sumter, Shaw AFB	33 54	80 22	169	22	25	95/77	92/76	90/75	21	79	78	77		W	100.0	15.4
SOUTH DAKOTA																
Aberdeen AP	45 27	98 26	1296	−19	−15	94/73	91/72	88/70	27	77	75	73	NNW 8	S	102.3	−28.1
Brookings	44 18	96 48	1637	−17	−13	95/73	92/72	89/71	25	77	75	73			97.8	−26.5
Huron AP	44 23	98 13	1281	−18	−14	96/73	93/72	89/71	28	77	75	73	NNW 8	S	101.5	−25.8
Mitchell	43 41	98 01	1346	−15	−10	96/72	93/71	90/70	28	76	75	73			103.0	−22.7
Pierre AP	44 23	100 17	1742	−15	−10	99/71	95/71	92/69	29	75	74	72	NW 11	SSE	105.7	−20.6
Rapid City AP	44 03	103 04	3162	−11	−7	95/66	92/65	89/65	28	71	69	67	NNW 10	SSE	100.9	−19.0
Sioux Falls AP	43 34	96 44	1418	−15	−11	94/73	91/72	88/71	24	76	75	73	NW 8	S	97.8	−26.5
Watertown AP	44 55	97 09	1738	−19	−15	94/73	91/72	88/71	26	76	75	73			100.8	−19.1
Yankton	42 55	97 23	1302	−13	−7	94/73	91/72	88/71	25	77	76	74				
TENNESSEE																
Athens	35 26	84 35	940	13	18	95/74	92/73	90/73	22	77	76	75				
Bristol-Tri City AP	36 29	82 24	1507	9	14	91/72	89/72	87/71	22	75	75	73	WNW 6	SW	97.2	9.8
Chattanooga AP	35 02	85 12	665	13	18	96/75	93/74	91/74	22	78	77	76	NNW 8	WSW	99.8	3.7
Clarksville	36 33	87 22	382	6	12	95/76	93/74	90/74	21	78	77	76				
Columbia	35 38	87 02	690	10	15	97/75	94/74	91/74	21	78	77	76				
Dyersburg	36 01	89 24	344	10	15	96/78	94/77	91/76	21	81	80	78				
Greeneville	36 04	82 50	1319	11	16	92/73	90/72	88/72	22	76	75	74	NE 8	W	95.6	0.8
Jackson AP	36 36	88 55	423	11	16	98/76	95/75	92/75	21	79	78	77	N 10	SW	99.2	6.6
Knoxville AP	35 49	83 59	980	13	19	94/74	92/73	90/73	21	77	76	75			96.0	7.0
Memphis AP	35 03	90 00	258	13	18	98/77	95/76	93/76	21	80	79	78			97.9	10.4
Murfreesboro	35 55	86 28	600	9	14	97/75	94/74	91/74	22	78	77	76			97.7	4.5
Nashville AP	36 07	86 41	590	9	14	97/75	94/74	91/74	21	78	77	76	NW 8	WSW		
Tullahoma	35 23	86 05	1067	8	13	96/74	93/73	91/73	22	77	76	75	NW 9	WSW	96.7	3.7

Col. 1	Col. 2		Col. 3		Col. 4	Col. 5		Col. 6			Col. 7	Col. 8			Col. 9		Col. 10	
State and Station [a]	Lat. °N		Long. °W		Elev. Feet	Winter,[b] °F Design Dry-Bulb		Summer,[c] °F Design Dry-Bulb and Mean Coincident Wet-Bulb			Mean Daily Range	Design Wet-Bulb			Prevailing Wind		Temp., °F Median of Annual Extr.	
	°	'	°	'		99%	97.5%	1%	2.5%	5%		1%	2.5%	5%	Winter Knots[d]	Summer	Max.	Min.
TEXAS																		
Abilene AP	32	25	99	41	1784	15	20	101/71	99/71	97/71	22	75	74	74	N 12	SSE	103.6	10.4
Alice AP	27	44	98	02	180	31	34	100/78	98/77	95/77	20	82	81	79		S	104.9	24.8
Amarillo AP	35	14	101	42	3604	6	11	98/67	95/67	93/67	26	71	70	70	N 11	S	100.8	0.9
Austin AP	30	18	97	42	597	24	28	100/74	98/74	97/74	22	78	77	77	N 11	S	101.6	19.7
Bay City	29	00	95	58	50	29	33	96/77	94/77	92/77	16	80	79	79				
Beaumont	29	57	94	01	16	27	31	95/79	93/78	91/78	19	81	80	80			99.7	23.5
Beeville	28	22	97	40	190	30	33	99/78	97/77	95/77	18	82	81	79	N 9	SSE	103.1	22.5
Big Spring AP	32	18	101	27	2598	16	20	100/69	97/69	95/69	26	74	73	72			105.3	10.7
Brownsville AP	25	54	97	26	19	35	39	94/77	93/77	92/77	18	80	79	79	NNW 13	SE	98.1	30.1
Brownwood	31	48	98	57	1386	18	22	101/73	99/73	96/73	22	77	76	75	N 9	S	105.3	13.0
Bryan AP	30	40	96	33	276	24	29	98/76	96/76	94/76	20	79	78	78				
Corpus Christi AP	27	46	97	30	41	31	35	95/78	94/78	92/78	19	80	80	79	N 12	SSE	97.0	27.2
Corsicana	32	05	96	28	425	20	25	100/75	98/75	96/75	21	79	78	77			104.2	15.2
Dallas AP	32	51	96	51	481	18	22	102/75	100/75	97/75	20	78	78	77	N 11	S		
Del Rio, Laughlin AFB	29	22	100	47	1081	26	31	98/73	98/73	97/73	24	79	77	76			103.8	23.0
Denton	33	12	97	06	630	17	22	101/74	99/74	97/74	22	78	77	76			104.5	11.8
Eagle Pass	28	52	100	32	884	27	32	101/73	99/73	98/73	24	78	78	77	NNW 9	ESE	107.7	22.1
El Paso AP	31	48	106	24	3918	20	24	100/64	98/64	96/64	27	69	68	68	N 7	S	103.0	15.7
Fort Worth AP	32	50	97	03	537	17	22	101/74	99/74	97/74	22	78	77	76	NW 11	S	103.2	13.5
Galveston AP	29	18	94	48	7	31	36	90/79	89/79	88/78	10	81	80	80	N 15	S	93.9	27.5
Greenville	33	04	96	03	535	17	22	101/74	99/74	97/74	21	78	77	77	NNW 10	SSE	103.6	11.7
Harlingen	26	14	97	39	35	35	39	96/77	94/77	93/77	19	80	79	79	NNW 11	S	102.3	29.3
Houston AP	29	58	95	21	96	27	32	96/77	94/77	92/77	18	80	79	79				
Houston Co	29	59	95	22	108	28	33	97/77	95/77	93/77	18	80	79	79			99.0	23.5
Huntsville	30	43	95	33	494	22	27	100/75	98/75	96/75	20	78	78	77			100.8	18.7
Killeen, Robert Gray AAF	31	05	97	41	850	20	25	99/73	97/73	95/73	22	77	76	75				
Lamesa	32	42	101	56	2965	13	17	99/69	96/69	94/69	26	73	72	71				
Laredo AFB	27	32	99	27	512	32	36	102/73	101/73	99/74	23	78	78	77	N 8	SE	105.5	8.9
Longview	32	28	94	44	330	19	24	99/76	97/76	95/76	20	80	79	78				
Lubbock AP	33	39	101	49	3254	10	15	98/69	96/69	94/69	26	73	72	71	NNE 10	SSE		
Lufkin AP	31	25	94	48	277	25	29	99/76	97/76	94/76	20	80	79	78	NNW 12	S		
McAllen	26	12	98	13	122	35	39	97/77	95/77	94/77	21	80	79	79				
Midland AP	31	57	102	11	2851	16	21	100/69	98/69	96/69	26	73	72	71	NE 9	SSE	103.6	10.8

3.23

TABLE 3.1 Climatic Design Criteria (Continued)

Col. 1 State and Station[a]	Col. 2 Lat. °'N	Col. 3 Long. °'W	Col. 4 Elev. Feet	Winter[b] Col. 5 Design Dry-Bulb 99%	97.5%	Summer[c] Col. 6 Design Dry-Bulb and Mean Coincident Wet-Bulb 1%	2.5%	5%	Col. 7 Mean Daily Range	Col. 8 Design Wet-Bulb 1%	2.5%	5%	Prevailing Wind Col. 9 Winter (Knots[d])	Summer	Col. 10 Median of Annual Extr. Max.	Min.
Mineral Wells AP	32 47	98 04	930	17	22	101/74	99/74	97/74	22	78	77	76			101.2	16.3
Palestine Co	31 47	95 38	600	23	27	100/76	98/76	96/76	20	79	79	78				
Pampa	35 32	100 59	3250	7	12	99/67	96/67	94/67	26	71	70	70				
Pecos	31 25	103 30	2610	16	21	100/69	98/69	96/69	27	73	72	71				
Plainview	34 11	101 42	3370	8	13	98/68	96/68	94/68	26	72	71	70			102.7	3.1
Port Arthur AP	29 57	94 01	16	27	31	95/79	93/78	91/78	19	81	80	80	N 9	S	97.7	24.0
San Angelo, Goodfellow AFB	31 26	100 24	1877	18	22	101/71	99/71	97/70	24	75	74	73	NNE 10	SSE		
San Antonio AP	29 32	98 28	788	25	30	99/72	97/73	96/73	19	77	76	76	N 8	SSE	101.3	21.1
Sherman, Perrin AFB	33 43	96 40	763	15	20	100/75	98/75	95/74	22	78	77	76	N 10	S	103.0	11.9
Snyder	32 43	100 55	2325	13	18	100/70	98/70	96/70	26	74	73	72				
Temple	31 06	97 21	700	22	27	100/74	99/74	97/74	22	78	77	77				
Tyler AP	32 21	95 16	530	19	24	99/76	97/76	95/76	21	80	79	78	NNE 23	S		
Vernon	34 10	99 18	1212	13	17	102/73	100/73	97/73	24	77	76	75				
Victoria AP	28 51	96 55	104	29	32	98/78	96/77	94/77	18	82	81	79			101.4	23.4
Waco AP	31 37	97 13	500	21	26	101/75	99/75	97/75	22	78	78	77				
Wichita Falls AP	33 58	98 29	994	14	18	103/73	101/73	98/73	24	77	76	75	NNW 12	S		
UTAH																
Cedar City AP	37 42	113 06	5617	-2	5	93/60	91/60	89/59	32	65	63	62	SE 5	SW	95.5	-7.8
Logan	41 45	111 49	4785	-3	2	93/62	91/61	88/60	33	65	64	63				
Moab	38 36	109 36	3965	6	11	100/60	98/60	96/60	30	65	64	63				
Ogden AP	41 12	112 01	4455	1	5	93/63	91/61	88/61	33	66	65	64	S 6	SW	99.5	-3.9
Price	39 37	110 50	5580	-2	2	93/60	91/60	89/59	33	65	63	62				
Provo	40 13	111 43	4448	1	6	98/62	96/62	94/61	32	66	65	64	SE 5	SW		
Richfield	38 46	112 05	5270	-2	5	93/60	91/60	89/59	34	65	63	62			98.1	-10.5
St George Co	37 02	113 31	2900	14	21	103/65	101/65	99/64	33	70	68	67			109.3	11.1
Salt Lake City AP	40 46	111 58	4220	3	8	97/62	95/62	92/61	32	66	65	64	SSE 6	N	99.4	-0.1
Vernal AP	40 27	109 31	5280	-5	0	91/61	89/60	86/59	32	64	63	62				
VERMONT																
Barre	44 12	72 31	600	-16	-11	84/71	81/69	78/68	23	73	71	70			92.4	-16.9
Burlington AP	44 28	73 09	332	-12	-7	88/72	85/70	82/69	23	74	72	71	E 7	SSW	92.5	-17.5
Rutland	43 36	72 58	620	-13	-8	87/72	84/70	81/69	23	74	72	71				

Col. 1 State and Station[a]	Col. 2 Lat. °N	Col. 3 Long. °W	Col. 4 Elev. Feet	Col. 5 Winter[b] Design Dry-Bulb 99%	97.5%	Col. 6 Summer[c] Design Dry-Bulb and Mean Coincident Wet-Bulb 1%	2.5%	5%	Col. 7 Mean Daily Range	Col. 8 Design Wet-Bulb 1%	2.5%	5%	Col. 9 Prevailing Wind Winter	Knots[d]	Summer	Col. 10 Temp. Median of Annual Extr. Max.	Min.
VIRGINIA																	
Charlottesville	38 02	78 31	870	14	18	94/74	91/74	88/73	23	77	76	75	NE	7	SW	97.4	8.0
Danville AP	36 34	79 20	590	16	16	94/74	92/73	90/73	21	77	76	75				100.1	9.2
Fredericksburg	38 18	77 28	100	14	14	96/76	93/75	90/74	21	78	77	76					
Harrisonburg	38 27	78 54	1370	12	16	93/72	91/72	88/71	23	75	74	73					
Lynchburg AP	37 20	79 12	916	12	16	93/77	90/74	88/73	21	77	76	75	NE	7	SW	97.2	7.6
Norfolk AP	36 54	76 12	22	20	22	93/77	91/76	89/76	18	79	78	77	NW	10	SW	97.2	15.3
Petersburg	37 11	77 31	194	14	17	95/76	92/76	90/75	20	79	78	77					
Richmond AP	37 30	77 20	164	14	17	95/76	92/76	90/75	21	79	78	77	N	6	SW	97.9	9.6
Roanoke AP	37 19	79 58	1193	12	16	93/72	91/72	88/71	23	75	74	73	NW	9	SW		
Staunton	38 16	78 54	1201	12	16	93/72	91/72	88/71	23	75	74	73	NW	9	SW	95.9	2.5
Winchester	39 12	78 10	760	6	10	93/75	90/74	88/74	21	77	76	75				97.3	3.7
WASHINGTON																	
Aberdeen	46 59	123 49	12	25	28	80/65	77/62	73/61	16	65	63	62	ESE	6	NNW	91.9	19.3
Bellingham AP	48 48	122 32	158	10	15	81/67	77/65	74/63	19	68	65	63	NNE	15	WSW	87.4	10.3
Bremerton	47 34	122 40	162	21	25	82/65	78/64	75/62	20	66	64	63	E	8	N		
Ellensburg AP	47 02	120 31	1735	2	6	94/65	91/64	87/62	34	66	65	63					15.2
Everett, Paine AFB	47 55	122 17	596	21	25	80/65	76/64	73/62	20	67	64	63	ESE	6	NNW	84.9	
Kennewick	46 13	119 08	392	5	11	99/68	96/67	92/66	30	70	68	67				103.4	2.0
Longview	46 10	122 56	12	19	24	88/68	85/67	81/65	30	69	67	66	ESE	8	NW	96.0	14.8
Moses Lake, Larson AFB	47 12	119 19	1185	1	7	97/66	94/65	90/63	32	67	66	64	N	8	SSW		
Olympia AP	46 58	122 54	215	16	22	87/66	83/65	79/64	32	67	66	64	NE	4	NE		
Port Angeles	48 07	123 26	99	24	27	72/62	69/61	67/60	18	64	62	61				83.5	19.4
Seattle-Boeing Field	47 32	122 18	23	21	26	84/68	81/66	77/65	24	69	67	65					
Seattle Co	47 39	122 18	20	22	27	85/68	82/66	78/65	19	67	67	65	N	7	N	90.2	22.0
Seattle-Tacoma AP	47 27	122 18	400	21	26	84/65	80/64	76/62	22	66	64	63	E	9	N	90.1	19.9
Spokane AP	47 38	117 31	2357	−6	2	93/64	90/63	87/62	28	65	64	62	NE	6	SW	98.8	−4.9
Tacoma, McChord AFB	47 15	122 30	100	19	24	86/66	82/65	79/63	22	68	66	64	S	5	NNE	89.4	18.8
Walla Walla AP	46 06	118 17	1206	0	7	97/67	94/66	90/65	27	69	67	66	W	5	W	103.0	3.8
Wenatchee	47 25	120 19	632	7	11	99/67	96/66	92/64	32	68	67	65					
Yakima AP	46 34	120 32	1052	−2	5	96/65	93/65	89/63	36	68	66	65	W	5	NW	101.1	1.0
WEST VIRGINIA																	
Beckley	37 47	81 07	2504	−2	4	83/71	81/69	79/69	22	73	71	70	WNW	9	WNW		
Bluefield AP	37 18	81 13	2867	−2	4	83/71	81/69	79/69	22	73	71	70					
Charleston AP	38 22	81 36	939	7	11	92/74	90/73	87/72	20	76	75	74	SW	8	SW	97.2	2.9

TABLE 3.1 Climatic Design Criteria (Continued)

Col. 1	Col. 2	Col. 3	Col. 4	Winter,[b] °F Col. 5		Summer,[c] °F Col. 6			Col. 7	Col. 8			Prevailing Wind Col. 9		Temp., °F Col. 10	
				Design Dry-Bulb		Design Dry-Bulb and Mean Coincident Wet-Bulb			Mean Daily	Design Wet-Bulb			Winter	Summer	Median of Annual Extr.	
State and Station[a]	Lat. °N	Long. °W	Elev. Feet	99%	97.5%	1%	2.5%	5%	Range	1%	2.5%	5%	Knots[d]		Max.	Min.
Clarksburg	39 16	80 21	977	6	10	92/74	90/73	87/72	21	76	75	74	WNW 9	WNW	90.6	-7.3
Elkins AP	38 53	79 51	1948	1	6	86/72	84/70	82/70	22	74	72	71	W 6	SW	97.1	2.1
Huntington Co	38 25	82 30	565	5	10	94/76	91/74	89/73	22	78	77	75			99.0	1.1
Martinsburg AP	39 24	77 59	556	6	10	93/75	90/74	88/74	21	77	76	75	WNW 10	W		
Morgantown AP	39 39	79 55	1240	4	8	90/74	87/73	85/73	22	76	75	74			95.9	0.7
Parkersburg Co	39 16	81 34	615	7	11	93/75	90/74	88/73	21	77	76	75	WSW 7	WSW		
Wheeling	40 07	80 42	665	1	5	89/72	86/71	84/70	21	74	73	72	WSW 10	WSW	97.5	-0.6
WISCONSIN																
Appleton	44 15	88 23	730	-14	-9	89/74	86/72	83/71	23	76	74	72			94.6	-16.2
Ashland	46 34	90 58	650	-21	-16	85/70	82/68	79/66	23	72	70	68			94.1	-26.8
Beloit	42 30	89 02	780	-7	-3	92/75	90/75	88/74	24	78	77	75				
Eau Claire AP	44 52	91 29	888	-15	-11	92/75	89/73	86/71	23	77	75	73				
Fond Du Lac	43 48	88 27	760	-12	-8	89/74	86/72	84/71	23	76	74	72			96.0	-17.7
Green Bay AP	44 29	88 08	682	-13	-9	88/74	85/72	83/71	23	76	74	72	W 8	SW	94.3	-17.9
La Crosse AP	43 52	91 15	651	-13	-9	91/75	88/73	85/72	22	77	75	74	NW 10	S	95.7	-21.3
Madison AP	43 08	89 20	858	-11	-7	91/74	88/73	85/71	22	76	75	73	NW 8	SW	93.6	-16.8
Manitowoc	44 06	87 41	660	-11	-7	89/74	86/72	83/71	21	76	74	72			94.1	-13.7
Marinette	45 06	87 38	605	-15	-11	87/73	84/71	82/70	20	75	73	71			95.9	-15.8
Milwaukee AP	42 57	87 54	672	-8	-4	90/74	87/73	84/71	21	76	74	73	WNW 10	SSW		
Racine	42 43	87 51	730	-6	-2	91/75	88/73	85/72	21	77	75	74				
Sheboygan	43 45	87 43	648	-10	-6	89/75	86/73	83/72	20	77	75	74			97.0	-12.4
Stevens Point	44 30	89 34	1079	-15	-11	92/75	89/73	86/71	23	77	75	73			95.3	-24.1
Waukesha	43 01	88 14	860	-9	-5	90/74	87/73	84/71	22	76	74	73			95.7	-14.3
Wausau AP	44 55	89 37	1196	-16	-12	91/74	88/72	85/70	23	76	74	72				
WYOMING																
Casper AP	42 55	106 28	5338	-11	-5	92/58	90/57	87/57	31	63	61	60	NE 10	SW	97.3	-20.9
Cheyenne	41 09	104 49	6126	-9	-1	89/58	86/58	84/57	30	63	62	60	N 11	WNW	92.5	-15.9
Cody AP	44 33	109 04	4990	-19	-13	89/60	86/60	83/59	32	64	63	61			97.4	-21.9
Evanston	41 16	110 57	6780	-9	-3	86/55	84/55	82/54	32	59	58	57			89.2	-21.2
Lander AP	42 49	108 44	5563	-16	-11	91/61	88/61	85/60	32	64	63	61	E 5	NW	94.9	-22.6
Laramie AP	41 19	105 41	7266	-14	-6	84/56	81/56	79/55	28	61	60	59				
Newcastle	43 51	104 13	4265	-17	-12	91/64	87/63	84/63	30	69	68	66			99.4	-19.0
Rawlins	41 48	107 12	6740	-12	-4	86/57	83/57	81/56	40	62	61	60				
Rock Springs AP	41 36	109 04	6745	-9	-3	86/55	84/55	82/54	32	59	58	57	WSW 10	W		
Sheridan AP	44 46	106 58	3964	-14	-8	94/62	91/62	88/61	32	66	65	63	NW 7	N	99.8	-23.6
Torrington	42 05	104 13	4098	-14	-8	94/62	91/62	88/61	30	66	65	63			101.1	-20.7

Col. 1 State and Station[a]	Col. 2 Lat. °N	Col. 3 Long. °W	Col. 4 Elev. Feet	Col. 5 Winter[b] Design Dry-Bulb 99%	97.5%	Col. 6 Summer[c] Design Dry-Bulb and Mean Coincident Wet-Bulb 1%	2.5%	5%	Col. 7 Mean Daily Range	Col. 8 Design Wet-Bulb 1%	2.5%	5%	Col. 9 Prevailing Wind Winter	Summer	Knots[d]	Col. 10 Median of Annual Extr. Max.	Min.
ALBERTA																	
Calgary AP	51 06	114 01	3540	-27	-23	84/63	81/61	79/60	25	65	63	62	NNW	SE	8		
Edmonton AP	53 34	113 31	2219	-29	-25	85/66	82/65	79/63	23	68	66	64	E	SE	9		
Grande Prairie AP	55 11	118 53	2190	-39	-33	83/64	80/63	78/61	23	66	64	62					
Jasper	52 53	118 04	3480	-31	-26	83/64	80/62	77/61	28	68	64	63					
Lethbridge AP	49 38	112 48	3018	-27	-22	90/65	87/64	84/63	28	68	66	65					
McMurray AP	56 39	111 13	1216	-41	-38	86/67	82/65	79/64	26	69	67	65					
Medicine Hat AP	50 01	110 43	2365	-29	-24	93/66	90/65	87/64	28	70	68	66					
Red Deer AP	52 11	113 54	2965	-31	-26	84/65	81/64	78/62	25	67	66	64					
BRITISH COLUMBIA																	
Dawson Creek	55 44	120 11	2164	-37	-33	82/64	79/63	76/61	26	66	64	62					
Fort Nelson AP	58 50	122 35	1230	-43	-40	84/64	81/63	78/62	23	67	65	64					
Kamloops Co	50 43	120 25	1133	-21	-15	94/66	91/65	88/64	29	68	66	65					
Nanaimo	49 11	123 58	230	16	20	83/67	80/65	77/64	21	68	66	65					
New Westminster	49 13	122 54	50	14	18	84/68	81/67	78/66	19	69	68	66					
Penticton AP	49 28	119 36	1121	0	4	92/68	89/67	87/66	31	70	68	67					
Prince George AP	53 53	122 41	2218	-33	-28	84/64	80/62	77/61	26	66	64	62	N	N	11		
Prince Rupert Co	54 17	130 22	170	-2	2	64/59	63/57	61/56	12	60	58	57					
Trail	49 08	117 44	1400	-5	0	92/66	89/65	86/64	33	68	67	66					
Vancouver AP	49 11	123 10	16	15	19	79/67	77/66	74/65	17	68	66	62					
Victoria Co	48 25	123 19	228	20	23	77/64	73/62	70/60	16	64	62	60	E	WNW	6		
MANITOBA																	
Brandon	49 52	99 59	1200	-30	-27	89/72	86/70	83/68	25	74	72	70	SE	S	11		
Churchill AP	58 45	94 04	155	-41	-39	81/66	77/64	74/62	18	67	65	63					
Dauphin AP	51 06	100 03	999	-31	-28	87/71	84/70	81/68	23	74	72	70					
Flin Flon	54 46	101 51	1098	-41	-37	84/68	81/66	79/65	19	70	68	67					
Portage La Prairie AP	49 54	98 16	867	-28	-24	88/73	86/72	83/70	22	76	74	71	W	W	8		
The Pas AP	53 58	101 06	894	-37	-33	85/68	82/67	79/66	20	71	69	68	W	W	8		
Winnipeg AP	49 54	97 14	786	-30	-27	89/73	86/71	84/70	22	75	73	71	W	S	8		

TABLE 3.1 Climatic Design Criteria (Continued)

Col. 1	Col. 2	Col. 3	Col. 4	Col. 5 Winter,[b] °F Design Dry-Bulb		Col. 6 Summer,[c] °F Design Dry-Bulb and Mean Coincident Wet-Bulb			Col. 7 Mean Daily Range	Col. 8 Design Wet-Bulb			Col. 9 Prevailing Wind (Knots[d])		Col. 10 Temp., °F Median of Annual Extr.	
State and Station[a]	Lat. ° ′ N	Long. ° ′ W	Elev. Feet	99%	97.5%	1%	2.5%	5%		1%	2.5%	5%	Winter	Summer	Max.	Min.
NEW BRUNSWICK																
Campbellton Co	48 00	66 40	25	−18	−14	85/68	82/67	79/66	21	70	68	66				
Chatham AP	47 01	65 27	112	−15	−10	89/69	85/68	82/67	22	71	69	68				
Edmundston Co	47 22	68 20	500	−21	−16	87/70	83/68	80/67	21	71	69	68				
Fredericton AP	45 52	66 32	74	−16	−11	89/71	85/69	82/68	23	73	71	69				
Moncton AP	46 07	64 41	248	−12	−8	85/70	82/69	79/67	23	71	69	68				
Saint John AP	45 19	65 53	352	−12	−8	80/67	77/65	75/64	19	69	67	65				
NEWFOUNDLAND																
Corner Brook	48 58	57 57	15	−5	0	76/64	73/63	71/62	17	66	64	63				
Gander AP	48 57	54 34	482	−5	−1	82/66	79/65	77/64	19	68	66	65	WNW 11	SW		
Goose Bay AP	53 19	60 25	144	−27	−24	85/66	81/64	77/63	19	67	66	64	N 9	SW		
St John's AP	47 37	52 45	463	3	7	77/66	75/65	73/64	18	68	66	65	N 20	WSW		
Stephenville AP	48 32	58 33	44	−3	4	76/65	74/64	71/63	14	67	65	64	WNW 10	S		
NORTHWEST TERRITORIES																
Fort Smith AP	60 01	111 58	665	−49	−45	85/66	81/64	78/63	24	68	66	64	NW 4	S		
Frobisher AP	63 45	68 33	68	−43	−41	66/53	63/51	59/50	14	54	52	51	NNW 9	NW		
Inuvik	68 18	133 29	200	−56	−53	79/62	77/60	75/59	21	63	61	60				
Resolute AP	74 43	94 59	209	−50	−47	57/48	54/46	51/45	10	49	47	46				
Yellowknife AP	62 28	114 27	682	−49	−46	79/62	77/61	74/60	16	64	62	61	SSE 7	S		
NOVA SCOTIA																
Amherst	45 49	64 13	65	−11	−6	84/69	81/68	79/67	21	71	69	68				
Halifax AP	44 39	63 34	83	1	5	79/66	76/65	74/64	16	68	66	65				
Kentville	45 03	64 36	40	−3	1	85/69	83/68	80/67	22	71	69	68				
New Glasgow	45 37	62 37	317	−9	−5	81/69	79/68	77/67	20	71	69	68				
Sydney AP	46 10	60 03	197	−1	3	82/69	80/68	77/66	19	70	68	67				
Truro Co	45 22	63 16	131	−8	−5	82/70	80/69	78/68	22	72	70	69				
Yarmouth AP	43 50	66 05	136	5	9	74/65	72/64	70/63	15	67	65	64	NW 11	S		

Col. 1 State and Station	Col. 2 Lat. °N (°)	(')	Col. 3 Long. °W (°)	(')	Col. 4 Elev. Feet	Col. 5 Winter °F Design Dry-Bulb 99%	97.5%	Col. 6 Summer °F Design Dry-Bulb and Mean Coincident Wet-Bulb 1%	2.5%	5%	Col. 7 Mean Daily Range	Col. 8 Design Wet-Bulb 1%	2.5%	5%	Col. 9 Prevailing Wind Winter (Knots)	Summer	Col. 10 Median of Annual Extr. Max.	Min.
ONTARIO																		
Belleville	44	09	77	24	250	−11	−7	86/73	84/72	82/71	20	75	74	73				
Chatham	42	24	82	12	600	0	3	89/74	87/73	85/72	19	76	75	74				
Cornwall	45	01	74	45	210	−13	−9	89/73	87/72	84/71	21	75	74	72				
Hamilton	43	16	79	54	303	−3	1	88/73	86/72	83/71	21	76	74	73				
Kapuskasing AP	49	25	82	28	752	−31	−28	86/70	83/69	80/67	23	72	70	69				
Kenora AP	49	48	94	22	1345	−32	−28	84/70	82/69	80/68	19	73	71	70				
Kingston	44	16	76	30	300	−11	−7	87/73	84/72	82/71	20	75	74	73				
Kitchener	43	26	80	30	1125	−6	−2	88/73	85/72	83/71	23	75	74	72				
London AP	43	02	81	09	912	−4	0	87/74	85/73	83/72	21	76	74	73				
North Bay AP	46	22	79	25	1210	−22	−18	84/68	81/67	79/66	20	71	70	68				
Oshawa	43	54	78	52	370	−6	−3	88/73	86/72	84/71	21	75	74	73				
Ottawa AP	45	19	75	40	413	−17	−13	90/72	87/71	84/70	21	75	73	72				
Owen Sound	44	34	80	55	597	−6	−2	84/71	82/70	80/69	21	73	72	70				
Peterborough	44	17	78	19	635	−13	−9	87/72	85/71	83/70	21	75	73	72				
St Catharines	43	11	79	14	325	−1	3	87/73	85/72	83/71	20	76	74	73				
Sarnia	42	58	82	22	625	0	3	88/73	86/72	84/71	19	76	74	73				
Sault Ste Marie AP	46	32	84	30	675	−7	−13	85/71	82/69	79/68	22	73	71	70				
Sudbury AP	46	37	80	48	1121	−22	−19	86/69	83/67	81/66	22	72	70	68				
Thunder Bay AP	48	22	89	19	644	−27	−24	85/70	83/68	80/67	24	72	70	68	W 8	W		
Timmins AP	48	34	81	22	965	−33	−29	87/69	84/68	81/66	25	72	70	68	N 10	SW		
Toronto AP	43	41	79	38	578	−5	−1	90/73	87/72	85/71	20	75	74	73				
Windsor AP	42	16	82	58	637	0	4	90/74	88/73	86/72	20	77	75	74				
PRINCE EDWARD ISLAND																		
Charlottetown AP	46	17	63	08	186	−7	−4	80/69	78/68	76/67	16	71	70	68				
Summerside AP	46	26	63	50	78	−8	−4	81/69	79/68	77/67	16	72	70	68				

TABLE 3.1 Climatic Design Criteria (Concluded)

Col. 1	Col. 2		Col. 3		Col. 4	Col. 5 (Winter,[b] °F)		Mean	Col. 6 (Summer,[c] °F) Design Dry-Bulb and Coincident Wet-Bulb			Col. 7	Col. 8 Design Wet-Bulb			Col. 9 Prevailing Wind		Col. 10
State and Station[a]	Lat. °N	'	Long. °W	'	Elev. Feet	Design Dry-Bulb 99%	97.5%	1%	2.5%	5%	Mean Daily Range	1%	2.5%	5%	Winter (Knots[d])	Summer	Median of Annual Extr. Max. / Min.
QUEBEC																	
Bagotville AP	48	20	71	00	536	−28	−23	87/70	83/68	80/67	21	72	70	68			
Chicoutimi	48	25	71	05	150	−26	−22	86/70	83/68	80/67	20	72	70	68			
Drummondville	45	53	72	29	270	−18	−14	88/72	85/71	82/69	21	75	73	71			
Granby	45	23	72	42	550	−19	−14	88/72	85/71	83/70	21	75	73	72			
Hull	45	25	75	44	200	−18	−14	90/72	87/71	84/70	21	75	73	72			
Megantic AP	45	35	70	52	1362	−20	−16	86/71	83/70	81/69	20	74	72	71			
Montreal AP	45	28	73	45	98	−16	−10	88/73	85/72	83/71	17	75	74	72			
Quebec AP	46	48	71	23	245	−19	−14	87/72	84/70	81/68	20	74	72	70			
Rimouski	48	27	68	32	117	−16	−12	83/68	79/66	76/65	18	71	69	67			
St Jean	45	18	73	16	129	−15	−11	88/73	86/72	84/71	20	75	74	72			
St Jerome	45	48	74	01	556	−17	−13	88/73	86/71	83/70	23	75	73	72			
Sept. Iles AP	50	13	66	16	190	−26	−21	76/63	73/61	70/60	17	67	65	63			
Shawinigan	46	34	72	43	306	−18	−14	86/72	84/70	82/69	21	74	73	71			
Sherbrooke Co	45	24	71	54	595	−25	−21	86/72	84/71	81/69	20	74	73	71			
Thetford Mines	46	04	71	19	1020	−19	−14	87/71	84/70	81/69	21	74	72	71			
Trois Rivieres	46	21	72	35	50	−17	−13	88/72	85/70	82/69	23	74	72	71			
Val D'or AP	48	03	77	47	1108	−32	−27	85/70	83/68	80/67	22	72	70	68			
Valleyfield	45	16	74	06	150	−14	−10	89/73	86/72	84/71	20	75	74	72			
SASKATCHEWAN																	
Estevan AP	49	04	103	00	1884	−30	−25	92/70	89/68	86/67	26	72	70	69			
Moose Jaw AP	50	20	105	33	1857	−29	−25	93/69	89/67	86/66	27	71	69	68			
North Battleford AP	52	46	108	15	1796	−33	−30	88/67	85/66	82/65	23	69	68	66			
Prince Albert AP	53	13	105	41	1414	−42	−35	87/67	84/66	81/65	25	70	68	67			
Regina AP	50	26	104	40	1884	−33	−29	91/69	88/68	84/67	26	72	70	68			
Saskatoon AP	52	10	106	41	1645	−35	−31	89/68	86/66	83/65	26	70	68	67			
Swift Current AP	50	17	107	41	2677	−28	−25	93/68	90/66	87/65	25	70	69	67			
Yorkton AP	51	16	102	28	1653	−35	−30	87/69	84/68	80/66	23	72	70	68			
YUKON TERRITORY																	
Whitehorse AP	60	43	135	04	2289	−46	−43	80/59	77/58	74/56	22	61	59	58	NW 5	SE	

Chapter 4

BUILDING HEAT TRANSMISSION SURFACES

In Chapter 2, the concept of heat transfer for homogeneous objects and surfaces was introduced. This chapter begins the first step in calculating the actual heating and cooling loads for a building. In this chapter, a simplified procedure for estimating the overall conductance through actual typical building surfaces is presented.

The rate of heat transfer per unit area through any given surface is a function of the temperature difference between the two sides and the overall thermal conductance through the composite surface. In order to estimate the heat loss or gain for a building, it is first necessary to determine an overall heat transfer coefficient for each type of surface and the area of each surface exposed to a temperature difference.

4.1 OVERALL HEAT TRANSFER COEFFICIENT

Frequently, in HVAC design, it becomes necessary to calculate the heat transferred though a composite building section rather than through a single element, as has been the case discussed in previous examples. In order to calculate the overall heat transfer through a composite section, a term called the **overall heat transfer coefficient (U value)** has been developed. The overall heat transfer coefficient (U value) is a single combined coefficient for conduction through all materials and the convection heat transfer through the surface film. As an example, suppose a composite brick wall, shown in Figure 4.1, had an outside temperature of T_o and inside temperature of T_i.

The U value for the composite wall, shown in Figure 4.1, is a combination of the outside convection coefficient, the thermal conductance of the brick, the thermal conductance of the insulation, the conductance of the drywall, and the inside convection coefficient. The heat transferred through a composite wall can be predicted by the following equation:

$$q = UA \, (T_i - T_o) \tag{4.1}$$

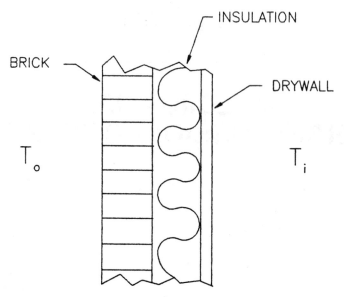

FIGURE 4.1   Composite Brick Wall

where

q = the heat transfer rate in Btu/hr•ft²•°F

U = the overall heat transfer coefficient in Btu/hr•ft²•°F

A = wall surface area in ft²

T_i = inside air temperature in °F

T_o = outside air temperature in °F

The overall heat transfer coefficient is a measure of the ability of a composite surface to transmit heat when a temperature difference exists between the two sides.

4.2   U *VALUE CALCULATION PROCEDURE*

As the conductance of a material is an indication of the ease with which heat is conducted through the material, the reciprocal of the conductance (the resistance) is a measure of the extent to which a material resists the flow of heat. The resistance to heat flow through a material is frequently referred to as the "*R* value" of the material. Mathematically, the thermal resistance of a material was given by Equation 2.6 as

$$R = \frac{1}{C}$$

Conductances, thermal resistances, and the thermal conductivity of many common building materials are presented in Appendix B.

TABLE 4.1 Air Film Resistances

Position of Surface	Direction of Heat Flow	Resistance
STILL AIR		
Horizontal	Upward	0.6l
Horizontal	Downward	0.92
Vertical	Horizontal	0.68
Air Space	Any	1.01
MOVING AIR		
15 mph wind (winter)	Any	0.17
7.5 mph wind (summer)	Any	0.25

In order to determine the U value of a composite building surface, it is first necessary to determine the thermal resistance of each element, starting with the convection resistance, or film resistance for air, on either side of the surface. Film resistances for air are given in Table 4.1

After the appropriate film resistances have been selected from Table 4.1, proceed through a cross section of the heat transfer surface, tabulating the thermal resistances for each element of the cross section. The equation for the U value of a composite surface is given by Equation 4.2.

$$U = \frac{1}{R_1 + R_2 + R_3 + \cdots + R_n} \tag{4.2}$$

Example 4.1 Suppose a frame wall had wooden shingles on the outside with felt roll, $\frac{1}{2}$ in plywood, $3\frac{1}{2}$ in fiberglass batt insulation, and $\frac{1}{2}$ in gypsum board on the inside. Assume there is a 15 mph wind outside. What is the overall heat transfer coefficient (U Value) of the wall?

From Table 4.1 and Appendix B, the following thermal resistances are obtained for each element in the cross section of the wall.

Element	R
Outside air film (15 mph)	0.17
Wooden shingles	0.87
Felt building paper	0.06
$\frac{1}{2}$ in Plywood	0.63
$3\frac{1}{2}$ in fiberglass insulation	11.83
$\frac{1}{2}$ in gypsum board	0.45
Inside air film (horizontal heat flow)	0.68
	$R_t = 14.69$

$$U = \frac{1}{R_t} = \frac{1}{14.69} = 0.07 \text{ Btu/hr•ft}^2\text{•°F}$$

In Example 4.1, the resultant heat transfer for each square foot of wall area would be 0.07 Btu/hr for each °F temperature difference.

It should be noted that the thermal resistances listed in Appendix B are for a material of a specified thickness. These values may be used directly when the calculation involves materials of the same thickness. Occasionally, materials of different thickness from those listed are used. Occasionally, no thickness is tabulated for a material and no corresponding resistance is given. In these cases, it is necessary to use Equation 2.7 to calculate the resistance of the material for the desired thickness. Substituting Equation 2.7 into Equation 2.6 results in Equation 4.3.

$$R = \frac{1}{C} = x\left(\frac{1}{k}\right)$$

(4.3)

where

     k = thermal conductivity in Btu/hr•ft²•°F•in

     x = thickness in inches

     C = conductance of a material

It is important to ensure that values with the proper units be used when applying Equation 4.3.

Example 4.2   Suppose a masonry wall, with a 15 mph wind outside, is composed of 4 in face brick, an air space, 8 in hollow light weight concrete block, $2\frac{1}{2}$ in expanded of polystyrene insulation, and $\frac{3}{4}$ in of plaster. What is the U value of the wall?

Since the $2\frac{1}{2}$ in polystyrene insulation does not have a resistance value listed, it is necessary to use Equation 4.3.

For 1 in polystyrene, Appendix B lists a thermal conductivity of 0.02 Btu•ft/hr•ft²•°F.

$$R = x\left(\frac{1}{k}\right) = (2.5 \text{ in})\left(\frac{1 \text{ ft}}{12 \text{ in}}\right)\left(\frac{1}{0.02}\right) = 10.42$$

Element	R
Outside air film	0.17
4 in face brick	0.44
Air space (Table 4.1)	1.01
8 in hollow lightweight concrete block	2.00
$2\frac{1}{2}$ in polystyrene insulation	10.42
$\frac{3}{4}$ in plaster	0.47
Inside air film (horizontal heat flow)	0.68
	$R_t = 15.19$

This is the "R value" of the wall.

$$U = \frac{1}{R_t} = \frac{1}{15.19} = 0.07 \text{ Btu/hr•ft}^2\text{•°F}$$

Example 4.3   A roof/ceiling consists of builtup roofing, 3 in preformed roof insulation, a metal deck, a 12 in air space, and acoustical ceiling tile. What is the U value of the roof/ceiling assembly?

Element	R
Outside air film (15 mph)	0.17
$^3/_8$ in builtup roofing	0.33
3 in preformed roof insulation	8.33
20 gage corrugated metal deck	0
12 in air space (Table 4.1)	1.01
$^1/_2$ in acoustical tile	1.26
Inside air film (vertical heat flow)	0.61
	$R_t = 11.71$

$$U = \frac{1}{R_t} = \frac{1}{11.71} = 0.09 \text{ Btu/hr•ft}^2\text{•°F}$$

Note that in this case, the air space in the ceiling was a "dead" air space. That is, the air was not moving or being exchanged with the room air. Frequently in HVAC applications, the ceiling space is used as an air plenum to return air from the room back to the air handling unit. In that case, the space above the ceiling becomes part of the conditioned space and the resistances for the air space and the ceiling tile would not be included in the calculation of the U value for the roof/ceiling assembly.

Note also that the thermal resistance for the metal deck is zero. This is due to the fact that metal is an excellent conductor of heat and offers little resistance to the flow of heat. For this reason, the thermal resistance of metal can be assumed as negligible.

In HVAC calculations, some types of surfaces, such as glass, are used very frequently. Since surfaces such as glass and various types of doors are used so frequently, U values have been tabulated for these surfaces and are tabulated in Tables 4.2, 4.3 and 4.4.

4.3   THERMAL MASS

In addition to the thermal resistance of a surface, it is frequently necessary to calculate its thermal mass, or weight. Densities for common building materials are presented in Appendix B, with the materials' thermal conductance or resistance. Since the calculation procedures are similar, and the data obtained from the same tables, it is convenient to calculate the surface weight along with the U value.

Section 2.3.4, Transient Heat Transfer, discussed the effect of thermal heat storage of materials and non-steady state heat transfer. The heat storage capacity of a material is closely related to its density or weight. Generally speaking, the greater the density or weight of a material, the greater its capacity to store heat. For this reason, it is usually necessary to determine the weight of an exterior surface.

Buildings are frequently classified based on their thermal mass or weight. Thermal mass of a substance or structure is actually the mass times the specific heat. A building with brick and block walls and with thick concrete floors is classified as heavy construction. A wooden frame building with wooden floors would be classified as light construction.

Classifications may be based on judgment; however, it is advisable to calculate the weight of surfaces. The mass or weight, per unit area of a surface, is the best way to classify a surface by weight. The weight per unit area is usually based on 1 ft^2 of surface area.

TABLE 4.2   Shading Coefficients for Single Glass with Indoor Shading by Venetian Blinds or Roller Shades

Description	Exterior Vertical Panels				Exterior Horizontal Panels (Skylights)	
	Summer		Winter		Summer	Winter
	No Indoor Shade	Indoor Shade	No Indoor Shade	Indoor Shade		
Flat Glass						
Single Glass	1.04	0.81	1.10	0.83	0.83	1.23
Insulating Glass, Double						
3/16 in. air space	0.65	0.58	0.62	0.52	0.57	0.70
1/4 in. air space	0.61	0.55	0.58	0.48	0.54	0.65
1/2 in. air space	0.56	0.52	0.49	0.42	0.49	0.59
1/2 in. air space low emittance coating						
$e = 0.20$	0.38	0.37	0.32	0.30	0.36	0.48
$e = 0.40$	0.45	0.44	0.38	0.35	0.42	0.52
$e = 0.60$	0.51	0.48	0.43	0.38	0.46	0.56
Insulating Glass, Triple						
1/4 in. air space	0.44	0.40	0.39	0.31		
1/2 in. air space	0.39	0.36	0.31	0.26		
Storm Windows						
1 in. to 4 in. air spaces	0.50	0.48	0.50	0.42		
Plastic Bubbles						
Single Walled					0.80	1.15
Double Walled					0.46	0.70

The procedure to determine the weight of a surface is quite straightforward. The weight, per square foot of surface area, of each layer of material is determined by first calculating the volume per square foot for each layer and then multiplying it by the mass density of the material. The weight of each layer is then added together to get the total weight of the surface per square foot.

Example 4.4   What is the thermal mass or weight, per square foot of area, of a lightweight 80 lb concrete floor that is 8 in thick?

From Appendix B, a density of 80 lb/ft³ is obtained for light weight concrete that is 8 in thick. The volume, per square foot of surface area, is calculated and then multiplied by the density of the material. Therefore, the weight, per square foot of floor surface area, is determined as follows:

$$\text{Weight} = (8 \text{ in})\left(\frac{1 \text{ ft}}{12 \text{ in}}\right)(80 \text{ lb/ft}^3) = 53.33 \text{ lb/ft}^2$$

Buildings are often classified according to the weight of their floors. The classifications are

TABLE 4.3 Coefficients of Transmission (U) for Wood Doors, Btu/hr•ft²•°F

| Door Thickness, in. | Description | Winter | | | Summer |
		No Storm Door	Wood Storm Door	Metal Storm Door	No Storm Door
1-3/8	Hollow core flush door	0.47	0.30	0.32	0.45
1-3/8	Solid core flush door	0.39	0.26	0.28	0.38
1-3/8	Panel door, 7/16-in. panels	0.57	0.33	0.37	0.54
1-3/4	Hollow core flush door	0.46	0.29	0.32	0.44
	with single glazing	0.56	0.33	0.36	0.54
1-3/4	Solid core flush door	0.33	0.28	0.25	0.32
	with single glazing	0.46	0.29	0.32	0.44
	with insulating glass	0.37	0.25	0.27	0.36
1-3/4	Panel door, 7/16-in. panels	0.54	0.32	0.36	0.52
	with single glazing	0.67	0.36	0.41	0.63
	with insulating glass	0.50	0.31	0.34	0.48
1-3/4	Panel door, 1-1/8-in. panels	0.39	0.26	0.28	0.38
	with single glazing	0.61	0.34	0.38	0.58
	with insulating glass	0.44	0.28	0.31	0.42
2-1/4	Solid core flush door	0.27	0.20	0.21	0.26
	with single glazing	0.41	0.27	0.29	0.40
	with insulating glass	0.33	0.23	0.25	0.32

TABLE 4.4 Coefficients of Transmission (U) for Steel Doors, Btu/hr•ft²•°F

| Door Thickness, in. | Description | Winter | | | Summer |
		No Storm Door	Wood Storm Door	Metal Storm Door	No Storm Door
1-3/4	Solid urethane foam core with thermal break	0.19	0.16	0.17	0.18
1-3/4	Solid urethane foam core without thermal break	0.40	-	-	0.39

0 to 64 lb/ft², lightweight

65 to 125 lb/ft², medium weight

Over 125 lb/ft², heavy weight

Example 4.5   What is the thermal mass or weight of the wall in Example 4.2? From the tables in Appendix B.

	Wt/ft²
Outside Air Film	0
4 in face brick; $(4 \text{ in})\left(\dfrac{1 \text{ ft}}{12 \text{ in}}\right)\left(130\dfrac{\text{lb}}{\text{ft}^3}\right) =$	43.3
Air space	0
8 in hollow lt. wt. con block; $(8 \text{ in})\left(\dfrac{1 \text{ ft}}{12 \text{ in}}\right)\left(45\dfrac{\text{lb}}{\text{ft}^3}\right) =$	30.0
2.5 in polystyrene insulation; $(2.5 \text{ in})\left(\dfrac{1 \text{ ft}}{12 \text{ in}}\right)\left(1.8\dfrac{\text{lb}}{\text{ft}^3}\right) =$	0.4
$^3/_4$ in plaster; $(0.75 \text{ in})\left(\dfrac{1 \text{ ft}}{12 \text{ in}}\right)\left(50\dfrac{\text{lb}}{\text{ft}^3}\right) =$	3.1
Inside air film	0
	76.8 lb/ft²

The weight of the wall is 76 lb/ft² of surface area. The wall is a medium weight surface.

4.4   HEAT TRANSMISSION AREAS

After the different types of heat transfer surfaces have been determined, and the respective *U* value calculated for each, it is necessary to determine the net area through which heat transfer will occur. Heat transfer will occur through any surface that encloses a space, is heated and/or cooled, and the outdoors or an unconditioned space. Examples of an unconditioned space would be an unheated attached garage or a crawl space under a floor of a conditioned space.

The rate of heat transfer through each surface is a function of its *U* value, the temperature difference across the surface, and the net area of the surface. In order to determine the net area of a heat transfer surface, it is first necessary to determine the gross area of the surface. The gross area of a wall would consist of all opaque wall areas, window areas, and door areas where the surfaces are exposed to outdoor conditions, or an unconditioned space that is adjacent to a heated and/or mechanically cooled space. This would also include interstitial areas between two conditioned spaces. In a similar fashion, the gross roof/ceiling area would include the entire roof and skylights. The net area of each surface is then determined by subtracting the areas of all windows, doors, and other surfaces with different *U* values.

For an entire building, or heating/cooling zone of a building, the net area is determined for each type of surface, and for each direction that surface faces. When calculating net heat transfer areas, it is necessary to establish the compass directions, on a plan of the space, and determine which direction each vertical surface is facing. Net areas must be calculated for horizontal heat transfer surfaces.

Chapter 5

INFILTRATION AND VENTILATION

One of the largest sources of heating and cooling loads for an HVAC system is outdoor air. Outside air enters and exits a building by one of two means: infiltration/exfiltration and forced ventilation.

Infiltration and exfiltration, or just infiltration for short, is the random uncontrolled flow, or leakage of air, into and out of a building. Since infiltration is an uncontrolled exchange of conditioned air inside the building with unconditioned air from outside, it is important to keep infiltration to a minimum for two reasons: First, it is undesirable to allow unconditioned air directly into a conditioned space because this could result in uncomfortable conditions for the occupants. Second, the rate of air exchange is uncontrolled, dependent on outdoor conditions such as wind conditions and temperature. In multi-story buildings, infiltration can also be the result of natural buoyancy forces caused by temperature differences between the inside and outside air (stack effect). Infiltration can be controlled by the HVAC system, by some extent, by forced ventilation. Infiltration can be reduced by drawing more air into the building through the HVAC system than is being mechanically exhausted, thereby pressurizing the building.

The need for outdoor ventilation air in buildings and the effects on indoor air pollution were briefly discussed in Chapter 1. Frequently, the most effective way to control indoor air quality, for any given building, is by the introduction of outdoor air into the building. The amount of outdoor air must be controlled, however. Excessive outdoor air ventilation rates can result in higher operating costs, as well as HVAC systems with excessive heating and cooling capacity.

5.1 VENTILATION

The primary need for outdoor ventilation air is to dilute indoor air contaminants. Outdoor air is used to control contaminants, such as carbon dioxide, carbon monoxide, odors, formaldehyde, radon, and tobacco smoke. Outdoor ventilation air may also be used to control indoor air humidity. There is no practical way to remove all the contaminants from indoor air, especially carbon dioxide. Air filters and cleaners can be used to remove particulate matter from the air, but are not effective for gaseous contaminants. Dilution with outdoor air is the most practical method for controlling indoor air contaminants.

Building mechanical codes set minimum outside air ventilation rates for occupied buildings. The codes govern ventilation of spaces within buildings intended for human occupancy. The most common code is the Building Officials and Code Administration (BOCA).

When discussing ventilation rates, it is necessary to understand the component flows in an HVAC system. When air is circulated within a building, air distributed to the various rooms and areas by the air handling unit is called supply air ventilation, or just supply air. Air returned back to the air handling unit for heating and/or cooling or recirculation is called return air. Return air is normally mixed with air drawn in from outside in the air handling unit. The supply air is a mixture of conditional return air and outside air. Exhaust air is air that is drawn out of a building mechanically by an exhaust fan. The return air rate is equal to the supply air rate, less the exhaust air rate. The outside air ventilation rate should be equal to or greater than the exhaust air rate.

The BOCA code provides three criteria upon which the ventilation rates for various areas of occupied buildings are determined. The first criteria is a minimum outdoor air ventilation rate per occupant. Presently, BOCA code requires a minimum outdoor ventilation rate of 15 ft^3/min per person. A second criteria prescribes the minimum supply air ventilation rate per occupant for various types of occupancies such as offices, theaters, etc. The code limits the maximum percentage of supply air which may be recirculated air, depending upon the efficiency of the air filtration of the HVAC system. The balance of the supply air must be from outside air. The third criteria of the BOCA Code prescribes the minimum amount of air that must be exhausted from spaces such as toilet rooms, locker rooms, janitors' closets, etc. Of course, the amount of air that is exhausted must be replaced by the supply air system.

When determining the amount of outside air required to be brought in, or made up, by the HVAC system for a building, the designer must examine all three criteria listed above to determine which criteria governs. The criteria that would result in the greatest amount of ventilation is the one that will govern.

In addition to consulting local building codes for ventilation requirements, the designer should also consult the latest issue of ASHRAE Standard 62, "Ventilation for Acceptable Indoor Air Quality." Although ASHRAE Standard 62 is a guide and not a code at the present time, it is quickly becoming the most accepted guideline for ventilation requirements. In addition to providing guides for minimum outdoor air, the standard provides procedures for ensuring adequate air quality in multiple spaces served by common air systems. ASHRAE Standard 62, as well as code requirements for ventilation, are discussed in greater detail in Chapter 11.

5.2 STACK EFFECT

Infiltration and exfiltration are caused by two driving pressures, wind forces and thermal buoyancy (stack effect) due to temperature differences between indoor air and outdoor air. The temperature differences result in pressure differences due to the difference in density between the warm air and cold air. Since warm air is less dense than colder air, the warm air tends to rise. In winter, the warm air tends to rise to the top of the building, creating a greater pressure at the top and a negative pressure at the bottom. This causes air to exfiltrate through openings and cracks near the top of a tall building and infiltrate at the bottom. A neutral pressure point is reached somewhere near the middle.

Usually the stack effect is not considered significant for buildings under four stories. For buildings above four stories, the stack effect is only significant in the winter when the indoor/outdoor temperature differences are the greatest. Another factor that tends to mitigate the stack effect is that buildings are not completely open between floors to allow the free flow of air from one level to another. This creates an additional resistance to air flow inside the building reducing the stack effect. In buildings where the stack effect is considered significant, a rule of thumb for estimating the indoor/outdoor pressure difference due to stack effect is to assume a pressure difference of 0.001 in of water per story per °F difference. For more detailed calculations of the pressure differences due to the stack effect, the reader should consult References 1 and 4.

5.3 WIND EFFECT

When wind flows over and around an object, such as a building, a relative high pressure is formed outside the windward side of the building and a relative low pressure is formed outside the leeward side. Pressures on the other sides of the building tend to be neutral, depending on the angle of the wind direction and the shape of the building. The pressure differences may vary significantly with time, due to air turbulence and changes in wind direction. The relative high pressure on the windward side of the building tends to cause infiltration into the building, while the low pressure causes a corresponding exfiltration on the leeward side.

5.4 COMBINED WIND AND STACK EFFECTS

Cracks occur at random fractures in building materials, at the interface of similar or dissimilar materials, and whenever one or more moveable surfaces meet with another surface (e.g., windows, doors, etc.). The number and size of the cracks depend on the type of construction, the workmanship during construction and manufacture of building components, and on building maintenance after construction.

Air flow through cracks and other openings in the building envelope are not linearly dependent upon the individual pressure differences caused by the different effects discussed above. Air flows caused by each pressure source should not be added together to get a net infiltration rate. In order to calculate the infiltration rate, it is necessary to determine the net pressure difference across a surface, then calculate a net infiltration rate. The net pressure difference, for any given location in a building, may be calculated using Equation 5.1.

$$\Delta P = \Delta P_s + \Delta P_w + \Delta P_p \tag{5.1}$$

where

ΔP = The combined net pressure difference

ΔP_s = Pressure difference due to the stack effect

ΔP_w = Pressure difference due to the wind effect

ΔP_p = Pressure difference due to pressurization of the building by the HVAC system

Each term in Equation 5.1 can be positive or negative depending on the location within the building and outdoor conditions.

A combined net infiltration rate can then be calculated with Equation 5.2.

$$Q = C (\Delta P)^n \tag{5.2}$$

where

Q = infiltration/exfiltration air flow rate

ΔP = the net pressure difference

n = flow exponent dependent upon the size and type of crack

C = pressure coefficient for the crack

The values for calculating pressure differences and infiltration have been determined experimentally for various building components and configurations. The reader is referred to References 1 and 4 for these detailed calculation methods.

5.5   SIMPLIFIED INFILTRATION CALCULATION METHODS

The most common method for calculating infiltration rates through building openings is the "crack method." The method uses tabulated values for infiltration rates in ft³/min per linear foot of crack length for various types of openings such as windows and doors. These tabulated values assume that the wind is perpendicular to the surface; that is, the wind is blowing directly at the surface. The tables also assume that all cracks around the opening are the same size. Table 5.1 lists infiltration rates for various types of windows and for various qualities of installation or maintenance. Table 5.2 lists infiltration rates for various types of doors for various qualities of installation or maintenance.

When calculating the rate of infiltration, it is first necessary to determine the crack length for each opening in the surface on the windward side of the building. During winter, this is usually the north side of the building; however, weather data should be consulted to determine the direction of the prevailing winds. In order to determine the required crack length to be used in calculating the rate of infiltration, it is necessary to determine the crack length for both the windward and leeward sides of the building. If air flows into the windward side of the building, an equivalent amount will flow out of the building, mostly from the cracks in the leeward side. If the cracks in the other sides of the building are significantly less than on the windward side, the rate of infiltration could be significantly less, since the building will tend to pressurize. The amount of crack used for calculating infiltration should not be less than ¹/₂ of the total crack length for the windward side of the room or building.

For small buildings, primarily residences, the air change method may be used to estimate the rate of infiltration. This is a very simplified rule of thumb method used for quick estimates and is only applicable to small buildings such as houses. It is based on empirical experience and is applicable to average conditions for average construction. The method assumes that the air inside the house will be exchanged with air from outside a given number of times per hour on average. The rate at which air infiltrates a building may be determined by multiplying the number of air changes per hour by the total volume of the building. The rate of infiltration can be calculated using Equation 5.3 and Table 5.3.

$$Q = AC \times V \times \frac{1 \text{ hr}}{60 \text{ min}} = AC \times V \times 0.0167 \qquad (5.3)$$

where

Q = the infiltration rate in ft³/min

AC = air changes per hour

V = volume of the house in ft³

TABLE 5.1 Infiltration Through Windows

Window Type	Infiltration (ft³/min•ft of crack)	
	No Strip	With Strip
Wood Sash, Double Hung		
Average fit	0.65	0.4
Poor fit	1.85	0.57
Poor fit w/stn sash	0.93	0.29
Metal Sash, Double Hung	1.23	0.53
Casement		
Average fit	0.55	—
Poor fit	0.87	—
Vertically Pivoted	2.4	

TABLE 5.2 Infiltration Through Doors

Door Type	Infiltration (ft³/min•ft of crack)
Wooden or Metal	
Good fit w/strip	0.9
Good fit w/o strip	1.8
Poor fit w/o strip	3.7
Glass Door	
Good fit	9.6
Average fit	14.0
Poor fit	19.0
Roll Up Door	9.6

TABLE 5.3   Infiltration Air Changes

	Air Changes	
Room Type	Windows w/o Strip	Windows with Strip or Storm Sash
No ext windows or doors	0.5	0.3
Windows or doors, 1 side	1	0.7
Windows or doors, 2 sides	1.5	1
Windows or doors, 3 sides or more	2	1.3

Heat losses and heat gains due to ventilation and infiltration are discussed in Chapters 6, 8, and 9.

Chapter 6

HEATING LOADS

The purpose of a building heating system is to maintain a desired indoor air temperature and indoor comfort conditions, despite the loss of heat to the outdoors. The amount of heat loss from a building is directly related to the difference between the indoor and outdoor temperatures. A heating load for a building occurs as the result of heat loss, or transfer, from inside the building to outside.

Heat loss from a building or space occurs by two methods:

1. The transmission of heat through the building exterior surfaces such as walls, roofs, etc.

2. Through infiltration and/or ventilation which exchanges warm indoor air with colder outside air.

The purpose of the heating system is to maintain the desired indoor temperature, typically 70°F. In order to do this, it must provide the amount of heat, equal to the heat loss from the space, at the desired indoor temperature. Figure 6.1 shows a typical heat balance for a space requiring heat.

In order for the temperature of the heated space to remain constant at the desired set point, the heat supplied by the system must be equal to the combined transmission, ventilation, and infiltration heat losses. The heat loss (combined heat losses) for a space is determined based upon an assumed constant outdoor winter design temperature. The winter design temperature is based on the climatic conditions for the location of the building and on the intended usage of the building. Climatic conditions and recommended design temperatures were discussed in Chapter 3, Climatic Conditions.

The heating load for a space is usually determined assuming that there is no internal heat contribution from internal heat sources such as lights, people, equipment, etc. The peak heating load usually occurs at night when the outside temperatures are the lowest. At night, the building usage is usually quite low and internal heat sources may not be occurring. If a credit for some internal heat source is taken, the heat source (heat gain) must be relatively constant. That is, the heat source must occur 24 hr per day.

The credit for the thermal storage of the building is usually not considered, although it can have an effect. In buildings with large thermal mass and large daytime internal heat gains, the heating load may be mitigated to some extent.

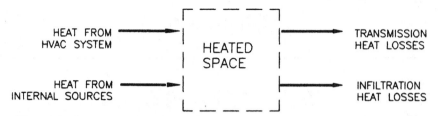

FIGURE 6.1   Heated Space Heat Balance

6.1   TRANSMISSION HEAT LOSSES

In order for the transmission of heat to occur through a surface, a temperature difference must exist across the surface. As discussed in Section 2.3.1, Conduction Heat Transfer, heat will be transmitted through a substance any time there is a temperature difference between the two sides. The amount of heat transferred is a function of the thermal conductivity of the surface, the temperature difference between the two sides, and the area of the surface.

Therefore, the transmission of heat, from the inside to the outside (a heat loss), will occur through all exterior surfaces of a building and through all surfaces between heated spaces and unheated spaces. An unheated space is any enclosed space adjacent to a heated space that does not have heat directly supplied to it. Examples of unheated spaces are attached unheated garages, attics, crawl spaces, and sun porches. The temperature of the unheated space will always be between the temperature of the heated space and the outdoor air temperature.

Example 6.1   An exterior wall is 10 ft long and 8 ft high with a 4 ft × 4 ft window. The wall has an overall heat transfer coefficient (U value) of 0.05 Btu/hr•ft²•°F. The window is a double pane type with a U value of 0.55 Btu/hr•ft²•°F. Assume that the outside air temperature is -3°F. If the wall and window were the only exterior surfaces in the room, how much heat would have to be provided to maintain an indoor air temperature of 72°F? Assume that the infiltration is negligible.

Referring to Section 4.4, Heat Transmission Areas, the net surface area for each surface is calculated.

$$\text{Window Area} = 4 \times 4 = 16 \text{ ft}^2$$
$$\text{Net Wall Area} = (8 \times 10) - 16 = 64 \text{ ft}^2$$

Using Equation 4.1, the heat loss through each surface is calculated.

$$\text{Wall loss: } q = (0.05)(64)[72 - (-3)] = 240 \text{ Btu/hr}$$
$$\text{Glass loss: } q = (0.55)(16)[72 - (-3)] = 660 \text{ Btu/hr}$$

Since the infiltration is assumed negligible, the amount of heat necessary to maintain 72°F in the room then is

$$q = 240 + 660 = 900 \text{ Btu/hr}$$

Transmission heat loss may also occur through surfaces that are indirectly exposed to outdoor temperatures such as floor slabs and underground walls of basements. In the case of floor slabs and underground walls, the earth acts as an insulator between the inside and outside air. Since the earth can store heat, it acts as a "heat sink" and causes a relatively constant rate of heat loss. Tables 6.1, 6.2, and 6.3 may be used to calculate heat losses for various below grade surfaces. Note that the heat loss for floor slabs is a function of the exposed perimeter of the slab and not the area of the slab.

TABLE 6.1 Heat Loss Coefficients for Floor Slab Construction (Btu/hr•°F•ft of perimeter)

		Degree Days (65°F Base)		
Construction	Insulationᵃ	2950	5350	7433
8-in block wall,	Uninsulated	0.62	0.68	0.72
brick facing	R = 5.4 from edge to footer	0.48	0.50	0.56
4-in block wall,	Uninsulated	0.80	0.84	0.93
brick facing	R = 5.4 from edge to footer	0.47	0.49	0.54
Metal stud wall,	Uninsulated	1.15	1.20	1.34
stucco	R = 5.4 from edge to footer	0.51	0.53	0.58
Poured concrete wall	Uninsulated	1.84	2.12	2.73
with duct near perimeterᵇ	R = 5.4 from edge to footer, 3 ft under floor	0.64	0.72	0.90

ᵃR value units in °F•ft²•hr/Btu•in.
ᵇWeighted average temperature of the heating duct was assumed at 110°F during the heating season (outdoor air temperature less than 65°F).
Copyright 1989 by the American Society of Heating, Refrigerating and Air Conditioning Engineers, Inc., from the Fundamentals Handbook. Used by permission.

TABLE 6.2 Heat Loss Below Grade Basement Walls

Depth, ft	Path Length Through Soil, ft	Heat Loss, Btu/hr•ft•°F						
		Uninsulated		R = 4.17		R = 8.34		R = 12.5
0–1	0.68	0.410		0.152		0.093		0.067
1–2	2.27	0.222	0.632	0.116	0.268	0.079	0.172	0.059	0.126
2–3	3.88	0.155	0.787	0.094	0.362	0.068	0.240	0.053	0.179
3–4	5.52	0.119	0.906	0.079	0.441	0.060	0.300	0.048	0.227
4–5	7.05	0.096	1.002	0.069	0.510	0.053	0.353	0.044	0.271
5–6	8.65	0.079	1.081	0.060	0.570	0.048	0.401	0.040	0.311
6–7	10.28	0.069	1.150	0.054	0.624	0.044	0.445	0.037	0.348

Copyright 1989 by the American Society of Heating, Refrigerating and Air Conditioning Engineers, Inc., from the Fundamentals Handbook. Used by permission.

TABLE 6.3 Heat Loss Through Basement Floors (Btu/hr•ft³•°F)

Depth of Foundation Wall Below Grade	Shortest Width of House, ft			
	20	24	28	32
5 ft	0.032	0.029	0.026	0.023
6 ft	0.030	0.027	0.025	0.022
7 ft	0.029	0.026	0.023	0.021

Copyright 1989 by the American Society of Heating, Refrigerating and Air Conditioning Engineers, Inc., from the Fundamentals Handbook. Used by permission.

The heat loss for a floor slab is determined by multiplying the heat loss for the floor slab, obtained from Table 6.1, by the exposed perimeter of the floor slab and the indoor-outdoor temperature difference. The Degree Days (65°F Base) needed may be obtained from Appendix C, Reference 5, or a similar source. It may be necessary to interpolate between values listed in Table 6.1 for locations with degree days between the listed values.

Implicit in the above heat loss calculation methods is the assumption of steady state heat transfer, as opposed to non-steady state, which was discussed in Section 2.3.4, Transient Heat Transfer. Since the outdoor air temperature fluctuates relatively slowly over time in this case, it is permissible to assume steady state heat conduction through exterior surfaces when calculating heat losses. Frequently, daily change in the winter high temperature and low temperature (daily range) results in a temperature change rate of less than 1°F/hr. This temperature change is gradual enough to assume steady state conditions when calculating heat losses for buildings.

6.2   INFILTRATION AND VENTILATION HEAT LOSSES

Whenever colder outside air is introduced into a building, sensible heat from the heating system must be provided to raise the temperature of the outside air to that of the inside air. The addition of latent heat is usually not considered unless the relative humidity, in the space, is to be maintained at some desired set point. Latent heat, in the form of moisture, may be added with a humidifier.

Applying the First Law of Thermodynamics for a gas, Equation 2.4 gave the amount of heat required to raise the temperature of a substance. In a situation where flow exists, the mass may be given in a mass flow rate, such as pounds per hour (lb/hr), or, as is common in HVAC design, cubic feet per minute (ft³/min). Since calculations involving the flow of air are so common in HVAC applications, Equation 2.4 can be modified for the flow of air. Air, at normal room temperatures, has a density of 0.075 lb/ft³ and a specific heat of 0.24 Btu/lb. Equation 2.4 may be written as follows:

$$q = Q(0.075 \text{ lb/ft}^3)(0.24 \text{ Btu/lb})(60 \text{ min/hr})(T_2 - T_1) \qquad (6.1)$$

where

q = the sensible heat added or removed from the air in Btu/hr

Q = the rate of air flow in ft³/min

$T_2 - T_1$ = the temperature change of the air in °F

Equation 6.1 may be simplified by using a constant for the air density specific heat and time conversion and may be written as

$$q = (1.08)Q(T_2 - T_1) \text{ for heating} \qquad (6.1a)$$
$$q = (1.10)Q(T_2 - T_1) \text{ for cooling} \qquad (6.1b)$$

Methods for determining the rate of infiltration were discussed in Chapter 5. Equation 6.1 (a) and (b) may be used to determine the heat loss, or gain, due to infiltration/ventilation, respectively.

Example 6.2 Assume that air infiltrates into a space at the rate of 100 ft³/min and an equivalent amount of air exfiltrates. The room air temperature is maintained at 70°F while the outside air is at 5°F. What is the sensible heat loss due to the infiltration?

Using Equation 6.1a,

$$q = (1.08)(100)(70 - 5) = 7020 \text{ Btu/hr}$$

As discussed in Chapter 5, it is frequently necessary, by code, to introduce outside air into a building via the HVAC system. This air must also be heated by the HVAC system. In the case of infiltration, it is necessary to determine how much heat must be supplied to a room or space, in order to offset the heat loss due to infiltration. The infiltration heat loss, in addition to the transmission heat loss for the space, determines how much air and at what temperature it must be supplied to the space.

In the case of outdoor ventilation air, the heat load is not a part of the individual space or room heat load, but is a part of the HVAC system load. Outside air brought into the building by the HVAC system is mixed with recirculated, or return air, heated or cooled, and supplied to the spaces at the desired temperature. In order to determine the sensible heat necessary to raise the total supply air from the temperature of the outside air/return air mixture, it is first necessary to determine the temperature of the mixed air. Equation 6.1a may be used to determine the heat required to raise the supply air temperature to the desired temperature. The mixed air temperature may be determined using Equation 6.2.

$$T_m = \frac{(Q_{OA})(T_{OA}) + (Q_{RA})(T_{RA})}{Q_{SA}} \tag{6.2}$$

where

T_m = the temperature of the mixed return air and outside air in °F

Q_{OA} = volumetric flow rate of the outside air in ft³/min

T_{OA} = outside air temperature in °F

Q_{RA} = volumetric flow rate of the return air in ft³/min

T_{RA} = the return air temperature in °F (usually the room temp)

Q_{SA} = volumetric flow rate of the supply air in ft³/min (return air + outside air)

The heat required by the HVAC system using Equation 6.1a is

$$q = (1.08)(Q_{sa})(T_{sa} - T_m) \tag{6.1c}$$

The desired supply air flow rate may be determined by code requirements as discussed in Chapter 5 or, as is usually the case, by the cooling load as discussed in Chapter 8. Usually the required air flow rate to a room is determined by the cooling load. Given a certain air flow rate into a space, and the desired room temperature, the required supply air temperature, for a given heat load, may be determined using Equation 6.3. Equation 6.3 is obtained by rearranging Equation 6.1a and solving for the supply air temperature.

$$T_{sa} = \frac{q}{T_r + (1.08)(Q_{sa})} \tag{6.3}$$

where

T_{sa} = required supply air temperature in °F

T_r = the desired room set point temperature in °F

q = total room heat load including transmission and infiltration heat losses in Btu/hr

Q_{sa} = volumetric flow rate of supply air in ft³/min

Example 6.3   A room which is to be maintained at 70°F has a combined heat loss of 3000 Btu/hr with 100 ft³/min of air supplied to it. At what temperature must the air be supplied in order to maintain 70° in the room?

Using Equation 6.3

$$T_{sa} = \frac{3000 \text{ Btu/hr}}{70 + (1.08)(100 \text{ ft}^3/\text{min})} = 98°F$$

6.3   HEATING LOAD CALCULATION PROCEDURE

A building heating system must be capable of providing sufficient heat, in order to maintain the desired inside set point temperature. The amount of heat required is equal to the combined heat losses for the space, or entire building, at the desired inside temperature and the outside design winter conditions.

In order to determine the required capacity of a heating system, it is necessary to estimate the maximum probable heat loss for each room or space to be heated. The combined heat losses include all transmission heat losses through exterior surfaces, through underground surfaces, and all infiltration heat losses. The ventilation heat loss may be added to the combined space heat losses to get the total combined heat loss. From the total combined heat loss or load, the capacity of the heating system may be determined.

The general procedure for calculating the design heat load for a space, or a building, is as follows:

1. Select the outdoor design weather conditions based upon the location and usage of the building.

2. Determine the desired indoor set point temperature.

3. Calculate the overall heat transfer coefficient for each heat transfer surface as described in Section 4.2.

4. Calculate the net heat transfer area of each exterior surface and surface adjacent to an unheated space as described in Section 4.4.

5. Calculate or estimate the rate of infiltration for the space or building as suggested in Chapter 5.

6. Determine the required outdoor ventilation rate, if any. See Chapters 5 and 11.

7. Compute the transmission heat loss for each exterior surface and surface adjacent to an unconditioned space of the space or building. See Section 6.1.

8. Compute the heat losses from below grade surfaces, such as floor slabs and basement surfaces. Refer to Tables 6.1, 6.2, and 6.3.

9. Compute the heat losses due to infiltration using Equation 6.1a.

10. Compute the heat losses due to outdoor ventilation (if any) using Equation 6.1a.

11. Sum all transmission, below grade, infiltration and ventilation heat losses to get the grand total heat loss for the space or building.

12. Allow for a "pick-up" load if night set back will be used. It is estimated that an additional 40% heating capacity should be provided for each 10°F night set back.

13. Select the heating system capacity equal to or slightly greater than the grand total heat loss or the "pick-up" load, whichever is greater.

Example 6.4 A house in St. Louis, MO has north-south walls that are 30 ft long and east-west walls that are 40 ft long as shown in Figure 6.2. All walls are 8 ft high. There are two 4 ft × 4 ft double pane windows in the east and west walls. The west wall also has a 3 ft × 7 ft wooden door. The floor slab is a slab-on-grade with no insulation.

Assume that the Degree Days (65°F Base) obtained from tabulated data in Reference 5 were equal to 5350.

The roof is flat. The west wall is a frame wall, while the other three are masonry. The inside temperature is to be maintained at 70°F, except at night, when it will be set back to 60°F. Assume that the infiltration rate of $1/2$ air change per hour was calculated by the methods given in Chapter 5.

Assume the following overall heat transfer coefficients were calculated by the methods given in Chapter 4.

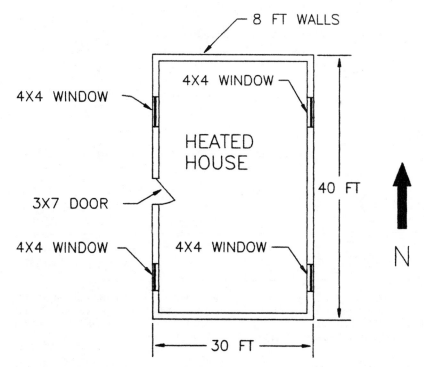

FIGURE 6.2 House Plan

West wall (frame):	$U = 0.076$ Btu/hr•ft²•°F
East, south and west walls:	$U = 0.09$
Roof:	$U = 0.08$
Glass windows:	$U = 0.56$
Door:	$U = 0.64$

What is the heat loss for the building at 70°F and what size furnace should it have?

Based on the criteria discussed in Chapter 3, an outside temperature is selected from Table 3-1. The outside design temperature selected for St. Louis, MO is 3°F.

The net wall areas are determined by calculating the gross wall area and subtracting the area of all other surfaces in the wall. The net wall areas in square feet are calculated as follows:

	N	E	S	W
Gross wall area (length × width)	240	320	240	320
Window area	0	32	0	32
Door area	0	0	0	21
Net wall areas	240	288	240	267
Roof area = (40 ft)(30 ft) = 1200 ft²				
Volume = (1200 ft²)(8 ft) = 9600 ft³				
Floor perimeter (west wall) = 40 ft				
Floor perimeter (south, east, north walls) = 30 + 40 + 30 = 100 ft				

Equation 5.3 is used to calculate the rate of infiltration as follows:

$$Q = \frac{(AC/hr)(volume)}{60} = \frac{(0.5AC/hr)(9600 \text{ ft}^3)}{60 \text{ min/hr}} = 80 \text{ ft}^3/min$$

The transmission losses are calculated using Equation 4.1 as follows:

$$q = UA(T_i - T_o)$$

Therefore:

Walls:		
West:	$(0.076)(267)(70 - 3) =$	1360 Btu/hr
North:	$(0.09)(240)(70 - 3) =$	1447
East:	$(0.09)(288)(70 - 3) =$	1737
South:	$(0.09)(240)(70 - 3) =$	1447
Windows:		
West:	$(0.56)(32)(70 - 3) =$	1201 Btu/hr
East:	$(0.56)(32)(70 - 3) =$	1201
Door:	$(0.64)(21)(70 - 3) =$	900 Btu/hr
Roof:	$(0.08)(1200)(70 - 3) =$	6432 Btu/hr

Referring to Table 6.1, the west frame wall is similar to a metal stud wall, and the other three masonry walls are similar to an 8 in block wall. There is no floor insulation. Heat loss factors for the frame and masonry walls are selected as 1.20 Btu/ft•°F and 0.68 Btu/ft•°F, respectively. The transmission heat loss for the floor is as follows:

West floor perimeter: $(1.20 \text{ Btu/ft} \cdot {}^\circ\text{F})(40 \text{ ft})(70 - 3) = 3216 \text{ Btu/hr}$

Other three walls perimeter: $(0.68 \text{ Btu/ft} \cdot {}^\circ\text{F})(100 \text{ ft})(70 - 3) = 4556$

Transmission subtotal = 23,497 Btu/hr

Equation 6.1a, and the rate of infiltration calculated above, is used to calculate the infiltration heat loss as follows:

Infiltration: $(1.08)(80 \text{ ft}^3/\text{min})(70 - 3){}^\circ\text{F} = 5789 \text{ Btu/hr}$

Grand total heat loss = 23,497 + 5789 = 29,286 Btu/hr

The grand total heat loss is the heat that would be required to maintain the inside temperature at 70°F with an outside air temperature of 3°F

Since the temperature is set back at night, an additional capacity of 40% for each 10°F of set back should be provided for pick-up load.

Allowance for pick-up = 0.4(29,286) = 11,714 Btu/hr

The required total heating capacity = 29,286 + 11,714 = 41,000 Btu/hr

Therefore, the furnace should have a heating capacity of at least 41,000 Btu/hr (12 kW) output.

HEAT LOSS CALCULATIONS

BUILDING: **ROOM:**

ELEMENT	TYPE	U X TEMP. DIFF. (BTUH/SQ FT)	AREA (SQ FT)	LOSS (BTUH)
TRANSMISSION LOSS				
WALL				
WALL				
WALL				
GLASS				
GLASS				
GLASS				
DOOR				
DOOR				
ROOF/CEILING				
ROOF/CEILING				
SKYLIGHT				
FLOOR				
			SUBTOTAL	

SLAB—ON—GRADE FLOOR			
LENGTH (FT)	FACTOR (BTUH/FT)	LOSS	

TRANSMISSION TOTAL (BTUH)	

INFILTRATION LOSS	
LOSS = 1.08 X °F (TD) X CFM	

SPACE TOTAL LOSS (BTUH)	

FIGURE 6.3   Blank Heat Loss Calculation Form

Chapter 7

EXTERNAL HEAT GAINS AND COOLING LOADS

The purpose of a building cooling system is to maintain a desired indoor temperature, and sometimes relative humidity level, despite the addition of heat to the space from both outdoor conditions and indoor sources. Unlike heating loads, the space cooling load is not just a function of the difference between the indoor and outdoor air dry bulb temperatures. The cooling load for a space or a building is the result of many complex variables. Heat gain to the space occurs from many sources, both inside and outside of the space.

As was discussed in Section 2.2.2 Forms of Heat, heat may take one of two forms, sensible heat or latent heat. Sensible heat is heat that is associated with a temperature change of a substance, such as air. Latent heat is heat that is associated with a phase change of a substance, such as condensation or evaporation. In the case of cooling loads, it is heat that is associated with the evaporation or condensation of moisture or water. For a cooling system, it is heat that is added to the space by evaporation of moisture (such as perspiration) and must be removed as the moisture or, latent heat, is condensed out of the air.

As previously stated, the total heat gain for a space, or a building, results from many sources. Basically, the sources of heat gains may be divided into two major categories, internal heat gains and external heat gains. Internal heat gains result from sources inside the space itself, such as from lights, people, equipment, etc. External heat gains are from sources outside the space, or building, and result from weather conditions. External heat gains include heat transmission through exterior walls, solar heat gains through windows, and from the introduction of outside air into the space in the form of infiltrated warm air. Figure 7.1 depicts a typical heat balance for a space requiring cooling.

This chapter will deal only with external heat gains. Internal heat gains will be discussed in Chapter 8.

Before discussing external heat gains, some definitions of terminology are in order. The distinction should be understood between the instantaneous heat gain for a space, the cooling load for a space, and the heat extraction rate from the space by the cooling system.

The instantaneous space heat gain is the rate at which heat is being generated in the space, or is entering the space, at any given instant in time. In Section 2.3.4, Transient Heat Transfer, the effect of building mass and thermal storage on heat gains were discussed.

The cooling load is the actual heat transfer to the air in the space at any given time. The space cooling load is different from the space instantaneous heat gain, due to the

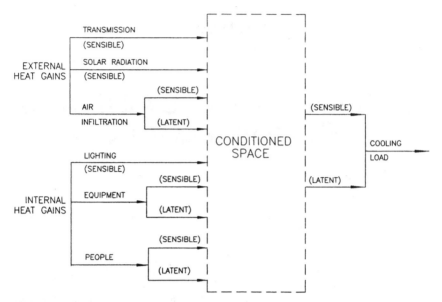

FIGURE 7.1   A Typical Heat Balance for a Space Requiring Cooling

thermal storage characteristics of the space itself. The cooling load is the rate at which heat must be removed from the space to maintain the room dry bulb and wet bulb temperatures at set values.

The heat extraction rate is the rate at which heat is actually removed from the space by the HVAC system. Due to the nature of HVAC systems and temperature control systems, the rate of heat removal is intermittent, causing "swings" in room air temperatures. Most thermostats have a "dead band" temperature range. When the room air temperature is in this range (usually 70°F to 78°F), neither heating nor cooling is required. Only when the room air temperature rises above the cooling set point temperature (78°F) does the space receive cooling or have heat removed. Heat is removed until the space air temperature drops below the set point temperature. This results in a non-constant or varying heat extraction rate which is frequently different from the actual cooling load at any given time.

7.1   EXTERNAL GAIN CALCULATION METHODS

The external heat gain (external gain) to a space or building is heat gain from outdoors, primarily through external surfaces. It includes heat transmission through walls, roofs, doors, and windows. External gains include solar heat gain through glass windows and glass doors (fenestration). External gains can include the addition of heat to a space by the infiltration of outside air into the space.

Unlike heat losses, heat gains through exterior surfaces may not be assumed to be steady state. Solar radiation on exterior surfaces, and the thermal storage effect of the surfaces, are the major reasons why external heat gains must be treated as non-steady state heat transfer. The effect of thermal storage was discussed in Section 2.3.4, Transient Heat Transfer.

ASHRAE has developed two calculation methods which take thermal storage and transient heat conduction into account for sunlit exterior surfaces. One method, introduced by ASHRAE in 1972, is the Total Equivalent Temperature Differential (TETD) Method. In the

TETD Method, various components of space heat gain are added together to obtain an instantaneous rate of space heat gain. The space heat gain is converted to an instantaneous cooling load by the Time Averaging (TA) technique of averaging the radiant heat gain with related values for previous hours. Space heat gains are separated into radiant and convective gains. The convection portion is assumed to be an instantaneous cooling load, while the radiant gain is time averaged to get a cooling load. The number of hours the radiant gain is averaged is based upon the thermal mass of the space and exterior surfaces.

The second calculation method is the Transfer Function Method (TFM), which uses a series of "weighing factors" (coefficients of room transfer functions). The TFM weighing factors use time averaging to approximate the heat gain for the present hour and for several previous hours in order to account for thermal storage of the space and determine the resultant cooling load. Since the actual calculation procedure for the TFM is so complicated that a computer would be necessary to perform the calculations, a simplified method suitable for manual calculation was developed by ASHRAE. The simplified method uses **Cooling Load Temperature Differences (CLTD)** for heat gains through sunlit walls and roofs and for heat conduction through glass (fenestration). The method also uses **Cooling Load Factors (CLF)** for solar gains through glass and for internal heat sources. Both CLTDs and CLFs take into account non-steady state heat transfer caused by thermal storage of the building mass.

CLTDs were calculated and tabulated for various exterior surfaces for three types of room space construction: light, medium, and heavy thermal characteristics. Tables for the various types of wall and roof construction are presented in this chapter. It is assumed that the heat flow through a similar wall, or roof, can be determined by multiplying the surface area and overall heat transfer coefficient (U value) by the tabulated CLTD. The tabulated values are based on an inside air temperature of 78°F and air outside design average daily temperature of 85°F. The tabulated values must be adjusted for other indoor air temperatures and other outdoor daily average temperatures. The tabulated values must be adjusted or corrected for latitude, surface color, and for the month in which the calculations are being performed. A correction factor must be applied to roofs with various types of attic ventilation.

The methods for manually calculating heat gains and cooling loads in this book are based on the **CLTD/CLF Method** developed by ASHRAE. The procedures presented herein are simplified and abridged in order to improve the ease of understanding the procedures. You should understand that the calculation of cooling loads for a building involves considerable judgment, and that the calculations are more of an art than a science. These procedures will only produce a "good estimate" of the cooling load for a building. There are a number of limitations involved with the CLTD/CLF Method. Most of the limitations are presented herein. The reader is referred to Reference 1 for a complete discussion of the CLTD/CLF Method.

7.2 COOLING LOADS FOR SUNLIT EXTERIOR SURFACES

The Cooling Load Temperature Difference (CLTD) Method may be used to calculate the cooling load for exterior surfaces exposed to sunlight and outdoor air temperatures, such as walls and roofs. It is not intended for shaded walls and roofs. The CLTD Method accounts for unsteady thermal storage affects caused by varying exterior conditions, such as sunlight and temperature changes. The thermal mass, or weight, of a surface and a procedure to calculate the weight of a surface was discussed in Section 4.3, Thermal Mass.

Surfaces that have a high mass, such as concrete or brick walls, will have a lower peak space cooling load for a given rate of heat gain. The time when the peak cooling load occurs will be delayed and the cooling load will be spread over a longer period of time. Surfaces that have the same U value (heat transfer coefficient) but a light mass, such as a curtain wall or a frame wall, will have a higher peak cooling load, peaking sooner with a shorter period for the cooling load. In an attempt to account for the thermal mass of a surface and transient heat gains, the CLTD is an equivalent steady state temperature difference that would result in the same cooling load as the surface in question. Representative surface types have been selected and classified, or grouped by their weight and U value.

ASHRAE has classified walls by thermal characteristics into seven groups, Groups A through G. Roofs are first divided according to whether the building has or does not have a suspended ceiling. They are then grouped according to their thermal characteristics in thirteen different groups, Groups 1 through 13.

The effect of sunlight on an exterior surface is a function of location (latitude) of the surface, the orientation (direction the surface is facing), the time of year, and the time of day. Obviously, the solar heat gain for an east facing wall, at 9:00 a.m., will be much greater than at 6:00 p.m. In addition to the time of day, the peak heat gain is a function of the day of the month, the season of the year, and the location of the building with respect to the earth's equator (latitude). Finally, the peak heat gain is a function of the orientation of the surface (horizontal, vertical, sloping, east facing, south facing, etc.). The amount and intensity of solar radiation is a function of all of the above variables.

In order to account for the effect of varying solar radiation on a surface and the mass of a surface, the **Sol-air temperature** concept was developed. The Sol-air temperature is that temperature of the outdoor air which, in the absence of all radiation exchanges, would result in the same rate of heat gain to the surface as would exist with the actual combination of incident solar radiation, radiant energy exchange with the sky, and other outdoor surroundings and convective air. The cooling load temperature difference (CLTD) is similar to the Sol-air temperature in that it attempts to predict a temperature difference and cooling load for exterior surfaces that would result in the presence of solar radiation. It also attempts to account for the thermal mass of a surface and variation of the cooling load with time.

In its simplified form, Equation 7.1 may be used to predict the cooling load for a given surface.

$$q = UA \, (\text{CLTD}) \qquad\qquad (7.1)$$

where

q = the cooling load in Btu/hr

U = the overall heat transfer coefficient in Btu/hr•ft^2•°F

A = the area of surface in ft^2

CLTD = the temperature difference which gives the cooling load at the designated time for the given surface type

The CLTD for actual situations must be corrected for individual situations, as will be discussed in the following sections.

In order to properly use the CLTD Method, the time of day at which the peak load occurs must be estimated. As will be discussed in the next section, this requires considerable judgment because different surfaces in a conditioned space frequently peak at different times. As an example, an east wall, a west wall, and a roof will each have a different time at which the peak CLTD occurs. In addition, rooms with different orientations in the building may each peak at a different time. The actual peak load for a cooling system serving several rooms may occur at a different time than any individual room.

7.2.1 Cooling Loads for Sunlit Walls

ASHRAE has classified, or grouped, exterior walls according to their construction weight and overall heat transfer coefficient (U value). There are seven different wall groups, Group A through Group G. Table 7.1, Wall Construction Groups, lists the seven

TABLE 7.1 Wall Construction Groups

Group No.	Description of Construction	Weight (lb/ft^2)	U-Value (Btu/h·ft^2·°F)
4-in. Face brick + (brick)			
C	Air space + 4-in. face brick	83	0.358
D	4-in. common brick	90	0.415
C	1-in. insulation or air space + 4-in. common brick	90	0.174-0.301
B	2-in. insulation + 4-in. common brick	88	0.111
B	8-in. common brick	130	0.302
A	Insulation or air space + 8-in. common brick	130	0.154-0.243
4-in. Face brick + (heavyweight concrete)			
C	Air space + 2-in. concrete	94	0.350
B	2-in. insulation + 4-in. concrete	97	0.116
A	Air space or insulation + 8-in. or more concrete	143-190	0.110-0.112
4-in. Face brick + (light or heavyweight concrete block)			
E	4-in. block	62	0.319
D	Air space or insulation + 4-in. block	62	0.153-0.246
D	8-in. block	70	0.274
C	Air space or 1-in. insulation + 6-in. or 8-in. block	73-89	0.221-0.275
B	2-in. insulation + 8-in. block	89	0.096-0.107
4-in. Face brick + (clay tile)			
D	4-in. tile	71	0.381
D	Air space + 4-in. tile	71	0.281
C	Insulation + 4-in. tile	71	0.169
C	8-in. tile	96	0.275
B	Air space or 1-in. insulation + 8-in. tile	96	0.142-0.221
A	2-in. insulation + 8-in. tile	97	0.097
Heavyweight concrete wall + (finish)			
E	4-in. concrete	63	0.585
D	4-in. concrete + 1-in. or 2-in. insulation	63	0.119-0.200
C	2-in. insulation + 4-in. concrete	63	0.119
C	8-in. concrete	109	0.490
B	8-in. concrete + 1-in. or 2-in. insulation	110	0.115-0.187
A	2-in. insulation + 8-in. concrete	110	0.115
B	12-in. concrete	156	0.421
A	12-in. concrete + insulation	156	0.113
Light and heavyweight concrete block + (finish)			
F	4-in. block + air space/insulation	29	0.161-0.263
E	2-in. insulation + 4-in. block	29-37	0.105-0.114
E	8-in. block	47-51	0.294-0.402
D	8-in. block + air space/insulation	41-57	0.149-0.173
Clay tile + (finish)			
F	4-in. tile	39	0.419
F	4-in. tile + air space	39	0.303
E	4-in. tile + 1-in. insulation	39	0.175
D	2-in. insulation + 4-in. tile	40	0.110
D	8-in. tile	63	0.296
C	8-in. tile + air space/1-in. insulation	63	0.151-0.231
B	2-in. insulation + 8-in. tile	63	0.099
Metal curtain wall			
G	With/without air space + 1- to 3-in. insulation	5-6	0.091-0.230
Frame wall			
G	1-in. to 3-in. insulation	16	0.081-0.178

wall groups with typical representative wall construction types and typical U values for each group.

Referring to Table 7.1, it can be observed that Group A walls have a relatively high mass or weight, while Group G walls are lightweight. The construction types and U values are for comparison with actual exterior walls when cooling load calculations are being made.

The first step in the cooling load calculation procedure is to select a wall group from Table 7.1 that is similar in weight to the wall for which load calculations are being made. After a similar wall construction group is selected, the appropriate cooling load temperature difference (CLTD) may be obtained from Table 7.2, Cooling Load Temperature Differences for Sunlit Walls. Referring to Table 7.2, it can be observed that CLTD values are listed for the seven different wall groups, depending upon the direction the wall is facing (orientation) and the time of day. The times listed are solar times. Notice that, for any given orientation, the heavier Group A walls have a lower peak CLTD and that occurs at a later hour than the light weight Group G walls.

The cooling load temperature differences listed in Table 7.2 are based upon several assumed conditions. They are:

a. The solar radiation is for 40° north latitude on July 21.

b. The surface is assumed to be dark in color.

c. The inside air temperature is 78°F.

d. The outside air temperature is 95°F with an outdoor mean temperature of 85°F and a daily range (variance) of 21°F.

e. The outside air film resistance is 0.333 hr•ft²•°F/Btu.

f. The inside air film resistance is 0.685 hr•ft²•°F/Btu.

If the design conditions for the building under consideration differ from the above conditions, the CLTD values must be modified or corrected before using Equation 7.1 to calculate the cooling load for the surface.

The following equation may be used to correct the CLTD values obtained from Table 7.2.

$$\text{CLTD}_c = (\text{CLTD} + \text{LM})k + (78 - T_i) + (T_{oa} - 85) \qquad (7.2)$$

where

CLTD_c = the corrected cooling load temperature difference.

CLTD = the cooling load temperature difference obtained from Table 7.2.

   LM = the latitude and month correction for the given surface location and calculation month. LM is obtained from Table 7.3 CLTD correction for latitude and month.

   k = the color adjustment factor. (k =1 if the surface is a dark color and k = 0.5 if the surface is permanently light colored and is in a rural area where there is little smoke.)

   T_i = the indoor design air temperature in °F.

   T_{oa} = the average outside air temperature on the design day. T_{oa} = the outside design temperature minus one half the daily range.

Example 7.1   Assume there is a 50 ft² wall, west facing and located at 24° north latitude on July 21. The wall has a 4 in medium weight hollow concrete block exterior, has a 1 in

TABLE 7.2 Cooling Load Temperature Differences for Sunlit Walls

North Latitude Wall Facing	0100	0200	0300	0400	0500	0600	0700	0800	0900	1000	1100	1200	1300	1400	1500	1600	1700	1800	1900	2000	2100	2200	2300	2400	Maximum CLTD	Minimum CLTD	Maximum CLTD	Difference CLTD
Group A Walls																												
N	14	14	14	13	13	12	11	10	10	10	10	11	11	12	13	13	14	14	14	14	14	14	14	14	2	10	14	4
NE	19	19	18	17	16	15	15	15	15	16	17	18	19	19	20	20	20	20	20	20	20	20	19	19	22	15	20	5
E	24	23	22	21	20	19	18	18	18	18	19	20	21	22	23	23	24	24	25	25	25	25	24	23	22	18	25	7
SE	24	23	22	21	20	19	18	18	18	18	18	19	20	21	22	22	23	24	24	24	24	24	23	23	22	18	24	6
S	20	20	19	18	18	17	16	15	14	14	14	14	15	16	17	17	18	19	20	20	20	20	20	20	23	14	20	6
SW	25	25	24	23	22	21	20	19	18	18	17	17	18	19	20	21	22	23	24	24	25	25	25	26	24	17	25	8
W	27	27	26	25	24	23	22	21	20	19	18	18	19	19	20	21	22	23	24	25	26	26	27	27	1	18	27	8
NW	21	21	20	20	19	18	17	16	16	15	14	14	15	15	16	17	18	19	20	20	21	21	21	21	1	14	21	7
Group B Walls																												
N	15	14	13	12	11	11	10	9	9	9	10	11	12	13	13	14	14	14	15	15	15	15	15	15	24	8	15	7
NE	19	18	16	15	14	13	13	15	17	18	18	17	16	16	17	19	20	21	21	21	21	21	20	20	21	12	21	9
E	23	21	20	18	17	15	15	19	22	24	25	22	20	19	20	22	24	25	26	27	27	27	25	24	20	15	27	12
SE	23	22	21	19	17	16	16	18	21	23	24	23	21	20	20	21	23	24	25	26	26	26	25	24	21	14	26	12
S	21	20	19	18	16	15	14	12	11	11	12	14	16	18	20	21	22	22	22	22	22	22	22	21	23	11	22	11
SW	27	26	25	24	22	20	19	17	15	14	14	15	17	20	23	25	27	28	28	28	28	28	28	30	24	13	28	15
W	29	28	27	26	24	22	21	19	18	16	15	15	17	18	20	22	24	26	27	29	30	30	29	32	24	14	30	16
NW	23	22	21	20	18	17	15	14	13	12	11	11	12	13	14	14	15	17	19	21	22	23	23	23	24	11	23	9
Group C Walls																												
N	15	14	13	11	11	10	10	8	8	8	9	11	12	13	14	15	15	15	16	17	17	17	16	16	22	7	17	10
NE	19	17	16	14	13	11	11	17	20	23	24	24	20	17	16	18	21	23	24	26	26	24	23	20	20	10	23	13
E	22	21	19	17	15	14	14	20	25	30	32	32	30	27	24	22	23	25	27	28	27	26	24	24	18	12	30	18
SE	21	21	19	17	15	13	13	19	24	29	32	31	26	24	22	21	22	24	25	26	25	24	24	24	19	12	29	17
S	19	19	17	15	14	12	12	11	12	16	20	24	26	26	24	20	20	21	25	27	25	24	22	22	21	9	26	17
SW	29	27	25	22	20	18	18	17	15	14	16	20	24	28	33	33	33	33	33	33	33	31	31	31	22	11	33	22
W	31	29	27	24	22	20	20	18	16	15	14	15	18	22	26	30	33	34	35	35	33	32	34	34	22	12	35	23
NW	25	23	21	19	18	16	14	13	12	11	10	11	12	14	18	22	25	27	27	27	27	27	27	27	22	10	27	17
Group D Walls																												
N	15	13	12	10	9	8	7	6	7	8	10	12	13	15	16	17	18	19	19	19	18	16	16	16	21	6	19	13
NE	17	15	13	11	10	8	8	17	20	22	23	22	20	18	18	20	22	24	24	22	20	20	18	18	19	7	25	18
E	19	17	15	13	11	9	8	20	27	30	32	30	26	24	22	24	26	28	30	30	26	24	22	22	16	8	32	25
SE	20	17	15	13	11	9	8	17	22	26	29	31	29	25	22	22	24	26	27	27	26	24	22	22	17	8	32	24
S	19	17	15	13	11	9	7	9	12	16	20	26	29	29	24	20	20	24	27	29	27	24	22	21	19	6	29	23
SW	28	25	22	20	17	14	12	10	10	12	16	22	29	34	37	38	38	38	37	35	34	33	31	31	21	8	38	30
W	31	27	24	21	19	16	12	10	9	11	14	18	24	31	37	40	41	41	41	40	38	37	34	34	21	9	41	32
NW	25	22	19	17	14	12	9	7	8	9	10	12	14	18	22	27	31	32	32	30	27	32	30	27	22	7	32	25

7.7

TABLE 7.2 Cooling Load Temperature Differences for Sunlit Walls (Concluded)

	Solar Time, h																								Hr of Maximum CLTD	Minimum CLTD	Maximum CLTD	Difference CLTD
	0100	0200	0300	0400	0500	0600	0700	0800	0900	1000	1100	1200	1300	1400	1500	1600	1700	1800	1900	2000	2100	2200	2300	2400				
Group E Walls																												
N	12	10	9	8	7	6	6	4	4	6	9	11	13	15	17	19	20	20	19	20	18	16	14	12	20	4	22	19
NE	13	11	9	8	7	6	20	26	33	36	38	37	34	32	31	30	29	28	25	22	19	17	15	13	16	5	26	22
E	14	12	10	8	7	6	26	38	45	43	39	36	34	33	32	32	31	30	28	25	22	20	17	15	13	5	38	33
SE	15	12	10	8	7	6	17	28	36	41	43	41	37	34	33	33	31	30	28	26	23	20	17	15	15	5	37	32
S	15	12	10	8	6	5	5	7	9	13	19	24	29	32	34	34	33	31	27	24	20	17	15	12	17	3	34	31
SW	22	18	15	12	10	8	7	6	8	11	14	17	20	24	29	35	40	43	44	45	40	35	29	24	19	6	45	40
W	25	21	17	14	12	10	8	7	8	9	11	14	17	20	24	29	34	38	43	49	45	40	34	29	20	6	49	43
NW	20	17	14	11	9	7	6	5	6	8	10	13	16	20	26	32	37	42	38	36	32	28	24	20	20	5	38	33
Group F Walls																												
N	8	6	5	3	2	1	4	6	7	9	11	14	17	19	21	22	23	24	24	23	20	16	13	11	19	1	23	23
NE	9	7	5	4	2	1	14	23	28	30	29	27	27	27	26	26	24	24	21	19	16	15	13	11	11	1	30	29
E	10	7	6	4	3	2	17	31	38	44	45	43	39	34	32	31	30	30	27	24	21	18	15	12	11	2	45	43
SE	10	8	6	5	3	2	10	19	28	36	41	43	42	39	36	33	31	30	28	25	21	18	15	13	12	2	43	41
S	10	8	6	5	3	1	3	7	13	20	27	34	38	39	38	36	33	31	27	24	20	17	15	12	13	1	39	38
SW	15	12	9	6	4	2	3	5	8	11	14	17	22	29	37	44	50	53	52	45	37	29	23	18	16	2	53	48
W	17	13	10	8	6	3	4	6	8	10	12	14	17	20	26	35	45	54	60	60	54	43	34	27	18	3	60	57
NW	14	10	8	6	4	2	3	5	6	8	10	13	15	20	27	35	43	46	46	42	37	31	25	20	19	2	46	44
Group G Walls																												
N	3	2	1	0	-1	-1	12	15	18	21	23	25	26	25	24	25	26	26	24	21	18	15	11	8	18	-1	26	27
NE	3	2	1	0	-1	-1	27	36	39	35	30	28	27	26	25	24	22	22	20	18	15	13	9	6	9	-1	39	40
E	4	2	1	0	-1	11	31	47	54	51	42	35	33	31	30	29	27	24	21	19	15	12	8	5	10	-1	55	56
SE	4	2	1	0	-1	5	18	32	42	48	51	48	42	37	33	30	27	25	22	20	17	13	8	5	11	-1	51	52
S	4	2	1	0	-1	-1	1	5	12	22	31	39	45	46	43	37	30	25	21	17	13	10	7	5	14	-1	46	47
SW	6	4	2	1	0	-1	2	5	8	12	16	22	26	31	39	50	59	63	61	52	37	26	17	11	16	0	63	63
W	6	3	2	1	0	-1	2	5	8	11	15	19	22	27	37	56	67	72	67	56	41	29	17	11	17	0	72	71
NW	5	3	2	0	-1	0	2	5	8	11	15	18	21	27	37	47	55	55	48	41	37	25	17	10	18	0	55	55

(1) *Direct Application of the Table Without Adjustments:*

Values in this table were calculated using the same conditions for walls as outlined for the roof CLTD table, Table __ . These values may be used for all normal air-conditioning estimates usually without correction (except as noted below) when the load is calculated for the hottest weather.

For totally shaded walls use the North orientation values.

(2) *Adjustments to Table Values:*

The following equation makes adjustments for conditions other than those listed in Note (1).

$$CLTD_{corr} = (CLTD + LM) K + (78 - t_R) + (t_o - 85)$$

where

CLTD is from Table __ . at the wall orientation.

LM is the latitude-month correction from Table 32.

K is a color adjustment factor applied after first making month-latitude adjustment

K = 1.0 if dark colored or light in an industrial area

K = 0.83 if permanently medium-colored (rural area)

K = 0.65 if permanently light-colored (rural area)

Credit should not be taken for wall color other than dark except where permanence of color is established by experience, as in rural areas or where there is little smoke.

Colors:

Light — Cream

Medium — Medium blue, medium green, bright red, light brown, unpainted wood and natural color concrete

Dark — Dark blue, red, brown and green

$(78 - t_R)$ is indoor design temperature correction

$(t_o - 85)$ is outdoor design temperature correction, where t_o is the average outside temperature on design day.

TABLE 7.3 CLTD Correction for Latitude and Month

Lat.	Month	N	NNE NNW	NE NW	ENE WNW	E W	ESE WSW	SE SW	SSE SSW	S	HOR
0	Dec	-3	-5	-5	-5	-2	0	3	6	9	-1
	Jan/Nov	-3	-5	-4	-4	-1	0	2	4	7	-1
	Feb/Oct	-3	-2	-2	-2	-1	-1	0	-1	0	0
	Mar/Sept	-3	0	1	-1	-1	-3	-3	-5	-8	0
	Apr/Aug	5	4	3	0	-2	-5	-6	-8	-8	-2
	May/Jul	10	7	5	0	-3	-7	-8	-9	-8	-4
	Jun	12	9	5	0	-3	-7	-9	-10	-8	-5
8	Dec	-4	-6	-6	-6	-3	0	4	8	12	-5
	Jan/Nov	-3	-5	-6	-5	-2	0	3	6	10	-4
	Feb/Oct	-3	-4	-3	-3	-1	-1	1	2	4	-1
	Mar/Sept	-3	-2	-1	-1	-1	-2	-2	-3	-4	0
	Apr/Aug	2	2	2	0	-1	-4	-5	-7	-7	-1
	May/Jul	7	5	4	0	-2	-5	-7	-9	-7	-2
	Jun	9	6	4	0	-2	-6	-8	-9	-7	-2
16	Dec	-4	-6	-8	-8	-4	-1	4	9	13	-9
	Jan/Nov	-4	-6	-7	-7	-4	-1	4	8	12	-7
	Feb/Oct	-3	-5	-5	-4	-2	0	2	5	7	-4
	Mar/Sept	-3	-3	-2	-2	-1	-1	0	0	0	-1
	Apr/Aug	-1	0	-1	-1	-1	-3	-3	-5	-6	0
	May/Jul	4	3	3	0	-1	-4	-5	-7	-7	0
	Jun	6	4	4	1	-1	-4	-6	-8	0	-7
24	Dec	-5	-7	-9	-10	-7	-3	3	9	13	-13
	Jan/Nov	-4	-6	-8	-9	-6	-3	9	3	13	-11
	Feb/Oct	-4	-5	-6	-6	-3	-1	3	7	10	-7
	Mar/Sept	-3	-4	-3	-3	-1	-1	1	2	4	-3
	Apr/Aug	-2	-1	0	-1	-1	-2	-1	-2	-3	0
	May/Jul	1	2	2	0	0	-3	-3	-5	-6	1
	Jun	3	3	3	1	0	-3	-4	-6	-6	1
32	Dec	-5	-7	-10	-11	-8	-5	2	9	12	-17
	Jan/Nov	-5	-7	-9	-11	-8	-15	-4	2	9	12
	Feb/Oct	-4	-6	-7	-8	-4	-2	4	8	11	-10
	Mar/Sept	-3	-4	-4	-4	-2	-1	3	5	7	-5
	Apr/Aug	-2	-2	-1	-2	0	-1	0	1	1	-1
	May/Jul	1	1	1	0	0	-1	-1	-3	-3	1
	Jun	1	2	2	1	0	-2	-2	-4	-4	2
40	Dec	-6	-8	-10	-13	-10	-7	0	7	10	-21
	Jan/Nov	-5	-7	-10	-12	-9	-6	1	8	11	-19
	Feb/Oct	-5	-7	-8	-9	-6	-3	3	8	12	-14
	Mar/Sept	-4	-5	-5	-6	-3	-1	4	7	10	-8
	Apr/Aug	-2	-3	-2	-2	0	0	2	3	4	-3
	May/Jul	0	0	0	0	0	0	0	0	1	1
	Jun	1	1	1	0	1	0	0	-1	-1	2
48	Dec	-6	-8	-11	-14	-13	-10	-3	2	6	-25
	Jan/Nov	-6	-8	-11	-13	-11	-8	-1	5	8	-24
	Feb/Oct	-5	-7	-10	-11	-8	-5	1	8	11	-18
	Mar/Sept	-4	-6	-6	-7	-4	-1	4	8	11	-11
	Apr/Aug	-3	-3	-3	-3	-1	0	4	6	7	-5
	May/Jul	0	-1	0	0	1	1	3	3	4	0
	Jun	1	1	2	1	2	1	2	2	3	2
56	Dec	-7	-9	-12	-16	-16	-14	-9	-5	-3	-28
	Jan/Nov	-6	-8	-11	-15	-14	-12	-6	-1	2	-27
	Feb/Oct	-6	-8	-10	-12	-10	-7	0	6	9	-22
	Mar/Sept	-5	-6	-7	-8	-5	-2	4	8	12	-15
	Apr/Aug	-3	-4	-4	-4	-1	1	5	7	9	-8
	May/Jul	0	0	0	0	2	2	5	6	7	-2
	Jun	2	1	2	1	3	3	4	5	6	1
64	Dec	-7	-9	-12	-16	-17	-18	-16	-14	-12	-30
	Jan/Nov	-7	-9	-12	-16	-16	-16	-13	-10	-8	-29
	Feb/Oct	-6	-8	-11	-14	-13	10	-4	1	4	-26
	Mar/Sept	-5	-7	-9	-10	-7	-4	2	7	11	-20
	Apr/Aug	-3	-4	-4	-4	-1	1	5	9	11	-11
	May/Jul	1	0	1	0	3	4	6	8	10	-3
	Jun	2	2	2	2	4	4	6	7	9	0

air space, has 1 in expanded polystyrene insulation, and has a ³/₄ in gypsum board interior finish. The outside design dry bulb temperature is 95°F, the daily range is 20°F, and the inside design dry bulb temperature is 75°F. What is the peak cooling load for the wall and at what hour does it occur?

First, it is necessary to calculate the U value and weight of the wall. Referring to Appendix B

Element	R	Wt/ft²
Outside Air Film	0.25	0
4 in M.W. Hollow Concrete Block	1.11	25.33
1 in Air Space	0.89	0
1 in Expanded Polystyrene	4.16	0.15
³/₄ in Gypsum Board	0.67	3.13
Inside Air Film	0.68	0
	7.76	28.61

$$U = \frac{1}{7.76} = 0.13 \text{ Btu/hr•ft}^2\text{•°F}$$

$$\text{Weight} = 28.61 \text{ lb/ft}^2$$

Referring to Table 7.1, a Type F wall with 4 in L.W. or H.W. concrete block + air space/insulation + finish is similar. The representative wall has a weight of 29 lb/ft² and a U value between 0.161 and 0.263.

Therefore, the calculations will be based on a Type F wall with a U value of 0.13. Assume the wall is dark since it is not permanently light colored.

From Table 7.2, it is determined that a Group F wall has a peak CLTD = 60°F and it occurs at hour 1900. The CLTD must now be corrected for the latitude/month, outside design temperature, and inside design temperature. From Table 7.3, the correction for 24° north latitude and July 21 is 0.

For an outside design dry bulb temperature of 95°F and a daily range of 20°F, the average daily dry bulb temperature is

$$T_{oa} = 95 - \frac{20}{2} = 85°F$$

Using Equation 7.2 for an inside design dry bulb temperature of 75°F, the corrected CLTD is

$$\text{CLTD}_c = (60 + 0)(1) + (78 - 75) + (85 - 85) = 63°F$$

Using Equation 7.1, the peak cooling load is

$$q = (0.13)(50)(63) = 409 \text{ Btu/hr}$$

When calculating cooling loads for spaces, considerable judgment is required. First, an exterior surface must be selected from the tables that has a similar weight and U value to the design surface. In all cases, the actual U value of the design surface should be used in Equation 7.1 to calculate the cooling load for the surface.

Second, if there is more than one exterior surface, or heat source for that matter, considerable judgment must be used to select the hour when the peak load for the space, or room, occurs. For this reason, it is frequently necessary to calculate cooling loads for several different hours, in order to determine the actual peak cooling load for a space and the time at which it occurs.

Example 7.2 A room has two 50 ft² walls that have the same construction as the wall described in Example 7.1. One wall faces east and one faces south. Assume the same location and design temperatures as Example 7.1. What is the peak cooling load for the space and when does it occur?

From Table 7.2, CLTD values and times are obtained as follows:

	1200	1400	1600
East Wall	45	39	34
South Wall	20	34	39

It can be seen that the east wall has a peak CLTD of 45°F at hour 1200 and then the CLTD decreases. The south wall CLTD continues to increase after hour 1200 until it peaks at a value of 39 at hour 1600. Since the two surfaces peak at different times, it will be necessary to calculate the cooling load for several different hours.

For the *east wall* using Equation 7.2, the corrected CLTD values are as follows:

at hour 1200

$CLTD_c = (45 + 0)(1) + (78 - 75) + (85 - 85) = 48°F$

at hour 1400

$CLTD_c = (39 + 0)(1) + (78 - 75) + (85 - 85) = 42°F$

at hour 1600

$CLTD_c = (34 + 0)(1) + (78 - 75) + (85 - 85) = 37°F$

For the *west wall*, the corrected CLTD values are:

at hour 1200

$CLTD_c = (20 + 0)(1) + (78 - 75) + (85 - 85) = 23°F$

at hour 1400

$CLTD_c = (34 + 0)(1) + (78 - 75) + (85 - 85) = 37°F$

at hour 1600

$CLTD_c = (39 + 0)(1) + (78 - 75) + (85 - 85) = 42°F$

The total external cooling load for the space is the sum of the cooling loads for each exterior surface. The cooling load at each hour is as follows:

at hour 1200

$q = (0.13)(50)(48) + (0.13)(50)(23) = 461.5$ Btu/hr

at hour 1400

$q = (0.13)(50)(42) + (0.13)(50)(37) = 513.5$ Btu/hr

at hour 1600

$q = (0.13)(50)(37) + (0.13)(50)(42) = 513.5$ Btu/hr

In this case, the peak load occurs twice, once at hour 1400 and again at hour 1600. The peak design exterior cooling load is 513.5 or 514 Btu/hr.

The above is a simplified example of a space with multiple heat gains and cooling loads. For an actual room, there are usually many heat gains, with peak cooling loads occurring at different times. This is one of the reasons why it is necessary to use considerable

TABLE 7.4   CLTD for Group A Walls with Additional Insulation

N	NE	E	SE	S	SW	W	NW
11	17	22	21	17	21	22	17

judgment when using the CLTD/CLF method to calculate cooling loads. This will become increasingly evident as additional sources of heat gain are discussed in this and the next chapter.

It may also be observed, in the above example, that the same correction factors were used over again for each hour for each wall. Since multiple calculations are frequently necessary, it is usually easier to calculate the correction factors for each surface separately, and then apply them to the hourly CLTD values.

For walls that have a U value greater than those listed in Table 7.1 for a given wall weight, the group selection must be adjusted. For each 7.0 increase in the R value (or a U value that is 14% lower) due to the added insulation, use the previous alphabetic, or heavier, wall group letter. For example, if a given wall is similar in actual weight to a Group C wall but has insulation that is $R7$ greater, then the calculations should be based on Group B CLTD values. If the given design wall is already similar to a Group A wall and has insulation that is $R7$, or greater than that listed for a Group A wall, then a CLTD should be used from Table 7.4.

7.2.2   Cooling Loads for Sunlit Roofs

The procedure used to calculate cooling loads for sunlit roofs is very similar to the procedure for sunlit walls. In a manner similar to sunlit walls, ASHRAE has grouped roofs by construction type or weight and also according to whether there is a suspended ceiling or not.

The roof types, along with the weight of the representative U values and the cooling loads, are given in Table 7.5, Cooling Load Temperature Differences for Flat Roofs. As in the case of sunlit walls, the CLTD values from Table 7.5 may be used to calculate cooling loads for sunlit roofs.

The same assumed conditions for the wall cooling load temperature differences apply to the roof cooling load temperature differences. Therefore, the roof CLTD values must be corrected. In addition to the conditions assumed for the walls, the representative roof types have the following assumed conditions:

a. The roof does not have a forced ventilation system, such as a fan in the space, or attic, between the ceiling and roof. The space is essentially a "dead air space."

b. The roof is flat.

Although roofs are not flat in many cases, there is no correction for an inclined roof. Using the values for a flat roof should not introduce significant errors, since the flat roof is probably the worst case situation. During the summer months, the sun is almost directly overhead and the solar radiation would fall directly on a flat roof.

The equation correcting the CLTD values listed is identical to Equation 7.2 for sunlit walls, except for an attic ventilation factor. The following equation may be used to correct the CLTD values obtained from Table 7.5.

$$\text{CLTD}_c = [(\text{CLTD} + LM)K + (78 - T_i) + (T_{oa} - 85)]f \qquad (7.3)$$

where

f = a factor for attic ventilation after all other adjustments have been made.

f = 1.0 for no ventilation and f = 0.75 for positive ventilation.

All other correction variables in Equation 7.3 are the same as for Equation 7.2. Notice that when using Table 7.3 to correct for the latitude and month, the column headed HOR should be used for roofs.

As in the case of walls, the U value in Table 7.5 is for comparison only. The actual U value of the roof, for the building being designed, should be used in Equation 7.1.

Example 7.3 Assume that a dark 200 ft² flat roof, similar to the one described in Example 4.3, is on a building in Chicago, Illinois on August 21. The weight of the roof has been calculated to be 8.72 lb/ft² and has a U value of 0.09 Btu/hr•ft²•°F. The attic/air space has no positive ventilation and a room below is to be maintained at 75°F. What is the peak cooling load for the roof and at what hour does it occur?

From Table 7.5, a Roof Number 1, with a suspended ceiling, would be chosen as the closest matching roof. The peak CLTD for the representative roof is 78°F which occurs at hour 15.

From Table 3.1, the 2½% climatic design conditions for Chicago, Illinois are 91°F db/74°F wb. The daily temperature range is 20°F and the latitude for Chicago is 42°N.

$$\text{Therefore } T_{oa} = 91 - \frac{20}{2} = 81°F$$

Using Equation 7.3 to correct the CLTD,

$$\text{CLTD} = \{[78 + (-3)](1) + (78 - 75) + (81 - 85)\}(1) = 74°F$$

Using Equation 7.1 and the calculated U value for the roof, the peak cooling load from the roof is

$$q = (0.09)(200)(74) = 1332 \text{ Btu/hr at hour 15}$$

In the above example, the air space, between the roof deck and the suspended ceiling, was a "dead air space," with little or no ventilation or air movement. Frequently, the space between the roof deck and suspended ceiling is used as a **return air plenum**. That is, the space above the ceiling is used to return the air from the occupied spaces to the air handling unit in lieu of ductwork. For spaces with ceiling return air plenums, the roof cooling load calculations should be based on the roof classification without a suspended ceiling. Not all of the roof cooling load becomes a cooling load for the room below the ceiling, though. Much of the cooling load becomes a return air cooling load for the HVAC system, but not a load for the room. This is an important distinction since the air supplied to the room should be based on the actual calculated room cooling load.

7.3   HEAT GAINS FOR WINDOWS AND GLASS

In the previous sections of this chapter, the effect of solar radiation on opaque surfaces, such as walls and roofs, was discussed. The resultant effect on the cooling load for a building was also discussed. The effect of solar radiation on glass areas, or fenestration areas, can have an even more significant effect on the cooling load for a building.

The space cooling load that results from solar radiation on fenestration areas is dependent upon a number of variables including:

TABLE 7.5 Cooling Load Temperature Differences for Flat Roofs

Roof No	Description of Construction	Weight, lb/ft²	U-value, Btu/h·ft²·°F	1	2	3	4	5	6	7	8	9	10	11	12	13	14	15	16	17	18	19	20	21	22	23	24	Hour of Maxi-mum CLTD	Mini-mum CLTD	Maxi-mum CLTD	Differ-ence CLTD
	Without Suspended Ceiling																														
1	Steel sheet with 1-in. (or 2-in.) insulation	7 (8)	0.213 (0.124)	1	-2	-3	-3	-5	-3	6	19	34	49	61	71	78	79	77	70	59	45	30	18	12	8	5	3	14	-5	79	84
2	1-in. wood with 1-in. insulation	8	0.170	6	3	0	-1	-3	-3	-2	4	14	27	39	52	62	70	74	74	70	62	51	38	28	20	14	9	16	-3	74	77
3	4-in. lightweight concrete	18	0.213	9	5	2	0	-2	-3	-3	1	9	20	32	44	55	64	70	73	71	66	57	45	34	25	18	13	16	-3	73	76
4	2-in. heavyweight concrete with 1-in. (or 2-in.) insulation	29 (0.122)	0.206	12	8	5	3	0	-1	-1	3	11	20	30	41	51	59	65	66	62	54	45	36	29	22	17		16	-1	67	68
5	1-in. wood with 2-in. insulation	9	0.109	3	0	-3	-4	-5	-7	-6	-3	5	16	27	39	49	57	63	64	62	57	48	37	26	18	11	7	16	-7	64	71
6	6-in. lightweight concrete	24	0.158	22	17	13	9	6	3	1	1	3	7	15	23	33	43	51	58	62	64	62	57	50	42	35	28	18	1	64	63
7	2.5-in. wood with 1-in. ins.	13	0.130	29	24	20	16	13	10	7	6	6	9	13	20	27	34	42	48	53	55	56	54	49	44	39	34	19	6	56	50
8	8-in. lightweight concrete	31	0.126	35	30	26	22	18	14	11	9	7	7	9	13	19	25	33	39	46	50	53	54	53	49	45	40	20	7	54	47
9	4-in. heavyweight concrete with 1-in. (or 2-in.) insulation	52 (52)	0.200 (0.120)	25	22	18	15	12	9	8	8	10	14	20	26	33	40	46	50	53	53	52	48	43	38	34	30	18	8	53	45
10	2.5-in. wood with 2-in. ins.	13	0.093	30	26	23	19	16	13	10	9	8	9	13	17	23	29	36	41	46	49	51	50	47	43	39	35	19	8	51	43
11	Roof terrace system	75	0.106	34	31	28	25	22	19	16	14	13	13	15	18	22	26	31	36	40	44	45	46	45	43	40	37	20	13	46	33
12	6-in. heavyweight concrete with 1-in. (or 2-in.) insulation	75 (75)	0.192 (0.117)	31	28	25	22	20	17	15	14	14	16	18	22	26	31	36	40	43	45	46	44	42	40	37	34	19	14	45	31
13	4-in. wood with 1-in. (or 2-in.) insulation	17 (18)	0.106 (0.078)	38	36	33	30	28	25	22	20	18	17	16	17	18	21	24	28	32	36	39	41	43	43	42	40	22	16	43	27
	With Suspended Ceiling																														
1	Steel Sheet with 1-in. (or 2-in.) insulation	9 (10)	0.134 (0.092)	2	0	-2	-3	-4	-4	-1	9	23	37	50	62	71	77	78	74	67	56	42	28	18	12	8	5	15	-4	78	82
2	1-in. wood with 1-in. ins.	10	0.115	20	15	11	8	5	3	2	3	7	13	21	30	40	48	55	60	62	61	58	51	44	37	30	25	17	2	62	60
3	4-in. lightweight concrete	20	0.134	19	14	10	7	4	2	0	0	4	10	19	29	39	48	56	62	65	64	61	54	46	38	30	24	17	0	65	65
4	2-in. heavyweight concrete with 1-in. insulation	30	0.131	28	25	23	20	17	15	13	13	14	16	20	25	30	35	39	43	46	47	46	44	41	38	35	32	18	13	47	34

No.	Description	Mass (lb/ft²)	U (Btu/h·ft²·°F)	1	2	3	4	5	6	7	8	9	10	11	12	13	14	15	16	17	18	19	20	21	22	23	24
5	1-in. wood with 2-in. ins.	10	0.083	25	20	16	13	10	7	5	5	7	12	18	25	33	41	48	53	57	57	56	52	46	40	34	29
6	6-in. lightweight concrete	26	0.109	32	28	23	19	16	13	10	8	7	8	11	16	22	29	36	42	48	52	54	54	52	47	42	37
7	2.5-in. wood with 1-in. insulation	15	0.096	34	31	29	26	23	21	18	16	15	15	17	20	25	29	34	38	41	44	44	43	41	38	34	31
8	8-in. lightweight concrete	33	0.093	39	36	33	29	26	23	20	18	15	14	15	17	20	25	29	34	38	42	45	46	46	44	42	39
9	4-in. heavyweight concrete with 1-in. (or 2-in.) ins.	53 (54)	0.128 (0.090)	30	29	27	26	25	24	22	21	20	20	21	24	27	29	32	34	36	37	37	36	34	33	32	32
10	2.5-in. wood with 2-in. ins.	15	0.072	35	33	30	28	26	24	22	20	18	18	20	21	24	27	30	33	35	38	38	38	37	36	34	33
11	Roof terrace system	77	0.082	30	29	28	27	26	25	23	22	22	22	24	25	27	29	31	32	33	34	34	33	33	32	31	31
12	6-in. heavyweight concrete with 1-in. (or 2-in) insulation	77 (77)	0.125 (0.088)	29	28	27	25	25	26	25	24	23	22	24	25	28	30	32	33	34	34	34	34	33	32	31	31
13	4-in. wood with 1-in (or 2-in.) insulation	19 (20)	0.082 (0.064)	35	34	33	32	29	27	26	24	23	21	20	22	24	29	34	37	39	41	42	41	40	37	34	33

(1) Direct Application of Table Without Adjustments:

Values were calculated using the following conditions:

- Dark flat surface roof ("dark" for solar radiation absorption)
- Indoor temperature of 78°F
- Outdoor maximum temperature of 95°F with outdoor mean temperature of 85°F and an outdoor daily range of 21°F
- Solar radiation typical of 40 deg North latitude on July 21
- Outside surface resistance, $R_o = 0.333$ ft²·°F·h/Btu
- Without and with suspended ceiling, but no attic fans or return air ducts in suspended ceiling space
- Inside surface resistance, $R_i = 0.685$ ft²·°F·h/Btu

(2) Adjustments to Table Values:

The following equation makes adjustments for deviations of design and solar conditions from those listed in (1) above.

$$CLTD_{corr} = [(CLTD + LM) K + (78 - t_R) + (t_o - 85)] f$$

where CLTD is from this table

(a) LM is latitude-month correction from Table for a horizontal surface.
(b) K is a color adjustment factor applied after first making month-latitude adjustments. Credit should not be taken for a light-colored roof except where permanence of light color is established by experience, as in rural areas or where there is little smoke.
  K = 1.0 if dark colored or light in an industrial area
  K = 0.5 if permanently light-colored (rural area)
(c) $(78 - t_R)$ is indoor design temperature correction
(d) $(t_o - 85)$ is outdoor design temperature correction, where t_o is the average outside temperature on design day

(e) f is a factor for attic fan and or ducts above ceiling applied after all other adjustments have been made
  $f = 1.0$ no attic or ducts
  $f = 0.75$ positive ventilation
Values in Table were calculated without and with suspended ceiling, but make no allowances for positive ventilation or return ducts through the space. If ceiling is insulated and fan is used between ceiling and roof, CLTD may be reduced 25% ($f = 0.75$). Analyze use of the suspended ceiling space for a return air plenum or with return air ducts separately.

(3) Roof Constructions Not Listed in Table:

The U-Values listed are only guides. The actual value of U as obtained from tables such as Table   or as calculated for the actual roof construction should be used.

An actual roof construction not in this table would be thermally similar to a roof in the table, if it has similar mass, lb/ft², and similar heat capacity, Btu/ft²·°F. In this case, use the CLTD from this table as corrected by Note (2) above.

(4) Additional Insulation:

For each R-7 increase in R-value from insulation added to the roof structure, use a CLTD for a roof whose weight and heat capacity are approximately the same, but whose CLTD has a maximum value 2 h later. If this is not possible, because a roof with longest time lag has already been selected, use an effective CLTD in cooling load calculation equal to 29°F.

Example: A flat roof without suspended ceiling has mass = 18.0 lb/ft², $U = 0.20$ Btu/h·ft²·°F, and heat capacity = 9.5 Btu/ft²·°F.
Use $CLTD_{uncorr}$ from Roof No. 13, to obtain $CLTD_{corr}$ and use the actual U value to calculate $q/A = U (CLTD_{corr}) = 0.20 (CLTD_{corr})$.

a. The size of the fenestration area.

b. The direction the fenestration is facing or its orientation.

c. Shading, both exterior and interior.

d. Solar radiation intensity and angle of incidence.

e. The type of glazing material and number of layers.

f. The time of day and the day of the year.

g. The mass of the space.

Any solar radiation falling on a surface, such as glass, will be reflected, absorbed or transmitted through the glass as shown in Figure 2.5. The relative proportions of reflected, absorbed and transmitted energy are primarily a function of the glazing material and the angle of incidence of the radiation. Figure 7.2 shows typical proportions of reflected, absorbed and transmitted energy for sunlit glass.

The total amount and intensity of solar radiation incident on a surface, such as a window, is a function of the angle of incidence that the radiation strikes the glazing surface and the extent to which the surface is shaded by any external objects. The angle of incidence is a function of the surface orientation, location, time of day and day of the year.

7.3.1 Angle of Incidence for Solar Radiation

The earth rotates on a nearly north/south axis, once each day, or every 24 hr. The earth also moves in a slightly elliptical orbit around the sun, called the elliptical plane.

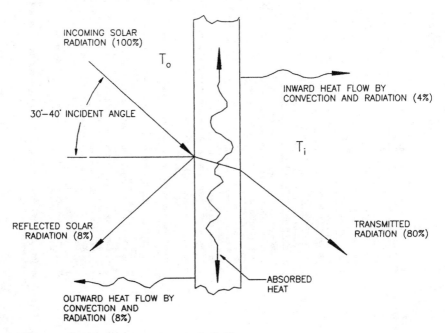

FIGURE 7.2   Instantaneous Heat Balance for Sunlit Glass

As a result of the earth's motion about its own axis and about the sun, the incident solar radiation, and the angle of incidence of that radiation, is constantly changing. These changes, in intensity and angle, have a significant effect on the resultant solar heat gain for surfaces.

Since the sun is so far away from the earth, it may be considered a point source of radiation. Thus, the relationship of the sun, with respect to a surface, may be described by angles and trigonometric relationships. The primary angles that determine the **angle of incidence** θ of solar radiation, on any given surface, are the **solar altitude** β of the sun, the **solar azimuth** ϕ, of the sun, the **surface-solar azimuth** γ, and the **surface tilt angle** α. These angles are shown graphically in Figures 7.3 and 7.4.

For a vertical surface, the surface-solar azimuth γ is the angle measured in the horizontal plane, between the projection of the sun's rays on that plane and a line perpendicular to the surface. The angle can be positive or negative, with a surface directly facing south having $\gamma = 0$. The solar azimuth angle ϕ is the angle in the horizontal plane, measured between south and the projection of the sun's rays on that plane. The angle can be positive or negative, with ϕ being positive for west of south or afternoon. When the sun is due south of a surface, $\phi = 0$.

The solar altitude β (the sun's altitude angle) is the angle between the sun's rays and the projection of a ray on a horizontal surface. It is simply the angle of the sun above the horizon and can vary from $\beta = 0°$ to $\beta = 90°$. It is particularly useful when calculating the effect of external shading. The angle of tilt α is the angle between the surface and the horizontal plane.

The **solar declination** δ is the angle between the sun's rays and the earth's equatorial plane. It is similar to the solar altitude β, except that the angle is measured between the sun's rays and the equatorial plane, instead of the horizontal plane. The solar declination may be approximated using Equation 7.4.

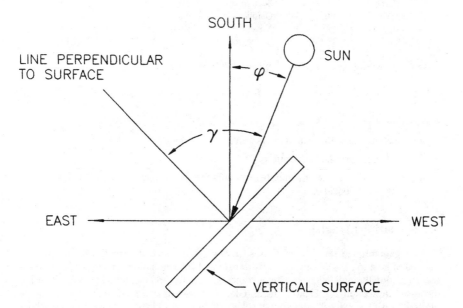

FIGURE 7.3   Surface-Solar Azimuth and Solar Azimuth

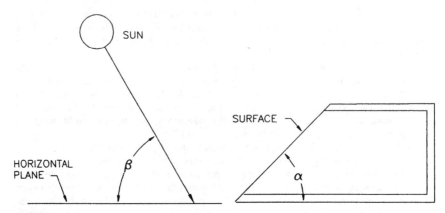

FIGURE 7.4   Solar Altitude and Surface Tilt Angle

$$\delta = 23.45 \sin \left[\frac{360° \, (284 + n)}{365} \right] \qquad (7.4)$$

where

  n = the day of the year

  Values of solar declination δ (in degrees), at mid-month are as follows:

January	−21.4	July	21.4
February	−14.0	August	14.0
March	−2.8	September	2.8
April	9.1	October	−9.1
May	18.6	November	−18.6
June	23.1	December	−23.1

The solar altitude β may be determined by using Equation 7.5.

$$\beta = \sin^{-1} (\sin \delta \sin L + \cos \delta \cos L \cos \omega) \qquad (7.5)$$

where

  L = the latitude with north being positive

  δ = the angle of solar declination

  ω = the hour angle, i.e., the number of hours from solar noon times 15° with mornings being positive

7.3.2   Glazing Materials

   Heat gain through glazing materials, such as glass, occurs by two methods. The primary method of heat transfer through glass is transmitted solar radiation. The second method is by conduction through the glass. Conduction heat transfer, and radiation heat transfer, were discussed in Section 2.3, Heat Transfer. Glazing materials can include clear and opaque plastics, as well as glass. The amount of solar radiation transmitted by a glazing material depends on the solar optical properties which include the wavelength of the radiation, the chemical composition of the material, coloring in the glazing material, and the angle of incidence θ of the radiation. One measure of a particular glazing material's ability to transmit solar radiation is the **index of refraction**. The index of refraction actually measures how much the path of light, passing through a material, is changed. The higher the index of refraction, the more a beam of light changes direction as it passes through the material, and the more it is reflected by the material.

An alternate method used to determine the ability of a particular glazing, or combination of glazing, to transmit solar radiation is the **shading coefficient (*SC*)**. The shading coefficient is defined as the ratio of solar gain through a given glazing system under a specific set of conditions, to the solar gain through a reference glass under the same conditions. Equation 7.6 is the equation for the shading coefficient.

$$SC = \frac{\text{solar heat gain of fenestration}}{\text{solar heat gain of reference glass}}$$

The reference glass is a single sheet of unshaded, clear, double strength glass (DSA) with a transmittance of 0.86, a reflectance of 0.08, and an absorptance of 0.06.

Typical shading coefficients (SC) are given in Table 7.6. The shading coefficient for any fenestration will increase the tabulated values when the inner surface film coefficient is increased and the outer coefficient decreased. The reverse is also true. Table 7.6 gives shading coefficients for outside film coefficients of 3 and 4 Btu/hr•ft²•°F. The larger film coefficient is for an outdoor wind speed of $7\frac{1}{2}$ mph, while the other is for lower wind speeds. The film convection coefficient affects the shading coefficient by affecting the amount of radiation that is absorbed by the glazing and then transferred away by convection.

The values presented in Table 7.6 are typical values. Whenever possible, the glazing manufacturer's literature should be consulted to get actual values for shading coefficients and *U* values.

7.3.3 Shading

The amount of solar heat gain to a space, through fenestration areas, is also directly related to the extent to which fenestration areas are shaded. Shading may be from an object external to the building, such as an overhang, or it may be internal, such as venetian blinds or shades.

External shading occurs when an object outside the building partially, or fully, blocks the sun's rays and shades the fenestration area. Such objects include overhangs, side fins, other buildings, or objects and vegetation, such as trees. External shading has a much greater effect on space heat gains and cooling loads than interior shading. It is also possible for an object to shade itself. For a vertical surface, such as a wall, whenever the surface-solar azimuth angle is greater than 90° and less than 180°, the surface will be shaded by itself. In effect, the surface "has its back to the sun." Surfaces that face directly north will always be shaded.

The most common form of external shading is the overhang. The amount, or fraction, of the fenestration surface that is shaded may be estimated by the **shading factor *S***. The shading factor is defined as the fraction of glazing area which is shaded by an external object. The shading factor for an overhang may be estimated by Equation 7.7.

$$S = \frac{1}{H}(W \tan \beta - D) \tag{7.7}$$

where

W = the width of the overhang as it extends out from the surface

D = the vertical distance from the top of the fenestration area to the underside of the overhang

β = solar altitude

H = the height of the glazing

Figure 7.5 graphically shows the variables for Equation 7.7

If S is equal to or less than 0, all of the fenestration area is in the sun. If S is equal to or greater than 1.0, the fenestration area is completely shaded. Any value of S between 0 and 1.0 is the fraction of the glazing that is shaded.

Equation 7.7 is only a good approximation for estimating the fraction of the glazing that is shaded. It does not take the length of the overhang or the fenestration into account. For more complex overhangs, as well as side fins, refer to Chapter 3 of Reference 1.

In addition to external shading, interior shading, such as blinds, drapes, or shades, is frequently employed for solar control. Interior shading has two effects. First, it reflects much of the solar radiation at the fenestration opening so the direct radiation never enters the space. Second, the interior shade prevents the solar radiation from being absorbed by the interior mass of the space itself. Instead, the interior shading absorbs much of the radiation, which in turn is convected to the air inside the space.

TABLE 7.6   Shading Coefficients for Single Glass and Insulating Glass

Type of Glass	Nominal Thickness[b]	Solar Trans.[b]	Shading Coefficient h_o=4.0	Shading Coefficient h_o=3.0
A. Single Glass				
Clear	1/8 in.	0.86	1.00	1.00
	1/4 in.	0.78	0.94	0.95
	3/8 in.	0.72	0.90	0.92
	1/2 in.	0.67	0.87	0.88
Heat Absorbing	1/8 in.	0.64	0.83	0.85
	1/4 in.	0.46	0.69	0.73
	3/8 in.	0.33	0.60	0.64
	1/2 in.	0.24	0.53	0.58
B. Insulating Glass				
Clear Out, Clear In	1/8 in.[c]	0.71[e]	0.88	0.85
Clear Out, Clear In	1/4 in.	0.61	0.81	0.82
Heat Absorbing[d] Out, Clear In	1/4 in.	0.36	0.55	0.58

[a]Refers to factory-fabricated units with $3/16$, $1/4$, or $1/2$-in air space or to prime windows plus storm sash.

[b]Refer to manufacturer's literature of values.

[c]Thickness of each pane of glass, not thickness of assembled unit.

[d]Refers to gray, bronze, and green tinted heat-absorbing float glass.

[e]Combined transmittance for assembled unit.

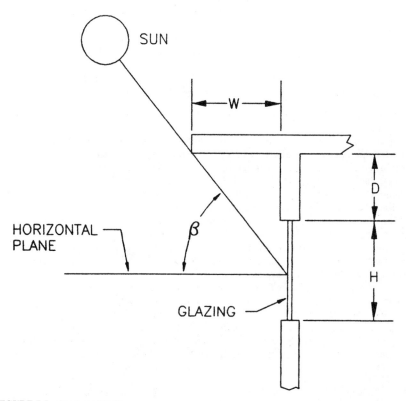

FIGURE 7.5 Overhang Shading

Since the solar radiation is absorbed by the interior shading and then convected to the room air, interior shades tend to be less effective in reducing solar heat gains than exterior shades. Table 7.7 gives typical shading coefficients for fenestration areas with venetian blinds and roller shades.

Shading coefficients of draperies, for single and multi-layer glazing, is a complex function of the color and weave of the drapery fabric. Whenever possible, shading coefficients for draperies should be obtained from the drapery or fabric manufacturer. If a manufacturer's data is not readily available, refer to Chapter 27 of Reference 1 or Chapter 3 of Reference 4.

7.3.4 Cooling Loads for Windows and Glass

Heat gain through a fenestration area to a space occurs by two methods, solar radiation and conduction heat transfer. Conduction heat transfer occurs as the result of the difference in temperature between indoors and outdoors, and due to radiant heat that has been absorbed by the glazing material itself. Equation 7.8 shows the components that make up the total heat flow through glass.

$$q_t = q_r + q_a + q_c \tag{7.8}$$

where

q_t = the total or net heat transmission through the glass in Btu/hr

q_r = the radiation heat transfer in Btu/hr

q_a = inward flow of absorbed radiation in Btu/hr

q_c = heat conducted through the glazing material in Btu/hr

Note that q_a and q_c may be positive or negative, i.e., the flow of heat may be into or out of the space.

In a manner similar to cooling loads for sunlit walls and roofs, ASHRAE has developed cooling load temperature differences (CLTD), and cooling load factors (CLF), to account for transient heat transfer and thermal storage of building materials. The effect of thermal storage was discussed in Section 2.3.4, Transient Heat Transfer. The Cooling Load Temperature/Cooling Load Factor Method, with the Transfer function method, was discussed in Section 7.1, External Heat Gain Calculation Methods.

The conduction heat gain, and resultant cooling load, is the sum of the absorbed radiant heat and the conducted heat due to the difference between indoor and outdoor temperatures. Equation 7.1 may be used to calculate the cooling load due to conduction through fenestration areas. Table 7.8 provides Cooling Load Temperature Differences for Conduction Through Glass to be used in Equation 7.1.

The CLTD values listed in Table 7.8 are valid for an indoor temperature of 78°F, an outdoor temperature between 93° and 102°F, and an outdoor daily range between 16° and 34°F, with an average daily temperature of 85°F. For conditions other than those listed above, the CLTD values must be corrected in accordance with the footnote provided with Table 7.8.

Referring again to Equation 7.8, the total heat transmission through a fenestration area includes radiation heat transfer. Again, ASHRAE has developed a method to account for

TABLE 7.7   Shading Coefficients for Single Glass with Indoor Shading by Venetian Blinds or Roller Shades

			Type of Shading				
			Venetian Blinds		Roller Shade		
					Opaque		Translucent
Type of Glass	Nominal Thickness[a], in.	Solar Transmittance[b]	Medium	Light	Dark	White	Light
Clear	3/32[c]	0.87 to 0.80	0.74[d] (0.63)[e]	0.67[d] (0.58)[e]	0.81	0.39	0.44
Clear	1/4 to 1/2	0.80 to 0.71					
Clear pattern	1/8 to 1/2	0.87 to 0.79					
Heat-absorbing pattern	1/8	—					
Tinted	3/16, 7/32	0.74, 0.71					
Heat-absorbing[f]	3/16, 1/4	0.46					
Heat-absorbing pattern	3/16, 1/4	—	0.57	0.53	0.45	0.30	0.36
Tinted	1/8, 7/32	0.59, 0.45					
Heat-absorbing or pattern	—	0.44 to 0.30	0.54	0.52	0.40	0.28	0.32
Heat-absorbing[f]	3/8	0.34					
Heat-absorbing or pattern		0.29 to 0.15					
	—	0.24	0.42	0.40	0.36	0.28	0.31
Reflective coated glass	S.C. = 0.30[g]		0.25	0.23			
	= 0.40		0.33	0.29			
	= 0.50		0.42	0.38			
	= 0.60		0.50	0.44			

[a]Refer to manufacturers' literature for values.
[b]For vertical blinds with opaque white and beige louvers in the tightly closed position, SC is 0.25 and 0.29 when used with glass of 0.71 to 0.80 transmittance.
[c]Typical residential glass thickness.
[d]From Van Dyck and Konen (1982), for 45° open venetian blinds, 35° solar incidence, and 35° profile angle.
[e]Values for closed venetian blinds. Use these values only when operation is automated for solar gain reduction (as opposed to dayfight use).
[f]Refers to gray, bronze, and green tinted heat-absorbing glass.
[g]SC for glass with no shading device.

TABLE 7.8 Cooling Load Temperature Differences for Glass

Solar Time, h	CLTD, °F	Solar Time, h	CLTD, °F
0100	1	1300	12
0200	0	1400	13
0300	−1	1500	14
0400	−2	1600	14
0500	−2	1700	13
0600	−2	1800	12
0700	−2	1900	10
0800	0	2000	8
0900	2	2100	6
1000	4	2200	4
1100	7	2300	3
1200	9	2400	2

Corrections: The values in the table were calculated for an inside temperature of 78°F and an outdoor maximum temperature of 95°F with an outdoor daily range of 21°F. The table remains approximately correct for other outdoor maximums 93 to 102°F and other outdoor daily ranges 16 to 34°F, provided the outdoor daily average temperature remains approximately 85°F. If the room air temperature is different from 78°F and/or the outdoor daily average temperature is different from 85°F

thermal storage by the building mass. For solar radiation, Cooling Load Factors (CLF) are used. Cooling Load Factors are different from Cooling Load Temperature Differences, in that they account for heat gains that are inside, or have directly entered, the space and are absorbed within the space itself. Such is the case for solar radiation which shines in through a window and is absorbed by floors, walls, and objects, and then released later as a cooling load on the space. The magnitude and time of the peak space cooling load are a function of the thermal mass of the space. Equation 7.9 may be used to calculate the radiant cooling load on a space, due to solar radiation through fenestration areas.

$$q_r = A(\text{SC})\,(\text{SHGF})\,(\text{CLF}) \tag{7.9}$$

where

A = the net glazing area of the fenestration area in ft²

SC = the shading coefficient

SHGF = the maximum solar heat gain factor in Btu/hr•ft² (from Table 7.9.)

CLF = Cooling Load Factor from Table 7.10, 7.11 or 7.12 for the appropriate room characteristics

Referring to Table 7.9, the values listed are maximum solar heat gain, for various fenestration orientations, at various north latitudes for each month. These values are typical values that take the solar angle of incidence into account. Whenever a fenestration area is fully, or partially shaded, the maximum solar heat gain, and CLF, for north facing fenestration should be applied to the shaded portion. This accounts for radiant heat gain due to indirect or reflected radiation. Note also that the CLF values in Table 7.10 are for glass **without internal shading** and **without carpeting** on the floors.

TABLE 7.9 Maximum Solar Heat Gain for Sunlit Glass (Btu/hr•ft²)

0° N Lat

	N	NNE/NNW	NE/NW	ENE/WNW	E/W	ESE/WSW	SE/SW	SSE/SSW	S	HOR
Jan.	34	34	88	177	234	254	235	182	118	296
Feb.	36	39	132	205	245	247	210	141	67	306
Mar.	38	87	170	223	242	223	170	87	38	303
Apr.	71	134	193	224	221	184	118	38	37	284
May	113	164	203	218	201	154	80	37	37	265
June	129	173	206	212	191	140	66	37	37	255
July	115	164	201	213	195	149	77	38	37	260
Aug.	75	134	187	216	212	175	112	39	38	276
Sept.	40	84	163	213	231	213	163	84	40	293
Oct.	37	40	129	199	236	238	202	135	66	299
Nov.	35	35	88	175	226	250	230	179	117	293
Dec.	34	34	71	164	226	253	240	196	138	288

4° N Lat

	N	NNE/NNW	NE/NW	ENE/WNW	E/W	ESE/WSW	SE/SW	SSE/SSW	S	HOR
Jan.	33	33	79	170	229	252	237	193	141	286
Feb.	35	35	123	199	242	248	215	152	88	301
Mar.	38	77	163	219	242	227	177	96	43	302
Apr.	55	125	189	223	223	190	126	43	38	287
May	93	154	200	215	206	161	89	38	38	272
June	110	164	202	215	200	156	73	38	38	263
July	96	154	197	214	196	181	85	39	38	267
Aug.	59	124	184	215	214	156	120	42	40	279
Sept.	39	75	156	209	231	216	170	93	44	293
Oct.	36	36	120	193	234	239	207	148	86	294
Nov.	34	34	79	168	226	248	232	190	139	284
Dec.	33	33	62	157	221	250	242	206	160	277

20° N Lat

	N	NNE/NNW	NE/NW	ENE/WNW	E/W	ESE/WSW	SE/SW	SSE/SSW	S	HOR
Jan.	29	29	48	138	201	243	253	233	214	232
Feb.	31	31	88	173	226	244	238	201	174	263
Mar.	34	49	132	200	237	236	206	152	115	284
Apr.	38	92	166	213	228	208	158	91	58	287
May	47	123	184	217	217	184	124	54	42	283
June	59	135	189	216	212	173	108	45	42	279
July	48	124	182	213	212	179	119	53	43	278
Aug.	40	91	162	206	220	200	152	88	57	280
Sept.	36	46	127	191	225	225	199	148	114	275
Oct.	32	32	87	167	217	236	231	196	170	258
Nov.	29	29	48	136	197	239	249	229	211	230
Dec.	27	27	35	122	187	238	254	241	226	217

24° N Lat

	N	NNE/NNW	NE/NW	ENE/WNW	E/W	ESE/WSW	SE/SW	SSE/SSW	S	HOR
Jan.	27	27	41	128	190	240	253	241	227	214
Feb.	30	30	80	165	220	244	243	213	192	249
Mar.	34	45	124	195	234	237	214	168	137	275
Apr.	37	88	159	209	228	212	169	107	75	283
May	43	117	178	214	218	190	132	67	46	282
June	55	127	184	214	213	179	117	55	43	279
July	45	116	176	210	212	185	129	65	46	278
Aug.	38	87	156	203	220	204	162	103	72	277
Sept.	35	42	119	185	222	225	206	163	134	266
Oct.	31	31	79	159	211	237	235	207	187	244
Nov.	27	27	42	126	187	236	249	237	224	213
Dec.	26	26	29	112	180	234	247	247	237	199

7.24

8° N Lat

	N	NNE/NNW	NE/NW	ENE/WNW	E/W	ESE/WSW	SE/SW	SSE/SSW	S	HOR
Jan.	32	32	71	163	224	250	242	203	162	275
Feb.	34	34	114	193	235	248	219	165	110	294
Mar.	37	67	156	215	241	230	184	110	55	300
Apr.	44	117	184	221	225	195	134	53	39	289
May	74	146	198	220	209	167	97	39	38	277
June	90	155	200	217	200	141	82	39	39	269
July	77	145	195	215	204	162	93	40	39	272
Aug.	47	117	179	214	216	186	128	51	41	276
Sep.	38	66	149	205	230	219	176	107	56	290
Oct.	35	35	112	187	231	239	211	160	108	288
Nov.	33	33	71	161	220	245	233	200	160	273
Dec.	31	31	55	149	215	246	247	215	179	265

12° N Lat

	N	NNE/NNW	NE/NW	ENE/WNW	E/W	ESE/WSW	SE/SW	SSE/SSW	S	HOR
Jan.	31	31	63	155	217	246	247	212	182	262
Feb.	34	34	105	186	235	248	226	177	133	286
Mar.	36	58	148	210	240	233	190	124	73	297
Apr.	40	108	178	220	227	200	142	64	40	290
May	60	139	194	220	212	173	106	40	40	280
June	75	149	198	217	204	161	90	40	40	274
July	63	139	191	215	207	168	102	41	41	275
Aug.	42	109	174	212	218	191	135	62	42	282
Sep.	37	57	142	201	229	222	182	121	73	287
Oct.	34	34	103	180	227	238	219	172	130	280
Nov.	32	32	63	153	214	241	243	209	179	260
Dec.	30	30	47	141	207	242	251	223	197	250

16° N Lat

	N	NNE/NNW	NE/NW	ENE/WNW	E/W	ESE/WSW	SE/SW	SSE/SSW	S	HOR
Jan.	30	30	55	147	210	244	251	223	199	248
Feb.	33	33	96	180	231	247	233	188	154	275
Mar.	35	53	140	205	239	235	197	138	93	291
Apr.	39	99	172	215	227	204	150	77	45	289
May	52	132	189	218	215	179	115	45	41	282
June	66	142	194	217	207	167	99	41	41	277
July	55	132	187	214	209	174	111	44	42	273
Aug.	41	100	168	209	219	196	143	74	46	282
Sep.	36	50	134	196	227	224	191	134	93	282
Oct.	33	33	95	174	223	237	225	183	150	270
Nov.	30	30	55	145	206	241	247	220	196	246
Dec.	29	29	41	132	198	241	254	233	212	234

28° N. Lat

	N (Shade)	NNE/NW	NE/NW	ENE/WNW	E/W	ESE/WSW	SE/SW	SSE/SSW	S	HOR
Jan.	25	25	35	117	183	235	251	247	238	196
Feb.	33	29	72	157	213	244	246	224	207	234
Mar.	37	41	116	189	231	237	221	182	157	265
Apr.	36	84	151	205	228	216	178	124	94	278
May	40	115	172	211	219	195	144	83	58	280
June	51	125	178	211	213	184	128	68	49	278
July	51	114	170	208	215	190	140	80	57	276
Aug.	38	83	149	199	220	207	172	120	91	272
Sep.	34	38	111	179	219	226	213	177	154	256
Oct.	30	30	71	151	204	236	238	217	202	229
Nov.	26	26	35	115	181	232	247	243	235	195
Dec.	24	24	24	99	172	227	248	251	246	179

32° N. Lat

	N (Shade)	NNE/NW	NE/NW	ENE/WNW	E/W	ESE/WSW	SE/SW	SSE/SSW	S	HOR
Jan.	24	24	29	105	175	229	249	250	246	176
Feb.	27	27	65	149	205	242	248	232	221	217
Mar.	32	37	107	183	227	237	227	195	176	252
Apr.	36	80	146	200	220	219	187	141	115	271
May	38	111	170	208	214	199	155	99	74	277
June	44	122	176	208	214	194	139	83	60	276
July	40	111	167	204	215	189	150	96	72	273
Aug.	37	79	141	195	210	210	181	136	111	265
Sep.	33	35	103	173	215	227	218	189	171	244
Oct.	28	28	63	143	195	234	239	225	215	213
Nov.	24	24	29	103	173	225	245	246	243	175
Dec.	22	22	22	84	162	218	218	252	252	158

36° N. Lat

	N (Shade)	NNE/NW	NE/NW	ENE/WNW	E/W	ESE/WSW	SE/SW	SSE/SSW	S	HOR
Jan.	22	22	24	90	166	219	247	250	252	155
Feb.	26	26	57	139	195	239	248	232	232	199
Mar.	30	53	99	176	223	238	232	206	192	238
Apr.	35	76	144	196	225	221	196	156	135	262
May	38	107	168	205	220	204	165	116	93	272
June	47	118	175	205	216	194	150	99	77	273
July	39	107	165	201	216	199	161	113	90	268
Aug.	36	75	138	190	218	212	189	151	131	257
Sep.	31	31	95	167	210	228	223	200	187	230
Oct.	27	27	56	133	187	230	239	231	225	195
Nov.	22	22	24	87	163	215	243	248	248	154
Dec.	20	20	20	69	151	204	241	253	254	136

TABLE 7.9 Maximum Solar Heat Gain for Sunlit Glass (Btu/hr·ft²) (Concluded)

40° N. Lat

	N (Shade)	NNE/NNW	NE/NW	ENE/WNW	E/W	ESE/WSW	SE/SW	SSE/SSW	S	HOR
Jan.	20	20	20	74	154	205	241	252	254	133
Feb.	24	24	50	129	186	234	246	244	241	180
Mar.	29	29	93	169	218	238	236	216	206	223
Apr.	34	71	140	190	224	223	203	170	154	252
May	37	102	165	202	220	208	175	133	113	265
June	48	113	172	205	216	199	161	116	95	267
July	38	102	163	198	216	203	170	129	109	262
Aug.	35	71	135	185	216	214	196	165	149	247
Sep.	30	30	87	160	203	227	226	209	200	215
Oct.	25	25	49	123	180	225	238	236	234	177
Nov.	20	20	20	73	151	201	237	248	250	132
Dec.	18	18	18	60	135	188	232	249	253	113

44° N. Lat

	N (Shade)	NNE/NNW	NE/NW	ENE/WNW	E/W	ESE/WSW	SE/SW	SSE/SSW	S	HOR
Jan.	17	17	17	64	138	189	232	248	252	109
Feb.	22	22	43	117	178	227	246	244	247	160
Mar.	27	27	87	162	211	236	238	224	218	206
Apr.	33	66	136	185	221	224	210	183	171	240
May	36	96	162	201	219	211	183	148	132	257
June	47	108	169	205	215	203	171	132	115	261
July	37	96	159	198	215	206	179	144	128	254
Aug.	34	66	132	180	214	215	202	177	165	236
Sep.	28	28	80	152	198	226	227	216	211	199
Oct.	23	23	42	111	171	217	237	240	239	157
Nov.	18	18	18	64	135	186	227	244	248	109
Dec.	15	15	15	49	115	175	217	240	246	89

60° N. Lat

	N (Shade)	NNE/NNW	NE/NW	ENE/WNW	E/W	ESE/WSW	SE/SW	SSE/SSW	S	HOR
Jan.	7	7	7	7	46	88	130	152	164	21
Feb.	13	13	13	58	118	168	204	225	231	68
Mar.	20	20	56	125	173	215	234	241	242	128
Apr.	27	59	118	168	206	222	225	220	218	178
May	43	98	149	192	212	220	211	198	194	208
June	58	110	162	197	213	215	202	186	181	217
July	44	97	147	189	208	215	206	193	190	207
Aug.	28	57	114	161	199	214	217	213	211	176
Sep.	21	21	50	115	160	202	222	229	231	123
Oct.	14	14	14	56	111	159	193	215	221	67
Nov.	7	7	7	7	45	86	127	148	160	22
Dec.	4	4	4	4	16	51	76	100	107	9

64° N. Lat

	N (Shade)	NNE/NNW	NE/NW	ENE/WNW	E/W	ESE/WSW	SE/SW	SSE/SSW	S	HOR
Jan.	3	3	3	3	15	45	67	89	96	8
Feb.	11	11	11	43	89	144	177	202	210	45
Mar.	18	18	47	113	159	203	226	236	239	105
Apr.	25	59	113	163	201	219	225	225	224	160
May	48	97	150	189	211	216	215	207	204	192
June	62	114	162	193	213	216	208	196	193	203
July	49	96	148	186	207	211	211	202	200	192
Aug.	27	58	109	157	193	211	217	217	217	159
Sept.	19	19	43	103	148	189	213	224	227	101
Oct.	11	11	11	40	83	135	167	191	199	46
Nov.	4	4	4	4	15	44	66	87	93	8
Dec.	0	0	0	0	1	5	11	14	15	1

48° N. Lat

	N (Shade)	NNE/ NNW	NE/ NW	ENE/ WNW	E/ W	ESE/ WSW	SE/ SW	SSE/ SSW	S	HOR
Jan.	15	15	15	53	118	175	216	239	245	85
Feb.	20	20	36	103	168	216	242	249	250	138
Mar.	26	26	80	154	204	234	239	232	228	188
Apr.	31	61	132	180	219	225	215	194	186	226
May	35	97	158	200	218	214	192	163	150	247
June	46	110	165	204	215	206	180	148	134	252
July	37	96	156	196	214	209	187	158	146	244
Aug.	33	61	128	174	211	216	208	188	180	223
Sep.	27	27	72	144	191	223	228	223	220	182
Oct.	21	21	35	96	161	207	233	241	242	136
Nov.	15	15	15	52	115	172	212	234	240	85
Dec.	13	13	13	36	91	156	195	225	233	65

52° N. Lat

	N (Shade)	NNE/ NNW	NE/ NW	ENE/ WNW	E/ W	ESE/ WSW	SE/ SW	SSE/ SSW	S	HOR
Jan.	13	13	13	39	92	155	193	222	230	62
Feb.	18	18	29	85	156	202	235	247	250	115
Mar.	24	24	73	145	196	230	239	238	236	169
Apr.	30	56	128	177	215	224	220	204	199	211
May	34	98	154	198	217	217	199	175	167	235
June	45	111	161	202	214	210	188	162	163	242
July	36	97	152	194	213	212	195	171	163	233
Aug.	32	56	124	169	208	216	212	197	193	208
Sep.	25	25	65	136	182	218	228	228	227	163
Oct.	19	19	28	80	148	192	225	238	240	114
Nov.	13	13	13	39	90	152	189	217	225	62
Dec.	10	10	10	19	73	127	172	199	209	42

56° N. Lat

	N (Shade)	NNE/ NNW	NE/ NW	ENE/ WNW	E/ W	ESE/ WSW	SE/ SW	SSE/ SSW	S	HOR
Jan.	10	10	10	21	74	126	169	194	205	40
Feb.	16	16	21	71	139	184	223	239	244	91
Mar.	22	22	65	136	185	224	238	241	241	149
Apr.	28	58	123	173	211	223	223	213	210	195
May	36	99	149	195	215	218	206	187	181	222
June	53	111	160	199	213	213	196	174	168	231
July	37	98	147	192	211	214	201	183	177	221
Aug.	30	56	119	165	203	216	215	205	203	193
Sep.	23	23	58	126	171	211	227	230	231	144
Oct.	16	16	20	68	132	176	213	229	234	91
Nov.	10	10	10	21	72	122	165	190	200	40
Dec.	7	7	7	7	47	92	135	159	171	23

TABLE 7.10   Cooling Load Factors for Glass without Shading

Fenestration Facing	Room Construction	Solar Time, h																							
		1	2	3	4	5	6	7	8	9	10	11	12	13	14	15	16	17	18	19	20	21	22	23	24
N (Shaded)	L	0.17	0.14	0.11	0.09	0.08	0.33	0.42	0.48	0.56	0.63	0.71	0.76	0.80	0.82	0.82	0.79	0.75	0.84	0.61	0.48	0.38	0.31	0.25	0.20
	M	0.23	0.20	0.18	0.16	0.14	0.34	0.41	0.46	0.53	0.59	0.65	0.70	0.73	0.75	0.76	0.74	0.75	0.79	0.61	0.50	0.42	0.36	0.31	0.27
	H	0.25	0.23	0.21	0.20	0.19	0.38	0.45	0.49	0.55	0.60	0.65	0.69	0.72	0.72	0.72	0.70	0.70	0.75	0.57	0.46	0.39	0.34	0.31	0.28
NNE	L	0.06	0.05	0.04	0.03	0.03	0.26	0.43	0.47	0.44	0.41	0.40	0.39	0.39	0.38	0.36	0.33	0.30	0.26	0.20	0.16	0.13	0.10	0.08	0.07
	M	0.09	0.08	0.07	0.06	0.06	0.24	0.38	0.42	0.39	0.37	0.37	0.36	0.36	0.34	0.33	0.30	0.27	0.24	0.20	0.17	0.14	0.12	0.11	0.10
	H	0.11	0.10	0.09	0.09	0.08	0.26	0.39	0.42	0.39	0.36	0.35	0.34	0.34	0.33	0.32	0.31	0.28	0.25	0.21	0.18	0.16	0.14	0.13	0.12
NE	L	0.04	0.04	0.03	0.02	0.02	0.23	0.41	0.51	0.51	0.45	0.39	0.36	0.33	0.31	0.28	0.26	0.23	0.19	0.15	0.12	0.10	0.08	0.06	0.05
	M	0.07	0.06	0.06	0.05	0.04	0.21	0.36	0.44	0.45	0.40	0.36	0.33	0.31	0.30	0.28	0.26	0.23	0.21	0.17	0.15	0.13	0.11	0.09	0.08
	H	0.09	0.08	0.08	0.07	0.07	0.23	0.37	0.44	0.44	0.39	0.34	0.31	0.29	0.27	0.26	0.24	0.22	0.20	0.17	0.14	0.13	0.12	0.11	0.10
ENE	L	0.04	0.03	0.03	0.02	0.02	0.21	0.40	0.52	0.57	0.53	0.45	0.39	0.34	0.31	0.28	0.25	0.22	0.18	0.14	0.12	0.09	0.08	0.06	0.05
	M	0.07	0.06	0.05	0.05	0.04	0.20	0.35	0.45	0.49	0.47	0.41	0.36	0.33	0.30	0.28	0.26	0.23	0.20	0.17	0.14	0.12	0.11	0.09	0.08
	H	0.09	0.09	0.08	0.07	0.07	0.22	0.36	0.46	0.49	0.45	0.38	0.33	0.30	0.27	0.25	0.23	0.21	0.19	0.16	0.14	0.13	0.12	0.11	0.10
E	L	0.04	0.03	0.03	0.02	0.02	0.19	0.37	0.51	0.57	0.57	0.50	0.42	0.37	0.32	0.29	0.25	0.22	0.19	0.15	0.12	0.10	0.08	0.06	0.05
	M	0.07	0.06	0.06	0.05	0.05	0.18	0.33	0.44	0.50	0.51	0.46	0.39	0.35	0.31	0.29	0.26	0.23	0.21	0.17	0.15	0.13	0.11	0.10	0.08
	H	0.09	0.09	0.08	0.08	0.07	0.20	0.34	0.45	0.49	0.49	0.43	0.36	0.32	0.29	0.26	0.24	0.22	0.19	0.17	0.15	0.13	0.12	0.11	0.10
ESE	L	0.05	0.04	0.03	0.03	0.02	0.17	0.34	0.49	0.58	0.61	0.57	0.48	0.41	0.36	0.32	0.28	0.24	0.20	0.16	0.13	0.10	0.09	0.07	0.06
	M	0.08	0.07	0.06	0.05	0.05	0.16	0.31	0.43	0.51	0.54	0.51	0.44	0.39	0.35	0.32	0.29	0.26	0.22	0.19	0.17	0.15	0.13	0.11	0.09
	H	0.10	0.09	0.09	0.08	0.08	0.19	0.32	0.43	0.50	0.52	0.49	0.41	0.36	0.32	0.29	0.26	0.24	0.21	0.18	0.16	0.14	0.13	0.12	0.11
SE	L	0.05	0.04	0.04	0.03	0.03	0.13	0.28	0.43	0.55	0.62	0.63	0.57	0.48	0.42	0.37	0.33	0.28	0.24	0.19	0.15	0.12	0.10	0.08	0.07
	M	0.09	0.08	0.07	0.06	0.05	0.14	0.26	0.38	0.48	0.54	0.56	0.51	0.45	0.40	0.36	0.33	0.29	0.25	0.21	0.18	0.16	0.14	0.12	0.10
	H	0.11	0.10	0.10	0.09	0.08	0.17	0.28	0.40	0.49	0.53	0.53	0.48	0.41	0.36	0.33	0.30	0.27	0.24	0.20	0.18	0.16	0.14	0.13	0.12
SSE	L	0.07	0.05	0.04	0.04	0.03	0.06	0.15	0.29	0.43	0.55	0.63	0.64	0.60	0.52	0.45	0.40	0.35	0.29	0.23	0.18	0.15	0.12	0.10	0.08
	M	0.11	0.09	0.08	0.07	0.06	0.08	0.16	0.26	0.38	0.48	0.55	0.57	0.54	0.48	0.43	0.39	0.35	0.30	0.25	0.21	0.18	0.16	0.14	0.12
	H	0.12	0.11	0.11	0.10	0.09	0.12	0.19	0.29	0.40	0.49	0.54	0.55	0.51	0.44	0.39	0.35	0.31	0.27	0.23	0.20	0.18	0.16	0.15	0.13
S	L	0.08	0.07	0.05	0.05	0.04	0.06	0.09	0.14	0.22	0.34	0.48	0.59	0.65	0.65	0.59	0.50	0.43	0.36	0.28	0.22	0.18	0.15	0.12	0.10
	M	0.12	0.11	0.09	0.08	0.07	0.08	0.11	0.14	0.21	0.31	0.42	0.52	0.57	0.58	0.53	0.47	0.41	0.36	0.29	0.25	0.21	0.18	0.16	0.14
	H	0.13	0.12	0.12	0.11	0.10	0.11	0.14	0.17	0.24	0.33	0.43	0.51	0.56	0.55	0.50	0.43	0.37	0.32	0.26	0.22	0.20	0.18	0.16	0.14
SSW	L	0.10	0.08	0.07	0.06	0.05	0.06	0.09	0.11	0.15	0.19	0.27	0.39	0.52	0.62	0.67	0.65	0.58	0.46	0.36	0.28	0.23	0.19	0.15	0.12
	M	0.14	0.12	0.11	0.09	0.08	0.09	0.11	0.13	0.15	0.19	0.25	0.35	0.46	0.55	0.59	0.59	0.53	0.44	0.35	0.30	0.25	0.22	0.19	0.16
	H	0.15	0.14	0.13	0.12	0.11	0.12	0.14	0.16	0.18	0.21	0.27	0.37	0.46	0.53	0.57	0.55	0.49	0.40	0.32	0.26	0.23	0.20	0.18	0.16
SW	L	0.12	0.10	0.08	0.06	0.05	0.06	0.08	0.10	0.12	0.14	0.16	0.24	0.36	0.49	0.60	0.66	0.66	0.58	0.43	0.33	0.27	0.22	0.18	0.15
	M	0.15	0.14	0.12	0.10	0.09	0.09	0.10	0.12	0.13	0.15	0.17	0.23	0.33	0.44	0.53	0.58	0.59	0.53	0.41	0.33	0.28	0.24	0.21	0.18
	H	0.15	0.14	0.13	0.12	0.11	0.12	0.13	0.14	0.16	0.17	0.19	0.25	0.34	0.44	0.52	0.56	0.56	0.49	0.37	0.30	0.25	0.21	0.19	0.17
WSW	L	0.12	0.10	0.08	0.07	0.05	0.06	0.07	0.09	0.10	0.12	0.13	0.17	0.26	0.40	0.52	0.62	0.66	0.61	0.44	0.34	0.27	0.22	0.18	0.15
	M	0.15	0.13	0.12	0.10	0.09	0.09	0.10	0.11	0.12	0.13	0.14	0.17	0.24	0.35	0.44	0.52	0.58	0.55	0.42	0.34	0.28	0.24	0.21	0.17
	H	0.15	0.14	0.13	0.12	0.11	0.11	0.12	0.13	0.14	0.15	0.16	0.19	0.26	0.36	0.46	0.53	0.56	0.51	0.38	0.30	0.25	0.21	0.19	0.17
W	L	0.12	0.10	0.08	0.06	0.05	0.05	0.07	0.08	0.10	0.11	0.12	0.14	0.20	0.32	0.45	0.57	0.64	0.61	0.44	0.34	0.27	0.22	0.18	0.14
	M	0.15	0.13	0.11	0.10	0.09	0.09	0.09	0.10	0.11	0.12	0.13	0.14	0.19	0.29	0.40	0.50	0.56	0.55	0.41	0.33	0.27	0.23	0.20	0.17
	H	0.14	0.13	0.12	0.11	0.10	0.11	0.12	0.13	0.14	0.15	0.16	0.17	0.24	0.34	0.44	0.51	0.55	0.51	0.38	0.30	0.24	0.21	0.19	0.16
WNW	L	0.12	0.10	0.08	0.06	0.05	0.05	0.07	0.09	0.10	0.12	0.13	0.15	0.17	0.26	0.40	0.53	0.63	0.62	0.44	0.34	0.27	0.22	0.18	0.14
	M	0.15	0.13	0.11	0.10	0.09	0.09	0.10	0.11	0.12	0.13	0.14	0.15	0.17	0.24	0.35	0.47	0.55	0.55	0.41	0.33	0.27	0.23	0.20	0.17
	H	0.14	0.13	0.12	0.11	0.10	0.11	0.12	0.13	0.14	0.15	0.16	0.17	0.18	0.25	0.36	0.46	0.53	0.52	0.38	0.30	0.24	0.20	0.19	0.16
NW	L	0.11	0.09	0.08	0.06	0.05	0.06	0.07	0.09	0.10	0.12	0.14	0.16	0.17	0.19	0.23	0.33	0.47	0.59	0.60	0.42	0.31	0.26	0.21	0.17
	M	0.14	0.12	0.11	0.09	0.08	0.09	0.10	0.11	0.13	0.14	0.16	0.17	0.18	0.21	0.30	0.42	0.51	0.54	0.39	0.32	0.26	0.22	0.19	0.16
	H	0.14	0.12	0.11	0.10	0.10	0.10	0.10	0.12	0.13	0.15	0.16	0.18	0.18	0.19	0.26	0.36	0.44	0.50	0.51	0.36	0.29	0.23	0.20	0.15
NNW	L	0.12	0.09	0.08	0.06	0.06	0.05	0.07	0.11	0.14	0.18	0.22	0.25	0.27	0.29	0.30	0.33	0.44	0.57	0.62	0.44	0.33	0.26	0.21	0.17
	M	0.15	0.13	0.11	0.10	0.09	0.10	0.12	0.15	0.17	0.20	0.23	0.25	0.26	0.28	0.28	0.31	0.39	0.51	0.56	0.41	0.33	0.27	0.23	0.18
	H	0.14	0.13	0.12	0.11	0.10	0.12	0.15	0.17	0.20	0.23	0.25	0.26	0.28	0.28	0.31	0.38	0.49	0.53	0.38	0.30	0.25	0.21	0.18	0.16
HOR	L	0.11	0.09	0.07	0.06	0.05	0.07	0.14	0.24	0.36	0.48	0.58	0.66	0.72	0.74	0.73	0.67	0.59	0.47	0.37	0.29	0.24	0.19	0.16	0.13
	M	0.16	0.14	0.12	0.11	0.09	0.11	0.16	0.24	0.33	0.43	0.52	0.59	0.64	0.67	0.66	0.62	0.56	0.47	0.38	0.32	0.28	0.24	0.21	0.18
	H	0.17	0.16	0.15	0.14	0.13	0.15	0.20	0.28	0.36	0.45	0.52	0.59	0.62	0.64	0.62	0.58	0.51	0.42	0.35	0.29	0.26	0.23	0.21	0.19

L=Light construction: frame exterior wall, 2-in. concrete floor slab, approximately 30 lb of material/ft^2 of floor area.

M= Medium construction: 4-in. concrete exterior wall, 4-in. concrete floor slab, approximately 70 lb of building material/ft^2 of floor area.

H=Heavy construction: 6-in. concrete exterior wall, 6-in. concrete floor slab, approximately 130 lb of building material/ft^2 of floor area.

For rooms that have carpeted floors, CLF values should be obtained from Table 7.11. Since the carpeting tends to shade and insulate the floor mass, the heat is converted to a cooling load much more quickly, and with a higher peak, than a bare or tiled floor. The carpeting prevents the floor from absorbing much of the radiant heat.

Similarly, for spaces that have interior shading on the fenestration areas, CLF values should be obtained from Table 7.12. The CLF values in Table 7.12 account for the fact that part of the incoming radiant energy is reflected back out through the glass, and part of the radiant energy is absorbed by the interior shade itself. Most of the radiant energy absorbed by the interior shade is converted to an instantaneous cooling load on the space.

Example 7.4   Assume a 4 ft × 4 ft west facing window, with two layers of clear glazing, and with $1/4$ in air space between the layers of glazing, in the exterior wall of a building located in St. Louis, MO as shown in the sketch below. Assume that there is a 2 ft wide

TABLE 7.11 Cooling Load Factors for Glass without Shading (Carpeted Floors)

Room Dir.	Mass	0100	0200	0300	0400	0500	0600	0700	0800	0900	1000	1100	1200	1300	1400	1500	1600	1700	1800	1900	2000	2100	2200	2300	2400
N	L	.00	.00	.00	.00	.01	.64	.73	.74	.81	.88	.95	.98	.98	.94	.88	.79	.79	.55	.31	.12	.04	.02	.01	.00
	M	.03	.02	.02	.02	.02	.64	.69	.69	.77	.84	.91	.94	.95	.91	.86	.79	.79	.56	.32	.16	.10	.07	.05	.04
	H	.10	.09	.08	.07	.07	.62	.64	.64	.71	.77	.83	.87	.88	.85	.81	.75	.76	.55	.34	.22	.17	.15	.13	.11
NE	L	.00	.00	.00	.00	.01	.51	.83	.88	.72	.47	.33	.27	.24	.23	.20	.18	.14	.09	.03	.01	.00	.00	.00	.00
	M	.01	.01	.00	.00	.01	.50	.78	.82	.67	.44	.32	.28	.26	.24	.22	.19	.15	.11	.05	.03	.02	.02	.01	.01
	H	.03	.03	.03	.02	.03	.47	.71	.72	.59	.40	.30	.27	.26	.25	.23	.20	.17	.13	.08	.06	.05	.05	.04	.04
E	L	.00	.00	.00	.00	.00	.42	.76	.91	.90	.75	.51	.30	.22	.18	.16	.13	.11	.07	.02	.01	.00	.00	.00	.00
	M	.01	.01	.00	.00	.01	.41	.72	.86	.84	.71	.48	.30	.24	.21	.18	.16	.13	.09	.04	.03	.02	.01	.01	.01
	H	.03	.03	.03	.02	.02	.39	.66	.76	.74	.63	.43	.29	.24	.22	.20	.18	.15	.12	.08	.06	.05	.05	.04	.04
SE	L	.00	.00	.00	.00	.00	.27	.58	.81	.93	.93	.81	.59	.37	.27	.21	.18	.14	.09	.03	.01	.00	.00	.00	.00
	M	.01	.01	.01	.00	.01	.26	.55	.77	.88	.87	.76	.56	.37	.29	.24	.20	.16	.11	.05	.04	.03	.02	.02	.01
	H	.04	.04	.03	.03	.03	.26	.51	.69	.78	.78	.68	.51	.35	.29	.25	.22	.19	.15	.09	.08	.07	.06	.05	.05
S	L	.00	.00	.00	.00	.00	.07	.15	.23	.39	.62	.82	.94	.93	.80	.59	.38	.26	.16	.06	.02	.01	.00	.00	.00
	M	.01	.01	.01	.01	.01	.07	.14	.22	.38	.59	.78	.88	.88	.76	.57	.38	.28	.18	.09	.06	.04	.03	.02	.02
	H	.05	.05	.04	.04	.03	.09	.15	.21	.35	.54	.70	.79	.79	.69	.52	.37	.29	.21	.13	.10	.09	.08	.07	.06
SW	L	.00	.00	.00	.00	.00	.04	.09	.13	.16	.19	.23	.39	.62	.82	.94	.94	.81	.54	.19	.07	.03	.01	.00	.00
	M	.02	.02	.01	.01	.01	.05	.09	.13	.16	.19	.22	.38	.60	.78	.89	.89	.77	.52	.20	.10	.07	.05	.04	.03
	H	.07	.06	.05	.05	.04	.07	.11	.14	.16	.18	.21	.35	.55	.71	.80	.79	.69	.48	.20	.14	.11	.10	.08	.07
W	L	.00	.00	.00	.00	.00	.03	.07	.10	.13	.15	.16	.18	.31	.55	.78	.92	.93	.73	.25	.10	.04	.01	.01	.00
	M	.02	.02	.01	.01	.01	.04	.07	.10	.13	.14	.16	.17	.30	.53	.74	.87	.88	.69	.24	.12	.07	.05	.04	.03
	H	.06	.06	.05	.04	.04	.06	.09	.11	.13	.15	.16	.17	.28	.49	.67	.78	.79	.62	.23	.14	.11	.09	.08	.07
NW	L	.00	.00	.00	.00	.00	.04	.09	.14	.17	.20	.22	.23	.24	.31	.53	.78	.92	.81	.28	.10	.04	.02	.01	.00
	M	.02	.02	.01	.01	.01	.05	.10	.13	.17	.19	.21	.22	.23	.30	.52	.75	.88	.77	.26	.12	.07	.05	.04	.03
	H	.06	.05	.05	.04	.04	.07	.11	.14	.17	.19	.20	.21	.22	.28	.48	.68	.79	.69	.23	.14	.10	.09	.08	.07
Hor.	L	.00	.00	.00	.00	.00	.08	.25	.45	.64	.80	.91	.97	.97	.91	.80	.64	.44	.23	.08	.03	.01	.00	.00	.00
	M	.02	.02	.01	.01	.01	.08	.24	.43	.60	.75	.86	.92	.92	.87	.77	.63	.45	.26	.12	.07	.05	.04	.03	.02
	H	.07	.06	.05	.05	.04	.11	.25	.41	.56	.68	.77	.83	.83	.80	.71	.59	.44	.28	.17	.13	.11	.10	.09	.08

Values of nominal 15 ft by 15 ft by 10 ft high space, with ceiling, and 50% or less glass in exposed surface at listed orientation.

L=Lightweight construction, such as 1-in. wood floor, Group G wall.

M= Mediumweight construction, such as 2 to 4-in. concrete floor, Group E wall.

H=Heavyweight construction, such as 6 to 8-in. concrete floor, Group C wall.

TABLE 7.12 Cooling Load Factors for Glass with Interior Shading

Fenestration Facing	0100	0200	0300	0400	0500	0600	0700	0800	0900	1000	1100	1200	1300	1400	1500	1600	1700	1800	1900	2000	2100	2200	2300	2400
N	0.08	0.07	0.06	0.06	0.07	0.73	0.66	0.65	0.73	0.80	0.86	0.89	0.89	0.86	0.82	0.75	0.78	0.91	0.24	0.18	0.15	0.13	0.11	0.10
NNE	0.03	0.03	0.02	0.02	0.03	0.64	0.77	0.62	0.42	0.37	0.37	0.37	0.36	0.35	0.32	0.28	0.23	0.17	0.08	0.07	0.06	0.05	0.04	0.04
NE	0.03	0.02	0.02	0.02	0.02	0.56	0.76	0.74	0.58	0.37	0.29	0.27	0.26	0.24	0.22	0.20	0.16	0.12	0.06	0.05	0.04	0.04	0.03	0.03
ENE	0.03	0.02	0.02	0.02	0.02	0.52	0.76	0.80	0.71	0.52	0.31	0.26	0.24	0.22	0.20	0.18	0.15	0.11	0.06	0.05	0.04	0.04	0.03	0.03
E	0.03	0.02	0.02	0.02	0.02	0.47	0.72	0.80	0.76	0.62	0.41	0.27	0.24	0.22	0.20	0.17	0.14	0.11	0.06	0.05	0.05	0.04	0.03	0.03
ESE	0.03	0.03	0.02	0.02	0.02	0.41	0.67	0.79	0.80	0.72	0.54	0.34	0.27	0.24	0.21	0.19	0.15	0.12	0.07	0.06	0.05	0.04	0.04	0.03
SE	0.03	0.03	0.02	0.02	0.02	0.30	0.57	0.74	0.81	0.79	0.68	0.49	0.33	0.28	0.25	0.22	0.18	0.13	0.08	0.07	0.06	0.05	0.04	0.04
SSE	0.04	0.03	0.03	0.03	0.02	0.12	0.31	0.54	0.72	0.81	0.81	0.71	0.54	0.38	0.32	0.27	0.22	0.16	0.09	0.08	0.07	0.06	0.05	0.04
S	0.04	0.04	0.03	0.03	0.03	0.09	0.16	0.23	0.38	0.58	0.75	0.83	0.80	0.68	0.50	0.35	0.27	0.19	0.11	0.09	0.08	0.07	0.06	0.05
SSW	0.05	0.04	0.04	0.03	0.03	0.09	0.14	0.18	0.22	0.27	0.43	0.63	0.78	0.84	0.80	0.66	0.46	0.25	0.13	0.11	0.09	0.08	0.07	0.06
SW	0.05	0.05	0.04	0.04	0.03	0.07	0.11	0.14	0.16	0.19	0.22	0.38	0.59	0.75	0.83	0.81	0.69	0.45	0.16	0.12	0.10	0.09	0.07	0.06
WSW	0.05	0.05	0.04	0.04	0.03	0.07	0.10	0.12	0.14	0.16	0.17	0.23	0.44	0.64	0.78	0.84	0.78	0.55	0.16	0.12	0.10	0.09	0.07	0.06
W	0.05	0.05	0.04	0.04	0.03	0.06	0.09	0.11	0.13	0.15	0.16	0.17	0.31	0.53	0.72	0.82	0.81	0.61	0.16	0.12	0.10	0.08	0.07	0.06
WNW	0.05	0.05	0.04	0.03	0.03	0.07	0.10	0.12	0.14	0.16	0.17	0.18	0.22	0.43	0.65	0.80	0.84	0.66	0.16	0.12	0.10	0.08	0.07	0.06
NW	0.05	0.04	0.04	0.03	0.03	0.07	0.11	0.14	0.17	0.19	0.20	0.21	0.22	0.30	0.52	0.73	0.82	0.69	0.16	0.12	0.10	0.08	0.07	0.06
NNW	0.05	0.05	0.04	0.03	0.03	0.11	0.17	0.22	0.26	0.30	0.32	0.33	0.34	0.34	0.39	0.61	0.82	0.76	0.17	0.12	0.10	0.08	0.07	0.06
HOR.	0.06	0.05	0.04	0.04	0.03	0.12	0.27	0.44	0.59	0.72	0.81	0.85	0.85	0.81	0.71	0.58	0.42	0.25	0.14	0.12	0.10	0.08	0.07	0.06

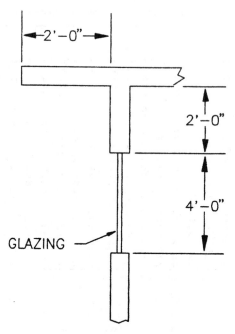

FIGURE 7.6   Window and Overhang for Example 7.4

overhang, located 2 ft above the top of the window, and that it is solar noon on June 15. Assume also that there is a 7½ mph wind outside, and the inside temperature is 78°F. The building floor is an uncarpeted, 6 in thick 120 lb/ft³ concrete floor.

What is the cooling load, due to the window? What would the cooling load be if carpeting were installed? What would the cooling load be if venetian blinds were installed?

By applying Equation 7.4, to calculate the solar angle of declination, or by using the Table in Section 7.3.1.

$$\delta = 23.1°$$

The latitude of approximately 38° is obtained from Table 3-1 for St. Louis, MO. Using Equation 7.5 to calculate the solar altitude for 38°N latitude

$$\beta = \sin^{-1}[\sin(23.1)\sin(38) + \cos(23.1)\cos(38)\cos(0)] = 68.4°F$$

Using Equation 7.7 to calculate the shading factor, or fraction

$$S = \frac{1}{4}[2\tan(68.4) - 2] = 0.76$$

$$\text{Window shaded area} = (4 \times 4)(0.76) = 12.2\ \text{ft}^2$$

$$\text{Unshaded area} = (4 \times 4)(1 - 0.76) = 3.8\ \text{ft}^2$$

For clear double glazing with a 7½ mph wind (h = 4.0) from Table 7.6

$$SC = 0.81$$

From Table 4.2 for double glazing

$$U = 0.61 \text{ Btu/hr•ft}^2\text{•°F}$$

From Table 7.8

$$\text{CLTD} = 9°F$$

Using Equation 7.1 to calculate the conduction heat gain

$$q_c = (0.61)(16)(9) = 87.8 \text{ Btu/hr}$$

Table 7-9 lists solar heat gain factors, for north latitude, in 4° increments. Since St. Louis, MO is at 38° north latitude and the nearest tabulated values are for 36° and 40°, the value for either latitude may be used. In this case, 40°N is chosen.

For a 6 in thick, 120 lb/ft³ concrete floor, assume heavy construction.

For the direct solar radiation, unshaded glass at 40°N latitude, west facing in June, from Table 7.9, the maximum solar heat gain is

$$\text{SHGF} = 216 \text{ Btu/hr•ft}^2$$

From Table 7.10, west fenestration, heavy construction, 12 noon solar time

$$\text{CLF} = 0.16$$
$$q_R = (3.8)(0.81)(216)(0.16) = 106.4 \text{ Btu/hr}$$

To calculate the indirect radiation, for the shaded part of the window, use the values for north facing fenestration.

$$\text{Maximum SHGF} = 48 \text{ Btu/hr•ft}^2$$
$$\text{CLF} = 0.69$$

Using Equation 7.9 to calculate the cooling load for the direct radiation

$$q_R = (12.2)\ (0.82)\ (48)\ (0.69) = 331.3 \text{ Btu/hr}$$

therefore, the total radiant and convective cooling load for the window is

$$q = 87.8 + 106.4 + 331.3 = 525.5 \text{ Btu/hr}$$

For a room with carpeting, the conduction cooling load through the glass would be the same.

$$q_c = 87.8 \text{ Btu/hr}$$

For the unshaded west glass, at solar noon, from Table 7.11

$$\text{CLF} = 0.17$$

Using Equation 7.9 to calculate the radiant cooling load again:

$$q_R = (3.8)(0.81)(216)(0.17) = 113.0 \text{ Btu/hr}$$

For the shaded west glass, solar noon, from Table 7.10

$$\text{CLF} = 0.87$$
$$q_R = (12.2)\ (0.81)\ (48)\ (0.87) = 412.7 \text{ Btu/hr}$$

the cooling load with carpeting is

$$q = 87.8 + 113.0 + 412.7 = 613.5 \text{ Btu/hr}$$

For a fenestration area with venetian blinds (indoor shade), from Table 4.2, $U = 0.55$ Btu/hr•ft²•°F. Using Equation 7.1, the conduction cooling load is

$$q_c = (0.55)\ (16)\ (9) = 79.2\ \text{Btu/hr}$$

For the unshaded west glass at solar noon from Table 7.12

$$CLF = 0.17$$
$$q_r = (3.8)\ (0.81)\ (216)(0.17) = 113.0\ \text{Btu/hr}$$

For the shaded area using the values for north from Table 7.12

$$CLF = 0.89$$
$$q_r = (12.2)\ (0.81)\ (48)\ (0.89) = 422.2\ \text{Btu/hr}$$

The total radiant and convective cooling load for the shaded window is

$$q = 79.2 + 113.0 + 422.2 = 614.4\ \text{Btu/hr}$$

7.4 COOLING LOADS FROM UNCONDITIONED SPACES

Cooling loads can result from heat transmission through surfaces adjacent to unconditioned spaces. Unconditioned spaces can include areas like garages, mechanical or electrical rooms, store rooms, boiler rooms, and kitchens that are not directly cooled.

Cooling loads for unconditioned spaces are essentially steady state load, and therefore, may be calculated in a manner similar to heat losses. Equation 4.1 may be used to calculate heat transmission through surfaces adjacent to unconditioned spaces. Table 7.13 lists typical recommended temperature differences between conditioned and unconditioned spaces.

TABLE 7.13   Design Temperature Differences (Extracted from Table 3.1, Reference 6)

Item	Temperature Difference* °F
Partitions	10
Partitions, or glass in partitions, adjacent to laundries, kitchens, or boiler rooms	25
Floors above unconditioned rooms	10
Floors on ground	0
Floors above basements	0
Floors above rooms or basements used as laundries, kitchens, or boiler rooms	35
Floors above vented spaces	17
Floors above unvented spaces	0
Ceilings with unconditioned rooms above	10
Ceilings with rooms above used as laundries, kitchens, etc.	20

*These temperature differences are based on the assumption that the air conditioning system is being designed to maintain an inside temperature 17°F lower than the outdoor temperature. For air conditioning systems designed to maintain a greater temperature difference than 17°F between the inside and outside, add to the values in the above table the difference between the assumed design temperature difference and 17°F.

For spaces other than those listed in Table 7.13, for which a temperature difference is not known, a temperature difference of 5°F may be assumed.

7.5 EXTERIOR COOLING LOAD CALCULATION PROCEDURE

In Sections 7.1 and 7.2, the need for experience and judgment, when calculating cooling loads, was briefly discussed. Most spaces have multiple heat sources that all go to make up the total peak cooling load. Each has a peak cooling load that frequently occurs at different times. For this reason, it is necessary to perform cooling load calculations for a number of different times or hours. Since calculations are frequently necessary for several different times, to determine the actual peak, an **External Cooling Load Calculation Form** has been provided at the end of this chapter to help organize the calculations.

When selecting the times for which cooling loads are to be calculated, it is usually necessary to use some judgment in order to determine which cooling loads are the dominant loads for the space. One method that may be used to determine which exterior loads are dominant is to calculate the $U \times A$ (overall heat transfer coefficient times the surface area) value for each exterior surface. The surface (or surfaces) with the largest $U \times A$ value are frequently the dominant loads. Another method is to look at exterior surfaces with large areas and light construction. Sometimes, those are the dominant loads.

A relatively large fenestration area will frequently dominate cooling loads for a space. Note that the maximum solar heat gain, for fenestration areas, can occur in the winter months. Finally, interior cooling loads, which are discussed in Chapter 8, should be examined to determine if they are the largest cooling load for the space.

Once the dominant space cooling load, or loads, are identified, the times that these loads peak are usually a good basis for selecting the times for which calculations should be made. Once a probable peak time is selected, it should be entered in the center column of the three time columns, on the External Cooling Load Calculation Form. The columns, to the left and right, are assigned times that are 1 or 2 hr different than the probable peak time, with the left column being earlier and the right column later. These then, are the hours for which calculations should be made, in order to estimate the actual peak space cooling load.

The procedure for estimating the peak external cooling load and the hour at which it occurs is as follows:

1. Select the interior design dry bulb and wet bulb temperatures or the dry bulb temperature and relative humidity.

2. Select the outdoor design conditions, from Table 3.1, for the appropriate building type and location. Select the design month or months.

3. Calculate the net surface area of each exterior surface, such as walls, roofs, fenestration areas, etc.

4. Calculate the U value and weight per ft^2 of all opaque exterior surfaces. Calculate a U value for surfaces adjacent to unconditioned spaces. Select an appropriate U value from Table 4.2 for each type of fenestration.

5. Select the appropriate Wall Group from Table 7.1, and/or Roof/Ceiling Group from Table 7.5, for each exterior wall and roof assembly.

6. Estimate the time at which the peak space cooling load will occur and select the times for which calculations should be made, based on the above suggestions.

7. Obtain the CLTD values, at the selected calculation hours, from Table 7.3 for Sunlit Walls, and from Table 7.5 for Sunlit Roofs.

8. Obtain the Latitude/Month Correction Factor (LM) from Table 7.3 and determine a color correction factor (if any). Determine the appropriate indoor, outdoor, and average outdoor temperature correction factors (if any). Determine the ventilation correction factor (if any), using Equation 7.2 for walls and/or Equation 7.3 for roofs. Calculate the corrected CLTD values at each calculation hour for each exterior wall and roof.

9. Using Equation 7.1, calculate the sensible cooling load for each opaque surface for each hour.

10. From Table 7.8, obtain CLTD values for the calculation times and multiply them by the fenestration area, to estimate the conduction cooling loads for glass.

11. Calculate the net area of all surfaces adjacent to unconditioned spaces. Using the recommended temperature differences in Table 7.13, or as recommended in Section 7.4, calculate the cooling loads for each surface adjacent to an unconditioned space.

12. Obtain shading coefficients (SC) for each type of fenestration area from Table 7.7.

13. Using the methods described in Section 7.3.3 and Equation 7.7, or in References 1 and 4, calculate the shaded and unshaded areas for fenestration area.

14. Obtain a maximum solar heat gain factor (SHGF) at each calculation time, for each fenestration orientation from Table 7.9, for the appropriate month and latitude. Use the values for north facing glass for all shaded areas of fenestration.

15. Obtain a Cooling Load Factor (CLF) for each calculation time, for each fenestration orientation from Tables 7.10, 7.11, or 7.12, for the appropriate design situation.

16. Using Equation 7.9, calculate the radiant cooling load for each fenestration area, for each calculation time.

17. Sum the individual cooling loads to get the exterior cooling load subtotal. These subtotals are then combined with the interior cooling loads to get the total peak cooling loads.

EXTERNAL COOLING LOADS

							CONDUCTION												
ITEM	TYPE	ORIENT	U	AREA	U X A	CLTD TABLE VALUE			ADJUST FOR LATITUDE & MONTH	COLOR CORRECT K	ADJUST FOR INSIDE & OUTSIDE TEMPS	CORR. CLTD			SENSIBLE COOLING LOAD				
						_HR	_HR	_HR				_HR	_HR	_HR	___HR	___HR	___HR		
ROOF																			
ROOF																			
ROOF																			
WALL																			
WALL																			
WALL																			
WALL																			
WALL																			
DOOR																			
DOOR																			
DOOR																			

ITEM	TYPE	U	AREA	U X A	TEMP. DIFF.	U X A X T
PARTITION						
PARTITION						
DOOR						
DOOR						
GLASS						
CEILING						
FLOOR						
SUM						

SOLAR		ORIENT	TOTAL AREA	SHADE COEFF.	MAX. SHGF	_HR		_HR		_HR	
						AREA	CLF	AREA	CLF	AREA	CLF
GLASS	UNSHADED										
	SHADED										
GLASS	UNSHADED										
	SHADED										
GLASS	UNSHADED										
	SHADED										
GLASS	UNSHADED										
	SHADED										

SUBTOTAL

Chapter 8

INTERNAL HEAT GAINS AND COOLING LOADS

Internal heat gains are gains that are generated within the conditioned space itself. Typical examples of internal heat gains include gains from people, lights, and energy consuming equipment located within the space.

Cooling loads from transient heat transfer, and the effect of thermal mass, were discussed in Chapters 4 and 7. You were also introduced to Cooling Load Temperature Differences (CLTD) and Cooling Load Factors (CLF) in Section 7.1.

ASHRAE has developed Cooling Load Factors (CLF), for internal heat gains, to account for the thermal storage of the building mass. Again, buildings have been classified into Light Mass, Medium Mass, and Heavy Mass construction groups. Cooling loads, and Cooling Load Factors, for interior heat gains are primarily a function of the length of time since the heat gain began, or entered the space, and the thermal mass of the space. As sensible heat is absorbed by the building mass, the temperature of the mass increases. As the temperature of the mass increases, it is able to absorb less heat, causing the cooling load to approach the instantaneous heat gain. After the heat gain stops, or leaves the space, the thermal mass continues to reject the stored heat to the space, causing a cooling load.

When calculating cooling loads that result from internal heat gains, a diversity factor is occasionally applied to the loads. Diversity of cooling loads results from not all of the cooling loads being on, or used, at any particular time. As an example, all the lights in a large building are rarely on at the same time. Some rooms may be occupied, while others are not. For large spaces or buildings, some lights may be on, while others are off. Diversity factors account for variable loads and can range from 70 to 100%, depending on the size and use of the building. The application of diversity factors requires some knowledge of the operation of a building and should be applied with considerable judgment. For small spaces, it is likely that all internal gains will be on and, therefore, a diversity factor of 100% should be used when calculating cooling loads for small spaces.

8.1   COOLING LOADS FROM PEOPLE

People (occupants) are one of the few sources of heat in a space that generate both sensible and latent heat. Sensible and latent heat were discussed in Section 2.2.2.

Heat is generated within a person's body by an oxidation process known as metabolism. Heat generated by humans is a function of the metabolic rate, which is the rate at which food is converted to energy in the form of heat. The metabolic rate is a function of the activity level of a person. The more active a person is, the more heat they generate and transfer to the space. The more sedentary a person is, the less they generate heat.

Sensible heat from occupants is transferred to a space as a result of the difference in temperature between the surrounding air and the body temperature. Sensible heat is also transferred by radiant heat exchange with the surroundings, to a lesser extent. Latent heat is transferred from people in the form of perspiration. The perspiration is then evaporated to the air in the space. Since all the latent heat is absorbed directly by the air as moisture, it is an instantaneous heat gain or cooling load for the space. Unlike sensible heat, there is no storage of latent heat by the building mass and furnishings. Therefore, the instantaneous latent heat gain is also a latent cooling load and no Cooling Load Factors need be applied. Instantaneous sensible and latent heat gains, for occupants and various activity levels, are given in Table 8.1.

Sensible heat gains, and the resultant cooling loads on a space, are a function of the number of occupants in the space, their activity level, the type of clothing worn by the occupants, and how long they have been in the space. The length of time the people have been in the space has an effect on the cooling loads since it also affects the thermal storage of the space as discussed earlier in this chapter.

The values given in Table 8.1 are for an average male, weighing 150 lb, that assumes a room air temperature of 75°F, and that assumes an occupancy of at least 3 hr. These values may be used, without the need for correction, unless there are significant differences between the actual, and assumed, conditions listed above. For a room dry bulb air temperature of 80°F, the total heat rejection would remain the same, but the sensible gain should be decreased approximately 20%, while the latent values are increased accordingly. The sex, weight, and age of a person will also affect the body heat rejection rate for any given activity level. An example of such a situation might occur when the majority of the occupants are small children. In such cases, it would be up to the designer's judgment to determine the magnitude of the adjustment, if any. Typically, the metabolic rate for females is about 85% of the values in Table 8.1 and about 75% for children.

The tabulated values for people eating in restaurants were increased, to account for the sensible and latent loads caused by the food served.

Animals also give off heat, just as humans do. The heat rejected by animals is directly proportional to their weight, for any given activity level. For sedentary activity, the total heat rejected per animal (sensible and latent) may be estimated by Equation 8.1.

$$q = 6.6 \ W^{0.75} \tag{8.1}$$

where

q = total heat rejected in Btu/hr

W = the weight of the animal in lb

In order to account for the thermal storage effect of the space, a Cooling Load Factor must be applied to the instantaneous sensible heat gain for occupants from Table 8.1. Equation 8.2 may be used to calculate the sensible cooling load for occupants.

TABLE 8.1 Heat Gain from People

| Degree of Activity | Typical Application | Metabolic Rate (Adult Male), Btu/hr | Average Adjusted Metabolic Rate*, Btu/hr | Room Dry-Bulb Temperature | | | | | | | | | |
| | | | | 82°F Btu/hr | | 80°F Btu/hr | | 78°F Btu/hr | | 75°F Btu/hr | | 70°F Btu/hr | |
				Sensible	Latent	Sensible	Latent	Sensible	Latent	Sensible	Latent	Sensible	Latent
Seated at rest	Theatre, Grade School	390	350	175	175	195	155	210	140	230	120	260	90
Seated, very light work	High School	450	400	180	220	195	205	215	185	240	160	275	125
Office worker	Office, Hotel, Apt., College	475	450	180	270	200	250	215	235	245	205	285	165
Standing, walking slowly	Dept., Retail or Variety Store	550	450	180	270	200	250	215	235	245	205	285	165
Walking, seated	Drug Store	550	500	180	320	200	300	220	280	255	245	290	210
Standing, walking slowly	Bank	550	500	180	320	200	300	220	280	255	245	290	210
Sedentary work	Restaurant	500	550	190	360	220	330	240	310	280	270	320	230
Light bench work	Factory, light work	800	750	190	560	200	530	245	505	295	455	365	385
Moderate dancing	Dance Hall	900	850	220	630	245	605	275	575	325	525	400	450
Walking, 3 mph	Factory, fairly heavy work	1000	1000	270	730	300	700	330	670	380	620	460	540
Heavy Work	Bowling alley, Factory	1500	1450	450	1000	465	985	485	965	525	925	605	845

Courtesy of Carrier Corp., Copyright 1972, McGraw-Hill. Used by permission.

Adjusted Metabolic Rate is the metabolic rate to be applied to a mixed group of people with a typical percent composition based on the following factors:

Metabolic rate, adult female=Metabolic rate, adult male × 0.85

Metabolic rate, children = Metabolic rate, adult male × 0.75

Restaurant—Values for this application include 60 Btu/hr for food per individual (30 Btu sensible and 30 Btu latent heat per hr).

Bowling—Assume one person per alley actually bowling and all others sitting, metabolic rate 400 Btu/hr; or standing, 550 Btu/hr.

$$q_o = (q_i)(n)(f)(\text{CLF}) \tag{8.2}$$

where

q_o = the sensible cooling load for occupants in Btu/hr

q_i = the instantaneous sensible heat gain for each occupant, from Table 8.1 or Equation 8.1, in Btu/hr

n = number of occupants

f = adjustment for age, sex, or weight (if any)

CLF = Cooling Load Factor from Table

Notice that the CLF values in Table 8.2 are a function of how long the occupants will be in the space and how long it has been since they entered the space.

Example 8.1   Assume there are ten boys playing basketball in a gymnasium, there are two adult female coaches, and one adult male referee. They have been in the gym for 2 hr and will be there for a total of 4 hr. The space temperature is 75°F. What are the sensible, latent, and total cooling loads?

Assume a heavy work level of activity for the children, with a 0.75 age adjustment factor.

Assume a standing, or walking, activity level, and a 0.85 sex adjustment factor for the female coaches.

Assume a 3 mph walking activity level for the male referee. No age or sex adjustment is required.

From Table 8.1 and using Equation 8.2, the cooling loads are as follows:

Occupants	Sensible Heat Gain	
Boys:	(525 Btu/hr•boy)(10 boys)(0.75 adj. fact)	= 3938 Btu/hr
Coaches:	(245)(2)(0.85)	= 416
Referee:	(380)(1)(1)	= 380
	Sensible subtotal	= 4734 Btu/hr

Occupants	Latent Heat Gain	
Boys:	(925 Btu/hr•boy)(10 boys)(0.75 adj. fact)	= 6938 Btu/hr
Coaches	(205)(2)(0.85)	= 349
Referee:	(620)(1)(1)	= 620
	Latent subtotal	= 7090 Btu/hr

From Table 8.2, the sensible CLF = 0.59

The cooling loads are

Sensible:	4734(0.59)	= 2,793
Latent:		= 7,907
Total:		= 10,700 Btu/hr

TABLE 8.2 Sensible Cooling Load Factors for People

Total hours in space	Hours After Each Entry Into Space																							
	1	2	3	4	5	6	7	8	9	10	11	12	13	14	15	16	17	18	19	20	21	22	23	24
2	0.49	0.58	0.17	0.13	0.10	0.08	0.07	0.06	0.05	0.04	0.04	0.03	0.03	0.02	0.02	0.02	0.02	0.01	0.01	0.01	0.01	0.01	0.01	0.01
4	0.49	0.59	0.66	0.71	0.27	0.21	0.16	0.14	0.11	0.10	0.08	0.07	0.06	0.06	0.05	0.04	0.04	0.03	0.03	0.03	0.02	0.02	0.02	0.01
6	0.50	0.60	0.67	0.72	0.76	0.79	0.34	0.26	0.21	0.18	0.15	0.13	0.11	0.10	0.08	0.07	0.06	0.06	0.05	0.04	0.04	0.03	0.03	0.03
8	0.51	0.61	0.67	0.72	0.76	0.80	0.82	0.84	0.38	0.30	0.25	0.21	0.18	0.15	0.13	0.12	0.10	0.09	0.08	0.07	0.06	0.05	0.05	0.04
10	0.53	0.62	0.69	0.74	0.77	0.80	0.83	0.85	0.87	0.89	0.42	0.34	0.28	0.23	0.20	0.17	0.15	0.13	0.11	0.10	0.09	0.08	0.07	0.06
12	0.55	0.64	0.70	0.75	0.79	0.81	0.84	0.86	0.88	0.89	0.91	0.92	0.45	0.36	0.30	0.25	0.21	0.19	0.16	0.14	0.12	0.11	0.09	0.08
14	0.58	0.66	0.72	0.77	0.80	0.83	0.85	0.87	0.89	0.90	0.91	0.92	0.93	0.94	0.47	0.38	0.31	0.26	0.23	0.20	0.17	0.15	0.13	0.11
16	0.62	0.70	0.75	0.79	0.82	0.85	0.87	0.88	0.90	0.91	0.92	0.93	0.94	0.95	0.95	0.96	0.49	0.39	0.33	0.28	0.24	0.20	0.18	0.16
18	0.66	0.74	0.79	0.82	0.85	0.87	0.89	0.90	0.92	0.93	0.94	0.94	0.95	0.96	0.96	0.97	0.97	0.97	0.50	0.40	0.33	0.26	0.24	0.21

CLF=1.0 for systems shut down at night and for high occupant densities such as in theaters and auditoriums.

8.2   COOLING LOADS FROM LIGHTS

Cooling loads from lighting fixtures represent one of the largest loads, either external or internal, for many types of buildings, such as offices, retail spaces, and any space with a large number of light fixtures. In office buildings, lighting levels can be 2 to 3 W/ft² of floor area. All the electrical energy supplied to a light fixture is eventually converted to heat.

The heat given off is either radiant (visible light), heat convected to the surrounding air, or heat conducted to adjacent building materials.

In a manner similar to heat gains for sunlit walls and roofs, the visible radiant energy (light) is absorbed by the space itself, and the furnishings within the space. The rate which electricity is converted to heat gain, by a light fixture, is a function of a number of variables, including the lamp type (incandescent, fluorescent, etc.), the type of fixture (recessed, pendant mounted, etc.), and the efficiency of the light fixture itself.

Lighting fixtures can be vented or unvented. A vented fixture is one that has air slots in it, enabling supply air, or return air, to pass through the slots. A typical recessed and unvented fixture is shown in Figure 8.1, while a typical recessed vented fixture is shown in Figure 8.2.

Incandescent lights convert about 10% of the power input to visible light, and about 90% is lost as radiant, conductive, and convective heat. Fluorescent lights convert about 25% of the power input to visible light, with the balance of the energy supplied lost as heat. Part of the energy loss, for a fluorescent fixture, is by the ballast, which amounts to about 20% of the energy supplied to the fixture.

All heat rejected by a light fixture is sensible. There is no latent heat. The instantaneous heat gain from lights may be calculated with Equation 8.3.

$$q_s = (P_l)(D)(\text{BF})(3.413) \tag{8.3}$$

where

q_s = the instantaneous heat gain in Btu/hr

P_l = lighting power input in W

D = a diversity or use factor to account for not all lights being on at once

BF = ballast factor. BF = 1.0 for incandescent light and BF = 1.2 for fluorescent

3.413 = conversion from W to Btu/hr

In Equation 8.3, note that the power is expressed in lighting power input. Sometimes, lighting fixture is described by the total power input, including losses. An example of this situation is a 200 W, four tube fluorescent fixture. The fixture has four, 40 W tubes that total 160 W. The other 40 W are for ballast losses. If the 200 W value was used for the fixture wattage in Equation 8.3, then BF = 1 since the ballast losses were already accounted for in the fixture rating.

The relative amount of heat that is either stored in the building mass and furnishings or converted to an instantaneous cooling load is dependent upon the type of fixture mounting, the space ventilation rate, and the building thermal mass. ASHRAE has developed Cooling Load Factors for lights, which are similar to those for people. In order to determine a CLF, the fixture type must be known, an approximate space ventilation rate must be known or estimated, the space construction weight must be estimated, the number of hours the lights have been on must be determined, and the total length of time the lights will be on must be known.

Table 8.3 lists an "a" classification, or coefficient, for lights. This classification is determined by the features of the room furnishings, the light fixture type, and the appropriate

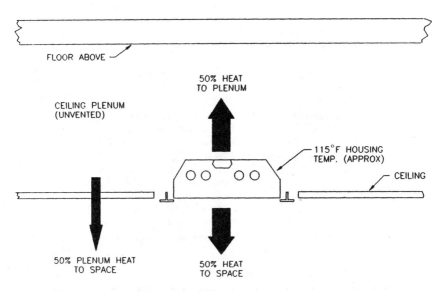

FIGURE 8.1 Typical Recessed Unvented Luminare Installation

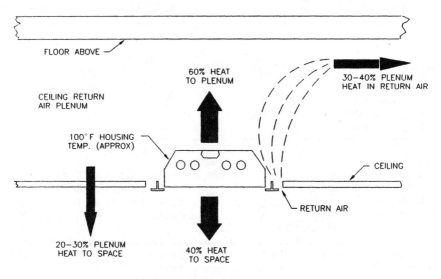

FIGURE 8.2 Typical Recessed Vented Luminare Installation

space ventilation rate. Table 8.4 lists a "b" classification or value for lights. This classification is determined by the envelope, or building construction, and the space ventilation rates.

After an "a" and "b" classification have been determined, from Tables 8.3 and 8.4, a Cooling Load Factor for the lights may then be determined from Table 8.5. Note that Table 8.5 is comprised of multiple tables, based on the length of time the lights are on.

TABLE 8.3   Design Values of "a" Coefficient Features of Room Furnishings, Light Fixtures, and Ventilation Rates

"a" Classification	Furnishings	Air Supply and Return	Type of Light Fixture
0.45	Heavyweight, simple furnishings, no carpet	Low rate, supply and return below ceiling	Recessed, not vented
0.55	Ordinary furniture, no carpet	Medium to high ventilation rate, supply and return below ceiling or through ceiling grille and space	Recessed, not vented
0.65	Ordinary furniture, with or without carpet	Medium to high ventilation rate or fan coil or induction type air conditioning terminal unit; supply through ceiling or wall diffuser; return around light fixtures and through ceiling space	Vented
0.75	Any type of furniture	Ducted returns through light fixtures	Vented or free hanging in air stream with ducted returns

TABLE 8.4   Classification for Different Envelope Construction and Room Air Circulation Rates

Room Envelope Construction	Mass of Floor Area, lb/ft^2	Room Air Circulation and Type of Supply and Return			
		Low	Medium	High	Very High
2 in Wood Floor	10	B	A	A	A
3 in Concrete Floor	40	B	B	B	A
6 in Concrete Floor	75	C	C	C	B
8 in Concrete Floor	120	D	D	C	C
12 in Concrete Floor	160	D	D	D	D

Example 8.2   Assume a room has four recessed, 200 W, fluorescent light fixtures in it. The fixtures are not the ventilated type. The lights were turned on at 8:00 a.m. and will be on until 6:00 p.m. The room has carpeting, over a 6 in concrete floor, with a desk, chair, and a bookcase. It is expected that the room will have a medium ventilation rate. What is the cooling load from the lights at 2:00 p.m.?

It is assumed that the wattage specified is the nominal power input per fixture. Therefore, the 200 W include a ballast factor. For a single room, the diversity factor is assumed to be 1.0. From Equation 8.3, the instantaneous heat gain is

$$q_i = (200)(1.0)(1.0)(3.413) = 683 \text{ Btu/hr}$$

Referring to Table 8.3, a = 0.55 is selected as the closest match, even though it is for no carpeting. It does match the ventilation rate and fixture type.

Referring to Table 8.4, b = c is chosen for a 6 in concrete floor, medium ventilation rate.

Since the lights were turned on at 8:00 a.m. and will be on until 6:00 p.m., the time the lights will be on is 10 hr. At 2:00 p.m., the lights have been on for 6 hr. Referring to Table 8.5, CLF = 0.75, the cooling load is

TABLE 8.5   Cooling Load Factors for Lights

Cooling Load Factors When Lights Are on for 8 Hours

"a" Coef- ficients	"b" Classi- fication	Number of hours after lights are turned on																							
		0	1	2	3	4	5	6	7	8	9	10	11	12	13	14	15	16	17	18	19	20	21	22	23
0.45	A	0.02	0.46	0.57	0.65	0.72	0.77	0.82	0.85	0.88	0.46	0.37	0.30	0.24	0.19	0.15	0.12	0.10	0.08	0.06	0.05	0.04	0.03	0.03	0.02
	B	0.07	0.51	0.56	0.61	0.65	0.68	0.71	0.74	0.77	0.34	0.31	0.28	0.23	0.22	0.20	0.18	0.16	0.15	0.13	0.12	0.11	0.10	0.09	0.08
	C	0.11	0.55	0.58	0.60	0.63	0.65	0.67	0.69	0.71	0.28	0.26	0.25	0.23	0.22	0.20	0.19	0.18	0.17	0.16	0.15	0.14	0.13	0.12	0.12
	D	0.14	0.58	0.60	0.61	0.62	0.63	0.64	0.65	0.66	0.22	0.22	0.21	0.20	0.20	0.19	0.19	0.18	0.18	0.17	0.16	0.16	0.16	0.15	0.15
0.55	A	0.01	0.56	0.65	0.72	0.77	0.82	0.85	0.88	0.90	0.37	0.30	0.24	0.19	0.16	0.13	0.10	0.08	0.07	0.05	0.04	0.03	0.03	0.02	0.02
	B	0.06	0.60	0.64	0.68	0.71	0.74	0.76	0.79	0.81	0.28	0.25	0.23	0.20	0.18	0.16	0.15	0.13	0.12	0.11	0.10	0.09	0.08	0.07	0.06
	C	0.09	0.63	0.66	0.68	0.70	0.71	0.73	0.75	0.76	0.23	0.21	0.20	0.19	0.18	0.17	0.16	0.15	0.14	0.13	0.12	0.11	0.11	0.10	0.10
	D	0.11	0.66	0.67	0.68	0.69	0.70	0.71	0.72	0.72	0.18	0.18	0.17	0.17	0.16	0.16	0.15	0.15	0.14	0.14	0.13	0.13	0.13	0.12	0.12
0.65	A	0.01	0.66	0.73	0.78	0.82	0.86	0.88	0.91	0.93	0.29	0.23	0.19	0.15	0.12	0.10	0.08	0.06	0.05	0.04	0.03	0.03	0.02	0.02	0.01
	B	0.04	0.69	0.72	0.75	0.77	0.80	0.82	0.84	0.85	0.22	0.19	0.18	0.14	0.14	0.13	0.12	0.10	0.09	0.08	0.08	0.07	0.06	0.06	0.05
	C	0.07	0.72	0.73	0.75	0.76	0.78	0.79	0.80	0.82	0.18	0.17	0.16	0.15	0.14	0.13	0.12	0.11	0.11	0.10	0.10	0.09	0.08	0.08	0.07
	D	0.09	0.73	0.74	0.75	0.76	0.77	0.77	0.78	0.79	0.14	0.14	0.13	0.13	0.13	0.12	0.12	0.11	0.11	0.11	0.10	0.10	0.10	0.10	0.09
0.75	A	0.01	0.76	0.80	0.84	0.87	0.90	0.92	0.93	0.95	0.21	0.17	0.13	0.11	0.09	0.07	0.06	0.05	0.04	0.03	0.02	0.02	0.02	0.01	0.01
	B	0.03	0.78	0.80	0.82	0.84	0.85	0.87	0.84	0.89	0.15	0.14	0.13	0.11	0.10	0.09	0.08	0.07	0.07	0.06	0.05	0.05	0.04	0.04	0.04
	C	0.05	0.80	0.81	0.82	0.83	0.84	0.85	0.86	0.87	0.13	0.12	0.11	0.10	0.10	0.09	0.09	0.08	0.08	0.07	0.07	0.06	0.06	0.06	0.05
	D	0.06	0.81	0.82	0.82	0.83	0.83	0.84	0.84	0.85	0.10	0.10	0.09	0.09	0.09	0.09	0.08	0.08	0.08	0.08	0.07	0.07	0.07	0.07	0.07

Cooling Load Factors When Lights Are on for 10 Hours

"a" Coef- ficients	"b" Classi- fication	Number of hours after lights are turned on																							
		0	1	2	3	4	5	6	7	8	9	10	11	12	13	14	15	16	17	18	19	20	21	22	23
0.45	A	0.03	0.47	0.58	0.66	0.73	0.78	0.82	0.86	0.88	0.91	0.93	0.49	0.39	0.32	0.26	0.21	0.17	0.13	0.11	0.09	0.07	0.06	0.05	0.04
	B	0.10	0.54	0.59	0.63	0.66	0.70	0.73	0.76	0.78	0.80	0.82	0.39	0.35	0.32	0.28	0.26	0.23	0.21	0.19	0.17	0.15	0.14	0.12	0.11
	C	0.15	0.59	0.61	0.64	0.66	0.68	0.70	0.72	0.73	0.75	0.76	0.33	0.31	0.29	0.27	0.26	0.24	0.23	0.21	0.20	0.19	0.18	0.17	0.16
	D	0.18	0.62	0.63	0.64	0.66	0.67	0.68	0.69	0.69	0.70	0.71	0.27	0.26	0.26	0.25	0.24	0.23	0.23	0.22	0.21	0.21	0.20	0.19	0.19
0.55	A	0.02	0.57	0.65	0.72	0.78	0.82	0.85	0.88	0.91	0.92	0.94	0.40	0.32	0.26	0.21	0.17	0.14	0.11	0.09	0.07	0.06	0.05	0.04	0.03
	B	0.08	0.62	0.66	0.69	0.73	0.75	0.78	0.80	0.82	0.84	0.85	0.32	0.29	0.26	0.23	0.21	0.19	0.17	0.15	0.14	0.12	0.11	0.10	0.09
	C	0.12	0.66	0.68	0.70	0.72	0.74	0.75	0.77	0.78	0.79	0.81	0.27	0.25	0.24	0.22	0.21	0.20	0.19	0.17	0.16	0.15	0.14	0.14	0.13
	D	0.15	0.69	0.70	0.71	0.72	0.73	0.75	0.74	0.75	0.76	0.76	0.22	0.23	0.21	0.20	0.19	0.19	0.18	0.18	0.16	0.17	0.16	0.14	0.15
0.65	A	0.02	0.66	0.73	0.78	0.83	0.86	0.89	0.91	0.93	0.94	0.95	0.31	0.25	0.20	0.16	0.13	0.11	0.08	0.07	0.05	0.04	0.04	0.03	0.02
	B	0.06	0.71	0.74	0.76	0.79	0.81	0.83	0.84	0.86	0.87	0.89	0.25	0.22	0.20	0.18	0.16	0.15	0.13	0.12	0.11	0.10	0.09	0.08	0.07
	C	0.09	0.74	0.75	0.77	0.78	0.80	0.81	0.82	0.83	0.84	0.85	0.21	0.20	0.18	0.17	0.16	0.15	0.14	0.14	0.13	0.12	0.11	0.11	0.10
	D	0.11	0.76	0.77	0.77	0.78	0.79	0.79	0.80	0.81	0.81	0.82	0.17	0.17	0.16	0.16.	0.15	0.15	0.14	0.14	0.14	0.13	0.13	0.12	0.12

TABLE 8.5 Cooling Load Factors for Lights (Continued)

Cooling Load Factors When Lights Are on for 10 Hours

"a" Coefficients	"b" Classification	Number of hours after lights are turned on																						
		0	1	2	3	4	5	6	7	8	9	10	11	12	13	14	15	16	17	18	19	20	21	22	23
0.75	A	0.01	0.76	0.84	0.84	0.88	0.90	0.92	0.93	0.95	0.96	0.97	0.22	0.18	0.14	0.12	0.09	0.08	0.06	0.05	0.04	0.03	0.03	0.02	0.02
	B	0.04	0.79	0.81	0.83	0.85	0.86	0.88	0.89	0.90	0.91	0.92	0.18	0.16	0.14	0.13	0.12	0.10	0.09	0.08	0.08	0.07	0.06	0.06	0.05
	C	0.07	0.81	0.82	0.83	0.84	0.85	0.86	0.87	0.88	0.89	0.89	0.15	0.14	0.13	0.12	0.12	0.11	0.10	0.10	0.09	0.09	0.08	0.08	0.07
	D	0.08	0.83	0.83	0.84	0.84	0.85	0.85	0.86	0.86	0.87	0.87	0.12	0.12	0.11	0.11	0.11	0.11	0.10	0.10	0.10	0.09	0.09	0.09	0.09

Cooling Load Factors When Lights Are on for 12 Hours

"a" Coefficients	"b" Classification	Number of hours after lights are turned on																						
		0	1	2	3	4	5	6	7	8	9	10	11	12	13	14	15	16	17	18	19	20	21	22	23
0.45	A	0.05	0.49	0.59	0.67	0.73	0.78	0.83	0.86	0.89	0.91	0.93	0.94	0.95	0.51	0.41	0.33	0.27	0.22	0.17	0.14	0.11	0.09	0.07	0.06
	B	0.13	0.57	0.61	0.65	0.69	0.72	0.75	0.77	0.79	0.82	0.83	0.85	0.87	0.43	0.39	0.35	0.31	0.28	0.25	0.23	0.21	0.18	0.17	0.15
	C	0.19	0.63	0.65	0.67	0.69	0.71	0.73	0.74	0.76	0.77	0.79	0.80	0.81	0.37	0.35	0.33	0.31	0.29	0.27	0.26	0.24	0.23	0.21	0.20
	D	0.22	0.66	0.67	0.68	0.69	0.70	0.72	0.72	0.73	0.74	0.75	0.76	0.76	0.32	0.31	0.30	0.29	0.28	0.27	0.26	0.26	0.25	0.24	0.23
0.55	A	0.04	0.58	0.66	0.73	0.78	0.82	0.86	0.89	0.91	0.93	0.94	0.95	0.96	0.42	0.34	0.27	0.22	0.18	0.14	0.11	0.09	0.07	0.06	0.05
	B	0.11	0.65	0.68	0.72	0.74	0.77	0.79	0.81	0.83	0.85	0.86	0.88	0.89	0.35	0.32	0.28	0.26	0.23	0.21	0.19	0.17	0.15	0.14	0.12
	C	0.15	0.69	0.71	0.73	0.75	0.76	0.78	0.79	0.80	0.81	0.83	0.84	0.85	0.30	0.29	0.27	0.25	0.24	0.22	0.21	0.20	0.19	0.17	0.16
	D	0.18	0.72	0.73	0.74	0.75	0.76	0.76	0.77	0.78	0.78	0.79	0.80	0.80	0.26	0.26	0.25	0.24	0.22	0.22	0.21	0.21	0.20	0.20	0.19
0.65	A	0.03	0.67	0.74	0.79	0.83	0.86	0.89	0.91	0.93	0.94	0.95	0.96	0.97	0.33	0.26	0.21	0.17	0.14	0.11	0.09	0.07	0.06	0.05	0.04
	B	0.09	0.73	0.75	0.78	0.80	0.82	0.84	0.85	0.87	0.88	0.89	0.90	0.91	0.27	0.25	0.22	0.20	0.18	0.15	0.15	0.13	0.12	0.11	0.10
	C	0.12	0.76	0.78	0.79	0.81	0.81	0.83	0.84	0.85	0.86	0.87	0.88	0.88	0.24	0.22	0.21	0.20	0.18	0.17	0.16	0.15	0.14	0.14	0.13
	D	0.14	0.79	0.79	0.80	0.80	0.81	0.82	0.82	0.83	0.83	0.84	0.84	0.85	0.20	0.20	0.19	0.18	0.18	0.17	0.16	0.16	0.16	0.15	0.15
0.75	A	0.02	0.77	0.81	0.85	0.88	0.90	0.92	0.94	0.95	0.96	0.97	0.97	0.98	0.23	0.19	0.15	0.12	0.10	0.08	0.06	0.05	0.04	0.03	0.03
	B	0.06	0.81	0.82	0.84	0.86	0.87	0.88	0.90	0.91	0.92	0.92	0.93	0.94	0.19	0.18	0.16	0.14	0.13	0.12	0.10	0.09	0.08	0.08	0.07
	C	0.09	0.83	0.84	0.85	0.86	0.87	0.88	0.89	0.89	0.90	0.90	0.91	0.91	0.17	0.16	0.15	0.14	0.13	0.12	0.12	0.11	0.10	0.10	0.09
	D	0.10	0.85	0.85	0.86	0.86	0.86	0.87	0.87	0.88	0.88	0.88	0.89	0.89	0.14	0.14	0.14	0.13	0.13	0.12	0.12	0.12	0.11	0.11	0.11

Cooling Load Factors When Lights Are on for 14 Hours

"a" Coefficients	"b" Classification	Number of hours after lights are turned on																						
		0	1	2	3	4	5	6	7	8	9	10	11	12	13	14	15	16	17	18	19	20	21	22	23
0.45	A	0.07	0.51	0.61	0.68	0.74	0.79	0.83	0.87	0.89	0.91	0.93	0.94	0.95	0.96	0.97	0.53	0.42	0.34	0.27	0.22	0.18	0.14	0.12	0.09
	B	0.18	0.61	0.65	0.68	0.72	0.74	0.77	0.79	0.81	0.83	0.85	0.86	0.88	0.89	0.90	0.46	0.41	0.37	0.34	0.30	0.27	0.24	0.22	0.20
	C	0.24	0.67	0.69	0.71	0.73	0.74	0.76	0.77	0.79	0.80	0.81	0.82	0.83	0.84	0.85	0.41	0.39	0.36	0.34	0.32	0.30	0.28	0.27	0.25
	D	0.26	0.71	0.72	0.72	0.73	0.74	0.75	0.76	0.77	0.78	0.78	0.79	0.80	0.80	0.80	0.36	0.35	0.34	0.33	0.32	0.31	0.30	0.29	0.28

TABLE 8.5 Cooling Load Factors for Lights (Concluded)

Cooling Load Factors When Lights Are on for 14 Hours

"a" Coefficients	"b" Classification	0	1	2	3	4	5	6	7	8	9	10	11	12	13	14	15	16	17	18	19	20	21	22	23
												Number of hours after lights are turned on													
0.55	A	0.06	0.69	0.68	0.74	0.79	0.83	0.86	0.89	0.91	0.93	0.94	0.95	0.96	0.97	0.98	0.43	0.35	0.28	0.22	0.18	0.15	0.12	0.09	0.08
	B	0.15	0.68	0.71	0.74	0.77	0.79	0.81	0.83	0.85	0.86	0.88	0.89	0.90	0.91	0.92	0.38	0.34	0.31	0.27	0.25	0.22	0.20	0.18	0.16
	C	0.19	0.73	0.75	0.76	0.78	0.79	0.80	0.81	0.83	0.84	0.85	0.86	0.86	0.87	0.88	0.34	0.32	0.30	0.28	0.26	0.25	0.23	0.22	0.21
	D	0.22	0.76	0.77	0.77	0.78	0.79	0.79	0.80	0.81	0.81	0.82	0.81	0.83	0.83	0.84	0.29	0.28	0.28	0.27	0.26	0.25	0.24	0.24	0.23
0.65	A	0.05	0.69	0.75	0.80	0.84	0.87	0.89	0.92	0.93	0.95	0.96	0.96	0.97	0.98	0.98	0.34	0.27	0.22	0.17	0.14	0.11	0.09	0.07	0.06
	B	0.11	0.75	0.78	0.80	0.82	0.64	0.85	0.87	0.88	0.89	0.90	0.91	0.92	0.93	0.94	0.29	0.26	0.24	0.21	0.19	0.17	0.16	0.14	0.13
	C	0.15	0.79	0.80	0.82	0.83	0.84	0.85	0.86	0.86	0.87	0.88	0.89	0.89	0.90	0.94	0.26	0.25	0.23	0.22	0.20	0.19	0.18	0.17	0.16
	D	0.17	0.81	0.82	0.82	0.83	0.83	0.84	0.84	0.85	0.85	0.86	0.86	0.87	0.87	0.87	0.23	0.22	0.21	0.21	0.20	0.20	0.19	0.18	0.18
0.75	A	0.03	0.78	0.82	0.86	0.88	0.91	0.92	0.94	0.95	0.96	0.97	0.97	0.98	0.98	0.99	0.24	0.19	0.16	0.12	0.10	0.08	0.07	0.05	0.04
	B	0.08	0.82	0.84	0.86	0.87	0.88	0.90	0.91	0.92	0.92	0.93	0.94	0.94	0.95	0.96	0.21	0.19	0.17	0.15	0.14	0.12	0.11	0.10	0.09
	C	0.11	0.85	0.86	0.87	0.88	0.88	0.89	0.90	0.90	0.91	0.91	0.92	0.92	0.93	0.93	0.19	0.18	0.17	0.16	0.15	0.14	0.13	0.12	0.11
	D	0.12	0.87	0.87	0.87	0.88	0.88	0.89	0.89	0.89	0.90	0.90	0.90	0.90	0.91	0.91	0.16	0.16	0.15	0.15	0.14	0.14	0.13	0.13	0.13

Cooling Load Factors When Lights Are on for 16 Hours

"a" Coefficients	"b" Classification	0	1	2	3	4	5	6	7	8	9	10	11	12	13	14	15	16	17	18	19	20	21	22	23
												Number of hours after lights are turned on													
0.45	A	0.12	0.54	0.63	0.70	0.76	0.81	0.85	0.88	0.90	0.92	0.94	0.95	0.96	0.97	0.97	0.98	0.98	0.54	0.43	0.35	0.28	0.23	0.18	0.15
	B	0.23	0.66	0.69	0.72	0.75	0.78	0.80	0.82	0.84	0.85	0.87	0.88	0.89	0.90	0.91	0.92	0.93	0.49	0.44	0.39	0.35	0.32	0.29	0.26
	C	0.29	0.72	0.74	0.75	0.77	0.78	0.80	0.81	0.82	0.83	0.84	0.85	0.86	0.87	0.88	0.88	0.89	0.45	0.42	0.39	0.37	0.35	0.33	0.31
	D	0.31	0.75	0.76	0.77	0.77	0.78	0.79	0.79	0.80	0.81	0.81	0.82	0.82	0.83	0.83	0.84	0.84	0.40	0.39	0.37	0.36	0.35	0.34	0.33
0.55	A	0.10	0.63	0.70	0.76	0.81	0.84	0.87	0.90	0.92	0.93	0.95	0.96	0.97	0.97	0.98	0.98	0.99	0.44	0.35	0.28	0.23	0.18	0.15	0.12
	B	0.19	0.72	0.75	0.77	0.80	0.82	0.84	0.85	0.87	0.88	0.89	0.90	0.91	0.92	0.93	0.94	0.94	0.40	0.36	0.32	0.29	0.26	0.24	0.21
	C	0.24	0.77	0.79	0.80	0.81	0.82	0.83	0.84	0.85	0.86	0.87	0.88	0.88	0.89	0.90	0.90	0.91	0.37	0.34	0.32	0.30	0.29	0.27	0.25
	D	0.26	0.80	0.80	0.81	0.83	0.82	0.83	0.84	0.84	0.85	0.85	0.86	0.86	0.86	0.87	0.87	0.87	0.33	0.32	0.31	0.30	0.30	0.28	0.27
0.65	A	0.07	0.71	0.77	0.81	0.85	0.88	0.90	0.92	0.94	0.95	0.96	0.97	0.97	0.98	0.98	0.99	0.99	0.34	0.27	0.22	0.18	0.14	0.12	0.09
	B	0.15	0.78	0.78	0.82	0.84	0.86	0.87	0.88	0.90	0.91	0.92	0.92	0.93	0.94	0.94	0.95	0.96	0.31	0.28	0.25	0.23	0.20	0.18	0.16
	C	0.18	0.82	0.83	0.84	0.85	0.86	0.87	0.88	0.89	0.89	0.90	0.90	0.91	0.92	0.92	0.93	0.93	0.28	0.27	0.25	0.24	0.22	0.21	0.20
	D	0.20	0.84	0.85	0.85	0.86	0.86	0.87	0.87	0.87	0.88	0.88	0.89	0.89	0.89	0.90	0.90	0.90	0.25	0.25	0.24	0.23	0.23	0.22	0.21
0.75	A	0.05	0.79	0.83	0.87	0.89	0.91	0.93	0.94	0.95	0.96	0.97	0.98	0.98	0.98	0.99	0.99	0.99	0.24	0.20	0.16	0.13	0.10	0.08	0.07
	B	0.11	0.85	0.86	0.87	0.89	0.90	0.91	0.92	0.93	0.93	0.94	0.95	0.96	0.96	0.96	0.96	0.97	0.22	0.20	0.18	0.16	0.15	0.13	0.12
	C	0.13	0.87	0.88	0.89	0.90	0.90	0.91	0.91	0.92	0.92	0.93	0.93	0.94	0.94	0.94	0.95	0.95	0.20	0.19	0.18	0.17	0.16	0.15	0.14
	D	0.14	0.89	0.89	0.89	0.90	0.90	0.90	0.91	0.91	0.91	0.91	0.92	0.92	0.92	0.92	0.93	0.93	0.18	0.18	0.17	0.17	0.16	0.16	0.15

CLF=1.0 when cooling system operates only during occupied hours or when lights are on 24 hr/day.
Copyright 1989 by the American Society of Heating, Refrigerating and Air Conditioning Engineers, Inc. from the ASHRAE Handbook—Fundamentals. Used by permission.

$$q_s = (\text{CLF})(q_i) = (0.75)(683) = 512 \text{ Btu/hr}$$

Refer to Section 8.5.2 for lighting cooling loads in buildings with return air plenums.

8.3 COOLING LOADS FOR EQUIPMENT AND APPLIANCES

Internal heat gains, and the resultant cooling loads, for equipment come from heat producing equipment located within the conditioned space. Equipment heat gains come from such items as office equipment (typewriters, copy machines, computers, etc.), appliances (ranges, stoves, ovens, etc.), and from motor driven equipment (pumps, compressors, machines, etc.) located in the conditioned space.

All of the above produce sensible heat only, except the kitchen appliances, which can also produce latent heat. In an effort to reduce the heat gain to a space, exhaust hoods are frequently placed over kitchen appliances to remove heat before it enters the space. For cooling load calculation purposes, appliances are classified according to whether they are located under an exhaust hood or not (hooded or unhooded). These classifications apply to all equipment, but most frequently apply to kitchen cooking appliances.

In a manner similar to people and lights, ASHRAE has developed Cooling Load Factors for equipment, and appliances, located within the conditioned space. Table 8.6 lists Cooling Load Factors for hooded appliances and Table 8.7 lists factors for unhooded appliances and equipment.

As in the case of Cooling Load Factors for lights, the CLF values for equipment are dependent upon the total time the appliances are operated and how long they have been on. The CLFs should be applied to the sensible portion of the heat gain only. They do not apply to latent heat gains. Equation 8.4 may be used to calculate the sensible cooling loads for equipment and appliances.

$$q = (q_i)(D)(\text{CLF}) \tag{8.4}$$

where

q = the sensible cooling load in Btu/hr

q_i = the instantaneous sensible heat gain in Btu/hr

D = a diversity or use factor

CLF = Cooling Load Factor from Table 8.6 or 8.7

Whenever possible, the manufacturer of the equipment should be contacted to get the actual instantaneous heat gain, or heat rejection rate. If that is not possible, the tables in the following sections may be used. This data represents average values for typical equipment. It may also be possible to check the equipment nameplate for heat rejection rates or power consumption rates.

8.3.1 Office Equipment

Table 8.8 lists typical heat gains for office equipment. Notice that some of the equipment includes a standby value, as well as a maximum power input. Equipment such as

TABLE 8.6 Cooling Load Factors for Appliances—Hooded

Sensible Heat Cooling Load Factors for Appliances—Hooded

Total Operational Hours	Hours after appliances are on																							
	1	2	3	4	5	6	7	8	9	10	11	12	13	14	15	16	17	18	19	20	21	22	23	24
2	0.27	0.40	0.25	0.18	0.14	0.11	0.09	0.08	0.07	0.06	0.05	0.04	0.04	0.03	0.03	0.03	0.02	0.02	0.02	0.02	0.01	0.01	0.01	0.01
4	0.28	0.41	0.51	0.59	0.39	0.30	0.24	0.19	0.16	0.14	0.12	0.10	0.09	0.08	0.07	0.06	0.05	0.05	0.04	0.04	0.03	0.03	0.02	0.02
6	0.29	0.42	0.52	0.59	0.65	0.70	0.48	0.37	0.30	0.25	0.21	0.18	0.16	0.14	0.12	0.11	0.09	0.08	0.07	0.06	0.05	0.05	0.04	0.04
8	0.31	0.44	0.54	0.61	0.66	0.71	0.75	0.78	0.55	0.43	0.35	0.30	0.25	0.22	0.19	0.16	0.14	0.13	0.11	0.10	0.08	0.07	0.06	0.06
10	0.33	0.46	0.55	0.62	0.68	0.72	0.76	0.79	0.81	0.84	0.60	0.48	0.39	0.33	0.28	0.24	0.21	0.18	0.16	0.14	0.12	0.11	0.09	0.08
12	0.36	0.49	0.58	0.64	0.69	0.74	0.77	0.80	0.82	0.85	0.87	0.88	0.64	0.51	0.42	0.36	0.31	0.26	0.23	0.20	0.18	0.15	0.13	0.12
14	0.40	0.52	0.61	0.67	0.72	0.76	0.79	0.82	0.84	0.86	0.88	0.89	0.91	0.92	0.67	0.54	0.45	0.38	0.32	0.28	0.24	0.21	0.19	0.16
16	0.45	0.57	0.65	0.70	0.75	0.78	0.81	0.84	0.86	0.87	0.89	0.90	0.92	0.93	0.94	0.94	0.69	0.56	0.46	0.39	0.34	0.29	0.25	0.22
18	0.52	0.63	0.70	0.75	0.79	0.82	0.84	0.86	0.88	0.89	0.91	0.92	0.93	0.94	0.95	0.95	0.96	0.96	0.71	0.58	0.48	0.41	0.35	0.30

TABLE 8.7 Cooling Load Factors for Appliances—Unhooded

Sensible Heat Cooling Load Factors for Appliances—Unhooded

Total Operational Hours	Hours after appliances are on																							
	1	2	3	4	5	6	7	8	9	10	11	12	13	14	15	16	17	18	19	20	21	22	23	24
2	0.56	0.64	0.15	0.11	0.08	0.07	0.06	0.05	0.04	0.04	0.03	0.03	0.02	0.02	0.02	0.02	0.01	0.01	0.01	0.01	0.01	0.01	0.01	0.01
4	0.57	0.65	0.71	0.75	0.23	0.18	0.14	0.12	0.10	0.08	0.07	0.06	0.05	0.05	0.04	0.04	0.03	0.03	0.02	0.02	0.02	0.02	0.01	0.01
6	0.57	0.65	0.71	0.76	0.79	0.82	0.29	0.22	0.18	0.15	0.13	0.11	0.10	0.08	0.07	0.06	0.06	0.05	0.04	0.04	0.03	0.03	0.03	0.02
8	0.58	0.66	0.72	0.76	0.80	0.82	0.85	0.87	0.33	0.26	0.21	0.18	0.15	0.13	0.11	0.10	0.09	0.08	0.07	0.06	0.05	0.04	0.04	0.03
10	0.60	0.68	0.73	0.77	0.81	0.83	0.85	0.87	0.89	0.90	0.36	0.29	0.24	0.20	0.17	0.15	0.13	0.11	0.10	0.08	0.07	0.07	0.06	0.05
12	0.62	0.69	0.73	0.79	0.82	0.84	0.86	0.88	0.89	0.91	0.92	0.93	0.94	0.31	0.25	0.21	0.18	0.16	0.14	0.12	0.11	0.09	0.08	0.07
14	0.64	0.71	0.76	0.80	0.83	0.85	0.87	0.89	0.90	0.92	0.93	0.93	0.94	0.95	0.40	0.32	0.27	0.23	0.19	0.17	0.15	0.13	0.11	0.10
16	0.67	0.74	0.79	0.82	0.85	0.87	0.89	0.90	0.91	0.92	0.93	0.94	0.95	0.96	0.96	0.97	0.42	0.34	0.28	0.24	0.20	0.18	0.15	0.13
18	0.71	0.78	0.82	0.85	0.87	0.89	0.90	0.92	0.93	0.94	0.94	0.95	0.96	0.96	0.97	0.97	0.97	0.98	0.43	0.35	0.29	0.24	0.21	0.18

TABLE 8.8 Heat Gain from Selected Office Equipment

Appliance	Size	Maximum Input		Standby Input		Recommended Rate of Heat Gain	
		Watts	Btu/h	Watts	Btu/h	Watts	Btu/h
Computer Devices							
Communication/transmission		1800-4600	6140-15700	1640-2810	5600-9600	1640-2810	5600-9600
Disk drives/mass storage		1000-10000	3400-34100	1000-6600	3400-22400	1000-6600	3400-22400
Microcomputer/wordprocessor	16-640 kbytes[a]	100-600	340-2050	90-530	300-1800	90-530	300-1800
Minicomputer		2200-6600	7500-15000	2200-6600	7500-15000	2200-6600	7500-15000
Printer (laser)	8 pages/min	870	3000	180	600	300	1000
Printer (Line, high speed)	5000-more pages/min	1000-5300	3400-18000	500-2550	2160-9040	730-3800	2500-13000
Tape drives		1200-6500	4100-22200	1000-4700	3500-15000	1000-4700	3500-15000
Terminal		90-200	300-700	80-180	270-600	80-180	270-600
Copiers/Typesetters							
Blue print		1150-12500	3900-42700	500-5000	1700-17000	1150-12500	3900-42700
Copiers (large)	30-67[a] copies/min.	5800-22500	1700-6600	5800-22500	900	3100	1700-6600

TABLE 8.8 Heat Gain from Selected Office Equipment (Concluded)

Appliance	Size	Maximum Input		Standby Input		Recommended Rate of Heat Gain	
		Watts	Btu/h	Watts	Btu/h	Watts	Btu/h
Copiers	6-30[a] copies/min.	1570-5800	460-1700	1570-5800	300-900	1000-3100	460-1700
Phototypesetter		1725	5900			1520	5200
Mailprocessing							
Inserting machine 3600-6800 pieces/h		600-3300	2000-11300			390-2150	1300-7300
Labeling machine 1500-30000 pieces/h		600-6600	2000-22500			390-4300	1300-14700
Miscellaneous							
Cash register		60	200			48	160
Cold food/beverage		1150-1920	3900-6600			575-960	1960-3280
Coffee maker	10 cup	1500	5120		sensible	1050	3580
					latent	450	1540
Microwave oven	1 ft³	600	2050			400	1360
Paper shredder		250-3000	850-10200			200-2420	680-8250
Water cooler	8 gal/h	700	2400			1750	6000

[a]Input is not proportional to capacity.
Copyright 1989 by the American Society of Heating, Refrigerating and Air Conditioning Engineers, Inc. from the Fundamentals Handbook. Used by permission.

copy machines are usually not operated at full power continuously, but are on standby much of the time they are on. The Recommended Rate of Heat Gain in the last column of Table 8.8 includes a diversity factor to account for standby time and full use time.

8.3.2   Kitchen Equipment

Most kitchen appliances produce both sensible and latent heat gains in a conditioned space. The majority of the latent heat gain is due to the loss of moisture from foods as they are heated and cooked. In addition to latent heat from cooking foods, gas burning appliances also produce latent heat as a part of the combustion process of the gas. A properly designed, and properly operating kitchen hood exhaust system, will remove most of the latent heat from appliances located under the hood before it enters the conditioned space.

Tables 8.9 and 8.10 list heat gains for unhooded electric, and unhooded gas burning, or steam heated kitchen appliances, respectively. If the appliances are located under a properly designed kitchen hood exhaust system, it may be assumed that all of the latent heat is removed, and that 50% of the sensible heat is removed by the exhaust system. Additional appliance heat gains may be found in Reference 1.

Equation 8.4 and the Cooling Load Factors in Tables 8.6 or 8.7 may be used to calculate cooling loads from kitchen appliances.

8.3.3   Motor Driven Equipment

Equipment driven by electric motors also produce sensible cooling loads in an air conditioned space. Examples of such equipment include lathes, printing presses, compressors, and pumps. Sometimes, the motor and the driven equipment are not both in the conditioned space. In that case, a distinction must be made according to whether only the motor, only the machinery, or both, are located within the conditioned space. Table 8.11 lists heat gains for motors, and motor driven equipment, for all three situations. Equation 8.4 and the Cooling Load Factors In Table 8.11 may be used to calculate the sensible cooling load.

8.4   INFILTRATION COOLING LOADS

Cooling loads caused by infiltration, or leakage, of outdoor air into a conditioned space are usually treated as internal loads, even though they are from an external source. Methods for estimating the amount of infiltration into, or out of, a space were discussed in Chapter 5.

Since the infiltration air mixes directly with conditioned air in the space, the heat gain becomes an immediate cooling load on the space. There is no thermal storage; therefore, it is not necessary to apply Cooling Load Temperature Difference or Cooling Load Factors.

The total cooling load from infiltration is both sensible and latent heat. The sensible heat gain is a result of the difference in the dry bulb temperatures between the indoor and outdoor air. Equation 6.1b may be used to calculate the sensible cooling load for infiltration air. The latent cooling load is a result of the difference in the moisture content between the indoor and outdoor air. Equation 8.5 may be used to calculate the latent cooling load for infiltration air.

TABLE 8.9 Heat Gain from Restaurant Appliances

Not Hooded* — Electric

Appliance	Overall Dimensions Less Legs and Handles (in)	Type of Control	Miscellaneous Data	MFR Max. Rating Btu/hr	Maintaining Rate, Btu/hr	Recom Heat Gain For Avg Use Sensible Heat, Btu/hr	Latent Heat, Btu/hr	Total Heat, Btu/hr
Coffee Brewer—½ gal		Man.		2,240	306	900	220	1,120
Warmer—½ gal		Man.		306	306	230	90	320
4 Coffee Brewing Units with 4½ gal Tank	20 × 30 × 26H	Auto.	Water heater—2000 W Brewers—2960 W	16,900		4,800	1,200	6,000
Coffee Urn—3 gal	15 Dia × 34H	Man.	Black Finish	11,900	3,000	2,600	1,700	4,300
3 gal	12 × 23 oval × 21H	Auto.	Nickel plated	15,300	2,600	2,200	1,500	3,700
5 gal	18 Dia × 37H	Auto.	Nickel plated	17,000	3,600	3,400	2,300	5,700
Doughnut Machine	22 × 22 × 57H	Auto.	Exhaust system to outdoors—½ hp motor	16,000		5,000		5,000
Egg Boiler	10 × 13 × 25H	Man.	Med. ht—550 W. Low ht—275 W	3,740		1,200	800	2,000
Food Warmer with Plate Warmer/ft² top surface		Auto.	Insulated, separate heating unit for each pot. Plate warmer in base	1,350	500	350	350	700
Food Warmer without Plate Warmer/ft² top surface		Auto.	Ditto, without plate warmer	1,020	400	200	350	550
Fry Kettle—11½ lb fat	12 Dia × 14H	Auto.		8,840	1,100	1,600	2,400	4,000
Fry Kettle—25 lb fat	16 × 18 × 12H	Auto.	Frying area 12 in × 14 in	23,800	2,000	3,800	5,700	9,500
Griddle, Frying	18 × 18 × 8H	Auto.	Frying top 18 in × 14 in	8,000	2,800	3,100	1,700	4,800
Grille, Meat	14 × 14 × 10H	Auto.	Cooking area 10 in × 12 in	10,200	1,900	3,900	2,100	6,000
Grille, Sandwich	13 × 14 × 10H	Auto.	Grill area 12 in × 12 in	5,600	1,900	2,700	700	3,400
Roll Warmer	26 × 17 × 13H	Auto.	One drawer	1,500	400	1,100	100	1,200
Toaster, Continuous	15 × 15 × 28H	Auto.	2 Slices wide—360 slices/hr	7,500	5,000	5,100	1,300	6,400
Toaster, Continuous	20 × 15 × 28H	Auto.	4 Slices wide—720 slices/hr	10,200	6,000	6,100	2,600	8,700
Toaster, Pop-Up	6 × 11 × 9H	Auto.	2 Slices	4,150	1,000	2,450	450	2,900
Waffle Iron	12 × 13 × 10H	Auto.	One waffle 7 in dia	2,480	600	1,100	750	1,850
Waffle Iron for Ice Cream Sandwich	14 × 13 × 10H	Auto.	12 Cakes, each 2½ in × 3¾ in	7,500	1,500	3,100	2,100	5,200

* If properly designed positive exhaust hood is used, multiply recommended value by 0.50.
Courtesy of Carrier Corp. Copyright 1965 McGraw-Hill, Inc. Used by permission.

TABLE 8.10 Heat Gain from Restaurant Appliances

Not Hooded* — Gas Burning and Steam Heated

Appliance	Overall Dimensions Less Legs and Handles (in)	Type of Control	Miscellaneous Data	MFR Max. Rating Btu/hr	Maintaining Rate Btu/hr	Recom Heat Gain For Avg Use		
						Sensible Heat Btu/hr	Latent Heat Btu/hr	Total Heat Btu/hr
Coffee Brewer—1/2 gal		Man.	Combination brewer and warmer	3,400		1,350	350	1,700
Warmer—1/2 gal		Man.		500	500	400	100	500
Coffee Brewer Unit with Tank	19 × 30 × 26H		4 Brewers and 4 1/2 gal tank			7,200	1,800	9,000
Coffee Urn—3 gal	15 in Dia × 34H	Auto.	Black Finish	3,200	3,900	2,900	2,900	5,800
—3 gal	12 × 23 oval × 21H	Auto.	Nickel Plated		3,400	2,500	2,500	5,000
—5 gal	18 Dia × 37H	Auto.	Nickel Plated		4,700	3,900	3,900	7,800
Food Warmer, Values/ft² top surface		Man.	Water bath type	2,000	900	850	450	1,300
Fry Kettle—15 lb fat	12 × 20 × 18H	Auto.	Frying area 10 in × 10 in	14,250	3,000	4,200	2,800	7,000
Fry Kettle—28 lb fat	15 × 35 × 11H	Auto.	Frying area 11 in × 16 in	24,000	4,500	7,200	4,800	12,000
Grill—Broil-O-Grill	22 × 14 × 17H	Man.	Insulated	37,000		14,400	3,600	18,000
Top Burner	(1.4 ft² grill surface)		22,000 Btu/hr					
Bottom Burner			15,000 Btu/hr					
Stoves, Short Order—Open Top. Values/ft² top surface		Man.	Ring type burners 12,000 to 22,000 Btu/ea	14,000		4,200	4,200	8,400
Stoves, Short Order—Closed Top. Values/ft² top surface		Man.	Ring type burners 10,000 to 12,000 Btu/ea	11,000		3,300	3,300	6,600
Toaster, Continuous	15 × 15 × 28H	Auto.	2 Slices wide—360 slices/hr	12,000	10,000	7,700	3,300	11,000

TABLE 8.10 Heat Gain from Restaurant Appliances (Concluded)

			Steam Heated			
Coffee Urn—3 gal	15 Dia × 34H	Auto.	Black finish	2,900	1,900	4,800
—3 gal	12 × 23 oval × 21H	Auto.	Nickel plated	2,400	1,600	4,000
—3 gal	18 Dia × 37H	Auto.	Nickel plated	3,400	2,300	5,700
Coffee Urn—3 gal	15 Dia × 34H	Man.	Black finish	3,100	3,100	6,200
—3 gal	12 × 23 oval × 21H	Man.	Nickel plated	2,600	2,600	5,200
—3 gal	18 Dia × 37H	Man.	Nickel plated	3,700	3,700	7,400
Food Warmer/ft² top surface		Auto.		400	500	900
Food Warmer/ft² ft top surface		Man.		450	1,150	1,500

* If properly designed positive exhaust hood is used, multiply recommended value by 0.50.
Courtesy of Carrier Corp. Copyright 1965 McGraw-Hill. Used by permission.

TABLE 8.11 Heat Gain from Electric Motors

Continuous Operating*

Nameplate or Brake Horsepower	Full Load Motor Efficiency %	Location of Equipment with Respect to Conditioned Space or Air Stream**		
		Motor In– Driven Machine In hp × 2545 % Eff	Motor In– Driven Machine In hp × 2545	Motor Out– Driven Machine Out hp × 2545 (1 − % Eff) % Eff
			Btu/hr	
$1/20$	40	320	130	190
$1/12$	49	430	210	220
$1/8$	55	580	320	260
$1/6$	60	710	430	280
$1/4$	64	1,000	640	360
$1/3$	66	1,290	850	440
$1/2$	70	1,820	1,280	540
$3/4$	72	2,680	1,930	750
1	79	3,200	2,540	680
$1^1/2$	80	4,770	3,820	950
2	80	6,380	5,100	1,280
3	81	9,450	7,650	1,800
5	82	15,600	12,800	2,800
$7^1/2$	85	22,500	19,100	3,400
10	85	30,000	25,500	4,500
15	86	44,500	38,200	6,300
20	87	58,500	51,000	7,500
25	88	72,400	63,600	8,800
30	89	85,800	76,400	9,400
40	89	115,000	102,000	13,000
50	89	143,000	127,000	16,000
60	89	172,000	153,000	19,000
75	90	212,000	191,000	21,000
100	90	284,000	255,000	29,000
125	90	354,000	318,000	36,000
150	91	420,000	382,000	38,000
200	91	560,000	510,000	50,000
250	91	700,000	636,000	64,000

*For intermittent operation, an appropriate usage factor should be used, preferably measured.
If motors are overloaded, and the amount of overloading is unknown, multiply the above heat gain factors by the following maximum service factors:

Maximum Service Factors						
Horsepower	$1/20–1/8$	$1/6–1/3$	$1/2–3/4$	1	$1^1/2–2$	3–250
AC Open Type	1.4	1.35	1.25	1.25	1.20	1.15
DC Open Type	—	—	—	1.15	1.15	1.15

No overload is available with enclosed motors.

**For a fan or pump in air conditioned space, exhausting air and pumping fluid to outside of space, use values in last column.
Courtesy of Carrier Corp. Copyright 1965, McGraw-Hill. Used by permission.

$$q_1 = 4840(\text{ft}^3/\text{min})(\Delta W) \qquad\qquad (8.5)$$

where

q_l = the latent cooling load in Btu/hr

4840 = a conversion factor for specific heat, mass flow and time

ft^3/min = infiltration air flow rate in ft^3/min

ΔW = the difference in humidity ratio between indoor and outdoor air in pounds of water vapor per pound of dry air

Although there are tables available which give humidity ratios (W) for air at various dry bulb and wet bulb temperatures, it is more convenient to obtain these values from a psychrometric chart. A complete discussion of the psychrometric chart is included in Chapter 9, with ventilation cooling load calculations.

Example 8.3 A room is to be maintained at 75°F dry bulb and 50% relative humidity. The outdoor air conditions are 94° dry bulb and 75°F wet bulb. If 100 ft³/min of outside air is infiltrating into the room, what are the sensible, latent, and total cooling loads from the infiltration? From Equation 6.1b, the sensible load may be calculated directly.

$$q_s = 1.10(100)(94 - 75) = 2090 \text{ Btu/hr}$$

From a psychrometric chart (see Chapter 9), the humidity ratios are as follows:

inside; $W = 0.0092$ lb water/lb dry air

outside; $W = 0.0144$ lb water/lb dry air

Using Equation 8.5, the latent load is

$$q_l = (4840)(100)(0.0144 - 0.0092) = 2517 \text{ Btu/hr}$$

and the total cooling load is

$$q_t = 2090 + 2517 = 4607 \text{ Btu/hr}$$

8.5   SYSTEM COOLING LOADS

System heat gains, or cooling loads, are loads that are internal cooling loads, but they apply to an entire air conditioning system, as opposed to air individual space. Examples of system loads are air handling unit fan motor heat gains, return air plenum gains, and duct heat gains.

All of these are loads on an entire system and not on an individual space. The minimum amount of cooling air supplied to a space in order to satisfy the space load should be based on the cooling load for that space (building codes may also set minimum air quantities). Therefore, it is necessary to distinguish between system (or return) loads and space (or room) loads.

8.5.1   Fan Motor Cooling Load

A fan and motor add energy to an air conditioning system in the form of pressure and motion of the air (See Section 2.1). Eventually, all the energy supplied to the circulated air becomes heat, due to friction losses, as the air flows through the ductwork, etc. In order to properly size an air conditioning system, it is necessary to account for this heat gain, particularly in a large system.

Section 8.3.3 discussed cooling loads for motor driven equipment and Table 8.11 listed heat gains for motors and machines. If the fan motor is in the air stream (or if the fan and motor are in the conditioned space), the values in Table 8.11, for "Motor In–Driven Machine In," should be used. If the fan and motor are outside the conditioned space, and the motor is not in the circulated air stream, the "Motor Out–Driven Machine In" values should be used.

Since a fan is not selected at this stage of the design process and the fan motor is not sized until after the cooling loads have been determined, it is usually necessary to estimate the size of the fan and motor for cooling load purposes. After the actual fan and motor have been selected, the calculations can be adjusted to reflect the actual values.

Fan brake horsepower is the power supplied directly to the fan itself. It does not take losses in motor efficiency, or fan/motor drive losses, into account. In order to get a "ball park" estimate of a system fan motor brake horsepower, Equation 8.6 may be used.

$$\text{Bhp} = \frac{QP_s}{6356 \, \eta_f} \tag{8.6}$$

where

Bhp = brake horsepower supplied to the fan in HP

Q = system air flow in ft³/min

P_s = system static pressure in in wg

6356 = a units conversion factor

η_f = the fan static efficiency

It is necessary to do a preliminary estimate for most of the variables in Equation 8.6. For a reasonably good ventilation rate in occupied spaces, ASHRAE recommends 4 to 12 air changes per hour for most spaces. The system air flow may then be approximated by Equation 8.7.

$$Q = C_a V \left(\frac{1 \text{ min}}{60 \text{ sec}} \right) \tag{8.7}$$

where

C_a = the space air changes per hour

V = the total space volume in ft³

60 = hours to minutes conversion

Unless a system is specifically designed otherwise, most air systems are low pressure—low velocity (less than 2 in. wg and less than 2000 ft/min). A value of 2 in wg may be assumed for P_s in Equation 8.6. The fan static efficiency η_f can vary widely, depending on the type and arrangement of the fan. A value of 50% is a reasonable average value to assume.

Since all the heat gain is absorbed by the circulated air stream, there is no thermal storage, and no CLTDs or CLFs need to be used. The values from Table 8.11 may be used directly for system cooling loads.

Example 8.4 An office has a total volume of 100,000 ft³. Assume a low velocity–low pressure system with the fan motor outside the air stream. The fan is located in an unconditioned penthouse mechanical room. What is the approximate system cooling load for the fan and motor?

Equation 8.7 may be used to estimate the system air flow. Assume 8 air changes per hour.

$$Q = (8)(100,000)\left(\frac{1}{60}\right) = 13,333 \text{ ft}^3/\text{min}$$

A fan static pressure of 2 in wg, and a fan static efficiency of 50%, is assumed. Equation 8.6 may be used to calculate the fan motor brake horsepower.

$$\text{Bhp} = \frac{(13333)(2)}{6356(0.5)} = 8.3 \text{ hp}$$

Assume a 10 hp motor, which is the next largest standard size motor.

Since the air handling unit is in a penthouse mechanical room and the motor is not in the air stream, the fan is in the conditioned space (the air stream) and the motor is out. From Table 8.11, the heat gain for the fan and motor is 25,500 Btu/hr, which is the cooling load.

8.5.2 Return Air Plenum Cooling Loads

Frequently, the space above a ceiling is used as a return air plenum. That is, the air is returned to the air handling system through the ceiling instead of through return air ducts. In this situation, the return air absorbs heat from the plenum as it is returned to the air handling unit. In this case these plenum loads should be considered system, or return loads, instead of cooling loads on the conditioned space.

A typical example in which part of a heat gain should be considered a plenum load is light fixtures in a ceiling with a return air plenum. Referring to Figures 8.1 and 8.2, 50% of the heat gain for unvented fixtures is directed up into the plenum with the balance to the space below. For vented light fixtures, 60% of the heat goes up to the plenum with 40% to the space below. When calculating cooling loads from lights, the effect of system or plenum loads should be taken into account.

8.5.3 Duct Heat Gains

Whenever supply air ductwork is routed through an unvented ceiling space (not a plenum) or when ductwork passes through an unconditioned space, the air in the duct may gain sensible heat, by conduction and convection, through the ductwork. The amount of heat gain is proportional to the duct surface area, the overall heat transfer coefficient (U value) of the duct insulation, and the temperature difference between the air inside the ductwork and the surrounding air. Overall heat transfer coefficients for ductwork and insulation are given in Table 8.12.

Equation 4.1 may then be used to calculate the sensible heat gain to the air in the ductwork. No CLTDs or CLFs need to be applied.

TABLE 8.12   Overall Heat Transfer Coefficient for Ductwork

Description of Ductwork	U
Sheet Metal, not insulated	1.18
1/2-in Thick Insulation Board With or Without Sheet Metal	0.38
1-in Thick Insulation Board With or Without Sheet Metal	0.22
1-in Thick Insulation Board With or Without Sheet Metal	0.15
2-in Thick Insulation Board With or Without Sheet Metal	0.12

Unless calculations are being made for an existing building, with an existing system, it is difficult to estimate the ductwork for a new system. As a rule of thumb, a sensible heat gain, of 1% of the space sensible cooling load may be assumed for duct heat gains.

8.6 INTERNAL COOLING LOAD CALCULATION PROCEDURE

In Section 7.5, the need for experience and judgment when calculating cooling loads was discussed. Methods for identifying the dominant cooling loads for a space and when they occur were also discussed. In addition to reviewing the external loads to see which loads may be most dominant, the internal loads must also be taken into account. In some buildings, such as office buildings, the lighting and equipment loads may be the dominant cooling loads, or have a significant effect on when the peak will occur. In some buildings, like theaters or sports areas, people may be the dominant load.

All cooling loads (internal and external) must be examined to determine when the space and/or system peak load might occur. Cooling load calculations for both interior and exterior loads should be made for the same times or hours. In order to help organize the calculations, an Internal Cooling Load Calculation Form has been provided at the end of this chapter. This form is to be used in conjunction with the External Cooling Load Form.

The procedure for estimating the various internal cooling loads and combining the internal loads with the external loads is as follows:

1. Identify the lighting type, electrical power input (and/or output) and the number of fixtures of each type. Calculate the total wattage for the lighting.

2. Based on which loads are judged to be the dominant loads (internal and external), select the times for which calculations will be made. Determine the total hours that the lights will be on.

3. Identify the lighting fixture type and estimate the ventilation rate for the space (low, medium or high). Based on the construction type, the ventilation rate and the fixture type, select an "a" value from Table 8.3, and a "b" classification from Table 8.4.

4. Based on the "a" value with the "b" classification, the total time the lights are on, and the hours corresponding to the calculation times, select Cooling Load Factors for each calculation time.

5. Multiply the lighting wattage by a ballast factor (if any). Note that the wattage rating for some fixtures includes the ballast. Multiply the total by 3.413 to convert from W to Btu/hr. This will give the instantaneous heat gain for the lights.

6. Determine what percentage of the lighting heat gain is a return gain (if any) and what percentage is a space heat gain. See Section 8.5 and Step 24.

7. Using the space lighting heat gain, calculate the lighting cooling load for each hour by multiplying the room lighting heat gain by the respective CLFs.

8. Enter the room lighting cooling loads for each calculation hour in the three right hand columns on the form. Repeat the procedure for each type of lighting fixture.

9. Estimate the number of occupants in the space for each calculation hour. Separate into groups according to activity level. Also, group by sex and age, if known.

10. Obtain a sensible and latent heat gain for each activity level group from Table 8.1.

11. Apply any age or sex adjustment, if necessary, in accordance with Section 8.1.

12. Estimate the total number of hours the occupants will be in the space, and how long it has been since they entered the space, for each calculation hour for each group. Select the appropriate CLFs for the calculation hours from Table 8.2.

13. To determine the sensible cooling load, multiply the number of occupants in each group by the sensible heat gain per occupant. Multiply the total for each group by the appropriate CLF for each hour to get the cooling load for each calculation time.

14. To get the latent cooling load, multiply the number of occupants in each group by the latent heat gain per occupant. The value should be the same for each calculation time if the occupancy and activity level remains constant for the time period.

15. Obtain the sensible and latent (if any) heat gains for each type and piece of equipment from the nameplate or manufacturer. If this information is not available, use the typical values listed in Tables 8.8, 8.9, 8.10, and/or 8.11.

16. To obtain the sensible cooling loads, determine how long the equipment will be on, and how long the equipment has been on, for each calculation hour. Select an appropriate CLF from Table 8.6 and 8.7. Multiply the equipment heat rejection rate by the watts to Btu/hr conversion (if necessary), and by the CLF, for each calculation hour.

17. Based on whether the equipment is hooded or not, the latent equipment gains may be summed to get the latent cooling load for the space. The value should be the same for each hour unless equipment is turned on and off during the calculation period.

18. Estimate the space infiltration rate in accordance with the methods in Chapter 5. Ventilation air should be handled separately, as discussed in Chapter 9.

19. Obtain the outdoor design conditions from Table 3.1.

20. Use Equation 6.1b, and the indoor/outdoor dry bulb temperature difference to calculate the sensible cooling load. The load should be the same for each hour unless there are significant changes in outdoor temperature during the calculation period.

21. Plot the indoor and outdoor design conditions on a psychrometric chart to get the indoor and outdoor humidity ratios. Refer to Chapter 9 for a discussion on psychrometric charts. Use Equation 8.5 to calculate the latent loads. They are also normally the same for each calculation hour.

22. Add all the internal sensible cooling loads to get an internal subtotal and add all of the internal latent loads to get an internal latent cooling load subtotal.

23. Bring the external sensible cooling load subtotal and the external latent subtotal forward for each calculation time. Add the external subtotals to the internal subtotals to get the space cooling load subtotals. These values may be used to determine the minimum air quantities supplied to the space, in order to satisfy the space cooling load.

24. Estimate any system or return cooling loads in accordance with Section 8.5 and Step 6 above.

25. Add the system or return loads to the space cooling loads, to get the Building Sensible Cooling Load (BSCL) and the Building Latent Cooling Load (BLCL) for each calculation hour. The hour that has the greatest total cooling load is when the peak load occurs, and the total is the peak building or space cooling load. These values, along with the ventilation loads, will be combined to get the Grand Total Loads. See Chapter 9.

INTERNAL AND TOTAL COOLING LOADS

INTERNAL SENSIBLE

	TYPE	TOTAL HOURS ON	START TIME	WATTS TO BTUH	BALLAST AND USE FACTOR	TOTAL GAIN	CLF		
							__HR	__HR	__HR
L I G H T S				3.413					
				3.413					
				3.413					

	HOURS IN SPACE	TIME ENTER	NUM. OF PEOPLE		GAIN/ PERSON	NUM. X GAIN	CLF		
P E O P L E									

	TYPE	HRS. ON	START TIME	HOODED YES/NO	SENSIBLE GAIN	CLF		
E Q U I P								
M I S C								

INFILTRATION–SENSIBLE	TEMP DIFF		
SCFM X 1.10 X T.D.			

INTERNAL SENSIBLE SUBTOTAL LOAD

EXTERNAL SUBTOTAL COOLING LOAD

SPACE SENSIBLE SUBTOTAL LOAD

INTERNAL LATENT LOAD	GAIN/ PERSON	__HR	__HR	__HR
PEOPLE				
INFILTRATION SCFM x 4840 x W. DIFF.				
EQUIPMENT				
			SUBTOTAL	

OASH: SCFM X 1.10 X TEMP. DIFF.

OALH: SCFM X 4840 X W. DIFF.

GRAND TOTAL COOLING LOAD

COOLING LOAD		
___HR	___HR	___HR

Chapter 9

HVAC PSYCHROMETRICS

Air conditioning involves the control of both the temperature and moisture content of the air in a conditioned space. The space temperature and moisture content are controlled by various thermodynamic processes provided by the HVAC system. Psychrometrics is the study of the thermodynamics processes of air and the moisture that is contained in air.

The proper control of the space temperature, as well as the humidity, has a direct effect on occupant comfort. Section 1.1 discussed some of the requirements for providing a comfortable indoor environment. The total cooling load for an air conditioning system is the sum of the space sensible and latent cooling loads, as well as sensible and latent loads from the introduction of outdoor ventilation air into the system. Air, as we know it, is actually a mixture of several substances, including gases and vapors. Dry air contains no moisture. Its two main components are gaseous nitrogen and gaseous oxygen. Nitrogen accounts for about 78% by volume and oxygen about 21%. Table 9.1 shows the gases that make up atmospheric air.

Atmospheric air contains small amounts of water in the form of vapor. The average amount of water vapor in air is about 0.08% by weight at 0°F, and about 1.56% at 70°F when the air is fully saturated with water vapor. Atmospheric air also contains some particulate matter, such as dust and pollen.

Standard atmospheric air is considered to have a temperature of 60°F, and a density of 0.075 lb/ft³ of dry air, at a pressure of 14.7 psia. Although these values may vary, standard air may be assumed for most HVAC calculations.

Since air is a mixture of dry air and water vapor, it is necessary to specify two independent properties in order to define the condition of air for any given pressure. These two properties are the temperature and the moisture content of the air. These two properties are necessary to define the condition, or **"state point,"** of the air. Methods for determining the state point of air for various thermodynamic and air conditioning processes will be discussed in the following sections of this chapter.

9.1 AIR AND WATER VAPOR MIXTURES

Since air is composed of a number of gases, it may be treated as an "ideal" or "perfect gas." A perfect gas is one in which the molecules are considered to be perfectly elastic, the volume occupied by the molecules is very small in comparison with the total volume, the

TABLE 9.1   Gaseous Composition of Dry Air

Gas	% by Weight	% by Volume
Nitrogen	75.47	78.03
Oxygen	23.19	20.99
Argon	1.29	.94
Carbon Dioxide	.05	.03
Hydrogen	.00	.01
Xenon, Krypton and other gas	minute portions	

attractive forces between adjacent molecules are very small, and the molecules move in random directions.

An ideal gas is also one that follows the **ideal gas law** (an equation of state):

$$PV = MRT \qquad (9.1)$$

where

P = the absolute pressure in psia

V = the volume in ft^3

M = the mass in lb mass

R = a gas constant

T = the absolute temperature in °R

Vapors, such as water vapor (moisture), do not follow the ideal gas law. This is the main difference between vapors and gases. Although vapors do not follow the ideal gas law, in some cases they come close to acting like ideal gases. Therefore, vapors may usually be treated as ideal gases with a reasonable degree of accuracy.

9.1.1   Partial Pressures of Mixtures

According to Dalton's Law of Partial Pressures, the pressure exerted by each gas in a mixture of gases is called the partial pressure of that gas. Equation 9.2 is Dalton's Law.

$$P_t = p_a + p_b + p_c \cdots \qquad (9.2)$$

where

P_t = the total pressure of the mixture

p_a = partial pressure exerted by gas A

p_b = partial pressure exerted by gas B, etc.

The total pressure of an air-vapor mixture is the sum of the partial pressures of each gas and the water vapor. The total barometric pressure of an air-vapor mixture can be expressed as

$$P_b = p_a + p_v \qquad (9.3)$$

where

P_b = the total barometric pressure

p_a = the partial pressure of dry air

p_v = the partial pressure of the water vapor in the mixture

9.1.2 Evaporation

All liquids have a tendency to evaporate. Some liquids, such as rubbing alcohol or gasoline, evaporate quickly. Other liquids, such as mercury, evaporate extremely slowly.

A liquid evaporates because there is always some of the liquid present in the air immediately above it as a vapor. This vapor, which is in contact with the liquid, has a pressure. This pressure is the equilibrium vapor pressure, or just vapor pressure.

When the partial pressure of the vapor and liquid are equal, evaporation does not take place. The number of molecules leaving the liquid and entering the vapor is equal to the number leaving the vapor and entering the liquid.

Evaporation occurs when some of the vapor diffuses or is blown away. When this happens, more of the liquid will turn to vapor in order to maintain the partial pressure of the vapor above the liquid. Evaporation will continue until the partial pressure of the vapor in the air is equal to the pressure of the liquid water. At this point evaporation will cease, the partial pressure of the vapor and the pressure of the water will be in equilibrium, and the air-vapor mixture will be "saturated."

The maximum amount of water vapor that air can hold in suspension at any given pressure is a function of its dry bulb temperature. The air-vapor mixture is said to be saturated when it can no longer hold any additional moisture at that temperature. The moisture that is in suspension is on the verge of condensing out.

The evaporation of water requires latent heat. As a result of the evaporation of water into the air, the water tends to cool as it surrenders the necessary heat for evaporation. This heat is called the latent heat of vaporization.

9.1.3 Condensation of Water Vapor

The condensation of water, from an air-vapor mixture, requires the same amount of latent heat as evaporation for the same set of conditions. The main differences are that the net flow of water molecules is out of the air-vapor mixture and the flow of latent heat is out of the air-vapor mixture.

In order for condensation to occur, the air-vapor mixture must be cooled to the saturation point. The saturation point occurs at the saturation temperature of the mixture. The **saturation temperature** (or **dew point temperature**) is the temperature at which the vapor begins to condense.

When an air-vapor mixture comes in contact with a cool surface, the vapor in the air will have a higher vapor pressure than the equivalent vapor pressure for the temperature of the cooler surface. This will result in a flow of water vapor molecules to the surface. As more molecules flow to the surface, they will combine and waterdroplets will begin to form on the surface. The sole criterion as to whether evaporation or condensation will take place is the relation between the dew point temperature of the air-vapor mixture, and the temperature of the water on a surface in contact with the air mixture.

9.1.4 Relative Humidity (ϕ)

Humidity is the moisture or water vapor occupying the same space as the air in an air-vapor mixture. Absolute humidity, or **humidity ratio** (W), is the weight of the water vapor per unit volume of dry air. It is usually expressed in pounds of moisture, per pound of dry air. There are 7000 grains of moisture in a pound of water.

The relative humidity of an air/vapor mixture is defined as the ratio of the pressure of the vapor in the mixture to the saturation pressure of the vapor at the same temperature or the temperature of the mixture. The relative humidity is expressed as a percentage and is given in Equation 9.4.

$$\phi = P_v/P_g \qquad\qquad (9.4)$$

where

P_v = pressure of the vapor in the air-vapor mixture

P_g = saturation pressure of the vapor

Relative humidity is an important measure of the amount of moisture in an air/vapor mixture for any given temperature. It is the most commonly used measurement for expressing the moisture content in air/vapor mixture. The higher the % of relative humidity, the higher the moisture content at a given temperature. If the temperature of the mixture is increased with no change in the moisture content of the mixture, relative humidity will decrease. This can be seen graphically on a psychrometric chart that will be discussed later in this chapter.

9.1.5   Dry Bulb and Wet Bulb Temperature

The dry bulb temperature of an air-water vapor mixture is simply the temperature that would be read from an ordinary thermometer. Changes in dry bulb temperature would indicate a change in sensible heat. The wet bulb temperature of an air-vapor mixture is a measure of the humidity level, or moisture content, of the mixture at a given dry bulb temperature. The dry bulb and wet bulb temperatures define the state or identify a state point of the mixture.

The wet bulb temperature of air is measured using a psychrometer. A psychrometer is a thermometer with a thin cloth (wick), wet with distilled water, covering the bulb of the thermometer (hence the term wet bulb temperature). Air is passed over the bulb and wick at a prescribed rate. If the water vapor in the air is not saturated, evaporation takes place, and latent heat is transferred to the moving air, with a resultant lowering of the water in the wick, until an equilibrium temperature is reached. The temperature at equilibrium is the wet bulb temperature of the air-vapor mixture. The difference between the dry bulb temperature and the wet bulb temperature is called the wet bulb depression. If the air-vapor mixture is at saturation, very little evaporation will take place. At this state, the dry bulb and wet bulb temperatures will be nearly equal. On the other hand, if the moisture in the mixture is low, there will be a high rate of evaporation and a lower wet bulb temperature will be measured.

A common device for measuring dry bulb and wet bulb temperatures is the sling psychrometer shown in Figure 9.1. The sling psychrometer has two thermometers one for measuring dry bulb temperature and the other for measuring wet bulb temperature. The thermometer for measuring the wet bulb temperature has a cloth wick on the bulb, which is wetted by dipping it in water. The thermometers are then whirled around to create the necessary air movement. The sling psychrometer is whirled until the two temperatures, the dry bulb and wet bulb, remain constant.

9.1.6   Enthalpy of Air-Vapor Mixtures

In Section 2.2.3, the thermal property of a substance known as enthalpy was discussed. **Enthalpy** (h) is a measure of the internal energy or total heat content of a substance.

The enthalpy of a mixture of perfect (ideal) gases, or of an air-vapor mixture, is the sum of the enthalpies of each constituent gas or vapor. Therefore, the enthalpy of an air-vapor mixture is actually the total heat of the mixture. The customary units for an air-vapor mixture are Btu/lb of dry air.

Since enthalpy is a measure of the total heat content of the air-vapor mixture, it is used extensively in analyzing HVAC processes as they add or remove sensible and latent heat to/from air-vapor mixtures.

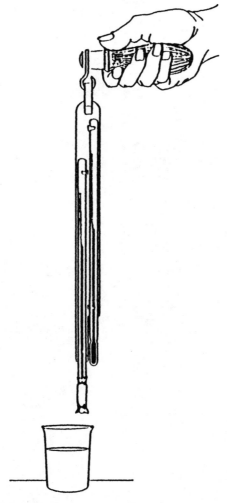

FIGURE 9.1 Sling Psychrometer

9.2 THE PSYCHROMETRIC CHART

Psychrometrics can be defined as the science involving the thermodynamic properties of moist air, and the effect of atmospheric moisture on materials and human comfort. A psychrometric chart is a plot of the psychrometric properties of moist air. A psychrometric chart is shown in Figure 9.2. It is used to graphically display the thermodynamic processes used in HVAC systems.

The chart is for the normal temperature range of HVAC processes (32°F to 105°F dry bulb) and for the standard barometric pressure of 14.7 psia (29.92 in mercury). If the total pressure of the air differs significantly from 14.7 psia, serious errors will result in the use of this particular chart. For other pressures, different charts should be used or corrections should be applied.

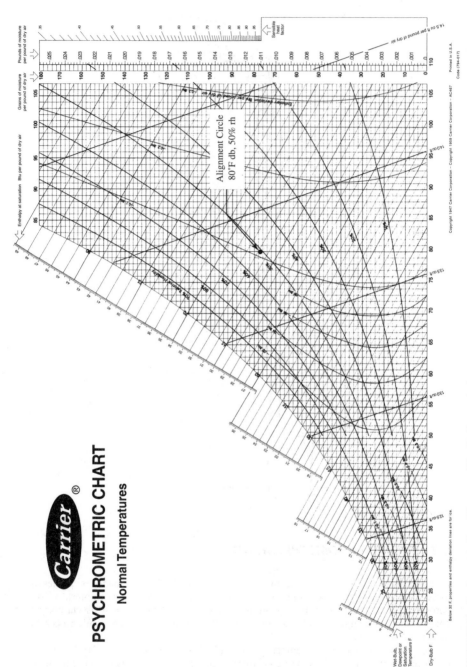

FIGURE 9.2 Psychrometric Chart (*Courtesy of Carrier Corporation*)

The chart is also useful in that if two properties of the air are known, such as the dry bulb and wet bulb temperatures, a state point has been defined and all the other properties for that state point may be obtained from the psychrometric chart. The state point would be located on the chart at the intersection of the two known properties.

A skeleton psychrometric chart that indicates where the various properties, scales, and features are located is shown in Figure 9.3.

Dry Bulb Temperature (Line 1): The temperature read on a standard thermometer in degrees Fahrenheit dry bulb (°Fdb) (see Section 9.1.5). Lines of constant dry bulb temperature are vertical straight lines on the chart and the temperature scale is across the bottom of the chart.

Wet Bulb Temperature (Line 2): The temperature read with a psychrometer (see Section 9.1.5). Lines of constant wet bulb temperature are straight lines that slope upward to the left. Wet bulb temperatures are given in degrees Fahrenheit wet bulb (°Fwb) and are read along the curve on the upper left side of the chart.

Humidity Ratio, W (Line 3): The ratio indicates the absolute moisture content of the air-vapor mixture, in grains of moisture per pound of dry air, or in Btu/lb of dry air (see Section 9.1.4). Lines of constant humidity ratio, or absolute moisture content, are horizontal lines. The humidity ratio is read to the right side of the chart.

Relative Humidity, ϕ (Line 4): Lines of constant relative humidity are the curved lines inside the chart that curve upward to the right (see Section 9.1.4).

Specific Volume (Line 5): The volume of the air-vapor mixture per unit weight is given in ft³/lb of dry air. The lines slope diagonally upward to the left.

Saturation Curve (Line 6): The saturation curve, or dew point, is a line at which the dry bulb and wet bulb temperatures are nearly equal, resulting in a saturated air-vapor mixture (see Sections 9.1.2 and 9.1.3). An air-vapor mixture must be cooled (have its dry bulb temperature lowered) to this point before any significant condensation will occur.

Enthalpy Scale (Scale 7): Lines of constant enthalpy of the air-vapor mixture.

Sensible Heat Ratio (Scale 8): The sensible heat ratio (SHR), or sensible heat factor, is the ratio of the sensible heat to the total (sensible and latent) heat in an HVAC process. SHR is read from the scale to the right of the chart.

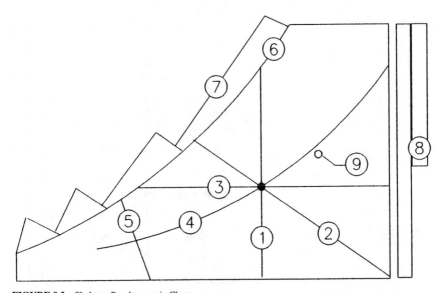

FIGURE 9.3 Skeleton Psychrometric Chart

Alignment Circle (Point 9): A point located inside the chart used in conjunction with SHR to plot various psychrometric processes. On the Carrier Psychrometric Chart, the alignment circle (or point) is located at 80°Fdb and 50% relative humidity (RH). Some other charts locate the alignment circle at 78°Fdb and 50% RH.

9.3   HVAC PROCESSES

Figure 9.4 graphically shows the typical processes used in HVAC systems on a skeleton psychrometric chart. The lines show the variation, in properties that the air-vapor mixture would pass through in going from one state point to another. State point 0 is assumed to be the beginning point of the process. The processes in actual HVAC systems are frequently more complex and involved than those shown in Figure 9.4. However, most HVAC processes can be analyzed by combinations of the basic processes. Any change in state point of the air-vapor mixture can be broken down into sensible and latent heat changes. These changes can be plotted on a psychrometric chart.

From this point on, the air-vapor mixture will just be referred to as air. Referring to Figure 9.4, the basic processes may be observed.

Humidification Only (state point _O_ to _A_) and Dehumidification (state point _O_ to _E_): These processes involve an increase (_O_ to _A_) or a decrease (_O_ to _E_) in the humidity ratio, with no change in the dry bulb temperature of the air. These are purely changes in latent heat with no change in sensible heat content.

Sensible Heating Only (state point _O_ to _C_) and Sensible Cooling Only (state point _O_ to _G_): These processes involve an increase (_O_ to _C_) and a decrease (_O_ to _G_) in the dry bulb temperature of the air. These are purely changes in sensible heat with no change in the moisture content. The humidity ratio (_W_) remains constant.

Combined Cooling and Dehumidification (state point _O_ to _F_): A result of the reduction in both dry bulb and wet bulb temperatures. Cooling coils normally perform a process like this, where both sensible and latent heat of the air decrease.

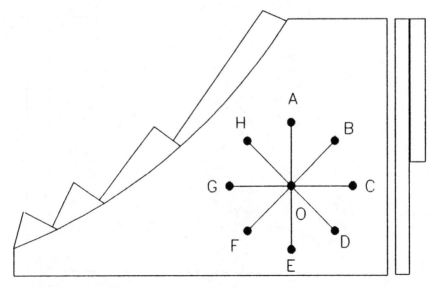

FIGURE 9.4   Basic HVAC Processes

Combination Heating and Humidification (state point *O* to *B*): A result of an increase in both dry bulb and wet bulb temperatures of the air. In this process, both sensible and latent heat are added to the air.

Evaporative Cooling Only (state point *O* to *H*): A process in which cooling and humidification occur. The wet bulb temperature of the air remains constant as the dry bulb temperature is reduced. This is an adiabatic (no net heat transfer) process. The process removes sensible heat, with the heat going to vaporize an additional amount of water added (added latent heat).

Chemical Dehumidification (state point *O* to *D*): A process in which moisture is removed from the air chemically. This is a constant enthalpy process, with the latent cooling nearly equal to the sensible heat increase.

Figure 9.5 shows a schematic of a typical HVAC system. A typical air conditioning unit includes a circulating fan, with a motor and a cooling coil, through which air is passed. The unit may or may not include a means of heating the circulated air. The cooling coil may be a chilled water coil or a direct expansion (DX) coil with a refrigerant.

The return air (ra) volume is usually less than the supply air volume, due to space exhaust, exfiltration out of the building, and relief air at the unit (air exhausted at the unit in order to provide an adequate outdoor air ventilation rate). The outdoor air (oa) volume is the difference between the supply air and return air volume. The outdoor air volume is usually set by codes (see Section 5.1). Equation 9.5 gives the relationship between the various air quantities.

$$Q_{sa} = Q_{ra} + Q_{oa} \qquad (9.5)$$

where

Q_{sa} = supply air volume in ft³/min

Q_{ra} = return air volume in ft³/min

Q_{oa} = outdoor air in volume in ft³/min

The minimum supply air volume may be determined by the methods described later in this chapter or by code requirements.

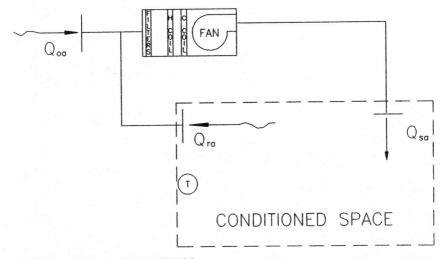

FIGURE 9.5 Schematic of Typical HVAC System

The sensible heat ratio (SHR), or factor (SHF), was discussed briefly in a previous section of this chapter. The room sensible heat factor (RSHF) is the ratio of the room sensible heat (cooling load) to the total room heat (sensible plus latent). Equation 9.6 is the equation for RSHF.

$$\text{RHSF} = \frac{q_{rs}}{q_{rs} + q_{rl}} = \frac{q_{rs}}{q_{rt}} \qquad (9.6)$$

where

q_{rs} = room sensible heat or cooling load in Btu/hr

q_{rl} = room latent heat or cooling load in Btu/hr

q_{rt} = room total heat or cooling load in Btu/hr

When plotted on a psychrometric chart, the slope of the RSHF line graphically shows the ratio of room sensible load to the room total load. A typical sensible heat factor line is shown between state points 3 and 4 in Figure 9.6.

When outdoor air is brought into an HVAC system, the system must remove the cooling loads due to the outdoor air as well as the space cooling loads. The Grand Total Cooling Load is the sum of the space (sensible and latent) cooling loads, as well as the outdoor (sensible and latent) loads. The Grand Total Sensible Heat Factor (GSHF) is the ratio of the total sensible heat (space plus outdoor) cooling load to the Grand Total Cooling Load. Equation 9.7 is the equation for GSHF.

$$\text{GSHF} = \frac{q_{st}}{q_{st} + q_{lt}} = \frac{q_{st}}{q_{gt}} \qquad (9.7)$$

where

q_{st} = total sensible load in Btu/hr

q_{lt} = total latent load in Btu/hr

q_{gt} = grand total cooling load in Btu/hr

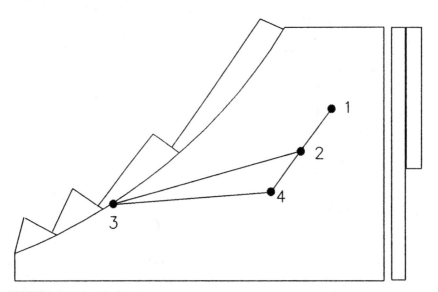

FIGURE 9.6   Typical Cooling Processes

When plotted on a psychrometric chart, the slope of the GSHF line graphically shows the ratio of the total system sensible loads to the Grand Total Cooling Loads. A typical GSHF line is shown between state points 2 and 3 in Figure 9.6.

Figure 9.6 shows the air state points usually found in an air cooling and dehumidifying process. As the conditions of the air pass from one state point to another, it is assumed that the process is really a series of state points that make up the interconnecting lines.

Ambient Outside Air Conditions (state point 1): This point is located, or described, by the design dry bulb and wet bulb temperatures for the location of the building. The point is located by the intersection of the dry bulb and wet bulb temperature lines. These values are obtained from climatic data (see Chapter 3).

Space Design Conditions (state point 4): These are the design conditions that the HVAC system will be designed to maintain within the conditioned space. These are usually a design dry bulb temperature and relative humidity. The point is plotted at the intersection of the dry bulb temperature and the percent relative humidity line.

Mixed Air Conditions (state point 2): The conditions at this point are a function of the outdoor air conditions, the return air conditions, and the quantities of outside and return air. The dry bulb temperature of the mixed air may be calculated with Equation 9.8.

$$T_{mdb} = \frac{Q_{oa}T_{oa} + Q_{ra}T_{ra}}{Q_{ra} + Q_{oa}} \tag{9.8}$$

where

T_{mdb} = mixed air temperature in °Fdb

Q_{oa} = outside air flow rate in ft³/min

T_{oa} = outside air dry bulb temperature in °Fdb

Q_{ra} = return air flow rate in ft³/min

T_{ra} = return air dry bulb temperature in °Fdb

It is usually necessary to estimate the return air temperature in Equation 9.8. Since the HVAC system should maintain the room design conditions in the space, the room design temperature may be assumed as a reasonable value of the return air temperature. If there are unusually large return cooling loads, it may be necessary to calculate a return air temperature, based on the return air flow rate and return cooling loads.

Since it is known that the mixed air conditions are between the space design conditions and the outdoor air conditions, the mixed air state point must be on a line between the indoor and outdoor conditions (state points). By locating the T_{mdb} value on the line between state points 1 and 4, state point 2 is defined and the mixed air wet bulb temperature (T_{mwb}) may be read from the psychrometric chart.

Leaving Air Conditions (state point 3): This point is, or is very near, the condition of the air as it comes out of the cooling coil. Notice that the point is at, or near, the saturation (dew) point. In order for any significant dehumidification to occur, the air must be cooled near the dew point temperature. As the air is cooled, it is cooled along the line between the mixed air conditions (state point 3). The slope of this line is parallel to the Grand Sensible Heat Factor (GSHF) of the mixed air.

When the conditioned air is supplied to the space, it is usually considered to be at, or near, the leaving air conditions (state point 3) of the cooling coil. As the air is supplied to the space, it heats up along the line from state point 3 to the room conditions at state point 4. Assuming that adequate air is supplied to handle the cooling loads, the line is parallel to the Room Sensible Heat Factor (RSHF). Methods for determining the required leaving air conditions and the required air flow rate are discussed later in this chapter.

Example 9.1 An air conditioning system is serving a space with a room sensible cooling load of 75,000 Btu/hr and latent cooling load of 5000 Btu/hr. Outdoor air is brought into the system, resulting in a sensible cooling load of 7500 Btu/hr and a latent cooling load of 4000 Btu/hr. What is the Room Sensible Heat Factor and the Grand Sensible Heat Factor?

From Equations 9.6 and 9.7

$$\text{RSHF} = \frac{75000}{75000 + 5000} = 0.94$$

$$\text{GSHF} = \frac{75000 + 7500}{75000 + 7500 + 5000 + 4000} = 0.90$$

Example 9.2 An air conditioning system is supplying 10,000 ft³/min of air to a space that is to be maintained at 75°Fdb and 50% RH. Assume that the return air dry bulb temperature is equal to the room air dry bulb temperature. Outside air is brought into the system at 95°Fdb and 76°Fwb at a rate of 2000 ft³/min. What are the conditions entering the cooling coil (the mixed air conditions), and what is the humidity ratio at each state point?

The outdoor air state point and indoor state points are plotted on a psychrometric chart and a straight line is drawn between the two points. From Equation 9.5, the return air flow is calculated as 8000 ft³/min. Equation 9.5 is used to calculate the return air flow as follows:

$$Q_{ra} = 10,000 \text{ ft}^3/\text{min} - 2000 \text{ ft}^3/\text{min} = 8000 \text{ ft}^3/\text{min}$$

Using Equation 9.8, the mixed air dry bulb temperature may be determined as follows:

$$T_{mdb} = \frac{8000(75) + 2000(95)}{10,000} = 79°\text{Fdb}$$

The mixed air wet bulb temperature may be determined by plotting 79°Fdb on the line between the indoor and outdoor state points, as shown in Figure 9.7.

From the psychrometric chart, the entering wet bulb temperature may be read as 65.5°Fwb (approx.). The humidity ratios for the state points may also be read directly from the right side of the psychrometric chart.

Outdoor: W = 0.0150 lb water/lb dry air

Mixed Air: W = 0.0106 lb water/lb dry air

Space Air: W = 0.0093 lb water/lb dry air

9.4 APPLICATION OF PSYCHROMETRIC TO COOLING LOAD CALCULATIONS

The final system cooling load calculations involve estimating the amount of air flow in dry air that will handle the space cooling loads, as well as the outdoor air cooling loads. The calculations also involve calculating the conditions, or state points, of the mixed air stream and the conditions leaving the cooling coil. All of this data is necessary to accurately size, select, and specify air conditioning equipment.

The requirements for determining the minimum outdoor air ventilation rate are usually determined by code requirements, as discussed in Chapter 5. Methods for calculating the space (room) sensible cooling load, space latent cooling load, return air sensible cooling load, building sensible cooling load, and building latent cooling load were discussed in Chapters 7 and 8. Section 8.4 presented methods for calculating cooling loads for outdoor air entering a conditioned space by infiltration. The same methods may be used to calculate cooling loads for outdoor ventilation air brought into an HVAC system.

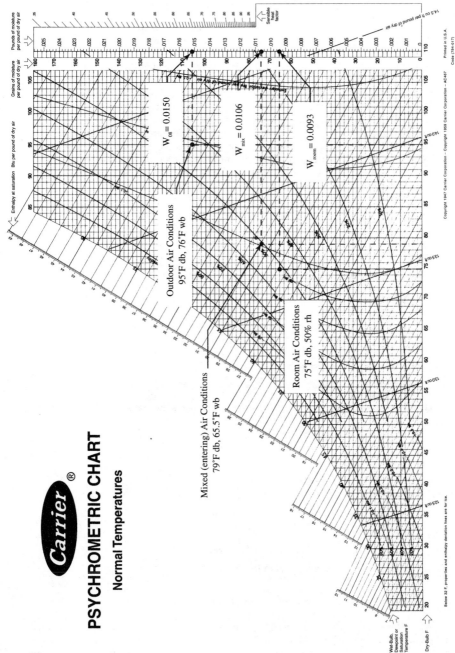

FIGURE 9.7   State Points for Example 9.2

9.4.1 Outdoor Air Cooling Load

The minimum amount of outdoor ventilation air is usually set by code requirements. In some cases, such as locker rooms or kitchens, the minimum outside ventilation air brought in by the HVAC system is determined by the amount of air exhausted from the space served by the HVAC system. The amount of outdoor air brought in by the HVAC system should almost always be equal to or greater than the exhaust air quantity for the space. Negative pressures in buildings should be avoided, to reduce infiltration and other operating problems.

The loads imposed by the introduction of unconditioned outdoor air into the space were discussed in Chapters 6 and 8. Just as infiltration loads must be handled by the HVAC system, so must the outdoor air ventilation loads. The main difference is that the infiltration heat gains are loads on the conditioned space while outdoor air loads are loads on the HVAC system. This chapter discusses the effect of outdoor air loads on the system.

Equation 6.1b may be used to calculate the outdoor air sensible heat and Equation 8.5 may be used to calculate the outdoor air latent heat.

$$q_{soa} = 1.1 \, Q_{oa} \, (T_{oa} - T_r) \tag{6.1b}$$

$$q_{loa} = 4840 \, Q_{oa} \, (W_{oa} - W_r) \tag{8.5}$$

where

q_{soa} = outside air sensible heat in Btu/hr

Q_{oa} = outdoor air flow in ft^3/min

T_{oa} = outside air temperature in °Fdb

T_r = room or space air temperature in °Fdb

Q_{loa} = outside air latent heat in Btu/hr

W_{oa} = outside air humidity ratio in lb water/lb dry air

W_r = room or space humidity ratio in lb water/lb dry air

The Grand Total Cooling Load is the sum of the space loads, return loads, and outdoor air loads.

$$q_{gt} = q_{rs} + q_{rl} + q_{ras} + q_{oas} + q_{oal} \tag{9.9}$$

where

q_{gt} = grand Total Cooling Load in Btu/hr

q_{rs} = room or space sensible cooling load in Btu/hr

q_{rl} = room or space latent cooling load in Btu/hr

q_{ras} = return air sensible cooling load in Btu/hr

q_{oas} = outside air sensible cooling load in Btu/hr

q_{oal} = outside air latent cooling load in Btu/hr

9.4.2 Bypass Factor and Effective Sensible Heat Factor

Whenever air is passed through a cooling coil, most of the air is affected by being cooled and usually dehumidified. It is inevitable, however, that some of the air passing through the coil will not be affected. The amount of air that passes through the coil unchanged is called bypassed air and the fraction of the total air flow bypassed is called the bypass factor. Typical bypass factors for various cooling applications are given below.

0.30 to 0.50:   Residential air conditioners, where cooling loads are not great and cooling is purely for comfort.

0.30 to 0.20:   Small commercial applications, usually using small packaged air conditioning units.

0.10 to 0.20:   Medium to large packaged air conditioning systems, such as packaged rooftop air conditioning units with standard cooling coils.

0.05 to 0.10:   Built-up air conditioning units with multiple row cooling coils specifically selected for the application. Used where fairly precise control of space temperature and humidity are required. Used also for moderately high latent cooling loads.

0 to 0.10:   Built-up air conditioning units with a many row cooling coils with close fin spacing. Used for process air conditioning, high latent cooling loads, and/or all outdoor air.

Cooling coils that have many rows and close fin spacing have a larger surface area for heat transfer from the flowing air to the cooling coil. These types of coils tend to have low bypass factors. Coils with comparatively wide fin spacing and few rows tend to have higher bypass factors. The velocity of the air passing through the coil also affects the bypass factor. Typical air face velocities through cooling coils range from 300 to 800 ft/min. Higher face velocities tend to result in higher bypass factors, while lower velocities result in lower bypass factors. The amount of coil surface area has a greater effect on bypass factors than does the face velocity.

The actual surface temperature of a cooling coil with chilled water, or with refrigerant, is not uniform, but varies with the location on the cooling coil. On a multi-row cooling coil, the surface nearest the entering air is in contact with the warmest air and, therefore, warmer than the average or mean surface temperature. The coil surface near the discharge side of the coil is in contact with air that has already been cooled to some extent. The coil surface temperatures, near the discharge side of the coil, tend to have a lower surface temperature.

In order to analyze the effect of a cooling coil on the air passed through the coil, it is necessary to assume an effective coil surface temperature which would produce the same leaving conditions as the actual non-uniform surface temperatures. For cooling and dehumidifying situations, the effective coil surface temperature is assumed to be where the GSHF line crosses the saturation, or dew point curve, on the psychrometric chart. This point is commonly called the **apparatus dew point temperature (adp)**.

Effective Sensible Heat Factor (ESHF) is the ratio of the effective room (space) sensible heat to the effective room total heat. The effective room sensible heat is the room sensible heat (cooling load), plus the outdoor ventilation air sensible heat that is assumed to be bypassed by the cooling coil. Equation 9.10 gives the effective room sensible heat.

$$q_{esh} = q_{rs} + (BF)q_{oas} \qquad (9.10)$$

where

q_{esh} = the effective room sensible heat in Btu/hr

q_{rs} = the room sensible heat in Btu/hr

BF = coil bypass factor

q_{oas} = outside air sensible heat in Btu/hr

The effective room latent heat may be found in a similar fashion by modifying Equation 9.10 as follows:

$$q_{erl} = q_{rl} + \text{BF}(q_{oal}) \tag{9.10a}$$

where

q_{erl} = effective room latent heat (Btu/hr)

q_{rl} = room latent heat (Btu/hr)

BF = coil bypass factor

q_{oal} = outside air latent heat (Btu/hr)

It is assumed that the outdoor air bypassed through the cooling coil becomes a cooling load on the conditioned space or room. This additional load affects the required cooling capacity of the HVAC system by increasing the amount of conditioned air that must be supplied to the space in order to handle the additional load. The Effective Sensible Heat Factor (ESHF) may be calculated using Equation 9.11.

$$\text{ESHF} = \frac{q_{esh}}{q_{esh} - q_{ehl}} = \frac{q_{rs} + (\text{BF})q_{oas}}{q_{rs} + (\text{BF})q_{oas} + q_{rt} + (\text{BF})q_{oal}} \tag{9.11}$$

Example 9.3   A packaged rooftop air conditioning unit is serving a space that is maintained at 75°Fdb/50% RH, has a sensible cooling load of 50,000 Btu/hr, and a latent cooling load of 2500 Btu/hr. The unit coil has a typical bypass factor of 0.15, without any return loads. Air is supplied to the space at the rate of 2300 ft³/min. Outdoor ventilation air is brought into the system at a rate of 460 ft³/min, at 94°Fdb/75°Fwb. What is the grand total heat and the ESHF for the space?

Using Equation 6.16, the outside air sensible heat is?

$$q_{soa} = 1.1(460)(94 - 75) = 9614 \text{ Btu/hr}$$

Plotting the room and outdoor air state points on a psychrometric chart, the indoor and outdoor humidity ratios may be determined. Using Equation 8.5, the outside air latent heat is?

$$q_{oal} = 4840(460)(0.0144 - 0.0092) = 11,577 \text{ Btu/hr}$$

The grand total cooling load may be calculated using Equation 9.9.

$$q_{gt} = 50000 + 2500 + 9614 + 11577 = 73,691 \text{ Btu/hr}$$

For a packaged rooftop air conditioning unit, a typical bypass factor is 0.15 for the cooling coil. The Effective Sensible Heat Factor may be calculated using Equation 9.11.

$$\text{ESHF} = \frac{50000 + (0.15)(9614)}{50000 + (0.15)(9614) + 2500 + (0.15)(11577)} = 0.92$$

9.4.3 The Conditioned Air Supply

The minimum amount of air supplied to a conditioned room or space is usually a function of the sensible cooling load in the space. During periods of low cooling loads, the minimum air flow may be based on ventilation code requirements.

For design purposes, the conditioned air supply to a space is based on the sensible cooling loads for the space, since cooling is usually controlled by a thermostat which senses the dry bulb temperature of the air in the space. In order to account for the outdoor air that is bypassed through the cooling coil and becomes a cooling load in the space, the conditioned air supply calculations for the space are based on the space, or room, effective sensible heat.

The total air flow for an entire system is determined in a similar fashion. A simplified one step method is presented as a calculation method for the conditioned air supply in this section. This involves the use of the effective total sensible heat (space and outdoor), the Grand Sensible Heat Factor (GSHF), the coil bypass factor, and the apparatus dew point temperature. The reader may consult References 1 and 7 for a more detailed discussion of the calculations for the conditioned air supply.

Since the GSHF is the relationship of the effective total sensible heat to the grand total heat, the GSHF describes how the air will be cooled and dehumidified as it passes through the cooling coil, losing both sensible and latent heat. When plotted on a psychrometric chart, the slope of the GSHF line graphically shows the relationship of the sensible heat to the total heat.

In order to determine the required apparatus dew point (adp) conditions, it is necessary to plot the ESHF line on a psychrometric chart. First, the ESHF must be calculated from the space and outside air cooling loads. The value of the ESHF is then plotted on the SHF scale on the right side of the psychrometric chart. A straight line is then drawn between the ESHF on the SHF scale and the alignment circle near the center of the chart. On the psychrometric chart shown in Figure 9.2, the alignment circle is located at 80°Fdb/50% RH. A point locating the room air conditions is then plotted on the psychrometric chart. A second line that passes through the room air state point is then drawn *parallel* to the first ESHF line. The second line is extended to where it intersects the saturation curve. The point where the second line intersects the saturation curve defines the required apparatus dew point (adp) conditions for the given loads. The adp defines the effective coil surface temperature. The line through the room's air state point and the adp describes how the air absorbs the sensible and latent heat after it enters the space.

The minimum amount of conditioned air that must be supplied to the space is based on the total effective sensible heat. Equation 9.12 may be used to calculate the minimum quantity of air that must be supplied in order to handle the total sensible cooling load.

$$Q_{da} = \frac{q_{esh}}{1.1(T_{rdb} - T_{adp})(1 - BF)} \tag{9.12}$$

where

Q_{da} = the rate of dry conditioned air supplied to the space in ft³/min

q_{esh} = the effective total sensible heat in Btu/hr

T_{edb} = the space or room design dry bulb temperature in °Fdb

T_{adp} = the apparatus dew point dry bulb temperature in °Fdb

BF = the cooling coil bypass factor

Since some of the outside air is bypassed when the air flows through the cooling coil, part of the outside air cooling load becomes a cooling load on the space by increasing the temperature of the air leaving the cooling coil. This outside air brought into the space is similar to outside air infiltration into the space. Equation 9.12 incorporates this additional load into the calculation with the bypass factor and effective sensible heat. This results in a coil discharge dry bulb temperature that is slightly higher than the adp temperature of the coil. Equation 9.13 may be used to calculate the leaving air dry bulb temperature.

$$T_{ldb} = T_{adp} + BF(T_{mdb} - T_{adp}) \tag{9.13}$$

where

T_{ldb} = leaving dry bulb temperature (°F)

T_{adp} = apparatus dew point temperature (°F)

BF = coil bypass factor

T_{mdb} = mixed air dry bulb temperature (°F)

In order to determine the leaving wet bulb temperature, the leaving dry bulb temperature from Equation 9.13 may be plotted on the GSHF line to determine the coil leaving air state point. The leaving air wet bulb temperature may be read directly from the psychrometric chart.

Example 9.4 A space served by a built-up air handling unit with a multi-row cooling coil has a sensible cooling load of 110,000 Btu/hr and a latent load of 5,500 Btu/hr. The coil has a bypass factor of 0.10. The space is to be maintained at 75°Fdb/50% RH. Outside air is brought in through the system at a rate of 1000 ft³/min and the conditions are 94°Fdb/75°Fwb.

What is the required total cooling capacity of the air conditioning unit? How much air must be supplied by the air handling unit to satisfy the cooling loads? What is the coil apparatus dew point temperature and what are the conditions of the air as it enters and leaves the cooling coil?

The indoor and outdoor state points are first plotted on the psychrometric chart and a straight line is drawn between them. The outside air sensible heat is calculated using Equation 6.1b. The indoor and outdoor humidity ratios are read from the psychrometric chart, and Equation 8.5 is used to calculate the latent cooling load. The state points are shown on a psychrometric chart in Figure 9.8.

$$q_{oas} = 1.1(1000)(94 - 75) = 20,900 \text{ Btu/hr}$$

$$q_{oal} = 4840(1000)(0.0146 - 0.0092) = 26,136$$

$$\text{(which may be rounded to 26,150 Btu/hr)}$$

The total required cooling capacity is calculated from Equation 9.9.

$$q_{gt} = 110,000 + 5500 + 20,900 + 26,150 = 162,500 \text{ Btu/hr}$$

The multi-row cooling coil has a bypass factor of 0.10. Equation 9.11 is used to calculate the ESHF.

$$\text{ESHF} = \frac{110,000 + .10(20,900)}{110,000 + 0.10(20,900) + 5,500 + .10(26,150)} = 0.93$$

The value of ESHF = 0.93 is then plotted on the Sensible Heat Factor Scale on the right side of the chart and a line drawn through the alignment circle. A second line, parallel to the first line, is drawn from the space design state point to the saturation curve to get the cooling coil adp temperature. See Figure 9.8.

$$T_{adp} = 54°\text{Fdb (approximate)}$$

The conditioned air supply rate may then be calculated using Equation 9.12.

$$Q_{da} = \frac{110,000 + 0.10(20,900)}{1.1(75 - 54)(1 - 0.10)} = 5,390 \text{ ft}^3/\text{min}$$

The mixed (entering) air dry bulb temperature may be estimated from Equations 9.5 and 9.8.

$$Q_{ra} = 5390 - 1000 = 4390 \text{ ft}^3/\text{min}$$

$$T_{mdb} = \frac{4390(75) + 1000(94)}{5390} = 78.5°\text{Fdb}$$

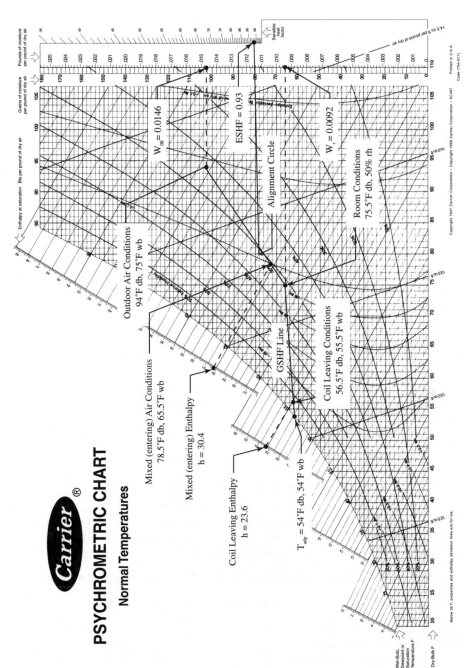

PSYCHROMETRIC CHART
Normal Temperatures

Copyright 1947 Carrier Corporation – Copyright 1959 Carrier Corporation – AC-687

Printed in U.S.A.
Code (784-017)

Copyright 1947 Carrier Corporation

Sensible heat factor

14.5 cu ft per pound of dry air

Pounds of moisture per pound of dry air

Grains of moisture per pound of dry air

Enthalpy at saturation Btu per pound of dry air

$W_{oa} = 0.0146$

$ESHF = 0.93$

Alignment Circle

$W_r = 0.0092$

Room Conditions
75.5°F db, 50% rh

Outdoor Air Conditions
94°F db, 75°F wb

Mixed (entering) Air Conditions
78.5°F db, 65.5°F wb

Mixed (entering) Enthalpy
h = 30.4

GSHF Line

Coil Leaving Conditions
56.5°F db, 55.5°F wb

Coil Leaving Enthalpy
h = 23.6

$T_{adp} = 54°F$ db, 54°F wb

Wet-Bulb, Dewpoint or Saturation Temperature F

Dry-Bulb F

Below 32 F, properties and enthalpy deviation lines are for ice.

FIGURE 9.8 State Points for Example 9.4

9.19

The entering dry bulb temperature is plotted on the line between the outdoor and indoor air state points, and the entering wet bulb temperature is read from the chart.

$$T_{mwb} = 65.5°Fwb$$

Equation 9.13 may then be used to calculate the leaving air dry bulb temperature.

$$T_{ldb} = 54 + 0.10(78.5 - 54) = 56.5°Fdb$$

A straight line is then drawn between the adp conditions and the entering air conditions, which represents the GSHF. The leaving dry bulb temperature is then plotted on the GSHF line, and the leaving wet bulb temperature is read from the psychrometric chart.

$$T_{lwb} = 55.5°Fwb$$

As a check of the calculations, the total cooling capacity and the sensible cooling capacity may be calculated by using the calculated air flow rate, entering air conditions, and leaving air conditions. Equation 9.14 may be used as an alternate method to calculate the total cooling capacity.

$$q_{gt} = 4.5 \, Q_{da} (h_e - h_l) \tag{9.14}$$

where

q_{gt} = grand total cooling load (Btu/hr)

Q_{da} = rate of conditioned air flow (ft^3/min)

h_e = enthalpy of the entering (mixed) air (Btu/lb dry air)

h_l = enthalpy of the leaving air (Btu/lb dry air)

4.5 = a constant to make units consistent

Example 9.5   Using the air quantity, air conditions, and enthalpies of the air as it enters and leaves the cooling coil, what is the total and sensible cooling capacities for the system described in Example 9.3?

From the enthalpy scale, on the upper left side of the psychrometric chart, and Equation 9.14,

$$q_{gt} = 4.5(5390)(30.4 - 23.6) = 164,935 \text{ Btu/hr}$$

From the entering and leaving dry bulb temperatures and Equation 6.1b,

$$q_S = 1.1(5390)(78.5 - 56.5) = 130,440 \text{ Btu/hr}$$

In the above examples, the ratio of the sensible heat to the total heat was fairly high. The cooling load was mostly sensible heat. In some applications, this is not necessarily the case. In many applications, there is a comparatively larger latent cooling load. Examples of an area with large latent loads might include a gymnasium with a large latent cooling load from many occupants. Another example might be a hospital operating room with an all outdoor air ventilation system.

In such instances, when the ESHF line is plotted on the psychrometric chart, the slope of the line is so steep that it either does not intersect the saturation curve at all, or it intersects the curve at a ridiculously low temperature. The normal leaving air dry bulb temperature range for comfort air conditioning is about 48 to 55°F. Lower temperatures are difficult to achieve without special application cooling coils. Higher leaving air temperatures will result in insufficient dehumidification.

In high latent cooling load applications, it is usually necessary to select a reasonable leaving air temperature and reheat the air before it enters the conditioned space. Reheating the air adds sensible heat to the leaving air, while the humidity ratio remains constant. It may also be necessary to provide a higher air flow rate, in order to handle the latent cooling load. Equation 9.15 may be used to estimate the amount of reheat required.

$$q_{rh} = \frac{\text{ESHF}[q_{rs} + q_{rl} + \text{BF}(q_{oas} + q_{oal})] - [q_{rs} + \text{BF}(q_{oas})]}{(1 - \text{ESHF})} \tag{9.15}$$

where

q_{rh} = heat required to reheat the dehumidified cooled air (Btu/hr)

ESHF = effective sensible heat factor

q_{rs} = room sensible cooling load (Btu/hr)

q_{rs} = room latent cooling load (Btu/hr)

q_{oas} = outdoor air sensible cooling load (Btu/hr)

q_{oal} = outdoor air latent cooling load (Btu/hr)

BF = coil bypass factor

Before using reheat in an HVAC system, you should consult the applicable energy codes for the project location. Some energy codes prohibit, or limit, the use of reheat.

Example 9.6 A gymnasium has a sensible cooling load of 110,000 Btu/hr and a latent load of 45,000 Btu/hr. The code requires 1000 ft³/min of outdoor air for the system. The space design conditions are 75°Fdb/50% RH and the outdoor conditions are 94°Fdb/75°Fwb. The cooling coil has a bypass factor of 0.10. What is the required total cooling capacity of the system, what is the required air flow rate, how much reheat is required (if any), and what are the coil entering, leaving, and reheat discharge state points?
From Example 9.4

q_{oas} = (1.1)(1000)(94 − 75) = 20,900 Btu/hr

q_{oal} = (4840)(1000)(0.0146 − 0.0092) = 26,150 Btu/hr (approx), therefore,

q_{gt} = 110,000 + 45,000 + 20,900 + 26,150 = 202,050 Btu/hr

The bypass factor is 0.10 and is used to calculate the ESHF.

$$\text{ESHF} = \frac{110,000 + 0.10(20,900)}{110,000 + 0.10(20,900) + 45,000 + 0.10(26,150)} = 0.70$$

Plotting the ESHF = 0.70 on the psychrometric chart, it crosses the saturation curve at 46°Fdb. This is below the range of a normal cooling coil. Assume a more reasonable apparatus dew point of 50°F. Using a T_{adp} = 50°Fdb and plotting a line on the psychrometric chart, results in ESHF = 0.75.

$$Q_{da} = \frac{110,000 + 0.10(20,900)}{1.1(75 - 50)(1 - 0.10)} = 4530 \text{ ft}^3/\text{min}$$

$$T_{mdb} = \frac{3530(75) + 1000(94)}{4530} = 79°\text{Fdb}$$

From the psychrometric chart:

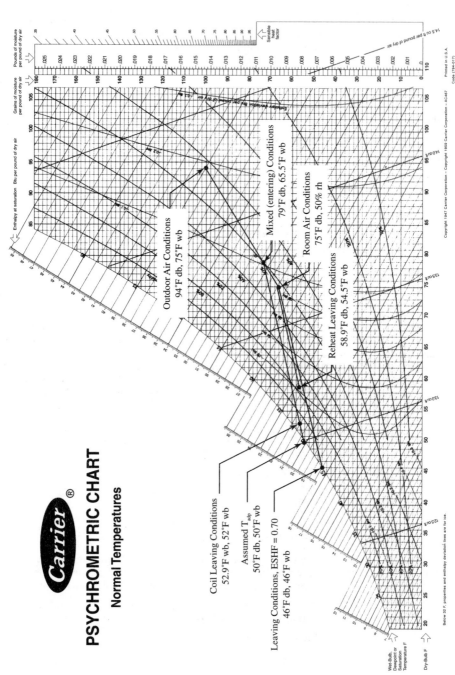

PSYCHROMETRIC CHART
Normal Temperatures

Outdoor Air Conditions
94°F db, 75°F wb

Mixed (entering) Conditions
79°F db, 65.5°F wb

Room Air Conditions
75°F db, 50% rh

Reheat Leaving Conditions
58.9°F db, 54.5°F wb

Coil Leaving Conditions
52.9°F wb, 52°F wb

Assumed T$_{adp}$
50°F db, 50°F wb

Leaving Conditions, ESHF = 0.70
46°F db, 46°F wb

Copyright 1947 Carrier Corporation – Copyright 1959 Carrier Corporation – AC-487

Printed in U.S.A.
Code (7844-017)

Below 32 F, properties and enthalpy deviation lines are for ice.

FIGURE 9.9 State Points for Example 9.6

$$T_{mwb} = 65.5°F$$
$$T_{ldb} = 50 + 0.10(79 - 50) = 52.9°Fdb$$
$$T_{lwb} = 52°Fwb$$

From Equation 9.15:

$$q_{rh} = \frac{0.75[110,000 + 45,000 + (0.10)(20,900 + 26,150)] - [110,000 - (0.10)(20,900)]}{(1 - 0.75)}$$

$$q_{rh} = 30,755 \text{ Btu/hr}$$

Equation 6.1a may be rearranged to calculate a temperature difference. Calculating the temperature rise through the reheat coil,

$$T_2 - T_1 = \frac{30,755}{1.08(4530)} = 6°Fdb$$

The conditions leaving the reheat coil are 58.9°Fdb and 54.5°Fwb.

This gets the supply air conditions close to the original ESHF = 0.7 line.

Checking the total cooling capacity by getting the entering and leaving enthalpy values from the chart,

$$q_{gt} = 4.5(4530)(30.5 - 21.6) = 181,425 \text{ Btu/hr}$$

Comparing this value to the previous total cooling load, this value is about 11% lower. The calculated air flow for the above design conditions will handle the sensible cooling load; however, it will not handle the latent cooling load. Therefore, it is necessary to determine a new air flow rate that will satisfy both the sensible and latent cooling loads.

Rearranging Equation 9.14 to solve for the air flow:

$$Q_{da} = \frac{202,050}{4.5(30.5 - 21.6)} = 5045 \text{ ft}^3/\text{min}$$

This air flow will handle the sensible and latent loads. Recalculating the reheat to get the required temperature rise,

$$q_{rh} = 1.1(5045)(6) = 33,300 \text{ Btu/hr}$$

The revised air flow rate will change the entering conditions slightly. It may be necessary to make several more calculations to get convergence.

9.5 HVAC PSYCHROMETRIC CALCULATION PROCEDURE

This section describes the steps necessary to determine the final cooling loads for a system and to specify and select the HVAC system for a space. It is assumed that you have already calculated the space cooling loads, in accordance with Chapters 7 and 8.

In order to perform the psychrometric calculations for a system, it is necessary to have a psychrometric chart and several straight edges to draw lines. Blank psychrometric charts can be obtained from HVAC equipment manufacturers and vendors. They are also available through ASHRAE.

The psychrometric calculation procedure is as follows:

1. Select the indoor design conditions for the project requirements and the outdoor air conditions for the project location from the climatic data in Chapter 3. Plot the state points on a psychrometric chart and draw a straight line between them.

2. Calculate the external, internal, and return cooling loads for the space, in accordance with the procedures in Chapters 7 and 8.

3. Determine the outdoor ventilation air requirements. Use Equations 6.1b and 8.5 to calculate the outdoor air sensible and latent cooling loads.

4. Calculate the Grand Total Cooling Load (q_{gt}), using Equation 9.9.

5. Assume a value for the cooling coil bypass factor based on the expected application and the guidelines listed in Section 9.4.2. Calculate The ESHF from Equation 9.11.

6. Locate the ESHF value on the SHF scale on the psychrometric chart. Draw a line from the ESHF on the SHF scale through the alignment circle near the center of the chart. (Other charts, such as the ASHRAE chart, use a slightly different method.) Draw a second line, beginning at the room state point, parallel to the first line, and extending to where it intersects the saturation curve on the chart. If the line does not intersect the saturation curve, or it intersects the curve at a point beyond the range of normal HVAC equipment, skip to Step 12.

7. The point where the second line intersects the saturation curve is the apparatus dew point (adp) of the cooling coil. Read the dry bulb temperature from the psychrometric chart.

8. Use Equation 9.12 to calculate the conditioned air supply for the space.

9. Use Equation 9.5 to calculate the return air quantity, and Equation 9.8 to calculate the cooling coil entering (mixed) air dry bulb temperature. Locate a point at the intersection of the entering dry bulb temperature and the line between the indoor and outdoor state points. Read the entering wet bulb temperature and the enthalpy from the chart. Draw a straight line from the entering air state point to the adp state point. This line represents the GSHF for the system.

10. Use Equation 9.13 to calculate the leaving dry bulb temperature. Locate a point at the intersection of the leaving dry bulb temperature and the GSHF line. Read the leaving wet bulb temperature and the enthalpy from the chart.

11. Use Equation 9.14 to calculate the total cooling load, based on the conditioned air supply, entering enthalpy and leaving enthalpy. As a check, compare this value to the value calculated in Step 4 above. If they are in reasonable agreement, the psychrometric calculations for the total cooling load, conditioned air supply, and state points are correct.

12. If the ESHF line through the room state point does not intersect the saturation curve at a reasonable value, assume a reasonable value for the coil adp. Draw a straight line from the room state point through the assumed adp. Using the SHF scale on the chart, determine an equivalent ESHF.

13. Calculate the conditioned air supply, entering conditions and leaving conditions, as described in Steps 8, 9, and 10 above, using the new ESHF line to locate the leaving conditions.

14. Using Equation 9.15, calculate the heat required for reheat (if any). Also calculate the resultant temperature rise in the air dry bulb temperature.

15. Draw a horizontal line from the cooling coil leaving conditions to the intersection of the reheat dry bulb temperature. This point locates the leaving conditions for the reheat coil. Read the reheat coil leaving wet bulb temperature from the chart.

Chapter 10

OVERVIEW OF HVAC SYSTEMS

Chapter 1 of this book discussed some of the major requirements of HVAC systems and some of the issues that should be considered when selecting an HVAC system. Basic descriptions of the system types are also included in Chapter 1. One major factor to consider when selecting an HVAC system is the total heating, ventilating, and air conditioning capacity requirement of the system or systems. Chapters 5 through 9 provide the most widely accepted method for determining the various system capacity requirements. This chapter provides you with an overview of the HVAC systems commonly used for heating, ventilating, and air conditioning of buildings today. Methods for providing the heating and cooling energy are also discussed. More detailed design considerations are discussed in the following chapters.

Basically, HVAC systems can be grouped into three major categories. The categories are Air Systems, Hydronic/Steam Systems, and Unitary or Packaged Systems. Practically all systems can be classified as being one of the major groups, or a combination of these groups.

In central air systems, heating and cooling is accomplished by heating and/or cooling air that is circulated to the conditioned space. A conditioned space, served by a system, may be considered a group of thermally similar areas which, when grouped together, are called a zone. Zoning was discussed in Chapter 1. Each space with a different exposure, or with a different magnitude of heating or cooling load, may be considered a zone. Each zone normally has a separate temperature control device (thermostat) and sometimes a humidity control device (humidistat). Zones have nearly uniform heating and/or cooling loads throughout the space. A zone is typically comprised of one or more rooms that are similar in their thermal load characteristics. A room implies an enclosed area that may or may not require a separate temperature control.

10.1 CENTRAL AIR SYSTEMS

Chapter 1 gave a brief description of a basic air handling system. Central air systems typically consist of a central air handling unit (AHU) and an air distribution system of ductwork and room air distribution devices. The AHU typically has a fan, filters, a cooling coil, and sometimes some sort of heating device. A typical AHU is shown in Figure 10.1.

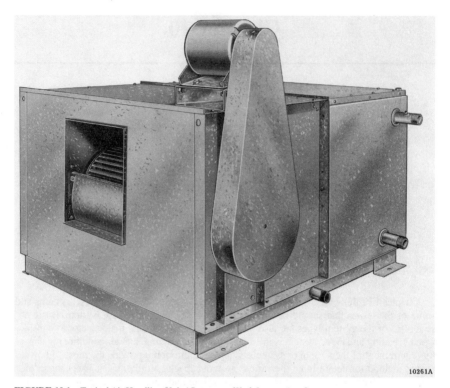

10261A

FIGURE 10.1   Typical Air Handling Unit (*Courtesy of York International*)

Heating devices can include hot water heating coils, steam coils, electric coils, gas furnaces, and oil furnaces. Cooling coils are usually chilled water type coils or direct expansion refrigerant coils. A typical water coil is shown in Figure 10.2. Heating and cooling coils are discussed in greater detail in Chapter 17, Air System Heating and Cooling.

Air handling units in central air systems are usually "built-up" units. Built-up units are comprised of various components that match the design requirements of the system. A built-up unit may have a chilled water cooling coil and an electric heating coil. Or, it may have a direct expansion (DX) cooling coil and a heating water coil. In a built-up unit, it is usually possible to mix and match fan types, coil types, filter types, etc.

Fans in built-up air handling units are usually centrifugal fans. A centrifugal fan is shown in Figure 10.3. Fans are discussed in Chapter 16, Fans and Central Air Systems.

Central air systems can basically be divided into three major types which describe how heating and cooling capacity control is achieved. The major types are constant volume/single path, multi-path, and variable air volume systems.

10.1.1 Constant Volume Single Path

The most basic type of central air HVAC systems is the constant volume single path system. As the name implies, the air flow rate is constant. There is only one path for the air to fol-

FIGURE 10.2 Water Heating or Cooling Coil (*Courtesy of York International*)

low as it leaves the AHU and is distributed to the conditioned space. This type of system is best suited for a single zone since the system is either in the heating mode or cooling mode at any given time. The system may be in the "dead band" mode in which it is neither heating nor cooling, just circulating air. It is not possible to simultaneously heat one zone and cool another with a single system. Typical applications for the constant volume single path systems are single family houses, large auditoriums with very few exterior surfaces, spaces with a single exterior exposure, and other spaces with fairly uniform heating and/or cooling loads. A schematic diagram of a constant volume single path system is shown in Figure 10.4.

Constant volume single path systems normally have an AHU with heating and/or cooling coils and ductwork in a series flow path. Sometimes the heating coil may be replaced with a gas or oil fired furnace, as in the case of a residential system. The heating or cooling is cycled on or off in response to a load in the conditioned space sensed by a thermostat.

A variation of the constant volume single path system is the constant volume reheat system. This system can provide either heating or cooling to multiple zones simultaneously, with a constant air volume system. A schematic diagram of a constant volume reheat system is shown in Figure 10.5.

In this type of system, air is circulated at a constant rate and leaves the AHU at a set temperature, usually about 55°F. Just before the air enters the conditioned zone, a duct

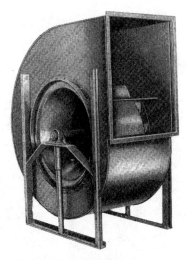

FIGURE 10.3   Centrifugal Fan (*Courtesy of the Trane Company*)

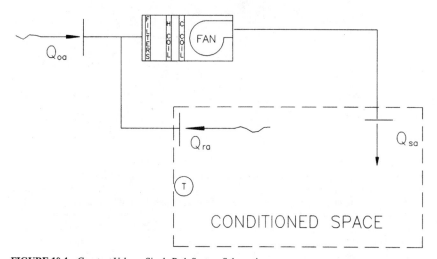

FIGURE 10.4   Constant Volume Single Path System Schematic

mounted, or air terminal mounted, reheat coil increases the temperature of the air entering the zone, in accordance with the zone heating or cooling load sensed by the zone thermostat. Reheat coils can be electric, steam, or heating water. A separate preheat coil may also be provided in the AHU to preheat the incoming outdoor air as necessary.

Constant volume reheat systems are rarely used on new installations since they are very energy inefficient. Air is first cooled to a set supply air temperature and then heated to meet the zone requirements. Some energy codes no longer allow constant volume reheat systems.

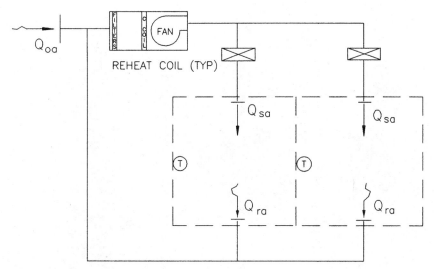

FIGURE 10.5   Constant Volume Reheat System Schematic

10.1.2   Multi-Path Systems

As the name implies, multi-path systems enable the conditioned air to follow multiple parallel paths to each zone. The temperature of the air supplied to each zone is a function of the heating or cooling load in that zone. Typically, the air handling unit has both heating and cooling coils, in order to provide either heated or cooled air to a zone as required. The two most common multi-path air systems are the multi-zone system and the dual duct system.

Multi-zone systems usually have an air handling unit with a cooling coil and multiple heating coils. A typical multi-zone air handling unit is shown in Figure 10.6.

Each zone has its own supply air duct from the AHU, enabling the air to flow through multiple parallel supply air ducts to the zones. Each zone has its own reheat coil inside the AHU. The zone reheat coil is controlled by a zone thermostat. Air is passed through the cooling coil in the AHU and is cooled to a predetermined temperature, usually about 55°F. The air flow is then divided inside the AHU into separate paths for each zone. Still inside the AHU, the air for each zone is then either passed through a reheat coil or bypassed around the reheat coil, according to the zone heating or cooling load. The proportion of air passed through or around the reheat coil is controlled by the zone thermostat.

Since the supply air is first cooled, then reheated or bypassed to provide the desired "mixed" air temperature, these systems are rather energy inefficient and energy wasteful. Very few multi-zone systems are installed in new buildings today. Most multi-zone systems are found in older buildings.

The second multi-path system in use today is the dual duct system. As the name implies, there are dual (two) ducts to each space or zone. One duct carries heated air, while the other carries cooled air. There can be a single or dual air handling units for the system. Either cooled air, heated air, or both are supplied to the space according to the thermal load in the space. In order to provide air to the space at the required temperature, a dual duct terminal is provided for each zone to mix the heated and cooled air in the required proportion. The dual duct terminal mixes the two air streams in accordance with the requirements of the zone thermostat. A typical dual duct terminal is shown in Figure 10.7.

FIGURE 10.6   Multi-Zone Air Handling Unit (*Courtesy of York International*)

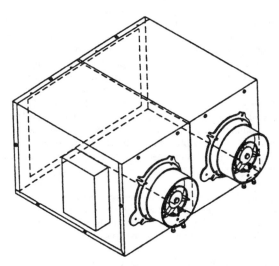

FIGURE 10.7   Dual Duct Terminal Unit (*Courtesy of the Trane Company*)

Notice that there are two air valves on the inlet of the terminal. One is for the cooled air and the other is for heated air. The air valves throttle each air flow according to the requirements of the space thermostat.

10.1.3 Variable Air Volume

Other than the constant volume single path system, the variable air volume (VAV) system is the most common type of air system installed in nonresidential buildings today. As the name implies, the rate of air flow in the system is not constant, but varies in accordance with the total sensible cooling load for the building. The supply air is usually cooled and supplied to the distribution ductwork at a constant temperature, although the air temperature may be reset in accordance with the cooling load. A schematic of a typical VAV system is shown in Figure 10.8.

As may be observed from Figure 10.8, the VAV system is also a single path system; however, it is not constant air volume. The amount of cooled air admitted to the conditioned space is a function of the sensible cooling load in the space, which is sensed by a thermostat. As the temperature of the space decreases, an air valve in a VAV terminal closes, reducing the air flow as the cooling load is satisfied.

Heat for the conditioned spaces is usually not provided by heating the supply air in the AHU. Frequently, heat is provided by baseboard radiators around the exterior perimeter of the conditioned space. Heat may also be provided in the VAV terminal with a reheat coil itself. A typical VAV terminal with a reheat coil is shown in Figure 10.9.

Notice that an air valve is on the inlet of the VAV terminal. The air valve throttles, or varies, the flow of conditioned air to the space in response to a space thermostat. For terminals with reheat, the reheat coil is usually located on the discharge of the terminal. The selection of various types of terminal units is discussed in Chapter 14.

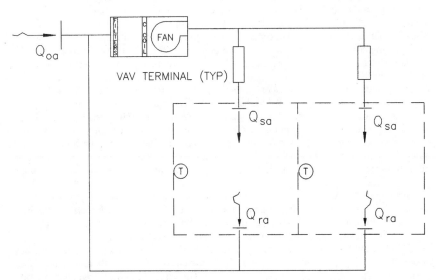

FIGURE 10.8 VAV System Schematic

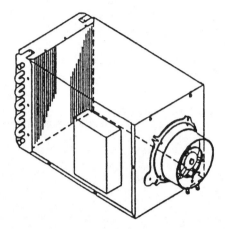

FIGURE 10.9 VAV Terminal with Hot Water Reheat (*Courtesy of the Trane Company*)

Since the quantity of air supplied to the spaces varies, it is good design practice to provide some method to control the air flow of the air handling unit fan. This is normally accomplished in one of three ways. The three methods are fan discharge dampers, fan inlet vanes, and variable frequency motor drives. Each method varies the flow of the AHU fan, in response to changes in static air pressure in the main supply air duct.

The first, and least desirable, are discharge dampers. Discharge dampers are automatically actuated dampers that are placed on the outlet, or discharge, of the AHU. The dampers open and close in response to the pressure in the main supply air duct. As the VAV terminals open and the pressure in the main duct drops, the discharge dampers open to allow more air from the fan into the ductwork. As the VAV terminals close and the duct pressure increases, the discharge dampers close. Discharge dampers are somewhat energy inefficient since the fan is rotating at full speed while the dampers block off the air flow. Using discharge dampers is a little like driving an automobile at full throttle while controlling the speed with the brakes. Discharge dampers may also result in undesirable high air pressures in the air handling unit.

A variation on discharge dampers is the bypass system. Air is simply bypassed around the conditioned space and returned to the AHU. Air is normally "dumped" into a return air plenum, or duct, by a motorized relief damper. The fan in the AHU is still rotating at full speed, fully loaded. Bypass systems are typically found on smaller commercial air systems.

A second, and more energy efficient, method to control the AHU fan air volume is with inlet vanes on the fan. These vanes are located at the air inlet on the fan and open or close in response to the pressure in the main supply air duct. Figure 10.10 shows a centrifugal fan with inlet vanes.

Since the inlet vanes restrict the amount of air entering the fan, the fan rotates in an unloaded condition, even though it is rotating at full speed. This reduces the load on the electric motor driving the fan and reduces the power consumed by the motor. Inlet vanes are more costly than discharge; however, they are also more effective.

The third, and most effective, way to control the air flow of a fan is a variable frequency motor drive. These drives vary the frequency of the electric power to the fan motor, in response to the pressure in the main supply air duct. The varying electrical frequency, in turn, causes the motor and fan to rotate at different speeds. When the air pressure in the duct drops, the fan speed increases to supply more air. When the air pressure increases, the fan

FIGURE 10.10   Fan with Inlet Vanes (*Courtesy of York International*)

speed decreases accordingly. Although variable frequency drives tend to be more expensive than the other methods to control air volume, they are also much more efficient. Frequently, the energy cost savings of the variable frequency drives justifies the additional first cost.

10.4.1   Air Distribution Systems

In central air systems, conditioned air is usually distributed throughout a building to the conditioned spaces via a network of ductwork and air distribution devices. The ductwork and devices between the air handling unit and the conditioned space are called the supply air system, while the devices and ductwork that return the air to the AHU are called the return air system.

Air distribution devices supply the air directly to, or return the air from, the conditioned space. Air devices include such things as ceiling air diffusers, air registers, and air grilles. In the design of HVAC systems, the location of the air devices in the conditioned space is important. They should be located so as to provide a uniform air distribution throughout the space. Supply air outlets and return inlets should not be located so close to each other that the air flow tends to "short circuit" from the outlet to the inlet. Air devices should also be located to avoid causing drafts within the space. However, the devices should be located so they provide good air motion within the space and so that there are no areas of stagnant air within the space. The selection of air distribution devices is discussed in Chapter 14.

When selecting air devices for a particular application, care should be taken to ensure that the device will not cause an objectionable noise level in the space at the design air flow rate.

Performance data for air devices normally includes NC (Noise Criteria) levels at various air flow rates. The higher the NC level, the higher the noise. Offices and similar spaces normally require a maximum NC level of around 35, while spaces such as restaurants or cocktail lounges can have an NC level as high as 45. Reference 3 and Chapter 12 list the recommended NC levels for various types of spaces. The Air Distribution Council provides standard methods for rating the performance of air supply and return devices. NC data is usually provided with the air device manufacturer's literature. Acoustics are discussed in Chapter 12.

In most central air systems, a network of ductwork is included to distribute the conditioned air from the AHU to the room devices. The ductwork system is actually two systems, a supply air system and a return air system. Sometimes, much of the return air ductwork above a ceiling is not included in lieu of a ceiling return air plenum. If a ceiling return air plenum is used, you should consult the applicable codes to determine which materials are permissible in return air plenums and for what applications return air plenums are allowed.

Supply air duct systems are usually classified according to their static air pressure and/or air velocity. Duct systems are classified as low pressure if the maximum static pressure is 2 in wg or less. They are classified as low velocity if the air velocity in the ductwork is 2000 fpm or less. Ductwork with pressures greater than 2 in wg and/or air velocities greater than 2000 ft/min is considered high pressure/high velocity. The Sheet Metal and Air Conditioning Contractors National Association (SMACNA) provide duct construction standards for ductwork handling air at various pressures and velocities. Duct systems design is discussed in Chapter 15.

There are a number of considerations involved when trying to make a decision of whether a supply air duct system should be low velocity/pressure or high velocity/pressure. Some of the major considerations include the amount of air to be circulated, the space available for ductwork, noise level, and installed cost. Designers should pay particular attention to noise levels for ductwork routed through quiet or sensitive areas and in high velocity systems.

10.2   HYDRONIC AND STEAM SYSTEMS

Hydronic systems are systems that provide heating and/or cooling to a space by circulating heated and/or chilled water. Heated water is usually generated by a boiler and circulated through a piping distribution system with a pump. Chilled water is usually produced by a chiller, a cooling tower, and circulating pumps. Chilled water and heating water systems are generally found in large buildings, building complexes, and buildings on a campus. Hydronic systems are generally very energy efficient; however, the need for a central plant to generate chilled and/or heating water is usually a large part of the cost of an HVAC system. Chillers, boilers, cooling towers, and related equipment are discussed later in this chapter.

In addition to the main central plant equipment, a hydronic system consists of one or more circulating pumps, piping, and some sort of heat exchange device. Heat exchange devices transfer heat between the circulated water and the air in the space, or the air supplied to the space, by an air handling unit. Heat exchange devices, which include hydronic heating/cooling coils, fan coil units, unit heaters, and radiators are discussed later in this chapter and in Chapter 17.

Hydronic systems can have several piping arrangements. Usually, there is supply piping that distributes the circulated water from the central plant to the heat exchange devices, and return piping that returns the water to the primary heating or cooling device in the central plant. Such an arrangement is called a two pipe system because there is a supply piping system and a return piping system. When the heat exchange devices use both heating water and chilled water, there may be a four pipe system. This type of system has one set of pipes for the chilled water and one set for the heating water. Sometimes, three pipe systems are used and the return piping is a common return for both chilled and heating water.

Hydronic systems have a number of advantages. First, since water has a much higher specific heat than air, the cross-sectional area of a pipe required to carry water is much less than the cross-sectional area for a duct carrying air for any given thermal load. Therefore, a hydronic system requires much less space in occupied areas of the building than does an air system. Hydronic systems also require less pumping horsepower than horsepower required to circulate air for any given thermal load. This can result in lower operating costs. Third, since the heat transfer device is frequently located in the conditioned space, hydronic systems are usually easier to zone if they are a four pipe system.

Steam systems are similar to hydronic systems except that steam is distributed as the heating medium in lieu of heated water. Obviously, steam systems are not directly used for cooling. Circulating pumps are not required for steam heating systems because the pressure of the steam itself is adequate for steam flow. Steam condensate (steam that has condensed back to water) is usually returned via a gravity drainage system. Occasionally, the condensate is pumped if gravity drainage is not possible.

Steam systems are somewhat different from hydronic systems in that the steam undergoes a phase change. As water is heated, it undergoes a phase change as it is converted to steam, usually in a boiler. Steam is converted back to water (condensate) in a heat exchange device, usually a steam heating coil or a steam radiator. Again, it undergoes a phase change. It is the phase change that releases the majority of the heat from steam. As an example, steam at 15 psig and 250°F will release about 945 Btu/lb of steam as it condenses from steam to condensate. The phase change occurs at a nearly constant temperature.

In order for the steam to condense in a heat exchange device, the steam itself must be kept inside the device until it condenses. After the steam becomes condensate, the condensate must be removed from the heat exchange device. This is usually accomplished with a steam trap. The trap will not allow steam to pass; however, it will allow condensate to pass. A steam trap is placed on the discharge piping of the heat exchange devices to hold the steam in while enabling the condensate to flow out.

10.2.1 Heating and Cooling Coils

Typically, heating coils use either heating water or steam as the heating medium, although electric heating coils are also used frequently. Cooling coils typically use chilled water or a refrigerant as the cooling medium. A typical water coil is shown in Figure 10.2. A typical direct expansion (DX) type refrigerant coil is shown in Figure 10.11.

In all finned coils, some of the air is passed through the coil and is not affected by the coil (it is not heated or cooled). The air that passes through the coil and is not affected is called bypass air. The percentage of bypassed air can range from as high as 30%, to as low as 2%. The amount of bypassed air depends on the air velocity, the coil fin spacing, and the number of coil rows. Typical bypass factors for coils are given in Section 9.4.2.

A coil is comprised of many fins (between which air flows), coil tubes, and headers. The fins increase the heat transfer area in contact with the passing air. Although much of the heat transfer from a coil is from the fins, the primary source of heat transfer is from round tubes, or pipes, that run between the coil headers at each end. The tubes are installed perpendicularly to the fins and air flow. If the fins run vertically, the tubes run horizontally. The number of tubes in the direction parallel to the air flow designate the number of rows of the coil.

The tubes are usually interconnected by return bends to form a serpentine arrangement. The fins are bonded to the tubes to provide good conduction heat transfer between the fluid in the tubes, the tubes themselves, and the fins. Water coils, steam coils, and refrigerant coils

FIGURE 10.11   DX Refrigerant Coil (*Courtesy of York International*)

usually have aluminum fins and copper tubes. A header is usually provided at one or both ends of the tubes to connect and to distribute the fluid to and from the tubes. Air vents and drains are usually provided on the headers of water coils.

DX refrigerant coils are a little more complicated than water coils. In a refrigerant coil, the refrigerant flows into the coil as a liquid, and then evaporates to vapor as it absorbs heat from the air. The refrigerant rate of flow must be properly metered and distributed to the tubes for uniform cooling throughout the coil. The flow of refrigerant into the cooling coil (or evaporator, as it is sometimes called) is controlled by a thermostatic expansion valve at the inlet to the coil. Direct expansion refrigerant systems are discussed later in this chapter.

Heating and cooling coils and their selection are discussed in Chapter 17.

10.2.2   Fan Coil Units and Unit Ventilators

A fan coil unit is, just as the name implies, a small unit with a fan and a chilled and/or heating water coil enclosed in a common cabinet. Fan coil units are usually small enough to be located in the space that they serve. The exterior cabinet frequently has a decorative finish. A typical fan coil unit is shown in Figure 10.12.

Fan coil units typically have a small fan, fan motor, filters, a single water coil, and controls mounted inside a decorative cabinet. Fan coil units can also have electric heating coil for

FIGURE 10.12 Fan Coil Unit (*Courtesy of the Trane Company*)

electrically heated units. A single water coil is usually used for both cooling and heating, even when the units are part of a four pipe system. A control valve, or valves, mounted on the inlet to the water coil, controls which set of pipes flow through the coil.

Fan coil units are usually located along exterior walls and can have limited outdoor air capabilities. The amount of outdoor air that a fan coil unit can handle is relatively small because the cooling coils are designed as "dry coils." Dry coils cannot handle very large latent cooling loads, as would be the case with higher quantities of outside air.

Air is normally discharged vertically through a grille in the top of the unit and returned near the bottom front of the cabinet. An outdoor air connection (if any) is usually provided in the back of the unit that is against the wall. A drain pan is usually provided inside the cabinet, under the coil, to collect condensate from latent cooling. Fan coil units, with bottom discharge and return, are available for installations above ceilings.

Unit ventilators are similar in construction to fan coil units, except that they are designed to handle much larger outdoor air quantities and higher latent cooling loads. It is possible for a unit ventilator to handle up to 100% outside air. Unit ventilators frequently include modulating return air, and outside air dampers to modulate the amount of outdoor air. Some unit ventilators have direct expansion cooling coils, as an option, in lieu of a chilled water coil.

10.2.3   Unit Heaters

As the name implies, unit heaters are small units that provide heating only. The heating medium for unit heaters is usually heating water or steam. Unit heaters can also have electric heating coils, or be gas fired with gas furnaces. A typical unit heater is shown in Figure 10.13.

A unit heater consists of a heating element (a coil or furnace) and a small fan enclosed in a single cabinet. They can be horizontal or vertical discharge. Unit heaters are usually installed to provide heating in unfinished spaces, such as warehouses and garages.

10.2.4   Radiators

Radiators are frequently used for heating in both finished and unfinished areas of buildings. They are usually located against exterior walls and especially under windows. Radiators can use steam, heating water, or electricity as the heating medium.

FIGURE 10.13   Unit Heater *(Courtesy of the Trane Company)*

Older buildings still use vertical sectional cast iron radiators. Although cast iron radiators are still in use today, they have been replaced by finned tube radiators in new buildings. The finned tube radiator has no moving parts. Steam and heating water radiators consist of a pipe, through which the heating medium flows, and metal fins bonded to the pipe. A finned tube element is shown in Figure 10.14.

The fins increase the heat transfer area of the heating element. Air flows through the fins by natural convection heat transfer. As the air is heated, it becomes less dense and rises. Air that rises is then replaced by cold air from below, which in turn is heated.

In finished areas, the finned tube elements are usually installed in decorative cabinets for aesthetic and safety reasons. Finned tube radiator cabinets are shown in Figure 10.15.

FIGURE 10.14 Finned Tube Element (*Courtesy of the Trane Company*)

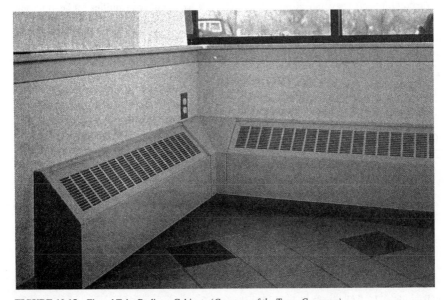

FIGURE 10.15 Finned Tube Radiator Cabinets (*Courtesy of the Trane Company*)

10.3   UNITARY PACKAGED HEATING AND COOLING UNITS

Unitary or packaged HVAC units are, just as their name implies, a complete HVAC unit in a single package or cabinet. Packaged units usually have a complete direct expansion cooling system, along with fans, and sometimes heating. The cooling system includes a refrigerant compressor, evaporator coil, condenser coil, and condenser fan. Heating is usually a gas furnace or electric coil, although heating water and steam coils are also available. Packaged units are also available with a heat pump cycle for heating.

Packaged equipment is quite popular because they usually have a lower first cost. Since everything is factory built and installed in a single cabinet, equipment costs are comparatively low. Installation costs are usually low because it is usually only necessary to set the unit, connect the utilities, and connect the distribution ductwork (if any).

Packaged units, however, are not capable of providing very precise control of temperature and humidity. These units are primarily for basic comfort heating and cooling, and are not suitable for process cooling or a situation with high latent cooling loads. Since these are packaged units, there are a limited number of options with the components. It is usually not possible to "build up" a custom unit for a special application.

10.3.1   Window and Through-the-Wall Units

Most people are familiar with window air conditioners and through-the-wall air conditioners. They are small capacity units, used primarily in residential applications, although they are commonly used in commercial applications for spot cooling and heating. Motel rooms frequently use through-the-wall units. Through-the-wall units are also referred to as packaged terminal air conditioners.

The units are usually completely self contained with all components in a single cabinet. Each unit is factory assembled with a direct expansion cooling system, and usually an electric coil for heat. They may also have reverse cycle heat pump capabilities for heating. The units are complete, including all temperature controls.

Window and through-the-wall units usually have a single utility connection that is electric power. The evaporator fan in these units has a low static pressure capability; therefore, they should not be connected to ductwork. Window and through-the-wall units are available with cooling capacities from $1/2$ to $7^1/2$ tons.

Window and through-the-wall units must be mounted on an outside wall to allow the condenser to reject heat to outdoors. These units usually have limited outdoor air ventilation capabilities.

10.3.2   Rooftop HVAC Units

As the name implies, rooftop HVAC units are packaged units, mounted on the roof of a building, and ducted down through the roof to the spaces below. Packaged rooftop units are also available with a horizontal duct discharge for grade mounted units. Rooftop units are usually installed on a roof curb which supports the unit and provides a weatherproof seal around the duct openings, through the roof. Rooftop units are completely factory assembled units, with a direct expansion refrigeration system heating system, fans, and controls in a single weatherproof cabinet. A typical rooftop unit is shown in Figure 10.16.

The most common forms of heating are gas furnaces and resistance electric heating coils. Units are also available with heating water coils and steam coils. Small and medium sized units are available as heat pumps.

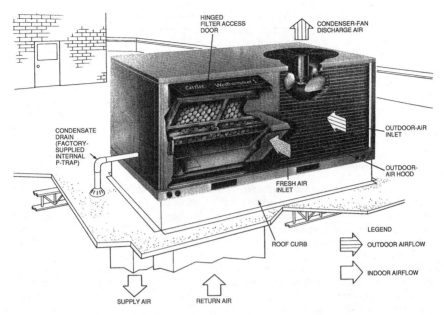

FIGURE 10.16   Rooftop HVAC Unit (*Courtesy of Carrier Corporation*)

Rooftop units are usually used in commercial applications, such as office buildings, retail spaces, and other spaces requiring purely comfort cooling. They are quite popular because of their low first cost and the fact that they do not use any space inside the building. Usually, the only required connections to the unit are the supply/return ductwork, electrical power, and gas (if applicable). One of the major disadvantages to rooftop units is that they can cause considerable noise and vibration in the spaces near the unit, particularly under the unit.

Rooftop units are available with cooling capacities ranging from as low as $2^{1}/_{2}$ tons, to as high as 125 tons or more in a single unit. Smaller units tend to be constant air volume, while larger units can be variable air volume. Small and medium sized units can use an air bypass arrangement to achieve variable air flow to the spaces.

10.4   HEATING METHODS AND EQUIPMENT

This section discusses the various methods, or sources of heat, and heating equipment for HVAC systems. Heat can be generated within the conditioned space itself, in an air handling unit serving the space, or in a central plant that may provide heat for the entire building. Central plants may also serve many buildings, such as a campus.

Heat generated within the space is usually resistance electric heat and occasionally natural gas or liquid petroleum gas. Electric baseboard heaters and electric coils in air terminal units or packaged terminal HVAC units are most frequently used in space electric heating methods. Gas fired unit heaters, and similar equipment, are the most common methods of providing gas heat from within the space. Of course, all gas fired equipment of this type must be properly vented.

Heat may also be provided to the conditioned space by an air handling unit serving the space. Again, resistance electric coils, natural gas, and propane gas are commonly used methods of generating heat. Heat pumps and fuel oil may also be used as a heating energy source. Heat pumps are discussed in Section 10.5. In order to use gas or fuel oil, a furnace is usually needed.

Finally, heat can be generated by a central plant in the form of heating water or steam. For hydronic or steam heating, a boiler is used to heat the circulated water or to generate steam. Boilers can be electric, gas fired, or fuel oil fired. Some boilers can use gas or fuel oil (dual fuel).

10.4.1 Furnaces

Furnaces are frequently used for heating when the heating source is a fossil fuel (chemical energy) such as natural gas, liquid petroleum (LP) gas, or fuel oil. Furnaces typically include an air circulation fan, air filters, controls, a burner, and a heat exchanger in a single cabinet. Oil fired furnaces frequently include a combustion air fan and motor (pressure atomizing burners). A typical furnace with an outdoor condensing unit (air conditioner) is shown in Figure 10.17.

Furnace controls usually include an ignition device (pilot light or igniter), a gas valve, a fan switch, and a temperature limit switch. The fan switch turns the fan on when the heat exchanger reaches a set temperature, and turns off when the heat exchanger drops below the set point temperature. If the heat exchanger temperatures become excessive, the temperature limit switch shuts the fuel off.

A heat exchanger is necessary for gas and oil fired furnaces to separate the combustion flue gases from the air circulated to the conditioned space. On older furnaces, air for the combustion process is drawn from the space in which the furnace is located. The flue gases are then discharged vertically through a flue pipe or chimney. On newer furnaces, a separate vent pipe is used to bring combustion air directly in from outdoors.

Furnaces are also used in commercial HVAC systems, usually in packaged rooftop equipment, and as duct furnaces for air handling units. Since it is difficult to provide zoning for

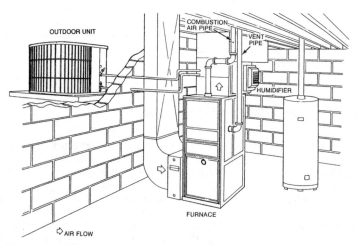

FIGURE 10.17   Residential Furnace and Air Conditioner (*Courtesy of Carrier Corporation*)

large systems with furnaces, space heating for large buildings is usually accomplished with either hydronic or electric units located in the conditioned space. Furnaces in large rooftop units are usually for morning warm-up, after night set-back.

Furnaces are usually rated according to their fuel consumption, or input, and their heating output capacity. The heating output in Btu/hr, divided by the fuel input heating value in Btu/hr gives an approximation of the furnace combustion efficiency. Older furnaces have a combustion efficiency around 70%, while newer furnaces can have efficiencies as high as 95%.

10.4.2 Electric Heating

The most common forms of resistance electric heating are electric baseboard heaters, electric coils in air handling units, and electric coils in air terminal units. Typical applications for electric heating coils include outdoor air preheat and electric reheat in air terminal units. Electric coils are ideal for preheat applications because they are not subject to freezing at low air temperatures. They are also well suited for air terminals because they can readily be located in the conditioned space, without the need for combustion venting or hydronic piping.

Another form of electric heating is the heat pump, which is discussed later in this chapter.

Electric heating coils produce heat as the result of the resistance to the flow of electrical energy. For larger coils, over about 5 kW, it is necessary to energize the coil, in stages, for temperature control. It is usually not advisable to have a large coil come on all at once. Safety controls for electric coils include primary and secondary thermal overload protection, over-current protection, disconnecting protection, and an air flow interlock. The air flow interlock ensures that the air is flowing when the coil is energized to prevent overheating.

All electric heating equipment should be approved by Underwriters Laboratory and the National Electrical Code.

10.4.3 Heating Boilers

Central heating plants use boilers as a source of heat to produce heating water or generate steam. The boilers can be gas fired, fuel oil fired, and occasionally electric. A typical gas fired, packaged steel boiler is shown in Figure 10.18.

Gas and oil fired boilers are similar to furnaces, in that they have a burner and a heat exchanger. Burners can be either natural draft or power burners with a combustion air fan. Heat exchangers can be classified as fire tube or water tube. Fire tube boilers have the hot combustion gases passing through the tubes surrounded by water. Water tube boilers, as the name implies, are the opposite of fire tube boilers. Water tube boilers have the water inside the tubes, and the hot combustion gases flow around the tubes containing the water.

Steam boilers and systems are classified according to the system steam pressure. Low pressure steam systems operate at steam pressures under 15 psig. Medium and high pressure steam systems operate at pressures greater than 15 psig. Steam boilers are typically available with steam capacities up to 50,000 lb of steam per hour.

Hydronic boilers and systems are classified according to their temperature and pressure. Low temperature/low pressure hydronic boilers operate up to 250°F and 160 psig pressure. Medium and high temperature/high pressure hydronic systems operate at pressures greater than 160 psig and temperatures greater than 250°F. Hydronic boilers are available with heating capacities up to 50,000,000 Btu/hr. Hydronic heating systems for buildings usually have a maximum operating temperature of 250°F or less, and are usually designed for a water temperature drop of 20°F to 40°F through a heating coil.

FIGURE 10.18   Packaged Steel Boiler (*Courtesy of Kewanee Manufacturing Co.*)

Both steam and heating water boilers are rated at their maximum working pressure and temperature, as determined by the American Society of Mechanical Engineers (ASME) Boiler Code Section.

Boilers are usually constructed of cast iron or steel. Cast iron boilers are constructed of individual sections of cast iron that can be bolted together to form the boiler. Steel boilers are usually packaged boilers that come from the factory completely assembled in a single welded steel jacket. Cast iron boilers range in capacity from 35,000 Btu/hr to 10,000,000 Btu/hr gross heat output. Packaged steel boilers start around 50,000 Btu/hr and go up to 50,000,000 Btu/hr gross heat output.

Operating and safety controls for boilers include temperature and pressure controls, temperature limit controls, high/low fuel pressure, flame failure fuel shut-off, pressure and temperature relief valve, and excess air control. Boilers must also have a "fuel train," which consists of regulators, gauges, valves, test cocks, pressure switches, etc. in the fuel piping to the boiler. Fuel trains are frequently factory furnished with the boiler.

10.5   COOLING AND REFRIGERATION SYSTEMS

Most cooling and refrigeration systems operate as direct expansion refrigeration systems, which are discussed in this section. Some chillers operate on the absorption refrigeration principle. Absorption chillers are discussed in Section 10.5.2.

Direct expansion (DX) systems are classified as direct systems by the ASHRAE Standard 15, Safety Code for Mechanical Refrigeration. A direct system is one in which the evaporator, or condenser, of the refrigerating system is in direct contact with the air to be cooled. Chillers are classified as indirect systems. An indirect system is one in which a secondary coolant (such as water) is cooled by the refrigerating system, which is then circulated to the air to be cooled.

When selecting HVAC equipment that has a refrigerating system, and uses a refrigerant, the designer should be aware that some popular refrigerants may no longer be available.

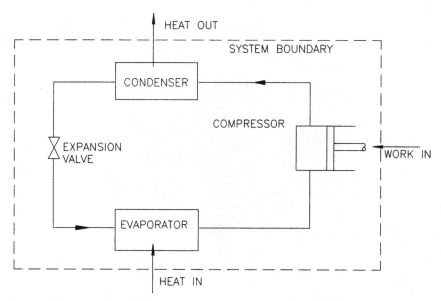

FIGURE 10.19   A Simple Refrigeration System Schematic

The Montreal Protocol, which was adopted in 1987, has banned the production of some refrigerants. Although refrigeration equipment may continue to operate with these refrigerants, the refrigerants may not be available when the equipment needs servicing. When designing a cooling system or a heat pump system, the designer should consult the latest issue of ASHRAE Standard 15 and the latest applicable codes.

10.5.1   Direct Expansion Systems

The most common system for producing a refrigerating effect is the direct expansion (DX) system. DX systems operate on a thermodynamic cycle that transfers heat from a low temperature region to a higher temperature region. Since heat only flows naturally from a high temperature region to a low temperature region, it is necessary to apply energy in the form of work to a refrigeration system. Work is provided by a compressor that compresses the refrigerant vapor. A simple DX refrigeration system is shown in Figure 10.19.

In the DX system, a refrigerant is circulated by the compressor. The refrigerant is a substance that will undergo a phase change (change from a gas to a liquid and back to a gas) as it goes through the cycle. The refrigerant enters the compressor as a gas and is compressed. Increasing the pressure of the gas causes its temperature to increase and it becomes a superheated vapor. The superheated refrigerant vapor then passes through a condenser where it rejects much of its heat. The heat is usually rejected directly to the atmosphere or to a circulated cooling fluid such as water. When the refrigerant comes out of the condenser, it is a saturated vapor at high pressure. (A high saturated substance is one in which vapor, or liquid phases, can be present at a given temperature and pressure.) The saturated vapor

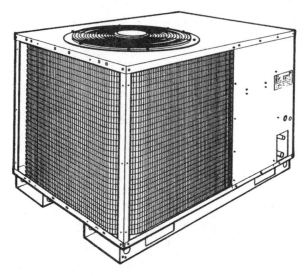

FIGURE 10.20   Residential Condensing Unit (*Courtesy of Carrier Corporation*)

refrigerant then passes through an expansion valve. The expansion valve reduces the pressure of the refrigerant, which also reduces its temperature. The refrigerant then becomes almost 100% saturated liquid at a much lower temperature. The saturated liquid refrigerant then passes through the evaporator. As the name implies, the liquid refrigerant evaporates and absorbs heat from the circulated air that is passed over the evaporator. The evaporator in an air conditioning system is the cooling coil. When the refrigerant comes out of the evaporator, it is again a vapor and the cycle repeats. Refrigeration for air systems is discussed in more detail in Chapter 17, Air System Heating and Cooling.

Figure 10.17 shows a residential heating and cooling system. This type of system is referred to as a split system. The evaporator coil and expansion valve are located in the furnace, or air handling unit, while the compressor and condenser coil are in the outdoor unit. Refrigerant piping connects the two units and carries the refrigerant between them. The outdoor unit is usually called a condensing unit. A residential condensing unit is shown in Figure 10.20.

Residential and small commercial units have cooling capacities ranging from two tons to about six tons. Large commercial condensing units can have cooling capacities over 150 tons. A large commercial condensing unit is shown in Figure 10.21.

When the outdoor unit contains the compressors, the condenser coils, and fans, it is referred to as a condensing unit. If the outdoor unit just has the condenser coils and fans with the compressors located elsewhere, the unit is referred to as a condenser.

A variation on the DX cooling system is the heat pump. A heat pump is used for heating a space, instead of cooling. It is almost identical to the cooling system, except the direction of heat exchange is reversed. In winter, heat is transferred from outdoors to the indoors by the refrigeration process. For a heat pump, the evaporator coil is outdoors and the condenser coil is indoors. Most heat pump systems function as a cooling system in summer and a heating system in winter. This is accomplished by a switching valve in the refrigeration piping that changes the function of the evaporator and condenser coils by changing the flow of the refrigerant.

FIGURE 10.21   Large Commercial Condensing Unit (*Courtesy of Carrier Corporation*)

10.5.2   Chilled Water Systems

Frequently, large buildings and multi-building campuses are cooled from a central plant. Central plants usually include one or more chillers to produce cooling, which is distributed to individual air handling, and fan coil units, by chilled water piping. Chilled water systems typically supply chilled water between 42°F and 50°F. If the water were any colder, it would require an anti-freezing solution and be called a brine. Brine systems are generally only used for process cooling, where lower temperatures are required. Typically, the water in HVAC systems is cooled by 8°F, 10°F, or 12°F as it passes through the chiller. As the chilled water passes through the chilled water coils in air handling units, the water temperature rises by an equivalent amount, before it is returned to the chiller. The chilled water is circulated throughout the system by one or more chilled water pumps.

Most chillers operate on the same principle as the DX systems. The main difference is that the evaporator removes heat from the circulated water instead of from the air. Electrically driven chillers have a DX refrigeration system and are usually reciprocating, rotary screw, or centrifugal compressor chillers. Reciprocating chillers use pistons inside cylinders to compress the refrigerant. Reciprocating chillers typically have cooling capacities from 20 to as high as 200 tons. Rotary screw chillers use a rotating helical screw to compress the refrigerant. Screw chillers have cooling capacities from 50 to 750 tons. Centrifugal chillers use a rotating wheel or impeller and centrifugal force to compress the refrigerant. A centrifugal chiller is shown in Figure 10.22.

FIGURE 10.22   Centrifugal Chiller (*Courtesy of York International*)

FIGURE 10.23   Cooling Tower (*Courtesy of Evapco, Inc.*)

Centrifugal chillers typically have cooling capacities ranging from 100 to 2000 tons in a single machine.

Chilled water may also be produced by absorption chillers, which do not operate on the DX cycle. Absorption chillers use heat, instead of work from an electrical motor, to produce a refrigerating effect. Most absorption chillers use a lithium bromide and water solution as the circulated refrigerant. The cooling cycle is based on the absorption of heat from the chilled water when the refrigerant (water) evaporates. The refrigerant solution in the chiller is maintained at a near perfect vacuum. The absorption cycle utilizes the affinity that lithium bromide has for water while under a vacuum.

The refrigerant (water) is cooled by latent cooling as it evaporates and is absorbed by a strong concentration of lithium bromide solution. As the lithium bromide solution becomes diluted by the water, it is transferred to another section, where it is heated. The heat causes the water to be driven off. The lithium bromide is then returned to the absorber to again absorb the water refrigerant. The water vapor that was driven out of the lithium bromide is condensed by cooling water and returned to the evaporator to repeat the cycle. Water is chilled as it is circulated through the tubes of the evaporator.

Smaller chillers, like reciprocating chillers, have condensers that are air cooled. Larger chillers, like centrifugal and absorption chillers, have condensers that are water cooled. Water cooled chillers reject heat to the atmosphere by a cooling tower, which is usually located outdoors. A typical cooling tower is shown in Figure 10.23.

The amount of heat rejected by a cooling tower is increased if the circulated water comes in contact with the outdoor air. Most cooling towers spray water into the outside air, as it is circulated through the cooling tower by fans. Spraying the water increases the evaporation of the water which increases the heat rejection rate. As the water evaporates into the air, latent heat is transferred from the water to the air.

Chapter 11

CODES AND STANDARDS FOR HVAC SYSTEM DESIGN

There are three major code organizations in the United States who set requirements for buildings through building codes. More than 90% of all local and state governments have adopted one of these code sets. These codes are the Building Officials and Code Administrators International (BOCA), International Conference of Building Officials (ICBO), and the Southern Building Code Congress International (SBCCI). Local fire protection districts also adopt and enforce the National Fire Codes, published by the National Fire Protection Association (NFPA). In general, the codes impose requirements on new construction and the renovation of existing facilities, for the protection of public health, safety, and welfare.

The building codes most directly related to the design of HVAC systems include the BOCA Mechanical Code, the Uniform Mechanical Code, and the mechanical sections of the Southern Building Code. The fire codes that most directly affect the design of HVAC systems are NFPA 54 National Fuel Gas Code, NFPA 90A Air Conditioning and Ventilating Systems, NFPA 90B Warm Air Heating Systems, and NFPA 101 Life Safety Code.

In addition to codes, which are enforceable by law, there are a number of design standards with which the designer should be familiar. In some cases, an entire standard has been adopted as part of a national or local code. The following standards, published by the American Society of Heating, Refrigerating, and Air Conditioning Engineers (ASHRAE), should be used as guidelines in the design of HVAC systems: Standard 15, *Safety Code for Mechanical Refrigeration*, Standard 55, *Thermal Environmental Conditions for Human Occupancy*, Standard 62, *Ventilation for Acceptable Indoor Air Quality*, Standard 90.1, *Energy Efficient Design of New Buildings, Except Low-Rise Residential Buildings*, and Standard 100.3, *Energy Conservation in Existing Buildings—Commercial*. In addition to ASHRAE Standards, you should also be familiar with *HVAC Duct Construction Standards, Metal and Flexible*, published by the Sheet Metal and Air Conditioning Contractors National Association, Inc. (SMACNA).

Whenever an HVAC system is being designed, the designer should always be aware of the codes that are in effect for the project location. The designer must also be aware of which version of the code is in effect. Frequently, it is not the latest edition of a published code. In addition to the above codes, local jurisdictions frequently enact additional requirements and restrictions, above and beyond the national codes.

This chapter provides you with an overview of codes and standards for HVAC design. It addresses only major requirements that are common to most codes. You should always consult local authorities having jurisdiction at the project location to determine actual code requirements for a particular project.

11.1 GENERAL CODE REQUIREMENTS FOR HVAC SYSTEM DESIGN

Codes for mechanical systems cover the design, installation, maintenance, and inspections of mechanical systems, including heating systems, ventilating systems, cooling systems, steam and hot water systems, water heaters, process piping, boilers and pressure vessels, mechanical refrigeration systems, and appliances using gas, as well as related accessories. Definitions of terms used in the codes are usually provided to help provide a clear understanding of the code. The designer should review the definitions before proceeding with a project.

Codes generally require that equipment must be installed and arranged to afford adequate access for inspection, maintenance, and repair. Codes also set minimum clearances for fuel burning equipment, or other heat producing equipment, and combustible materials. The designer should also refer to the equipment manufacturer's recommendations and instructions for equipment clearances.

Frequently, codes set minimum energy efficiency standards for energy consuming equipment and systems. Energy efficiency requirements vary between codes; however, the Energy Policy Act of 1992 requires that each state must certify that its energy code meets or exceeds the requirements listed in ASHRAE Standard 90.1. Although ASHRAE 90.1 is a standard and not a code, it is the basis for many state and local energy codes.

Some codes, such as the Uniform Mechanical Code, set minimum pipe sizes for the condensate drains from air conditioning equipment. These pipe sizes assume a minimum slope of $1/8$ in/ft, and are based on the nominal cooling capacity of the air conditioning unit in tons of refrigeration. Table 11.1 gives the required pipe sizes.

The pipe sizes also assume 20% outside air at 90°Fdb and 73°Fwb, and 80% room air at 75°Fdb and 62.5°Fwb. The pipe sizes should be adjusted accordingly for other design conditions.

All electrical work must be in compliance with the National Electrical Code (NFPA 70).

11.2 FIRE AND SMOKE CONTROL IN AIR SYSTEMS

Approved fire dampers must be provided whenever air ducts penetrate, pass through, or terminate in walls that have a fire resistance rating of 2 hr or more. Whenever ductwork passes through a fire rated floor, it must either be enclosed in a fire rated chase or have fire dampers in the vertical openings of each floor. Fire dampers are also required wherever the

TABLE 11.1 Minimum Condensate Pipe Sizes

Equipment Capacity (tons)	Minimum Pipe Diameter
Up to 20	$3/4$ in
21 to 40	1 in
41 to 90	$1 1/4$ in
91 to 125	$1 1/2$ in
126 to 250	2 in

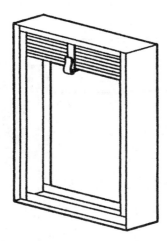

FIGURE 11.1 Curtain Type Fire Damper

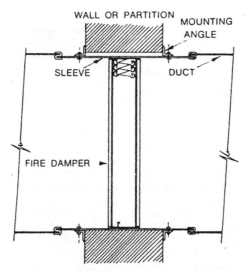

FIGURE 11.2 Typical Fire Damper Installation

ductwork passes through the walls of a fire rated chase. The purpose of a fire damper is to maintain the fire integrity of the wall or floor and prevent a fire from spreading from one area of a building to another.

A typical fire damper is shown in Figure 11.1 and the installation of a fire damper is shown in Figure 11.2.

The fire damper can be either a curtain type, as shown in Figure 11.1, or it can be a multi-blade type. In either case, the damper is spring loaded in the closed position, but is held open by a fusible link. The fusible link holds the damper open during normal operation. In the event of a fire, the fusible link melts and the springs force the damper closed.

Junction Box

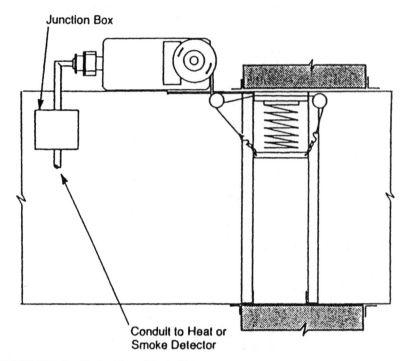

**Conduit to Heat or
Smoke Detector**

FIGURE 11.3   Combination Fire and Smoke Damper

The dampers are rated according to how long they can maintain their fire resistive integrity during a fire. Ratings for dampers are usually $1\frac{1}{2}$ hr and 3 hr. Fire dampers with a $1\frac{1}{2}$ hr rating are required in walls and floors with a 2 hr fire resistance rating. Fire dampers with a 3 hr rating are required in walls and floors that have a fire resistance rating of 3 or more hours. Occasionally, fire dampers are required in ceiling penetrations, such as air diffusers, if the ceiling is part of a fire rated ceiling and floor assembly.

All fire dampers must meet the requirements of, and be listed by, *UL 555 Fire Dampers and Ceiling Dampers*.

In addition to fire dampers, codes require that smoke dampers must be installed at, or adjacent to, the point where ductwork passes through a smoke barrier such as a wall. Frequently, the smoke barrier is also fire rated, in which case a combination fire and smoke damper is required. Figure 11.3 shows an typical installation of a combination fire and smoke damper.

Combination fire and smoke dampers may also be a curtain type or a multi-blade type. A combination fire and smoke damper is spring loaded closed with a fusible link holding the damper open that melts when the temperature reaches a set point. The damper may also be closed and opened by the operation of the damper actuator, which also holds the damper open during normal operation. Upon a signal from a fire alarm system, the actuator actuates and allows the springs to pull the damper to the closed position.

The purpose of a smoke damper is to prevent the spread of smoke from one area of a building to another in the event of a fire. Typically, the walls of corridors are both fire and smoke rated, since they must be used as a means of egress in the event of a fire. Figure 11.4 shows where fire and smoke dampers are required by code in a building. The numbers in the figure refer to the appropriate sections of NFPA 90A.

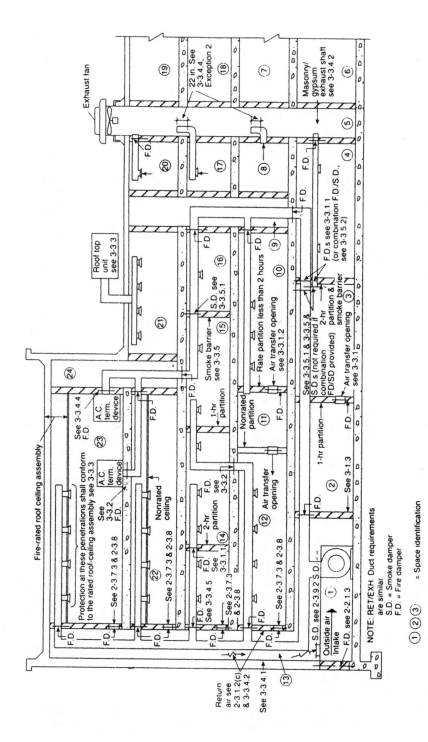

FIGURE 11.4 Standard 90A Fire Damper Applications

Fire codes also require smoke detectors in air handling systems. The purpose of these detectors is to shut the fan motor off if smoke is present in the air stream of the system. This prevents the spread of smoke throughout the building by the air system. At least one detector is required in all air systems that supply air greater than 2000 ft³/min. Different codes vary as to the required location of the single smoke detector in systems over 2000 ft³/min. Some codes require the detector in the return air, upstream of the air handling unit's air filters. Others require the detector downstream of the supply air fan. Controls for air systems are discussed in Chapter 20.

For air systems that circulate 15,000 ft³/min, or more, a smoke detector is required in the return air of the air handling unit and also in the return air duct for each story of a multistory building, prior to the duct connection to a common return air duct. Some codes allow an exception to the additional smoke detector in systems over 15,000 ft³/min, if the area served by the air distribution system is protected by a system of area smoke detectors.

All smoke detectors must be connected to a building fire alarm system, in addition to shutting the supply air fan off if they detect smoke.

11.3   VENTILATION REQUIREMENTS

In recent years, ventilation and indoor air quality have become a very important aspect of HVAC system design. People in the United States spend up to 90% of their time indoors. More than 70 million employees work indoors. Of that 70 million, it has been estimated that 21 million (about 30%) are exposed to poor indoor air quality (IAQ). It has also been estimated that there are more than 8000 chemicals in the air of most buildings. There are more than 13,000 if smoking is permitted. Outdoor air ventilation rates were reduced, with the advent of the energy crisis in the 1970s, in an attempt to reduce energy consumption in buildings. Until recently, some codes required that HVAC systems provide only 5 ft³/min of outside air, per person, in a space or building. Many studies have shown that ventilation systems are a major source of IAQ problems. 53% of the IAQ problems that the National Institute for Occupational Safety and Health (NIOSH) have investigated have been due to ventilation deficiencies. Increased outdoor air is an effective way to reduce indoor air pollution.

In 1989, the American Society of Heating, Refrigerating, and Air Conditioning Engineers (ASHRAE) introduced new guidelines for ventilation rates and building IAQ, with Standard 62–1989 Ventilation for Acceptable Indoor Air Quality. While this standard is not a code, enforceable by law, requiring compliance by HVAC designers and operators, it is becoming the standard by which system design and operations are judged. Standard 62–89 sets the minimum outdoor ventilation air rate at 15 ft³/min per person. The recommended outdoor ventilation rates increase, depending upon the occupancy of the space or building. The standard also provides guidelines for estimating the number of occupants for different building occupancies. Table 11.2 gives the ASHRAE 62–89 ventilation guidelines. The 1993 BOCA Mechanical Code increased the minimum outdoor air requirements from 5 ft³/min, to 15 ft³/min per person. Like many other codes, it has essentially adopted the same requirements as those recommended by ASHRAE Standard 62–1989. In addition to the outdoor air requirements, it sets minimum exhaust air requirements for areas such as public restrooms. The code requires that the HVAC system provide at least 75 ft³/min of exhaust air for each water closet, or urinal, in a public restroom.

11.4   AIR DISTRIBUTION

Many mechanical codes specify, or restrict, the manner in which air is distributed throughout a building by the HVAC system. This section discusses some of the require-

Table 11.2 Outdoor Air Requirements for Ventilation
Copyright 1989 by the American Society of Heating Refrigerating and Air Conditioning
Engineers, Inc. from ASHRAE Standard 62-89 Ventilation for Indoor Air Quality. Used by permission

OUTDOOR AIR REQUIREMENTS FOR VENTILATION*
COMMERCIAL FACILITIES (offices, stores, shops, hotels, sports facilities)

Application	Estimated Maximum** Occupancy P/1000 ft² or 100 m²	Outdoor Air Requirements				Comments
		cfm/ person	L/s· person	cfm/ft²	L/s·m²	
Dry Cleaners, Laundries						Dry-cleaning processes may require more air.
Commercial laundry	10	25	13			
Commercial dry cleaner	30	30	15			
Storage, pick up	30	35	18			
Coin-operated laundries	20	15	8			
Coin-operated dry cleaner	20	15	8			
Food and Beverage Service						
Dining rooms	70	20	10			
Cafeteria, fast food	100	20	10			
Bars, cocktail lounges	100	30	15			Supplementary smoke-removal equipment may be required.
Kitchens (cooking)	20	15	8			Makeup air for hood exhaust may require more ventilating air. The sum of the outdoor air and transfer air of acceptable quality from adjacent spaces shall be sufficient to provide an exhaust rate of not less than 1.5 cfm/ft² (7.5 L/s·m²).
Garages, Repair, Service Stations						
Enclosed parking garage				1.50	7.5	Distribution among people must consider worker location and concentration of running engines; stands where engines are run must incorporate systems for positive engine exhaust withdrawal. Contaminant sensors may be used to control ventilation.
Auto repair rooms				1.50	7.5	
Hotels, Motels, Resorts, Dormitories				cfm/room	L/s·room	Independent of room size.
Bedrooms				30	15	
Living rooms				30	15	
Baths				35	18	Installed capacity for intermittent use.
Lobbies	30	15	8			
Conference rooms	50	20	10			
Assembly rooms	120	15	8			
Dormitory sleeping areas	20	15	8			See also food and beverage services, merchandising, barber and beauty shops, garages.
Gambling casinos	120	30	15			Supplementary smoke-removal equipment may be required.
Offices						
Office space	7	20	10			Some office equipment may require local exhaust.
Reception areas	60	15	8			
Telecommunication centers and data entry areas	60	20	10			
Conference rooms	50	20	10			Supplementary smoke-removal equipment may be required.
Public Spaces				cfm/ft²	L/s·m²	
Corridors and utilities				0.05	0.25	
Public restrooms, cfm/wc or urinal		50	25			Mechanical exhaust with no recirculation is recommended.
Locker and dressing rooms				0.5	2.5	
Smoking lounge	70	60	30			Normally supplied by transfer air, local mechanical exhaust; with no recirculation recommended.
Elevators				1.00	5.0	Normally supplied by transfer air.

* Table prescribes supply rates of acceptable outdoor air required for acceptable indoor air quality. These values have been chosen to control CO₂ and other contaminants with an adequate margin of safety and to account for health variations among people, varied activity levels, and a moderate amount of smoking.

**Net occupiable space.

ments, and restrictions, common to most codes. The actual design of air distribution systems is covered in Chapters 14 and 15.

A plenum is defined as an enclosed portion of a building, or structure, which is designed to allow air movement and thereby serve as part of an air distribution system. All materials exposed to the air flow from the air HVAC system, such as in return or supply air plenums,

Table 11.2 Outdoor Air Requirements for Ventilation
Copyright 1989 by the American Society of Heating Refrigerating and Air Conditioning
Engineers, Inc. from ASHRAE Standard 62-89 Ventilation for Indoor Air Quality. Used by permission

OUTDOOR AIR REQUIREMENTS FOR VENTILATION* (Continued)
COMMERCIAL FACILITIES (offices, stores, shops, hotels, sports facilities)

Application	Estimated Maximum** Occupancy P/1000 ft² or 100 m²	Outdoor Air Requirements				Comments
		cfm/ person	L/s· person	cfm/ft²	L/s·m²	
Retail Stores, Sales Floors and Show Room Floors						
Basement and street	30			0.30	1.50	
Upper floors	20			0.20	1.00	
Storage rooms	15			0.15	0.75	
Dressing rooms				0.20	1.00	
Malls and arcades	20			0.20	1.00	
Shipping and receiving	10			0.15	0.75	
Warehouses	5			0.05	0.25	
Smoking lounge	70	60	30			Normally supplied by transfer air, local mechanical exhaust; exhaust with no recirculation recommended.
Specialty Shops						
Barber	25	15	8			
Beauty	25	25	13			
Reducing salons	20	15	8			
Florists	8	15	8			Ventilation to optimize plant growth may dictate requirements.
Clothiers, furniture				0.30	1.50	
Hardware, drugs, fabric	8	15	8			
Supermarkets	8	15	8			
Pet shops				1.00	5.00	
Sports and Amusement						
Spectator areas	150	15	8			When internal combustion engines are operated for maintenance of playing surfaces, increased ventilation rates may be required.
Game rooms	70	25	13			
Ice arenas (playing areas)				0.50	2.50	
Swimming pools (pool and deck area)				0.50	2.50	Higher values may be required for humidity control.
Playing floors (gymnasium)	30	20	10			
Ballrooms and discos	100	25	13			
Bowling alleys (seating areas)	70	25	13			
Theaters						Special ventilation will be needed to eliminate special stage effects (e.g., dry ice vapors, mists, etc.)
Ticket booths	60	20	10			
Lobbies	150	20	10			
Auditorium	150	15	8			
Stages, studios	70	15	8			
Transportation						Ventilation within vehicles may require special considerations.
Waiting rooms	100	15	8			
Platforms	100	15	8			
Vehicles	150	15	8			
Workrooms						
Meat processing	10	15	8			Spaces maintained at low temperatures (−10°F to +50°F, or −23°C to +10°C) are not covered by these requirements unless the occupancy is continuous. Ventilation from adjoining spaces is permissible. When the occupancy is intermittent, infiltration will normally exceed the ventilation requirement. (See Ref 18).

* Table prescribes supply rates of acceptable outdoor air required for acceptable indoor air quality. These values have been chosen to control CO_2 and other contaminants with an adequate margin of safety and to account for health variations among people, varied activity levels, and a moderate amount of smoking.

**Net occupiable space.

must be noncombustible, or have limited combustibility, and have a smoke developed index of 50 or less. Mechanical equipment rooms, used as return air plenums for air handling units, may not be used for storage, or be occupied. Insulation and other materials used in ductwork must not have a flame spread over 25 and a smoke developed rating that is no higher than 50 if they are inside the building. Weatherproof coatings for ductwork outside a building need not meet these requirements. Some codes restrict the use of PVC and CPVC

Table 11.2 Outdoor Air Requirements for Ventilation
Copyright 1989 by the American Society of Heating Refrigerating and Air Conditioning
Engineers, Inc. from ASHRAE Standard 62-89 Ventilation for Indoor Air Quality. Used by permission

OUTDOOR AIR REQUIREMENTS FOR VENTILATION* (*Concluded*)
COMMERCIAL FACILITIES (offices, stores, shops, hotels, sports facilities)

Application	Estimated Maximum** Occupancy P/1000 ft² or 100 m²	Outdoor Air Requirements				Comments
		cfm/ person	L/s· person	cfm/ft²	L/s·m²	
Photo studios	10	15	8			
Darkrooms	10			0.50	2.50	
Pharmacy	20	15	8			
Bank vaults	5	15	8			
Duplicating, printing				0.50	2.50	Installed equipment must incorporate positive exhaust and control (as required) of undesirable contaminants (toxic or otherwise).

INSTITUTIONAL FACILITIES

Application	Estimated Maximum Occupancy	cfm/ person	L/s· person	cfm/ft²	L/s·m²	Comments
Education						
Classroom	50	15	8			
Laboratories	30	20	10			Special contaminant control
Training shop	30	20	10			systems may be required for
Music rooms	50	15	8			processes or functions including
Libraries	20	15	8			laboratory animal occupancy.
Locker rooms				0.50	2.50	
Corridors				0.10	0.50	
Auditoriums	150	15	8			
Smoking lounges	70	60	30			Normally supplied by transfer air. Local mechanical exhaust with no recirculation recommended.
Hospitals, Nursing and Convalescent Homes						
Patient rooms	10	25	13			Special requirements or codes and
Medical procedure	20	15	8			pressure relationships may deter-
Operating rooms	20	30	15			mine minimum ventilation rates
Recovery and ICU	20	15	8			and filter efficiency. Procedures generating contaminants may require higher rates.
Autopsy rooms				0.50	2.50	Air shall not be recirculated into other spaces.
Physical Therapy	20	15	8			
Correctional Facilities						
Cells	20	20	10			
Dining halls	100	15	8			
Guard stations	40	15	8			

* Table prescribes supply rates of acceptable outdoor air required for acceptable indoor air quality. These values have been chosen to control CO_2 and other contaminants with an adequate margin of safety and to account for health variations among people, varied activity levels, and a moderate amount of smoking.

**Net occupiable space.

plastic piping in air plenums. The space between a corridor ceiling and the floor, or roof above, may be used as a return air plenum as long as the integrity of any fire and/or smoke rated surfaces are maintained. See Section 11.2.

Air may not be recirculated from any space in which flammable vapors or dust are present in quantities and concentrations that would cause hazardous conditions. Return air inlets must be located at least 3 in above the floor.

Exit passage ways, public corridors, stairs, and other exits may not be used as part of a supply air, return air, or exhaust air system serving other areas of a building. The only exception to this requirement is for toilet rooms, bathrooms, janitors' closets, and similar auxiliary spaces opening directly into a corridor. In these cases, exhaust air for the auxiliary space may be drawn in from a corridor.

Outside air intakes for HVAC systems must be located to avoid drawing in any combustible materials, flammable vapors, objectionable odors, or vehicle exhaust. They should also be located to minimize the hazard from fires that may occur in other nearby structures. The outside air intakes must prevent the introduction of undesirable material and debris by having screens made of corrosion resistant material that does not have a mesh larger than $1/2$ in. Outside air should also not be taken from any location closer than 10 ft from an appliance vent, plumbing vent, or the discharge from an exhaust fan, unless the outside air intake is at least 3 ft below the vent or discharge. The outside air inlet must be at least 10 ft above the surface of any abutting driveway, or at least 10 ft horizontally from a sidewalk, street, alley, or driveway.

Exhaust air discharge may not be directed onto walkways or sidewalks.

11.5   DUCT CONSTRUCTION FOR HVAC SYSTEMS

Guidelines and recommendations for the construction and installation of ductwork are contained in *HVAC Duct Construction Standards, Metal and Flexible*, published by the Sheet Metal and Air Conditioning Contractors National Association, Inc. (SMACNA) (Ref. 21). Ductwork is normally classified according to whether it is under a positive (supply) or negative (return or exhaust) pressure, the static pressure in the duct, and the velocity of the air in the duct. A low pressure air distribution system is one in which the static pressure in the ductwork is 2 in. wg or less. A medium pressure system is one in which the static pressure is greater than 2 in, up to 6 in. wg. A high pressure system is one that has a static pressure greater than 6 in. wg. Duct system design is covered in Chapter 15.

Ductwork can be either rigid or flexible. Rigid ductwork is usually constructed of galvanized steel, stainless steel, or fiberglass ductboard. Flexible ductwork (or air connector) may be metallic or non-metallic. Material thickness and construction standards are a function of the duct usage classification, material, air pressure, and air velocity. Guidelines for material thickness and construction requirements may be found in References 3 and 21.

Ductwork must be securely supported by metal hangers or brackets. The maximum allowable distance between supports for ductwork is 10 ft. The designer should also consult local codes for any requirements regarding seismic support, or restraint, of ductwork and other equipment.

In warm air furnace type systems, the minimum unobstructed free area of the supply air duct should be at least 2 in²•1000 Btu/hr output of the heating device. For heat pump heating systems, the free area should be at least 6 in²•Btu/hr output.

Flexible ductwork is frequently used as a connector, or "run-out," between air terminals and room air devices, or main trunk ducts and room air devices. Flexible ductwork is usually factory made and must be made in compliance with UL 181, *Standard for Factory-Made Air Ducts and Connectors*. Flexible air ducts may not be used where air temperatures will exceed 250°F, and may not be used as vertical air ducts that are more than two stories in height. The maximum allowable length for flexible ductwork is 14 ft. It may not pass through any wall, partition, etc. that has a fire rating of 1 hr or more. Flexible air ducts may not pass through floors.

Provisions must be made to prevent the formation of condensation on the exterior of any ductwork. Duct coverings and linings, as well as tape and adhesives, may not have a flame spread rating over 25, or a smoke developed rating of greater than 50. Duct coverings and linings must not flame, glow, smolder, or smoke when subjected to a flame. Duct coverings may not extend through walls or floors that are required to be fire stopped, or required to have a fire resistive rating.

11.6   FUEL BURNING HVAC SYSTEMS

Fuel burning HVAC systems include natural gas, propane, and fuel oil furnaces. Burners used in furnaces can be the atmospheric type, or power type. All combustion processes consume oxygen as part of the process. The oxygen is obtained from the oxygen in air. Atmospheric type burners draw air, at atmospheric pressure, into the combustion chamber, as a natural part of the combustion process. Power type burners use a fan, or blower, to provide air to the combustion chamber at a higher rate than atmospheric type, and at a pressure above atmospheric pressure. Power burners are normally used for equipment with relatively high heating capacities.

Since all combustion processes consume oxygen, they must have an adequate supply of combustion air. If the burner is inside a building, the HVAC system designer must ensure that there is an adequate supply of combustion air. In some cases, combustion air can be drawn from inside the building if the enclosed space is large enough to provide sufficient combustion air. In other cases, combustion air must be supplied from outside the building. Most codes determine the requirements for combustion air based on the fuel input to the burner in Btu/hr.

For burners inside a building where the combustion air is completely drawn from inside, there must be at least 40 ft^3 of room air volume for each 1000 Btu/hr of total fuel input. If the room where the burner is located does not meet the above criteria, openings may be provided to adjacent spaces, so the combined volume of all spaces meets the criteria. There must be two openings between the room with the burner and the adjacent spaces. One opening must be near the top of the room and one near the bottom, as shown in Figure 11.5. Each opening must have a minimum of 1 in^2 of free area per 1000 Btu/hr total input rating, with a minimum opening size of 100 in^2.

Combustion air may also be drawn in from outdoors. There must be one combustion air opening, near the top of the room, and one near the bottom, as shown in Figure 11.6. For ducted horizontal combustion air openings, each opening must have at least 1 in^2 of free area for each 2000 Btu/hr total input rating. For direct combustion air openings through a wall, each opening must have at least 1 in^2 of free area per 4000 Btu/hr total input rating. Outside combustion air may also be provided vertically, such as through a roof. Each open-

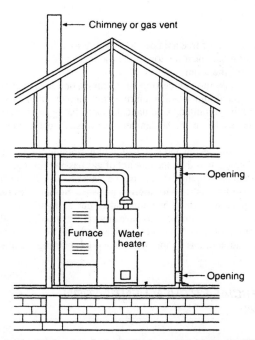

FIGURE 11.5 Combustion Air Openings for Inside Combustion Air (*Courtesy of National Fire Protection*)

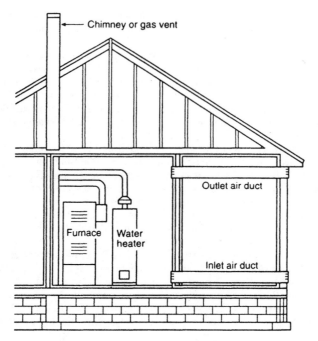

FIGURE 11.6 Combustion Air Openings for Outside Combustion (*Courtesy of National Fire Protection Association*)

ing must have at least 1 in² of free area per 4000 Btu/hr total input rating. One of the vertical openings must terminate near the top of the room. The other must be ducted to within 12 in above the floor of the room.

Some codes will not permit combustion air openings, or ducts, to pass through any construction that is fire rated and would require a fire damper. Manually operated volume dampers also may not be permitted. Automatic dampers that open when the burner fires, are permitted by some codes; however, the damper must be such that it opens if it loses control power.

Codes also regulate the design and installation of gas piping for gas burning equipment. The sizing of gas piping is based on the amount of gas passing through the pipe, the initial pressure of the gas, and the length of the gas piping. Most codes have several tables that give the minimum pipe size for gas flow for a given pressure. Some codes restrict, or prohibit, the routing of gas piping through air plenums and air ducts. If gas piping is permitted in an air plenum, it is usually not permissible to have any valves or unions within the plenum.

You should consult NFPA 54 National Fuel Gas Code, as well as local codes, regarding the design and installation of gas piping systems and gas burning equipment.

11.7 COMMERCIAL KITCHEN VENTILATION SYSTEMS

Codes regulate the design, construction, and installation of kitchen ventilation systems in order to minimize health and safety hazards. They usually regulate the amount of exhaust air, and make-up air, as well as air velocity in the ductwork, hood construction, duct construction, gas piping to the appliances under the hood, and kitchen hood fire suppression systems. Code requirements vary with various code jurisdictions. The designer should always review local code requirements before proceeding with the design of a kitchen ventilation system.

The cooking equipment is actually a source of contaminated air which must be removed from the conditioned space. Cooking equipment also generates both sensible and latent heat, which also must be removed from the conditioned space. Contaminated air is usually removed by placing the cooking equipment under a kitchen hood with an exhaust fan and ductwork. Make-up air to replenish the exhausted air may be provided from the kitchen itself or through the kitchen hood.

Commercial kitchen hoods must be constructed of either steel or stainless steel. All joints and seams in the hood must be welded liquid tight. Hoods must be tested and labeled in accordance with UL 710, *Exhaust Hoods for Commercial Cooking Equipment*. Canopy type hoods are hoods that completely cover the cooking equipment, whereas, non-canopy hoods do not completely cover the cooking equipment. Canopy hoods attached to a wall must exhaust at least 100 ft^3/min•ft^2 of hood area. Canopy hoods exposed on all sides must exhaust at least 150 ft^3/min•ft^2 of hood area. Non-canopy hoods must exhaust at least 300 ft^3/min•ft^2 of cooking surface.

Make-up air must be provided to replenish the air exhausted by a kitchen hood. The amount of make-up air must be approximately equal to the amount exhausted. The temperature of the make-up air must be within 10°F of the conditioned space. The make-up air temperature can be greater than a 10°F difference from the space if it will not decrease the comfort conditions in the space. Untempered make-up air may be used if the air is supplied to a hood that is designed to supply untempered air from under the hood itself. The designer should verify that local codes will permit the use of untempered air and, if so, that the hood is suitable for untempered make-up air.

The kitchen exhaust system must be independent of the building ventilation or exhaust system. It must have separate ductwork and an exhaust fan. The ductwork must be constructed of steel, or stainless steel, with all joints and seams welded liquid tight. Fiberglass duct board and flexible ductwork may not be used. The air velocity in the exhaust duct must be between 1500 and 2200 ft/min. Clean out openings must be provided every 20 ft in horizontal ductwork, or as otherwise required by code. Exhaust ductwork must terminate, or discharge, at least 10 ft above grade level. The fan motor of the exhaust fan must be located so it is outside the exhaust air stream.

Most codes also require a fire suppression system for kitchen hoods. The suppression system must discharge automatically in the event of a fire under the hood. The fire suppression system must also be such that it can be activated manually by "pull station." If the fire suppression system is activated, all fuel and electricity to the cooking equipment under the hood must automatically be shut off.

Chapter 12

ACOUSTICS, VIBRATION, AND SEISMIC DESIGN FOR HVAC SYSTEMS

While the main purpose of an HVAC system is to provide acceptable temperatures, humidities, and air quality inside a building, it must not produce objectionable levels of noise or vibration in the process. Other than complaints about temperature problems, more complaints are received by building managers about excessive HVAC system noise than from any other source. In recent years, the trend has been toward lighter building construction, which has resulted in greater sound transmission. People occupying the space also are having more sophisticated requirements. Studies have shown that people working in sound controlled spaces are more productive than those who are not working in a sound controlled space. As a result, there are economic benefits to providing a space without objectionable noise or vibration.

Most causes of excessive noise in HVAC systems are due to improper initial design, poor installation, and cost cutting without regard to noise and vibration. The greatest sources of noise problems in HVAC systems are chillers, air handling units, air terminal units, and room air devices, such as diffusers, grilles, and registers. Problems of excessive noise and vibration can be avoided if the fundamental principles of sound and vibration control are applied during the selection and design of an HVAC system.

The purpose of this chapter is to present the basics of acoustical and vibration design for HVAC systems, as well as basic seismic design. It will help the system designer to select and design a system that provides a comfortable and safe indoor environment. Some HVAC systems are inherently noisier than others. Noise and vibration design should begin during the schematic design and system selection phases, and continue through the entire design process.

In addition to not generating objectionable levels of noise or vibration, the system may require seismic restraints to prevent excessive movement in the event of an earthquake. Local codes usually specify the extent of seismic restraint for HVAC system equipment and components.

12.1   THE BASICS OF SOUND AND VIBRATION

A basic principle in the study of sound and vibration is that of wave motion. A common example of wave motion similar to sound is a situation of a stone tossed into a quiet pool of water. When the stone enters the water, it creates an ever-widening set of circular waves that spread across the surface from the point of impact. The water does not move as a whole, but a small part of the surface does move. Another example of wave motion, which is more akin to vibration, is the motion of a disturbance along a rope that is stretched horizontally. If there is only one wave, it is known as a single-pulse wave. If the waves are repeated many times, at equal time intervals, they are known as periodic waves. Still another example of wave motion is the motion that the prongs of a tuning fork set up in the air surrounding the prongs. Periodic waves are created in the air as the forward movement of a prong compresses the air in front of it while the backward movement rarefies the air. These conditions are transmitted outward from the fork as a wave disturbance. Upon entering the ear, these waves produce the sensation of sound, and the waves themselves are known as sound waves.

Therefore, sound may be defined as a variation in pressure due to vibration in an elastic fluid, such as air. Airborne sound is a variation of air pressure, with atmospheric pressure as the mean value. Since sound is transmitted by compression and expansion of the molecules of the material through which it is traveling, it cannot travel through a vacuum. The greater the density of the material, the faster the traveling speed of sound waves.

Vibration is similar to sound in that it involves wave motion; however, the media through which the waves travel are different. Sound waves usually travel through a fluid that is an elastic material, such as air or water. Vibration, on the other hand, involves the transmission of wave motion through an elastic solid or body. A material that is elastic is one which may be subjected to a stress that causes a deformation but will still be able to return to its original shape upon removal of the stress.

An example of a vibrating system is a weight attached to a spring which is dropped and is restrained from falling by a spring. As the weight falls, it converts kinetic energy of the falling weight to stored potential energy within the spring. The weight continues to fall until it is stopped by the spring. At this point, the stored energy in the spring accelerates the weight in an upward motion until all the spring's potential energy is converted back to kinetic energy, as the weight travels upward. The weight travels upward until gravity stops the upward motion, at which time the cycle begins to repeat itself.

The weight continues to cycle, or oscillate, until it is acted upon by some outside force or until all of the energy is dissipated due to friction in the motion of the system. The oscillations of sound or vibrations repeat themselves at a constant time rate called a period, and at a constant magnitude called the amplitude. A graph of a typical undampened oscillating system is shown in a sin wave, which is shown in Figure 12.1.

Note that the waves in Figure 12.1 repeat themselves at the same time period and at the same amplitude (maximum and minimum values). Frequently, the waves are dampened by some outside attenuation. A graph of a dampened oscillating system is shown in Figure 12.2.

In the case of sound, the dampening might occur as the result of some sound absorbing material. In the case of a vibrating system, the dampening might occur as the result of a device which absorbs energy (such as a shock absorber) or a device that limits amplitude, or travel (snubber), of the oscillations.

A cycle is one complete excursion of an oscillation. The frequency is the number of cycles (or complete oscillations) in a given unit of time, usually cycles/sec (Hz). The wavelength of the sound wave is the distance between regions of identical rarefaction or compression of the medium. The amplitude of motion is the maximum displacement beyond the normal or "rest" position of the element being considered.

The velocity of a longitudinal wave in a fluid is a function of the density and the modulus of elasticity of the medium. In a given medium, for fixed conditions, the speed of

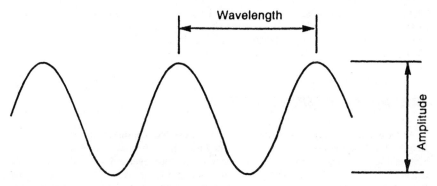

FIGURE 12.1   Undampened Oscillating System Sin Wave

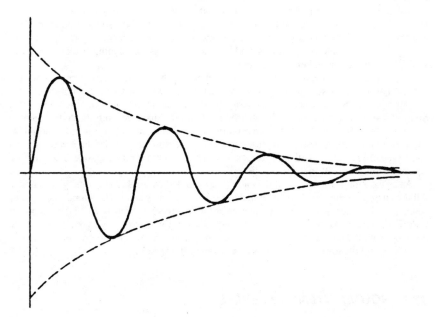

FIGURE 12.2   Dampened Oscillating System Sin Wave

sound (acoustic velocity) is a constant. The relationship between velocity, frequency, and wavelength is expressed by Equation 12.1.

$$c = \lambda f \tag{12.1}$$

where

   c = the speed of sound in ft/sec

   f = the frequency in cycles/sec (Hz)

   λ = the wavelength in ft

In fluids and most solids the wave or impulse expands spherically in a manner similar to an expanding soap bubble. The surface of the expanding sphere is called the "wave front." The energy of the original disturbance is diminishing at any point on the wave front as the square of the distance from the source.

12.2   SOUND REFERENCE LEVELS

In most cases, the absolute values obtained from sound measurements are of little use. In order for them to have any significance, they must be compared with some base or reference. Sound level is measured with respect to a reference level and is usually a ratio with respect to the reference level. The zero level of sound pressure is not a true absolute zero. It is not the absence of any pressure in excess of what would exist with no sound energy present in the medium. It is, however, something of an average "threshold of hearing," which is a pressure of about 3×10^{-9} lb/in^2 (psi) for humans.

Sound pressure is an audible disturbance within the atmosphere whose intensity is influenced by the surroundings and the distance from the source of the sound. Sound pressure is influenced by the surroundings and can be measured directly. However, sound power values must be calculated by making sound pressure readings and adding back any kind of source and environmental effects due to the surroundings.

Acoustic energy is the total energy of a given part of the transmitting medium minus the energy that would exist in the same part in the medium with no sound present. Sound intensity, or sound power, is acoustical energy emitted by the sound source measured in watts. Sound pressure is a function of the environment and the distance from the source. Sound power, on the other hand, is not a function of the environment and distance. Sound intensity is difficult to measure directly; however, sound pressure may readily be measured with some type of pressure sensitive device, such as a human ear or a microphone. The pressure squared is proportional to the intensity.

As an example, assume a sound source is surrounded by a sphere of some given radius. All the energy emitted by the source must pass through the sphere. The sound intensity, or sound power, is the power flow per unit area of the sphere, expressed as W/m^3. Sound power follows the inverse square law; i.e., the sound intensity varies inversely as the square of the distance from the source. The greater the distance from the source, the greater the area of the sphere and the lower the energy per unit area of the sphere.

12.3   SOUND INTENSITY LEVELS

The human ear is capable of hearing a wide range of sound levels. If a value of 1 were assigned to the lowest level, or threshold of hearing, a value of 1×10^{13} would be the value assigned to the sound level produced by a jet engine. That is a factor of 10 million. With such a wide range in values, it is almost impossible to have a linear scale to include measurements and values at both ends of the extremes. Since audible sound has such a wide range in values, it was necessary to develop a logarithmic scale to measure and describe sound intensity levels. For this reason, the sound pressure levels are given in decibels (dB), which are a dimensionless quantity. The decibel is a basic unit of measurement in acoustics. Since decibels are a dimensionless measurement, it is necessary to specify a reference level when using decibel scales.

Numerically, the decibel is ten times the base 10 logarithm of the ratio of two like quantities proportional to acoustical power or energy. A reference quantity is always implied, even if it is not noted. Just as ground elevations are measured relative to sea level as a datum, sound intensity levels are measured from a particular sound intensity chosen as an arbitrary reference standard. Equation 12.2 is the equation for sound intensity levels in dB.

$$N = 10 \log_{10} \left(\frac{I}{I_r} \right)$$
(12.2)

where

N = the sound intensity level in dB

I = the intensity level of the sound under consideration in W/cm²

I_r = the reference intensity of 10^{-16} W/cm²

The more intense the sound, the higher the number of decibels, which expresses its intensity level.

For comparison purposes, the effect of 1 dB can be observed by setting $N = 1$ and solving for I/I_r.

The result is $10^{0.1} = 1.26$, showing that a 1 dB increase in sound would result in a 26% increase in sound intensity. If the threshold of hearing has a sound intensity level of 0 dB, then a jet engine would have a sound intensity level of 140 dB.

Obviously, since sound intensity levels are logarithmic, decibel levels cannot just be added arithmetically. Two sources each producing 50 dB do not produce a combined sound level of 100 dB; rather, they produce 53 dB (since there are two 50 dB sources, $10 \log_{10} 2 = 3$, $50 + 3 = 53$ for the level of the two sources).

12.4 SOUND LEVEL RATINGS

Pitch is the physical response of the human ear to various sound frequencies. Low frequencies are identified as being low in pitch and high frequencies are referred to as being high in pitch. Human hearing encompasses a range of frequencies from about 16 cps (cycles per second) to almost 20,000 cps. Only at high sound pressure levels is the response reasonably flat throughout the frequency range. Low frequencies do not strongly affect the human ear; however, frequencies from 500 to 5000 cps are very noticeable. Humans tend to be very sensitive to sound in the middle and higher frequencies.

Loudness is strongly dependent upon pitch as well as on the energy level of the sound. Loudness of a sound is a physical response to sound pressure (intensity) and the frequency of the sound. Loudness is strongly dependent on pitch, as well as on the amount of energy of the sound. At any given frequency, or pitch, the loudness varies directly as the sound pressure and intensity; however, it does not vary linearly. Figure 12.3 shows a set of curves that are called "equal loudness contours."

The curves represent the sound pressure level necessary at each frequency to produce the same loudness response in the "average listener." The lower dashed curve represents the threshold of hearing for the human ear. It represents the sound pressure level necessary to produce the sensation of hearing in the average listener. The actual threshold varies by as much as ± 10 dB for "normal" people. The upper curve represents the "threshold of feeling" where the sound pressure level is so high that pain may be felt.

Since loudness is dependent on sound pressure and frequency, sounds have been subdivided into the audible range of frequencies called octaves or octave bands. Each octave band is a continuous range of frequencies identified by its center frequency. The center

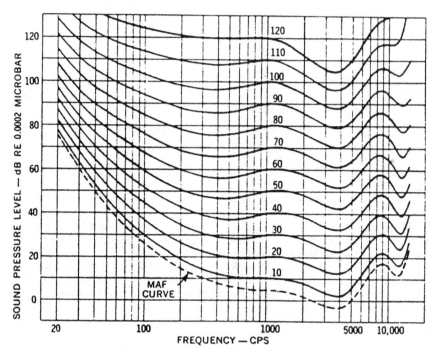

FIGURE 12.3   Equal Loudness Curves

TABLE 12.1   Octave Band Frequencies

Octave Band	Center Frequency	Band Edge Frequencies
1	63	44.6–88.5
2	125	88.5–177
3	250	177–354
4	500	354–707
5	1,000	707–1,414
6	2,000	1,414–2,830
7	4,000	2,830–5,650
8	8,000	5,650–11,300

frequency of a band is double that of the next lower band. For sound analysis purposes, the audible range of frequencies is normally broken down into eight octave bands. Table 12.1 gives the octave bands and frequency points used in HVAC system acoustics.

There have been a number of methods developed to relate sound pressure and frequency combinations to sound levels as perceived by the human ear. The most commonly used methods of expressing sound levels are:

A-B-C Weighting Networks

Phons

Sones

Noise Criteria (NC) Levels

Room Criteria (RC) Levels

The A-B-C Weighting Network method is a simple method for comparing the relative loudness of one noise with another. It is widely used since sound measurements may be performed with an instrument such as a hand-held sound meter. In an attempt to simulate the response of a human ear to both loudness and frequency, the sound meter typically has three scales, A, B and C. The "C" scale of the sound meter provides equal responses in all frequencies above 50 Hz. It gives an indication of ear response at high sound levels. The "A" scale simulates ear responses to sounds of low pressure levels, while the "B" scale simulates ear response to sounds of medium pressure levels.

The A-weighted sound level is frequently used in the design of HVAC systems and to measure HVAC system acoustics. The measurements, however, only measure the loudness of the sound without regard to the frequency distribution of the sound. Since both the sound and frequency of a sound are important in HVAC system acoustics, the A-weighted sound measurements are of limited use. They should not be used to set noise control or design objectives for HVAC systems.

Another frequently used measurement of loudness is the phon. Since the ear's perception of loudness is based upon both the intensity and frequency of the sound, the phon is a somewhat subjective method of expressing loudness. Two sounds, one with a frequency of 1000 cps and another sound at some other frequency, are compared. The intensity level of the second sound is adjusted until it is judged to be equally loud as the reference sound. The unit of loudness is then called the phon. Curves of equal loudness, or phons, that relate intensity and frequency may then be plotted on a graph, as shown in Figure 12.4.

The curves are lines of equal loudness for various intensity levels and frequencies, as judged by the "average observer." Note that the phon scale is not linear, but logarithmic. A sound that is 40 phons is not "twice as loud" as 20 phons. Although the phon scale covers a wide range, it does not fit a subjective loudness scale since it is logarithmic.

The sone is another frequently used method of measuring total loudness and expressing it as a single number. The sone is computed from an octave band analysis, based on equal loudness curves or contours. A sone may be defined as the loudness of a sound with a frequency of 1000 cps and a pressure of 0.02 microbars, which is 40 dB (re: 0.0002 microbars). This is also equal to the loudness level of 40 phons. The relationship between phons and sones is shown in Figure 12.5.

Notice that the phon scale is logarithmic, while the sone scale is linear. The sone scale is linear to the human ear. A sound that is 20 sones is twice as loud as a sound that is 10 sones. Sones are frequently used to rate non-ducted fans, such as propeller fans and power ventilators.

Humans find absolute silence very uncomfortable, except for very brief periods of time. People almost always prefer some background or "masking" sound. Privacy, rather than total silence, is what people actually prefer. Therefore, some background sound is not only acceptable, but is desirable. In addition to providing temperature, humidity, and ventilation control in buildings, HVAC systems frequently provide desirable background noise in buildings.

The acceptable levels of background noise for various types of spaces can be related to two principal characteristics of noise, which are speech interference level and loudness level. The speech interference level is governed by the amount of high frequency noise

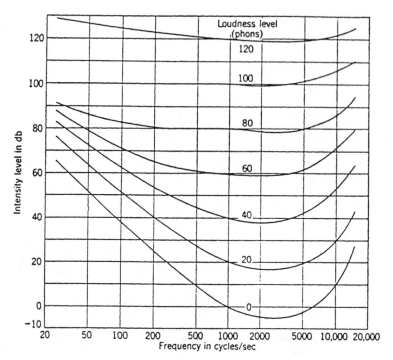

FIGURE 12.4 Equal Loudness Curves (Phons)

(between 600 and 4800 cps). The loudness level is a single number and is governed by both high and low frequency noise.

Noise Criteria (NC) Curves are derived by measuring various sound pressure levels at each of the eight octave bands. Each curve is assigned a value that represents average tolerances to sound levels over an entire frequency spectrum. The curves define limits that a noise source must not exceed, in order to achieve levels of occupant acceptance. NC Curves attempt to identify what sound levels of broad band noise are acceptable at each octave band. Figure 12.6 shows the NC Curves for the eight octave bands. The sound level at each octave band, along with a given NC Curve, would be judged about the same loudness anywhere along the curve.

NC Curves are frequently used to determine the noise level in a room that is generated by HVAC systems. By applying NC sound levels, it is possible to express acceptable sound levels for various types of spaces. The NC noise limits are to specify upper limits of the octave band levels caused by HVAC systems which are acceptable depending on the type of occupancy.

Recommended NC levels for various types of occupancies are given in Table 12.2. The room NC levels may, in turn, be used to set sound output levels for HVAC system equipment. HVAC system equipment manufacturers usually provide sound power data, by octave band, for their equipment.

The application of NC Curves is valid only for continuous, steady-state sound that has no obvious fluctuations in sound level with time. Since the shape of an NC Curve is not that of a well balanced bland sounding noise, NC Curves must be used with caution. A more preferable method of setting acceptable noise levels for HVAC systems are the **Room Criteria (RC) Curves.**

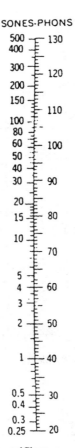

SONES-PHONS

FIGURE 12.5 Relationship between Sones and Phons

Room Criteria (RC) Curves are similar to NC Curves; however, the use of RC Curves is preferable to the use of NC Curves and the A-B-C Weighting Method. They are specifically intended for establishing HVAC system design criteria goals. One reason that RC Curves are preferable is that they address sounds at lower frequencies better than other methods of analyzing sound. The RC Curves cover a range of frequencies from 16 to 4000 Hz. The shape of the RC curve is a close approximation of a well balanced bland-sounding spectrum. RC Curves are shown in Figure 12.7.

The advantages of using RC Curves include:

A. They account for the influence of both spectrum shape and level on the subjective assessment of sound.

B. They include data in the octave bands centered at 31.5 Hz and 16 Hz.

C. They account for the ability of low-frequency acoustical energy to cause noticeable vibration in buildings of light construction.

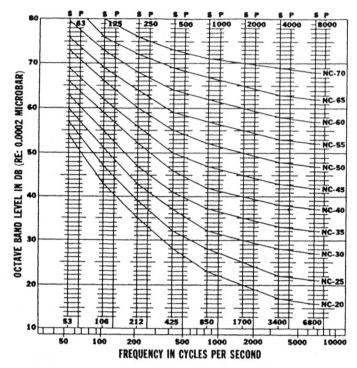

FIGURE 12.6   NC Criterion

TABLE 12.2   Acceptability Criteria for Steady State Background Noise

Type of Space	Recommended Criterion	
	Preferred	Alternative
Recording studios	RC 10-20(N)	NC 10-20
Concert and recital halls	RC 15-20(N)	NC 15-20
TV studios, music rooms	RC 20-25(N)	NC 20-25
Legitimate theaters	RC 20-25(N)	NC 20-25
Private residences	RC 25-30(N)	NC 25-30
Conference rooms	RC 25-30(N)	NC 25-30
Lecture rooms, classrooms	RC 25-30(N)	NC 25-30
Executive offices	RC 25-30(N)	NC 25-30
Private offices	RC 30-35(N)	NC 30-35
Churches	RC 30-35(N)	NC 30-35
Cinemas	RC 30-35(N)	NC 30-35
Apartments, hotel bedrooms	RC 30-35(N)	NC 30-35
Courtrooms	RC 35-40(N)	NC 35-40
Open-plan offices and schools	RC 35-40(N)	NC 35-40
Libraries	RC 35-40(N)	NC 35-40
Lobbies, public areas	RC 35-40(N)	NC 35-40
Restaurants	RC 40-45(N)	NC 40-45
Public offices (large)	RC 40-45(N)	NC 40-45

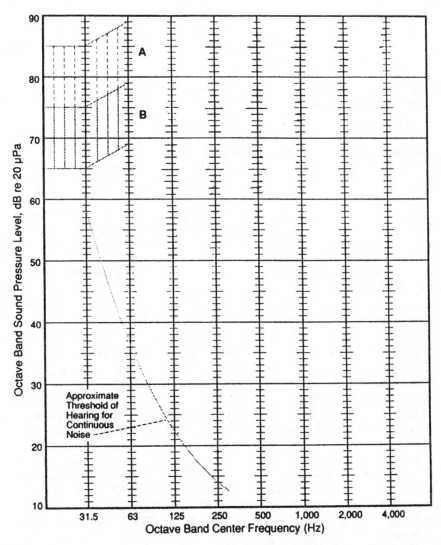

FIGURE 12.7 Room Criteria (RC) Curves   (Copyright 1993 by the American Society of Heating, Refrigerating and Air Conditioning Engineers, Inc. from the Fundamentals Handbook. Used by permission.)

In addition to assigning a numerical value to the level of sound, the RC method also considers the "quality" or type of sound. Sounds are assigned a letter designation which describes the quality of sound. The designations are as follows:

Hissy Spectrum (*H*): A sound is judged to be hissy if there is a spectral imbalance created by relatively high octave band levels at high frequencies.

Rumbly Spectrum (*R*): A sound is judged to be rumbly if there is a spectral imbalance created by relatively high octave band levels at low frequencies.

Neutral Spectrum (N): A sound is judged to be neutral if it is neither hissy nor rumbly. This results in an approximate balance between the relative loudness and the low frequency and high frequency regions of the spectrum over the range of recommended application (RC 25 to RC 50).

Rumbly Vibration Spectrum (RV): A sound is judged to cause rumbly vibration when it has the potential to produce perceptible vibration of lightweight walls and ceilings. This can result from high levels in the octave bands centered at 16.5 Hz and 31.5 Hz.

Recommended RC levels for various types of occupancy are shown in Table 12.2.

The lower value in Table 12.2 is intended for use as a goal in the design of an HVAC system, while the higher number is the recommended maximum limit. The alternative NC criterion may be used when the quality of the space usage is not demanding and background noise with a rumble or hiss can be tolerated, *as long as it is not too loud*. Whenever the NC criterion is used, there is an inherently higher possibility that the occupants will be dissatisfied with the background noise.

The shaded regions in octave bands from 16 Hz to 63 Hz of the RC Curves in Figure 12.7 indicate the octave band frequency and sound pressure level that may cause noticeable vibrations in the walls and ceilings of lightweight building construction. Sounds with frequencies and sound power levels in these areas should be avoided in occupied spaces.

The procedure for determining the noise caused by an HVAC system, using the RC rating method, is as follows:

1. Measure the octave band noise level in the 31.5 Hz, 63 Hz, 125 Hz, 250 Hz, 500 Hz, 1000 Hz, 2000 Hz, and 4000 Hz octave band frequency centers. If the possibility that acoustically induced vibration could occur, a measurement should also be made in the 16 Hz band.

2. Plot the octave band levels measured in Step 1 on the RC Curves in Figure 12.7.

3. For the octave band levels of 500 Hz, 1000 Hz, and 2000 Hz, measured in Step 1, calculate the arithmetic average in decibels. This value corresponds to the numerical value of the RC rating.

4. Using the plot of the data from Step 2, plot the arithmetic average calculated in Step 3 at 1000 Hz. Draw a line from this point with a −5 dB/octave slope. This line is called the *reference curve.*

5. Draw two straight lines that are parallel to the reference line in Step 4. The first line is drawn 5 dB above the reference curve, between 31.5 Hz and 500 Hz. The second line is drawn 3 dB above the reference curve between 1000 Hz and 4000 Hz. These two lines form the permissible limits of a neutral spectrum.

6. Assign a letter designating the quality of sound to the RC rating determined in Step 3. A sound quality letter is assigned as follows:

The letter N is assigned if the spectrum being evaluated falls entirely within the reference curve and the two parallel lines drawn in Step 5, i.e., between the reference curve and the permissible limits of a neutral spectrum between 31.5 Hz and 500 Hz, and between 1000 Hz and 4000 Hz.

The letter H is assigned if the spectrum being evaluated is above the line drawn 3 dB above the reference curve between 1000 Hz and 4000 Hz in Step 5, and does not exceed the octave band levels at 500 Hz and lower frequencies by more than 5 dB.

The letter R is assigned if the spectrum being evaluated is above the line drawn 5 dB above the permissible limit between 63 Hz and 500 Hz in Step 5, and it does not exceed the octave band levels at 1000 Hz and greater by more than 3 dB.

The letter *RV* is assigned if the spectrum being evaluated has an octave band in the 16 Hz and/or 31.5 Hz octave bands which is 70 dB or higher. Between 65 dB and 70 dB, acoustically induced vibration probably will not be a problem.

Example 12.1   The following four noises A, B, C and D were measured in a room.

	Octave Band Center Frequency, Hz								
Noise	16	31.5	63	125	250	500	1000	2000	4000
A	—	53	51	46	40	34	33	32	30
B	—	62	58	52	47	42	36	31	27
C	70	70	62	53	43	40	33	26	20
D	82	84	70	55	47	37	30	27	22

What is the RC rating and spectrum for each noise ?
The data for each of the four noises are plotted in Figures 12.8 A, B, C and D.
The results are:

Sound A: RC = 33 dB, Hissy (*H*)

Sound B: RC = 36 dB, Neutral (*N*)

Sound C: RC = 33 dB, Rumbly (*R*)

Sound D: RC = 31 dB, Rumbly Vibration (*RV*)

12.5   HVAC EQUIPMENT SOUND

Manufacturers of HVAC equipment, such as fans, usually can provide sound power data for their equipment. The data usually includes the sound power at each of the eight octave bands, although sones are sometimes listed for smaller fans. HVAC air system equipment, for which manufacturer's sound data is usually available, includes fans, variable air volume (VAV) terminals, fan-powered VAV terminals, grilles, registers, and diffusers. Sound power data is also available for packaged equipment, such as packaged rooftop air conditioners.

For larger pieces of equipment, like centrifugal fans in air handling units, the data is not usually published in the equipment catalogs. Since the sound power levels are a function of the operating characteristics, such as air flow, fan rotation speed, static pressure, etc., the data must be requested for each design application. In order to standardize sound power rating procedures for equipment, organizations, such as the Air Movement and Control Association (AMCA), have developed standard testing methods for various types of equipment. Equipment is frequently certified that it was tested in accordance with the AMCA standards.

Fans are a major source of noise in HVAC systems. Most noise generated by fans is due to high air velocities through the fan wheel and housing. Fan blade frequency, which is a function of the rotational speed of the fan, is a significant cause of fan noise. As a centrifugal fan wheel rotates, air is forced from the fan blades by centrifugal forces, and impacts against the cut-off at the fan discharge. As the air strikes the cut-off, a "thumping" noise is created each time a fan blade passes the cut-off.

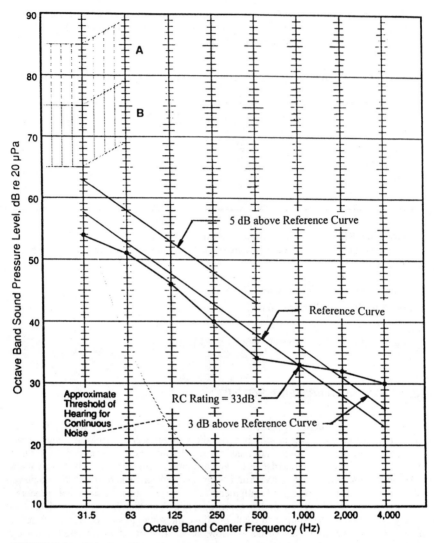

FIGURE 12.8A   Noise Data Plots

Fan noise can also be generated as a result of the point of operation on the fan curve. An oversized fan will be operating out of the normal design range and will approach a block tight condition. The air flow rate is relatively small, compared to the static pressure produced by the fan. This causes a large pressure differential between the fan inlet and discharge, which tends to cause the air to bubble back through the fan. Under this situation, the operation of the fan is unstable due to pressure pulsations which generate noise. At the opposite extreme, undersized fans operate far out on their performance curves at, or near, wide open air flow. This results in noise generated by high air velocities. It is important to properly select fans for their proper application. Fan selection is discussed later in Chapter 16.

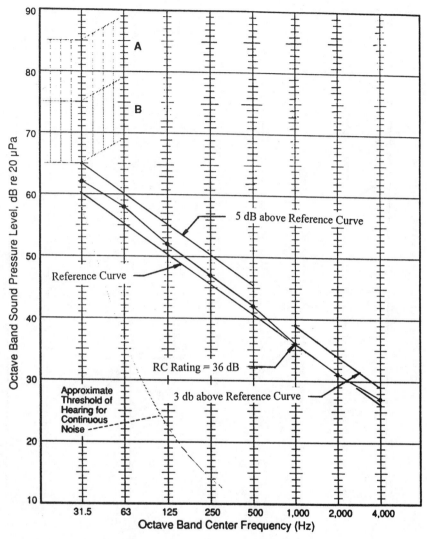

FIGURE 12.8B Noise Data Plots

Fan sound output level is a function of the type of fan, i.e. centrifugal, propeller, vane axial, etc. For any given fan, the sound output level is a function of the air flow, static pressure, and rotational speed (rpm). The fan motor and drive can also generate sound. Sound data for fans is usually given, assuming that there is no ductwork connected to the fan which will attenuate, or reduce, the sound.

For any given fan, the sound output level is also a function of the installation of the fan. The sound output level on the supply side of an air handling unit is frequently greater than the return side, due to the sound attenuation (reduction) caused by the unit casing, coils, etc. Fan noise can travel to an occupied space by a variety of methods. One common method is

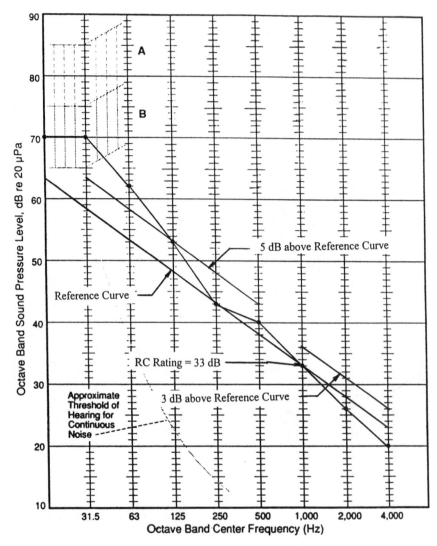

FIGURE 12.8C   Noise Data Plots

for the noise to be transmitted directly through the building structure, such as the walls of an equipment room, to an occupied space. Another common path for fan noise to enter an occupied space is noise that is transmitted by the supply air ductwork. Sound from the inlet of a fan can also be transmitted to an occupied space by the return air ductwork. The effect of ductwork on HVAC system noise is discussed later in this chapter.

In packaged HVAC equipment, such as rooftop air conditioners and condensing units, the compressors in the refrigeration system, as well as the fans, are a major source of sound and vibration. Noise and vibration are particularly problematic in equipment with cooling capacities of 20 tons and greater. In the space directly under the rooftop air conditioning

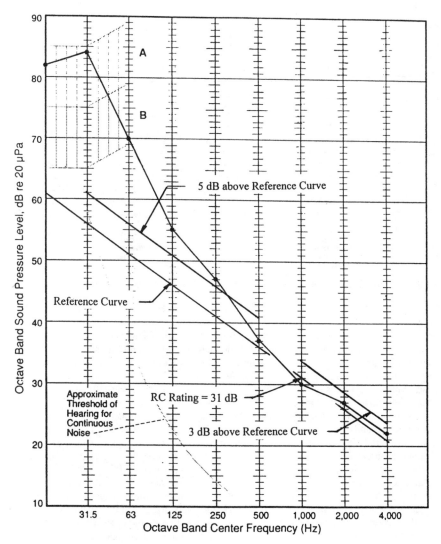

FIGURE 12.8D   Noise Data Plots

units, it is not unusual to hear a combination of air flow noise, compressor noise, and low frequency rumble. Figure 12.9 shows the typical sources and paths for sound from a rooftop air conditioning unit.

Paths through which sound enters a space include the supply air ductwork, the return air ductwork, and through the building structure. The sound paths through the ductwork include two components, duct breakout noise and duct borne noise. Duct breakout noise is sound that continues to travel in a straight line when the ductwork changes direction. Duct breakout noise typically occurs directly under a rooftop air conditioner at the first elbow in

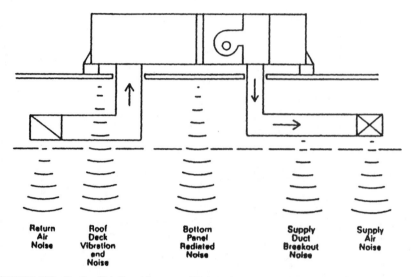

FIGURE 12.9   Rooftop Unit Sound Sources and Paths

the ductwork. The sound simply continues in a straight line as the ductwork changes direction. Duct borne noise is sound that travels down the ductwork and enters the space through an air diffuser, or register, just as the air does. Finally, structure borne noise is sound that results from the vibration of the structure itself. Lightweight metal deck roofs are particularly vulnerable to this type of problem. The structure borne noise usually results from the vibration caused by the fans and/or compressors.

In an attempt to reduce sound caused by roof mounted equipment, such as rooftop air conditioners and compressors, there are several design procedure rules of thumb that should be considered. For maximum sound attenuation (reduction), roof mounted equipment should be located over non-critical, non-occupied spaces, such as storerooms and toilet rooms. Slightly higher sound levels can be tolerated in these areas. It is also advisable to install duct liner in the supply and return air ductwork, especially up to the first change in direction. Whenever possible, round distribution ductwork should be used because round ductwork tends to naturally resist sound breakout. Ductwork may also be enclosed with drywall to reduce the sound transmitted from ductwork. In extreme cases, the ductwork may be enclosed, or encased, in lead lining. Finally, roof mounted vibrating equipment should be mounted on concrete slabs, or curbs, to add mass and reduce vibration, and roof mounted equipment should be mounted on spring mounted vibration isolation bases and curbs.

Variable air volume (VAV) terminal units and fan terminal units can be a source of HVAC system noise. When selecting air terminals, it is important to consider the sound generated by these units. Air terminals should be tested and certified to be in compliance with ARI Standard 880, "Industrial Standards for Air Terminals." Manufacturers usually provide sound levels produced by their terminals for the octave bands between 125 Hz and 4000 Hz.

Since these types of terminals are usually located in ceiling return air plenums above an occupied space, sound can easily be transmitted through a ceiling to a space. A well-sealed gypsum board ceiling will provide about 20 dB of sound reduction, but if a return air opening is nearby, the reduction may only be about 3 dB. The noise emitted by a terminal can either be duct borne noise from the HVAC system transmitted through the terminal or it can be noise generated within the terminal itself. The noise generated within the terminal can result from two sources, the fan and from throttling of the primary air. If the unit is a fan terminal unit, noise can be generated by the fan much like noise generated by fans in air handling units, as previously discussed. Noise can also be generated by the VAV throttling damper. As the damper closes, the velocity of the primary air passing through the terminal tends to increase, which may result in an increase in the generated noise.

Sound from a terminal unit may be either from the discharge of the unit or radiated sound. Discharge sound travels out the end of the unit and is transmitted to the space through the distribution ductwork, and out the air devices, such as air diffusers. Radiated sound is sound that "breaks out" from the unit casing, or return air port, of the terminal and passes through the ceiling to the space.

To reduce the noise that a terminal may cause in a space, the terminal should be selected for the proper design air flow and with the sound power output in mind. Table 12.3 shows typical performance data for a VAV terminal. As a general rule, terminal units should be located over areas that are less sensitive to sound and noise, such as corridors, storerooms, etc.

TABLE 12.3 Typical VAV Terminal Performance Data

Size	Air Flow Rate CFM	Min. ΔP_s Inches W.G.	$\Delta P_s = \Delta P_t = 1.0"$ W.G. Sound Power Levels, dB Octave Band (Hz)						$\Delta P_s = \Delta P_t = 1.5"$ W.G. Sound Power Levels, dB Octave Band (Hz)						$\Delta P_s = \Delta P_t = 3.0"$ W.G. Sound Power Levels, dB Octave Band (Hz)					
			125	250	500	1K	2K	4K	125	250	500	1K	2K	4K	125	250	500	1K	2K	4K
4	50	.05	—	—	27	—	—	—	—	36	28	—	—	—	—	40	31	—	—	—
	100	.19	46	41	34	26	27	25	48	43	36	27	27	26	51	46	39	28	28	28
	150	.41	50	44	39	36	38	35	52	46	41	37	38	36	55	50	44	37	39	38
	200	.74	53	47	42	43	46	42	55	49	44	44	46	43	58	53	47	44	47	45
5	75	.04	—	36	28	—	—	—	—	38	30	—	—	—	47	42	32	—	—	—
	100	.08	45	39	31	—	—	—	46	41	33	—	—	—	49	45	36	—	—	—
	200	.32	52	45	39	32	31	28	53	47	41	32	31	29	56	51	43	33	32	31
	300	.71	56	49	43	42	42	38	57	51	45	42	42	39	60	55	48	43	43	41
6	100	.03	—	37	28	—	—	—	45	40	30	—	—	—	48	43	33	—	—	—
	200	.13	50	44	36	22	18	16	52	46	38	23	19	18	55	50	41	23	20	20
	300	.30	54	48	41	32	29	26	56	50	42	32	30	28	59	53	45	33	31	30
	400	.54	57	50	44	39	37	33	59	52	45	39	38	35	62	56	48	40	39	37
7	150	.03	46	40	31	—	—	—	48	42	32	—	—	—	51	46	35	—	—	—
	300	.12	53	46	38	24	19	17	55	49	40	24	19	18	58	52	43	25	20	20
	450	.26	57	50	43	34	30	27	59	52	44	34	30	28	62	56	47	35	31	30
	600	.47	60	53	46	41	38	34	62	55	48	41	38	35	65	59	51	42	39	37
8	200	.04	48	42	32	—	—	—	50	44	33	—	—	—	53	48	36	—	—	—
	400	.14	55	48	39	24	—	—	57	50	41	25	—	17	60	54	44	25	19	19
	600	.32	59	52	44	34	29	25	61	54	46	34	29	27	64	58	49	35	30	29
	800	.56	62	55	47	41	37	32	64	57	49	41	37	34	67	60	52	42	38	36
9	300	.05	54	49	40	—	22	30	56	52	43	—	24	33	59	58	48	23	27	40
	500	.15	57	52	43	28	28	33	60	55	46	30	30	37	63	61	52	32	33	43
	700	.29	60	54	46	34	32	35	62	57	49	35	34	39	66	63	54	38	37	45
	900	.48	62	56	48	38	35	37	64	59	51	40	37	40	68	65	56	42	40	47
10	400	.08	56	50	42	23	26	32	58	54	45	24	28	36	62	59	50	27	31	43
	600	.18	59	53	44	30	31	35	61	56	47	31	33	39	65	62	53	34	36	45
	800	.32	61	55	47	35	34	37	63	58	50	36	36	41	67	64	55	38	39	47
	1000	.50	63	56	48	39	37	38	65	60	51	40	39	42	69	65	56	42	42	48
12	500	.05	58	52	43	24	29	36	60	55	46	25	31	39	64	61	51	28	34	46
	750	.11	61	54	46	31	34	38	63	58	49	32	36	42	67	63	54	35	39	48
	1000	.20	63	56	48	36	37	40	65	59	51	37	39	44	69	65	56	40	42	50
	1500	.46	66	59	51	42	42	42	68	62	54	44	44	46	72	68	59	47	47	53
14	700	.06	61	54	45	28	33	39	63	57	48	29	35	43	66	63	53	31	38	49
	1000	.13	63	56	47	34	38	41	66	59	50	35	39	45	69	65	55	38	43	52
	1500	.29	67	59	50	41	42	44	69	62	53	42	44	48	72	68	58	44	47	54
	2000	.50	69	60	52	46	46	46	71	64	55	47	48	49	74	69	60	49	51	56
16	1000	.05	64	56	47	32	38	43	66	59	50	33	40	46	69	65	55	36	43	53
	1500	.11	67	58	50	39	43	45	69	62	53	40	45	49	72	67	58	42	48	55
	2000	.20	69	60	52	44	46	47	71	64	55	45	48	51	75	69	60	47	51	57
	3000	.44	72	63	55	51	51	50	74	66	58	52	53	53	78	72	63	54	56	60
16 x 24	2000	.04	63	56	50	37	45	46	64	58	53	39	48	50	66	62	57	42	52	56
	3000	.10	69	61	54	42	49	50	70	63	56	44	52	53	73	67	61	48	56	59
	4000	.17	73	65	56	46	51	52	75	67	59	48	54	56	77	70	64	52	59	62
	6000	.40	80	70	60	52	55	55	81	72	63	54	58	59	83	75	68	58	62	65

The terminal unit should include a sound attentuation section on the discharge side of the terminal. In lieu of a sound attentuation section, a plenum that is lined with fiberglass duct liner can be provided on the discharge side of the unit for sound attentuation. Runout ductwork to diffusers should also be lined. When the terminal is located in a ceiling return air plenum, the terminal unit should be located as far as possible from the return air grille to reduce the passage of sound through the grille. The runout ductwork, downstream of the terminal unit, should be sized for low velocity and low static pressure. Duct sizing is discussed in more detail in Chapter 15.

Finally, ductwork and air devices such as diffusers, grilles, and registers can be a source of noise. In general, ductwork will tend to attenuate or reduce duct borne sound. The amount of attenuation is a function of the construction of the ductwork, the length of the ductwork, and whether the ductwork is lined. Air flow in ductwork can also generate sound. Air flow generates sound as it flows through elbows, vanes, and any location where air turbulence may occur. The sound generally has a broad spectrum. To reduce the sound generated by air flow in ductwork, it is advisable to design the ductwork to reduce flow turbulence and to keep air velocities below certain levels. Table 12.4 gives recommended air velocities for main trunk ducts for various applications.

These are recommended velocities that will reduce the possibility of sound generation within the ductwork. These velocities may be exceeded; however, additional steps to reduce sound may be necessary.

Air flowing through air devices in the room can also be a source of noise in the space. Air devices produce the greatest noise in the higher octave bands, primarily 1000 Hz to 4000 Hz. The amount of noise in an air device is primarily dependent on the air flow rate and the free area of the air device which determines the air velocity. Higher air velocities tend to generate higher noise levels. Most manufacturers of air devices include noise data with their catalog performance literature, giving noise output as related to the air flow rate, open free area of the device, and air pressure drop through the device. When comparing the noise data of various devices, it is important that the devices be tested and certified in accordance with the Air Diffusion Council Standard 1062 GRD 84, "Test Code for Grilles, Registers, and Diffusers."

To control the amount of noise caused by an air device in a room, the following rules of thumb should be considered.

TABLE 12.4   Recommended Main Duct Velocities

Application	Main duct felocities—fpm
Residences	600
Apartments	1000
Hotel bedrooms	1000
Hospital bedrooms	1000
Private offices	1200
Directors' rooms	1200
Libraries	1200
Theaters	800
Auditoriums	800
General offices	1500
High class restaurants	1500
High class stores	1500
Banks	1500
Average stores	1800
Cafeterias	1800
Industrial	2500

1. Select the air devices that have an NC or RC rating that are rated at least 5 dB below the desired room NC or RC level.

2. The ductwork connected to the air device should be straight for at least three equivalent duct diameters.

3. Air balancing dampers should not be attached to the collar of the air device. The damper should be located at least three equivalent duct diameters away from the air device.

12.6   VIBRATION

In addition to creating objectionable noise in buildings, HVAC systems can also cause excessive and objectionable vibrations. Both sound and vibration are associated with all equipment that have moving parts, including rotating equipment, such as fans, and reciprocating equipment, like compressors. Vibration occurs in materials such as machine parts or building structures, because they are elastic (if deformed, they will return to their original shape when the deforming force is removed). An inelastic material, such as a lump of putty, cannot support a vibration.

Vibration is similar to sound in that it involves periodic to-and-fro motion or wave motion. Vibration has an amplitude (maximum displacement), a period, and a frequency of oscillations. The period of vibration is the time of a single cycle and the frequency is the number of cycles occurring per unit of time. The natural frequency is the frequency of free vibration (periodic motion that continues after the original disturbance is removed). Forced vibration is motion that persists because of the continuation of a disturbing force. The amplitude of the vibration is the maximum displacement of the motion. In the study of vibrations, frequency is the most important aspect of vibration. If the forcing frequency becomes equal to the natural frequency of the system, resonance is said to occur and the frequency is the resonant frequency.

All rotating equipment operates at certain rotational speeds that result in vibrational frequencies that are "sympathetic" or are harmonic with the support structure of the equipment. This is referred to as the critical speed and is the speed at which resonance of the equipment and support system occurs. The critical speed results in resonant or harmonic vibrations. When harmonic vibrations occur, there is a marked increase in the amplitude of the vibrations. In order to keep the amount of vibration transmitted to a building by HVAC equipment to an acceptable level, it is necessary to provide vibration isolation for most rotating and/or reciprocating equipment. A rigidly mounted machine will transmit all of its vibratory forces to its supporting structure.

12.7   VIBRATION ISOLATION OF HVAC SYSTEMS

All machines, including HVAC equipment, are inherently vibration generators. It is impossible to design a machine that is completely vibration free. It is therefore necessary to provide support systems for HVAC equipment that minimize the transmission of the vibration to the base structure or building.

A large percentage of the vibrations to the base structure can be eliminated by the proper selection of vibration equipment and materials. Vibration isolators are resilient supports on which equipment is mounted to isolate them from the base structure. These supports may be made of some resilient material, such as neoprene, or, more frequently, they are springs.

The effectiveness of vibration isolators is usually expressed in terms of efficiency. The efficiency of an isolator is a measure of the percentage of vibration it absorbs, as opposed to the percentage that it transmits. The efficiency of the isolator is a function of the following:

- the spring constant or stiffness of the isolator.
- the dead load deflection of the isolator.
- the disturbing frequency of the forced vibration.
- the natural frequency of the vibration isolator.

The dead load deflection is a function of the equipment weight and the stiffness of the vibration isolator. The natural frequency of the vibration isolator is a function of the static deflection and is given by Equation 12.3.

$$f_n = 3.13\sqrt{\frac{1}{d}} \tag{12.3}$$

where

f_n = the natural frequency of the isolator in cps (Hz)

d = the static deflection of the isolator in in

The disturbing frequency of the forced vibration (f_d) may be obtained from the operating characteristics of the equipment. It is simply the rotational speed of the equipment in cycles per second. The rotational speed in rpm may be converted to cps simply by dividing by 60. For example, an 1800 rpm fan would have a frequency of 30 cps. Whenever the equipment has a variable speed, such as a VAV fan, the lowest rpm should be used for the disturbing frequency. If two rotating pieces of equipment are mounted on a common isolation base, the speed of the slower machine should be used. For example, if a 1725 rpm motor is used to drive a fan at 600 rpm, the speed of the fan should be used to determine the disturbing frequency. Using the natural frequency of a given isolator, the transmissibility of an isolator may be calculated using Equation 12.4.

$$\tau = \left[\left(\frac{f_d}{f_n}\right)^2 - 1\right]^{-1} \tag{12.4}$$

where

τ = the transmissibility of the isolator

f_d = disturbing frequency of the equipment in cps

f_n = the natural frequency of the isolator in cps

The efficiency of the isolator may be obtained from Equation 12.5.

$$\eta = (1 - \tau) \tag{12.5}$$

where

η = the efficiency of the isolator

It may be observed from Equation 12.4 that equipment with lower rotational speeds will result in higher transmission of vibration to the base structure. A larger static deflection of the isolator results in lower transmission of vibration. It should also be observed that when the disturbing frequency and the natural frequency of the isolator are almost equal, f_d/f_n approaches 1.0, and the transmissibility of the isolator increases to infinity and results in no isolation of the vibration. Obviously, this is a situation which must be avoided. In order to provide adequate vibration isolation for equipment, the following "rules of thumb" should be considered.

- provide at least three times (and preferably four or more times) as much deflection in the isolators as the dead load deflection of the supporting structure.
- ensure that equipment is not operated at rotational speeds that result in a frequency at, or near, the natural frequency of the isolator.
- insofar as possible, the mounting system should keep the equipment center of gravity within the plane of the isolators.
- the natural frequency of the vibration isolator should be less than $1/3$ of the driving frequency, in most applications, and less than $1/5$ in critical applications.
- consider providing a heavy rigid inertia block, or base, for the equipment which will minimize the effects of unequal weight distribution, add mass to the system to dampen vibrations, and stabilize the entire resiliently mounted assembly.
- provide housekeeping pads for the equipment which will add mass to the floor beneath the equipment and help dampen vibrations.

There are three basic types of vibration isolators used for HVAC equipment. They are pad type isolators, spring type isolators, and vibration isolation bases. Pad type isolators are relatively inexpensive and are usually made of rubber, neoprene, or some other type of resilient material. Figure 12.10 shows a typical neoprene pad type vibration isolator.

Pad type isolators normally provide about 0.2 in to 0.5 in of static deflection. They are suitable for small vibrations from small high-speed equipment in which the unbalance forces are relatively small. Depending upon the load, the material should have a certain minimum druometer number. These types of isolators should be used only in non-critical, non-sensitive areas.

Steel spring vibration isolators are the most commonly used type of vibration isolator used for HVAC equipment. A typical spring vibration isolator is shown in Figure 12.11.

Steel spring vibration isolators are suitable for static deflections of 5 in or more. They should be designed to have a low spring constant, i.e., they should be soft. This will cause the natural frequency to be much lower than the disturbing frequency. In order to provide suitable vibration isolation, the following rules of thumb should be considered when selecting spring vibration isolators.

- the outside diameter of the spring should be no less than 80% of the operating height of the spring, to ensure lateral stability.
- the spring should have remaining travel to solid of no less than 50% of the static deflection to reduce the possibility of "bottoming out."
- a $1/4$ in neoprene friction pad should be provided between the spring base and the floor or housekeeping pad to provide isolation from high frequency vibrations.

FIGURE 12.10 Typical Neoprene Pad Type Vibration Isolator

FIGURE 12.11   Spring Vibration Isolator

Occasionally, it is desirable to limit the deflection or motion of a spring isolator. It is usually desirable to limit the upward travel of a spring isolator when the weight is not on the isolator. This allows for easier installation of equipment on the isolators. Typical restrained spring vibration isolators are shown in Figure 12.12.

Restrained spring vibration isolators are also used for roof mounted equipment. The restraints provide safety stops in the event of high wind loads. Since the loads exerted by equipment are sometimes greater during startup, the restraints limit excessive deflections when the equipment is started. Restrained isolators are also beneficial as seismic equipment restraints, to reduce equipment movement in the event of an earthquake. Generally speaking, the selection criteria for restrained spring isolators are the same as for unrestrained spring isolators. Care should be taken to select restrained isolators with the proper limits of travel.

Vibration isolation bases are essentially floating base pads, mounted on vibration isolation springs. A typical vibration base is shown in Figure 12.13.

Vibration isolation bases are used to add mass to the equipment support. The additional mass tends to dampen and absorb vibrations. The base can be made of structural steel or it can be made of concrete. These types of isolators are frequently used to support equipment, such as pumps, although they may also be used to support rooftop air conditioners. When

FIGURE 12.12 Restrained Spring Vibration Isolators

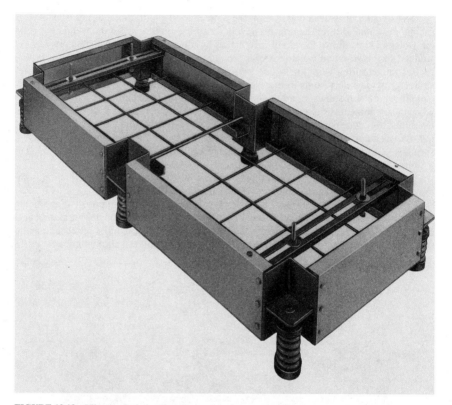

FIGURE 12.13   Vibration Isolation Base

a rooftop air conditioner is mounted on a lightweight, low mass roof, such as a metal deck, vibration isolation bases, or curbs, are frequently used. Vibration isolation bases also provide rigidity to the assembly and can help maintain proper alignment of the equipment, such as a fan and motor. When selecting vibration isolation bases, it is best to consult the equipment manufacturer and the vibration isolator manufacturer for recommendations regarding selection.

Finally, it is recommended that a flexible connection be provided between medium or large sized fans and the connecting ductwork. Flexible duct connectors are usually made of rubber or canvas. They allow for relative motion between the fan and ductwork without transmitting vibration to the ductwork. These connections are usually reasonably air tight, although significant air leakage can occur if they are not properly installed and maintained.

12.8   SEISMIC DESIGN CONSIDERATIONS FOR HVAC SYSTEMS

This section is intended as an introduction to seismic design for HVAC systems. In general, local building codes set the requirements for seismic restraints of mechanical equipment, as well as other building components. Nearly all seismic requirements in local codes are based on one of the three model codes developed by the International Conference Officials and Code Officials (ICBO), Building Officials and Code Administrators International (BOCAI), and the Southern Building Code Conference, Inc. (SBCCI).

The main intent of seismic restraints is to prevent non-structural items in a building, such as HVAC equipment, from becoming lethal missiles in the event of an earthquake. Earthquakes can cause equipment to break loose and fall from a ceiling or slide across a floor when the ground shakes a building and its contents. In addition to the lethal missile problem, it is also important to keep building mechanical systems in operation during and after an earthquake. The attachment (or seismic restraint) of equipment must have sufficient strength to transmit the forces during an earthquake, or it must provide sufficient isolation to allow for the induced motion of the equipment.

In order to properly select suitable seismic restraints for HVAC equipment, it is first necessary to determine the seismic zone in which the building is located. Figure 12.14 shows the seismic zones for the United States.

These zones give some measure of the severity of earthquakes that can be expected. In areas that have a 0 rating, there is little probability of a severe earthquake, whereas areas with a 4 rating (such as California) have a high probability of a severe earthquake.

There are two general methods for restraining equipment, such as HVAC and other mechanical equipment. The first method is to rigidly attach the equipment to the building structure to prevent any movement. In such an arrangement, the supporting members used to attach the equipment to the building structure must be strong enough to withstand the forces that would be expected in the event of an earthquake. Rigidly attaching equipment to the building structure has some disadvantages, as discussed in the section on Vibration Isolation. The second method is the isolation approach, in which equipment is provided with sufficient space and the support has sufficient flexibility to reduce the possibility of excessive motion of the equipment. This method usually employs restraints to limit the travel of the equipment, such as those shown in Figure 12.12.

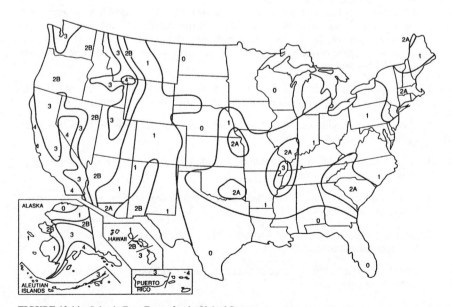

FIGURE 12.14   Seismic Zone Factor for the United States

TABLE 12.5   Damage to Mechanical Equipment during Past Earthquakes

Components	Recorded Damage
Rigidly mounted equipment such as boilers, chillers, generators, tanks	Generally perform well where there is no damage to structural base. Some shearing of attachment devices and corresponding horizontal displacement; tall tanks overturn; supports fail. Greatest damage is to equipment that rested on structural base without positive anchorage; overturning and horizontal movement sever connected lines and pipes.
Vibration-isolation-mounted fans, pumps, air handlers, etc.	Devices fail and cause equipment to fall. Some damage due to unrestrained shaking on vibration-isolation device. Suspended equipment fails more often than floor-mounted equipment.
Water, stream, sprinkler, gas, and waste piping	Large-diameter rigid piping fails at elbows and bends. Joint separations; hanger failures. Small-diameter piping performs better than larger piping due to bending without breaking. Single failures of hanger assemblies frequently causes progressive overloading and failures at other hangers and piping supports. Piping performs better in vertical runs where there are lateral restraints than in horizontal runs where there is no lateral bracing. Failures at building seismic joints due to differential movements.
Rectangular, square, and round ducts	Breakage most common at bends. Supporting yokes fail; long runs fail as a result of large-amplitude swaying.

Past experience with mechanical equipment systems in earthquakes has provided some indication of how equipment supported by both methods will react and the type of damage that can by expected. Table 12.5 provides some indication as to what can be expected from mechanical equipment in an earthquake.

When designing and selecting seismic restraints for HVAC equipment, it is important to consult the appropriate codes, the manufacturer of the restraints (if any), and an engineer experienced in seismic design. However, there are some guidelines which should be followed when designing HVAC systems that will reduce the problems caused by earthquakes. These guidelines include:

1. heavy mechanical equipment should not be mounted on the upper floors of tall buildings, unless the mounts are carefully selected and analyzed for earthquake resistance.

2. floor mounted equipment, with vibration isolation devices, should be bolted to the equipment base and to the structural slab.

3. lateral and vertical restraints should be provided for all isolated floor mounted equipment to restrain displacement.

4. resilient material should be provided on the contact surfaces of the restraining devices to reduce impact loads.

5. vibration isolation hangers for suspended equipment should be tightly installed against the supporting structural member.

6. cross bracing should be provided between the hanger rods on all four sides of suspended lightweight equipment.

7. horizontal ductwork should be supported as close as possible to the structural member.

8. long hangers and supports for ductwork should have lateral sway bracing.

9. all diffusers and registers should be secured to the ductwork with sheet-metal screws.

10. supports for tanks and heavy equipment must be designed with sufficient strength to withstand earthquake forces. Supports should be anchored to the floor or otherwise secured.

11. supports for elevated tanks or equipment should be sway braced and anchored to the structural slabs or walls.

12. piping should be anchored to only one structural system. Where structural systems change, and relative deflections might occur, flexible or moveable joints should be provided to allow for movement.

13. whenever possible, pipes should not cross the seismic joints of the building. If they must cross, it should be at the lowest floor possible and seismic piping joints should be provided.

14. flexible pipe connectors should be provided where rigidly supported piping connects to equipment with vibration isolation.

15. all horizontal piping $2^1/_2$ in and larger should be sway braced in both the longitudinal and transverse directions.

The above guidelines are general guides that may be used in the design of HVAC systems. Additional information on seismic design may be obtained from References 2 and 23.

Chapter 13

HUMAN COMFORT, INDOOR AIR QUALITY, AND VENTILATION

The main purpose of HVAC systems in most buildings is to provide a comfortable indoor environment for the occupants of the building. Comfort may be defined as any condition which, when changed, will make a person uncomfortable. Comfort in an indoor environment may be defined as the state of mind in which one acknowledges satisfaction with regard to the indoor environment.

Comfort, or satisfaction, with the indoor environmental conditions is a subjective and complex response to a number of interacting variables. Variables that are controllable by the HVAC system include the dry bulb temperature of the air, the relative humidity of the air, air cleanliness, and air motion. Comfort is also a function of several uncontrollable variables, which include activity level of the occupants, the type of clothing worn by the occupants themselves, and the temperature of surrounding surfaces, such as walls, floors, etc.

The function of the HVAC system is to maintain comfortable indoor conditions, despite a wide range of activities within the space, and despite changes in the outdoor weather conditions. The air conditioning system must provide a comfortable indoor space by controlling the air temperature, humidity, air movement, and cleanliness of the air in the occupied space. Since comfort is frequently a subjective judgment, the objective of an HVAC system is to provide an indoor environment in which at least 80% of the occupants find the indoor environment acceptable.

13.1 THE HUMAN BODY AS A HEAT ENGINE

A heat engine may be defined as a system that operates in a cycle that has net positive work and a net positive heat transfer. Just as a heat engine operates on a cycle, produces work, and produces heat, so does the human body.

Since food that is consumed by humans is primarily carbon and hydrogen, it may be considered as fuel for the human heat engine. When food is consumed, the energy contained in the food (fuel) is released by an oxidation process. Oxygen comes from the air that is brought into the body during respiration or breathing. The process of converting fuel into energy in the human body is called metabolism. An objective of the HVAC system is to remove excess heat that may be produced through metabolism in the body and help maintain the deep body temperature at 98.6°F.

Since the human body may be considered as a heat engine, which is a thermodynamic system, the First Law of Thermodynamics is applicable to the human body. (See Section 2.2.1, The First Law of Thermodynamics.) It is possible to perform a heat balance of the human body by treating the body as controlled volume. Basically, the energy balance would be: heat produced = heat loss + work performed + heat stored. Equation 13.1 expresses the First Law of Thermodynamics for a human body.

$$S = M \pm W \pm q_s \pm q_l \pm q_r \qquad (13.1)$$

where

S = the rate of heat storage within the body in Btu/hr

M = metabolism of the human body in Btu/hr

W = net work performed by the body in Btu/hr

q_s = sensible heat exchange in Btu/hr

q_l = latent heat exchange in Btu/hr

q_r = radiant heat exchange in Btu/hr

Note that each of the terms in Equation 13.1, with the exception of the metabolism, can be positive or negative (net energy gain or loss). In order for the body to be in equilibrium, i.e., not increase in or decrease in temperature, the values in Equation 13.1 must be such that $S = 0$. One of the main functions of the HVAC system is to produce conditions such that $S = 0$.

The energy generated within the human body by metabolism can vary widely and is a function of a number of variables. One of the main variables is activity level of the individual. Naturally, a higher activity level will result in a higher rate of metabolism. For an "average person," the basal metabolism is about 240 Btu/hr. The metabolic rate of individuals is also a function of body weight, age, sex and body conditioning. The metabolic rate is usually expressed in met units with one met unit defined as 18.4 Btu/hr·ft² (of body surface area). For an average sized male, a met unit is equivalent to about 360 Btu/hr. This represents the heat produced by an "average" sedentary man. Met rates for females are about 30% lower than for males.

The work term in Equation 13.1 is usually negative (energy out of the body), although it can be positive. Work was defined in Section 2.1.2 as a form of energy in transit, resulting from a force acting through a given distance. Work performed by the human body is usually in the form of exercise or physical exertion. As an example, a 200 lb person walking up a 5% grade at 3 mph (4.4 ft/sec) would lift a 200 lb weight a height of 0.22 ft every second. This would result in a work rate of 44 lb/ft·sec, or 204 Btu/hr. The work term in Equation 13.1 would be negative.

Sensible heat loss (or gain) is primarily a function of the difference between the body and the surrounding air. Sensible heat losses can occur as both connective and radiant heat exchanges. Radiant heat exchange will be discussed separately. Connective heat exchange can occur through the body's exterior skin or through exchange with respiration air. A significant amount of heat is lost through respiration since air is brought into direct contact with the lungs in the core of the body, which is at the core temperature of 98.6°F. This usually results in a maximum difference in the air temperature and the skin temperature. Sensible heat is also exchanged through the body's exterior skin surface. For bare skin, it is a function of the exposed surface area of the body. Since the human body is usually clothed, the actual exposed surface area of the body is usually fairly small in most cases. The clothing acts as an insulator or resistance to the flow of heat and moisture. Clothing insulation is usually measured in **clo** units. One clo unit has an equivalent thermal resistance of about 0.88 ft²/hr•°F•Btu. The thermal resistance for clothing is a function of the clothing material and thickness. Since clothing insulation cannot be measured for most design applications, clo values are available for various clothing ensembles. The reader may wish to consult Reference 1 for a more detailed discussion of clothing insulation.

Latent heat is transmitted from the human body through evaporation as a result of perspiration and also through respiration. When the surrounding air temperature reaches about 90°F, the sensible heat loss through the skin becomes negligible. As the air temperature rises above 98°F, the body no longer loses sensible heat, but begins to absorb sensible heat from the air. In order to keep the core temperature of the human body from increasing, the body rejects heat by evaporation through perspiration. The amount of evaporation is a function of the difference in the water vapor pressure at the skin and the ambient air. The vapor pressure of the ambient air is a function of the relative humidity of the air, which will be discussed later in this chapter.

The human body also transfers heat with its surroundings, through radiant heat exchange. A radiant heat exchange will occur between two objects whenever there is a temperature difference, and whenever the objects can "see" each other (are not shielded from each other). The amount of heat that is transferred between a human body and its surroundings is a function of the Mean Radiant Temperature (MRT), which is discussed in the next section.

Although providing a comfortable indoor environment for human occupants is a function of many complex variables, the amount of heat given off by a human body for cooling load calculation purposes is fairly well documented. You may refer to Reference 1 for additional information concerning cooling loads for occupants.

13.2 EFFECTIVE TEMPERATURE, MEAN RADIANT TEMPERATURE, AND COMFORT

The two most obvious indicators of comfort in a space under the control of the HVAC system are the space dry bulb temperature and the relative humidity in the space. However, the dry bulb temperature alone is not a true indicator of thermal comfort conditions in the space. The Effective Temperature and the Mean Radiant Temperature are more accurate indicators of thermal conditions in a space.

The dry bulb temperature of the ambient air in a space is the most obvious, and easily measured, indicator of the thermal conditions in the space. As previously discussed, the sensible heat exchange of a human body occurs by two modes, convection heat transfer and radiant heat transfer. The convection heat transfer is a function of the difference in temperature between the human body and the dry bulb temperature of the air, as well as the ambient air velocity. The radiant exchange, however, is not a function of the air temperature.

The radiant exchange between a human body and the space in which it is located is primarily a function of the difference in temperature of the body and the temperature of the surfaces that make up the room or enclosure. Since there are usually multiple surfaces in a space with which a body can exchange heat, the concept of a **Mean Radiant Temperature (MRT)** for a space has been developed. The MRT is the uniform temperature of an imaginary enclosure, in which radiant heat transfer from the human body equals the radiant heat transfer in the actual non-uniform enclosure.

Radiant heat transfer was discussed in Section 2.2.3, and Equation 2.11 gave the rate of heat exchange for radiant heat transfer. It may be observed from Equation 2.11 that the radiant heat exchange is a function of the difference between the fourth powers of the absolute temperatures of the surfaces exchanging heat. It may also be observed, from Equation 2.11, that the radiant heat exchange is a function of the shape, or configuration, factor between the two objects. The Mean Radiant Temperature takes into account the difference between the human body and the various shape factors (or angle factors) between the body and each surface in the space. The shape factor between the body and each space is a function of the location of the body within the space and the relative size of each surface in the space. Since the Mean Radiant Temperature is a function of the room configuration, a series of graphs have been developed to simplify the calculation process of the MRT. You may wish to refer to Reference 1 for more detail on the procedure for calculating the Mean Radiant Temperature of a space and for the shape factor graphs.

An indicator of thermal comfort that is closely related to the Mean Radiant Temperature is the **globe temperature**. The globe temperature of a space is often used to estimate the MRT for the space and is measured with a globe thermometer. A globe thermometer consists of a hollow copper sphere that is 6 in in diameter and has a coating of black paint on the outer surface. A precision thermometer is located inside the sphere at the center. Since the radiant heat exchange at the globe surface is balanced by the convective heat transfer, the globe surface reaches equilibrium, which approximates the Mean Radiant Temperature of the space.

Another more widely used indicator of thermal comfort is the **effective temperature** (**ET^***). Environmental comfort is strongly affected by the air dry bulb temperature and the relative humidity. The ET^* combines the temperature and humidity into a single index of environmental comfort. Two situations with the same effective temperature should result in the same response to comfort. In order to compare the ET^* in two different environments, they both must have the same air velocity in the space. Based upon various effective temperatures, ASHRAE has developed what is called a "comfort zone." The comfort zone prescribes acceptable ranges of temperature and humidity levels for both summer and winter conditions. The ASHRAE comfort zones are shown in Figure 13.1.

Notice that there are two zones, one for winter and one for summer. Any point that defines a temperature and humidity inside the winter comfort zone is considered to be a comfortable environment for winter applications. Similarly, any point inside the summer comfort zone would provide suitable comfort conditions in the summer. The chart is based on an occupant activity level of 1.2 met, typical summer and winter clothing, room air velocities of less than 50 ft/min, and altitudes from sea level to 7000 ft. The ASHRAE Comfort Chart and comfort zones are generally applicable to offices, homes, schools, retail spaces, theaters, and similar commercial type spaces. A more detailed discussion of the ASHRAE Comfort Chart and comfort zones is available in References 1 and 24.

13.3   HUMIDITY LEVELS

Humidity is the moisture or water vapor occupying the same space as the air in an air-vapor mixture. Absolute humidity, or humidity ratio, is the weight of water vapor per unit volume of dry air. The relative humidity of an air vapor mixture is defined as the ratio of the pressure of the vapor in the mixture to the saturation pressure of the vapor in the mixture at the same temperature, or temperature of the mixture. Equation 13.2 gives the relative humidity of an air-vapor mixture.

$$\phi = \frac{P_v}{P_g} \tag{13.2}$$

where

ϕ = the relative humidity of the mixture of the air and water vapor in %

P_v = the partial pressure of the vapor

P_g = the saturation pressure of the vapor at the temperature of the mixture.

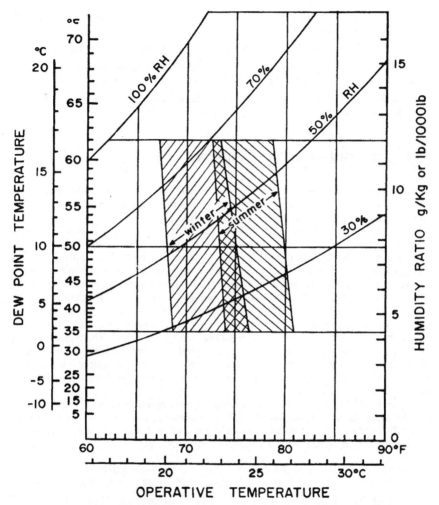

FIGURE 13.1   ASHRAE Comfort Chart

The previous section discussed the conditions for which the indoor environment may be considered comfortable. This included comfort zones, which specified various combinations of dry bulb temperatures and relative humidity levels. In addition to providing comfort conditions, there are additional considerations regarding the control of the humidity level in the space.

When designing an HVAC system, the following effects of relative humidity should be taken into consideration.

1. Maintaining the relative humidity in the space to a maximum of 30% in winter will reduce the possibility of condensation on the inside of windows and glass.

2. Relative humidities in the space may be significantly higher than the design conditions during part load conditions.

3. When space relative humidities are above 75%, conditions are favorable for the growth of bacteria, fungi, and mold.

4. Space humidities below 23% can increase the possibility of respiratory infections.

5. Space relative humidities below 25% and air temperatures below 70°F can result in static electricity and uncomfortable shocks.

6. Some hygroscopic materials, such as natural textile fibers, paper, wood, leather, and some foods, are affected by space relative humidity.

13.4 AIR MOVEMENT AND COMFORT

Air motion and velocity can also have a significant effect on an occupant's perception of comfort in a space. Air velocity, draftiness, and how the air is introduced into a space are all indices of comfort related to air movement in a space.

A draft may be defined as a noticeable air current. It results in an undesired local cooling of the human body, caused by the movement of air. Draftiness is one of the most significant annoyances in spaces such as offices. When people feel a draft, it usually causes them to want higher temperatures in the room, or even causes them to stop the ventilation system. Excessive air motion in a space can be uncomfortable enough when the air temperature is a constant. It is even more uncomfortable when the air temperature is fluctuating, especially when it is directed towards the back of an occupant's neck or at the ankles. Generally speaking, a velocity of 15 ft/min to 25 ft/min is considered still air. Air velocities at 65 ft/min would be considered drafty by most people. The type of activity in the space and the clothing also have an effect on the acceptable air velocities within a space. The physical arrangement of the space and how the air is introduced into the space will also have an effect on the perception of draftiness. Room air distribution is discussed in Chapter 14.

Typically, HVAC systems introduce cooling air into a space between 12°F and 30°F below the space set point temperature. Air is also introduced at velocities well above 15 ft/min. In order to reduce problems with air motion and draftiness, a good air distribution system should be designed with the following considerations:

1. Entrain enough room air with the space supply air such that, when the air reaches the occupied zone, it will be well mixed and warm enough to not be objectionable.

2. Air should be supplied to the space so the velocity will be reduced before it reaches the occupied zone.

3. The supply air should provide a turbulent eddying motion within the space.

4. Select air devices that do not create any objectionable noise within the space.

13.5 INDOOR AIR QUALITY (IAQ)

It is estimated that people in the United States spend 65% to 90% of their time indoors. More than 70 million employees work indoors. Of that 70 million, it is estimated that 21 million (about 30%) are exposed to poor indoor air quality. It has been estimated that there are more than 8000 chemicals in the air of most buildings, more than 13,000 if smoking is permitted. With so much time spent indoors exposed to so many chemicals, indoor air that occupants breathe has a significant effect on their comfort and health.

13.5.1 Indoor Air Quality, Comfort, and Health

Indoor air quality (IAQ) may be generally defined as "the nature of air that affects one's health and well being." Health is a state of complete physical, mental, and social well-being, and not just the absence of disease or infirmity. The effect of poor indoor air quality on occupants results in symptoms that are generally upper respiratory in nature. These include irritation of the eyes, nose, throat, and the upper respiratory system; headaches; skin irritation; dizziness; general fatigue; and respiratory infections. A notable indicator of poor indoor air quality is that occupants who complain about the above symptoms report that the symptoms go away, or simply lessen, after leaving the building.

The quality of air in an enclosed space is an indicator of how well the air satisfies the following two conditions:

1. The content of the oxygen and carbon dioxide must be within acceptable ranges to allow for the functioning of the respiratory system.

2. The concentrations of gases, vapors, aerosols, and particulate matter should be below levels that have deleterious effects, or that can be perceived as objectionable to the occupant.

Building IAQ is a result of a complex relationship between contamination sources in a building, the ventilation rate, and the dilution or removal of the contaminants. This complex relationship is further complicated by outdoor sources of contaminants in the outdoor ventilation air brought into the building to reduce the indoor pollutants.

There are three methods of controlling building air quality: source control, dilution, and removal. Only source control can eliminate exposure to a contaminant. Dilution and removal require the mixture of the contaminant with the air in the space before it can be controlled. Dilution and removal are the only two methods affected by the design and operation of the HVAC system. The ability of the HVAC system to control the contaminants in the space within acceptable levels is highly dependent on the air ventilation rate and air distribution patterns within the space. Ventilation for acceptable indoor air quality is discussed later in this chapter, and air distribution is discussed in Chapter 14.

13.5.2 Sources of Contamination in Buildings

Originally, it was believed that human occupants were the main source of pollutants for indoor air. This theory was put forth more than a century ago. Ever since then, ventilation requirements have primarily been based on the number of occupants in a space. Even today, ventilation codes are based on space occupancy. More recently, however, it has become evident that humans are not the only, or even primary, source of indoor air contamination. Emissions from contaminant sources in a building include building materials, consumer products, cleaning fluids, furnishings, combustion appliances and processes, biological growth from standing water or damp surfaces, as well as the human occupants. The HVAC system itself can also be a source of indoor air pollution, if the system is improperly designed and especially if it is improperly operated and maintained. Table 13.1 gives typical levels of pollution sources for buildings.

In any building, there is a large number of pollutants in the indoor air. As indicated above, the air quality of an indoor space is a function of the levels of many gases, vapors, aerosols, and particulate matter found in the air of a space. Some typical pollutants found in indoor air include the following:

Radon: a heavy radioactive gas that results from the decay of uranium which is known to cause lung cancer.

TABLE 13.1   Pollutant Test Results of Randomly Selected Buildings

Pollutant Source	% of Pollution Contributed
Outdoor Air	0%
Space Materials and Furnishings	36%
Space Occupants	26%
HVAC System	38%

Microbial Growth: microorganisms such as fungi, viruses, and bacteria that usually grow as the result of excessive moisture and standing water.

Volatile Organic Compounds (VOCs): gaseous chemical pollutants that are generally given off by building materials and furnishings.

Particulate Matter: airborne microscopic solid material such as dust, asbestos particles, pollen, fungus spores, insect parts, etc. Particulate matter and its control are discussed in Chapter 18.

Environmental Tobacco Smoke (ETS): gases and vapors given off as the result of the smoking of tobacco products. ETS is becoming less of an indoor air pollutant as smoking is banned in more buildings.

Carbon Dioxide: as part of the respiration process, humans absorb oxygen through the lungs and discharge carbon dioxide as they exhale. Spaces with large numbers of occupants can experience problems with high levels of carbon dioxide.

Combustion Products: can result from combustion appliances inside the building such as stoves, gas or oil fired water heaters, furnaces, and boilers. Combustion products from vehicle exhaust can also be entrained in outdoor ventilation air brought into a building by the HVAC system.

Other pollutants frequently found in building air include lead, nitrogen dioxide, carbon monoxide, ozone, sulfur dioxide, and various other chemicals.

Volatile organic compounds (VOCs) in indoor air, from building materials or furnishings, are a significant source of indoor air pollutants, most prevalent in new buildings, or when new furnishings are installed. They are commonly found in furniture, paint, adhesives, solvents, upholstery, draperies, carpeting, aerosol cans, clothing, construction materials, cleaners, deodorizers, copy machine toner, marker pens, and correction fluids. Table 13.2 lists building materials that emit VOCs along with the chemicals that are emitted. Table 13.3 lists building furnishings and equipment that emit VOCs along with the chemicals that are emitted.

Microbial growth is another significant source of indoor air pollution. The primary requirement for the growth of microorganisms is moisture, especially in standing water and areas that are constantly wet. Both of these situations can, and frequently do, occur in HVAC systems. Standing water can occur in cooling coil condensate drain pans in air handling units. If the drain pan does not have adequate slope for drainage, or if the piping is blocked, standing water can occur in the drain pan. The standing water can become stagnant and have microbial slime. As the fan circulates air through the cooling coil, over the drain pan, and to the occupied spaces, at least some of the microbial particulate is aerosolized into the air. It is important that HVAC systems have adequate drainage of condensed water in drain pans.

Microbial growth from moisture is also prevalent in humidification systems. Basically, there are three types of humidification systems: evaporative pan, steam injection, and water spray. The evaporative pan type consists of a pan with heated water or, in some cases, a pan with absorber plates. As air flows through the humidifier, moisture is absorbed from the pan or the absorbent plates. The steam injection humidifier consists of a set of nozzles on a man-

TABLE 13.2   Volatile Organic Compounds Emitted from Building Materials and Furnishings

Source	Pollutant Emitted	Source	Pollutant Emitted
Adhesives	Alcohols	Particle Board	Alcohols
	Aminies		Alkanes
	Benzene		Amines
	Formaldehyde		Benzene
	Terpenes		3-Carene
	Toluene		Formaldehyde
	Xylenes		Terpenes
			Toluene
Caulking	Alcohols	Tile and Linoleum	Acetates
Compounds	Alkanes	Floor Coverings	Alcohols
	Amines	Wall Coverings	Alkanes
	Benzene		Amines
	Formaldehyde		Benzene
	Methylketone		Formaldehyde
	Xylenes		Methyl Styrene
			Xylenes
Carpeting	Alcohols	Paints, Stains,	Acetates
	Alkanes	and Varnishes	Acrylates
	Formaldehyde		Alcohols
	4-Methylbenzene		Alkanes
	Styrene		Amines
			Benzenes
			Formaldehyde

ifold, or grid, located in the air stream, that inject steam directly into the air. In a water spray humidification system, water is sprayed through atomizing nozzles, directly into the air stream. The water spray results in a fine mist which is mixed into the air stream. For any type of humidification system, it is important that the amount of moisture put into the air be accurately controlled. Improperly controlled humidification systems can result in wet duct liners and standing water in ducts. If final filters are installed downstream of the humidifier, the filters can also become saturated with moisture. Pan type humidifiers, by their design, will have standing water exposed to the air stream. Finally, steam supplied from a boiler frequently contains chemicals necessary for corrosion control in the boiler and steam distribution system.

Another major source of indoor air pollution is pollution that is entrained in outdoor air as it is brought into the building by the HVAC system. This problem usually results from the location of the outdoor air intake itself. The outdoor air intake must be located so it is not in close proximity to sources of contamination such as vehicle loading docks, cooling towers, toilet exhaust outlets, kitchen exhaust outlets, plumbing vents, industrial equipment, and parking lots where motor vehicle emissions are generated. Other sources of contamination include debris in the air intake, bird nests in the air intake, standing water on a roof, and contamination from trash dumpsters.

13.6   VENTILATION FOR ACCEPTABLE INDOOR AIR QUALITY

In addition to providing acceptable temperature and humidity conditions in a building, it is also the function of an HVAC system to provide adequate ventilation resulting in

TABLE 13.3   Volatile Organic Compounds and Other Pollutants Emitted from Appliances, Office Equipment and Supplies

Source	Pollutant Emitted	Source	Pollutant Emitted
Ovens, Kerosene and Gas Heaters	Carbon monoxide	Computer, video display terminals	n-Butanol
	Nitrogen dioxide		Cresol
	Sulphur dioxide		Phthalates
	Polyaromatic hydrocarbons		Dodecamethyl cyclosiloxane
Duplicating machines	Ethanol		2-Ethoxyethyl acetate
	Methanol		Ethylbenzene
	Halogenated alkanes		3-Methylene-2-pentanone
Michrofiche developers	Ammonia		Ozone
Carbonless paper	Chlorobiphenyl		Penol
	Cyclohexane		Phosphoric acid
	Dibutyphthalate		Toluene
	Formaldehyde		Xylene
Paper forms	Acetaldehyde	Electrographic printers and photocopying machines	Amnonia
	Acetic acid		Benzaldehyde
	Acetone		Benzene
	Acrolein		Butyl methacrylate
	Alkanes		Carbon black
	Benzaldehyde		Ethyl benzene
	1,5-Dimethylcyclopentene		Halogenated alkanes
	2-Ethyl furane		Isopropanol
	Hexanal		Ozone
	Isopropanol		Methyl methacrylate
	Paper dust		Styrene
	Proprionaldehyde		Terpene
	1,2,2-Trichloroethane		Toluene
			Xylenes
		Typewriter correction fluid	Acetone
			1,1,1-Trichloroethane

acceptable indoor air quality. When selecting and designing an HVAC system, the designer should always be aware of any pollution causing material, systems, equipment, or processes within a building. The previous section discussed some of the pollutants, and their sources, commonly found in most buildings. This section discusses the methods for providing acceptable indoor air quality.

Most indoor air pollutants can easily be controlled, if adequate outdoor ventilation air is brought into the building by the HVAC system. Ventilation may be defined as the process of supplying or removing conditioned or unconditioned air by natural or mechanical means, to, or from, any enclosed space. Ventilation in a building serves many functions, including:

1. the replacement of oxygen consumed by the occupants during respiration

2. dilution and/or removal of body or other noxious odors

3. dilution and/or removal of indoor air contaminants

4. reduction in the levels of carbon dioxide

5. providing heating and/or cooling to the conditioned space

6. providing desired pressure relationships between spaces

The vitalizing quality of outdoor ventilation air is its oxygen content. Actually, the amount of the oxygen consumed by building occupants that must be replaced by outdoor ventilation air is relatively small. Far more outdoor ventilation air is required for the dilution and/or removal of impurities than to replenish the oxygen. A corresponding amount of air must be exhausted, or relieved, from the building if excessive pressure gradients between indoors and outdoors are to be avoided.

Most building codes set outdoor air ventilation requirements for buildings. Chapter 11 discussed the major codes that govern the design of HVAC systems. Most ventilation codes for building are based on ASHRAE Standard 62-1989, Ventilation for Acceptable Indoor Air Quality. Whenever an HVAC system is being designed, it is always important that the designer comply with the local codes as a minimum. In addition to complying with local codes, the design should also comply with the requirements of ASHRAE Standard 62, which is recognized as the most authoritative guide for ventilation and indoor air quality.

ASHRAE Standard 62 allows the HVAC system designer to use one of two methods for designing a system that provides adequate ventilation for acceptable indoor air quality. The two procedures suggested by ASHRAE 62-1989 are the Ventilation Rate Procedure and the Indoor Air Quality Procedure. An HVAC system designed and operated by either method should provide acceptable indoor air quality in a building.

13.6.1 Ventilation Rate Procedure

The Ventilation Rate Procedure is the most widely used method for achieving acceptable indoor air quality. It is a prescriptive procedure that provides an indirect method for controlling indoor air quality. It specifies the amount of outdoor air needed to dilute and remove contaminants for 91 different building or space applications. The procedure prescribes rates at which outdoor ventilation air must be delivered to a space and various means to condition that air.

Table 11.2 lists the outdoor air requirements for ventilation for various types of building occupancies. Under the Ventilation Rate Procedure, Table 11.2 prescribes the minimum outdoor air required, per occupant, for various types of facilities or occupancies. The procedure is easily applied by simply estimating the space occupancy and calculating the minimum outdoor air ventilation rate, using the ft^3/min•person factor from Table 11.2.

Example 13.1   How much outdoor ventilation air is required for a 5000 ft^2 office space? From Table 11.2, the estimated occupancy for an office is seven occupants per 1000 ft^2 and the outdoor air requirements are 20 ft^3/min•person.

$$\text{Occupancy} = (7 \text{ occupants}/1000 \text{ ft}^2)(5000 \text{ ft}^2) = 35 \text{ people}$$

$$\text{O.A. ventilation rate} = (35 \text{ people})(20 \text{ ft}^3/\text{min•person}) = 700 \text{ ft}^3/\text{min}$$

The application of the Ventilation Rate Procedure is fairly easy and straightforward for most applications. Additional calculations are required whenever multiple types of spaces are served by a common air system, and when spaces have variable occupancy levels.

Frequently, a single building air system serves areas of a building that have different ventilation and occupancy requirements. An example of such a situation would be an air system serving a building with conference rooms as well as regular office space. These spaces would have different ventilation requirements due to their difference in occupancy estimates and the variable occupancy of the conference rooms.

To determine the minimum outdoor air requirements for multiple types of spaces, or spaces, with different occupancy levels, ASHRAE 62 requires that Equation 13.3 be used to calculate the minimum outdoor air supplied by the system.

$$Y = \frac{X}{(1 + X + Z)} \qquad (13.3)$$

where

$Y = Q_{OT}/Q_{ST}$ = the corrected fraction of outdoor air in the system supply

$X = Q_{ON}/Q_{ST}$ = the uncorrected fraction of outdoor air in the system supply

$Z = Q_{OC}/Q_{SC}$ = the fraction of outdoor air in the critical space (the space with the greatest required fraction of outdoor air supplied to the space)

Q_{OT} = the corrected outdoor air flow rate

Q_{ON} = the sum of outdoor air flow rates for all branches on the system

Q_{OC} = the outdoor air flow rate required in the critical space

Q_{ST} = the total supply flow rate, i.e., the sum of all supply for all branches

Q_{SC} = the supply air flow rate in the critical space

Equation 13.3 is shown graphically in Figure 13.2.

Example 13.2 An office building has a total floor area of 10,000 ft², which is comprised of 9500 ft² of general office space and 500 ft² of conference room space. A common air handler serves the entire building. Assume a steady occupancy rate for the spaces. The air system supplies a total of 10,250 ft³/min (9500 ft³/min to the general office and 750 ft³/min to the conference room) based on the building cooling load. How much outside ventilation air should the HVAC system bring into the building in order to meet the requirements of ASHRAE Standard 62-89?

Referring to Table 11.2, the occupancy may be estimated for each space as follows:

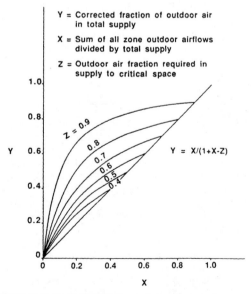

FIGURE 13.2 Ventilation Reduction in Multiple Spaces from a Common Source

$$\text{Office: } \left(\frac{7 \text{ people}}{1000 \text{ ft}^2}\right)(9500 \text{ ft}^2) = 66.5 \text{ people}$$

$$\text{Conference: } \left(\frac{50 \text{ people}}{1000 \text{ ft}^2}\right)(500 \text{ ft}^2) = 25 \text{ people}$$

The outside air requirements for each space are also obtained from Table 11.2 as follows:

Office: (20 ft³/min•person)(66.5 people) = 1330 ft³/min

Conference Room: (20 ft³/min•person)(25 people) = 500 ft³/min

Since the conference room has a higher people density than the general office space, it is the critical space.

$$Q_{SC} = 750 \text{ ft}^3/\text{min}$$

$$Q_{OC} = 500 \text{ ft}^3/\text{min}$$

The total system supply air flow is 10,250 ft³/min = Q_{ST}.
From the definition of the variables for Equation 13.3

$$X = \frac{Q_{OT}}{Q_{ST}} = \frac{(1330 + 500)}{10,250} = 0.1785$$

$$Z = \frac{Q_{OC}}{Q_{SC}} = \frac{500}{750} = 0.667$$

Using Equation 13.3

$$Y = \frac{X}{(1 + X + Z)} = \frac{0.1785}{(1 + 0.1785 - 0.667)} = 0.349$$

Solving for the outside air flow rate,

$$Q_{OT} = Y Q_{ST} = (0.349)(10,250) = 3,577 \text{ ft}^3/\text{min}$$

Another modification to the outdoor ventilation rates, required under the Ventilation Rate Procedures, may be applied to the spaces that have variable occupancy levels. Such spaces may have a frequently changing occupancy rate, or may be occupied only for relatively short periods of time. An example of the first type of variable occupancy would include a space such as a waiting room in a doctor's office. An example of the second type of variable occupancy would be a church sanctuary or a theater.

HVAC systems that provide outdoor ventilation air to spaces with intermittent or variable occupancy may have their outdoor air quantity adjusted by the use of dampers, or by starting and stopping the fan system to provide adequate dilution of contaminants within the space at all times. Such system adjustments may lag the actual occupancy in time. Preferably, the adjustments should lead the occupancy, depending on the source of contamination and the variation on occupancy. If the source of the contamination is from the occupants themselves, or from their activities, the occupancy and activity does not present a short-term health hazard, and if the contaminants are dissipated during unoccupied periods, the outdoor air supplied to the space may lag the occupancy. The length of time lag is a function of the amount of outdoor air supplied, the volume of the space, and the number of occupants. Figure 13.3 may be used to determine the maximum permissible time lag for the space.

Example 13.3 A conference room that seats 200 people has a volume of 32,000 ft³ and has an outdoor ventilation rate of 3000 ft³/min (15 ft³/min•person). How long may the outdoor air dampers of the air system remain closed after the conference room is occupied?

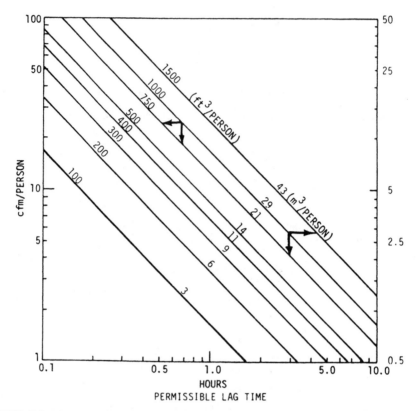

FIGURE 13.3   Maximum Permissible Ventilation Time Lag

$$\text{Volume per person} = \frac{32,000}{200} = 160 \text{ ft}^3/\text{person}$$

Locating the intersection of 15 ft³/min•person and 160 ft³/person on Figure 13.3, the maximum permissible time lag is 0.15 hr or 9 min.

When the contaminants are generated in the space or the conditioning system, and are independent of the occupants or their activities, the outdoor ventilation air should be supplied to the space prior to the entry of the occupants. This will allow acceptable conditions to exist at the start of occupancy in the space. Figure 13.4 may be used to determine the maximum required lead time for outdoor ventilation prior to occupancy.

Example 13.4   A telecommunications room has a large amount of equipment that emits VOCs into the room. The room has a volume of 40,000 ft³, has a variable occupancy of 300 people, and the outdoor air is supplied at 4000 ft³/min (15 ft³/min•person) when the room is occupied. When the room is occupied, the outdoor air damper is closed. What is the minimum time the air dampers should be opened prior to occupancy?

$$\text{Volume per person} = \frac{40,000}{300} = 133 \text{ ft}^3/\text{person}$$

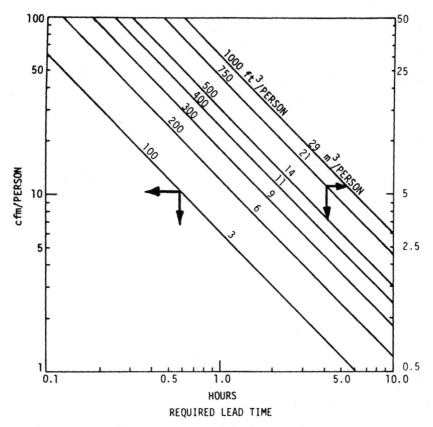

FIGURE 13.4   Minimum Ventilation Time Required Before Occupancy of Space

Locating the intersection of 15 ft³/min·person and 133 ft³/person on Figure 13.4, the minimum lead time is determined to be 0.6 hr or 36 min. This implies that the outdoor air supplied should lead the occupancy by 36 min, in order to have acceptable concentrations at the start of occupancy.

13.6.2   Indoor Air Quality Procedure

An alternate method to the Ventilation Rate Procedure for ensuring acceptable indoor air quality is the Indoor Air Quality (IAQ) Procedure. Since the amount of outdoor ventilation air that is brought into a building and conditioned can have a significant effect on energy consumption, the IAQ method was introduced to permit the use of advanced energy efficient means for controlling building air quality.

The IAQ procedure provides a performance based method for achieving acceptable air quality. The procedure allows the HVAC system designer to use any amount of outdoor air necessary, provided that it can be shown that the concentrations of indoor air contaminants would be below the recommended levels. ASHRAE Standard 62 lists the maximum allowable concentrations of pollutants in buildings used for human occupancy.

Under the IAQ procedure, the use of recirculated air in conjunction with proper filtration and air cleaning equipment may be used as an alternative to higher levels of outdoor ventilation air. In order to use the procedure, the designer must determine the rate at which various contaminants are emitted into the space, the filter/cleaner efficiency, and the ventilation effectiveness, as well as the supply air and outdoor air ventilation rates. Since the procedure is an involved procedure, dependent on space contaminant generation and filter efficiencies, the procedure is beyond the scope of this book. You may wish to refer to Reference 19 for additional information on the Indoor Air Quality Procedure.

Chapter 14

SPACE AIR DISTRIBUTION

In Chapter 13, the conditions in a space that are controlled by an HVAC system affecting human comfort were discussed. This chapter begins the discussion of how an HVAC system provides comfort in a space by the distribution of conditioned air in the space, and discusses some of the issues to be considered in the design of space air distribution systems. This chapter is primarily concerned with air movement within the conditioned space, although space temperature distribution is considered, since it is directly affected by the distribution of conditioned air within the space. The objective, in the design of the air distribution system, is to assure a proper combination of air temperature, humidity, and air motion in the occupied space to provide comfortable conditions.

Discomfort can be caused by a lack of uniform conditions within the space or by excessive fluctuations of conditions within the space. Discomfort may be the result of excessive room air temperature variations (horizontally, vertically, or both), excessive air motion (draft), and lack of air motion resulting in areas of stagnate air.

In order to provide comfortable conditions in an occupied space served by an HVAC system, it is first necessary to define what is considered the occupied space. The occupied space, or zone, is the volume in a room that is bounded by the vertical distance from 1 ft above the floor to 6 ft above the floor, and horizontally from 1 ft inside each wall. This is the space where almost all human activity takes place. Usually, the objective of the HVAC system is to only control the conditions within the occupied zone. Conditions outside the occupied zone can, and frequently do, vary considerably from normal comfort conditions.

Occupants of a space can usually sense air movement as low as 50 ft/min. An objective of the HVAC system is to provide air motion within the occupied zone between 20 and 100 ft/min. Areas with air velocities less than 20 ft/min are considered to be stagnate areas. Stagnate areas can result in large temperature variations and high contamination levels within the stagnate air volume. Air velocities above 100 ft/min in the occupied zone can result in draftiness. A draft is any localized feeling of coolness or warmth of any portion of the body due to both air movement and temperature. The optimum air velocity within the occupied zone is generally considered to be about 50 ft/min.

14.1 BASIC AIR MOTION
AND AIR DISTRIBUTION

An objective of an HVAC system is to introduce conditioned air into an occupied zone or space so the desired combination of room air temperature, humidity, and air motion occur in the occupied zone. Air is introduced into a space by various types of air devices and at various locations within the room. Air flowing from the air devices results in jets of air in the space. In order to achieve the desired conditions in the occupied zone, the performance of air jets at various conditions and locations must be considered.

14.1.1 Air Jets and Air Motion

Air discharged from an air device, such as a grille, register, or diffuser, results in a stream of air that is projected into the space. A jet of air is similar to a stream of water flowing from a garden hose. An air jet is an air stream that is discharged from an outlet with a significantly higher velocity than that of the surrounding air. The jet moves along a central axis until its terminal velocity is reduced to a value that equals the velocity of the surrounding ambient air.

The velocity of the air as it is discharged from an outlet is a function of the air flow rate and the free, or open, area of the outlet. The free area of an outlet is the total minimum area of the openings in the air outlet through which air can pass. The outlet velocity is the average velocity of the air emerging from the outlet, measured in the plane of the opening. The Continuity Equation (Equation 2.16) may be used to determine the outlet velocity for any given flow rate through an outlet with a given free area. If the density of the air is assumed to be constant, Equation 2.16 may be used to calculate the outlet velocity.

Unlike a jet of water, a jet of air does not flow in a confined stream tube. As the air flows in a jet stream, resistance from the room air (which generally is not moving or is moving slowly) will tend to reduce the velocity of the jet and expand it as it flows. The expansion of the flow stream as it flows is called the spread and is shown in Figure 14.1.

The boundaries of the air stream are referred to as the envelope of the air stream, and the distance that the air stream flows in the direction of discharge is called the **throw**. The terminal velocity is the air velocity at the end of the throw.

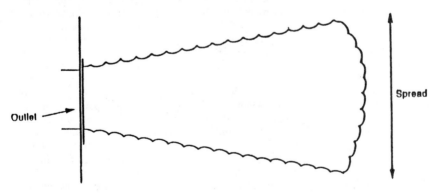

FIGURE 14.1   The Spread of an Isothermal Jet

As the jet flows and encounters air resistance from the still room air, it tends to draw the room air into the jet. The tendency of the jet to induce motion of the room air, in the direction of the jet flow, is known as **entrainment of the room air**. Entrainment is defined as the movement of room air into the jet caused by the air stream discharge from an outlet. Entrainment is also referred to as secondary air motion. The total air is the mixture of discharge air and entrained air, and the **entrainment ratio** is the total air divided by the amount of air discharged from the outlet.

14.1.2 Isothermal Free Jets

In order to study and understand the characteristics and performance of air jets, it is first necessary to study air jets in their simplest form. The simplest, most basic air jet is an isothermal jet. An isothermal jet is an air jet that is at the same temperature as the surrounding air into which the jet is discharged. Since the jet temperature and ambient air temperature are equal, buoyancy forces, due to differences in air density, need not be considered. A free jet is one that is not confined by any obstruction, such as solid surfaces or air motion from another source.

Figure 14.2 shows the flow pattern for a typical isothermal free jet. It may be observed from Figure 14.2 that the air jet expands as it flows. The expansion is due to the resistance of the still ambient air, and the angle at which the jet expands is known as the **angle of divergence**. The angle of divergence for an isothermal jet is typically about 10° to 12° from the centerline axis of the jet, or a total of about 20° to 24°. The air jet forms an expanding cone as it travels outward from the discharge point.

The flow of an isothermal jet can be divided into four distinct regions, or zones of expansion, with different characteristics. The four zones in an isothermal jet expansion are shown in Figure 14.3. The characteristics of the zones are as follows:

Zone 1, which is the first zone as the air is discharged, is a short core zone in which the maximum velocity of the air stream remains essentially constant. Zone 1 extends about 4 diameters, or widths, from the face of the outlet.

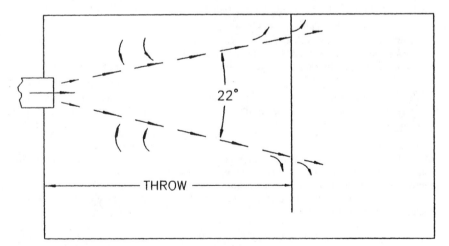

FIGURE 14.2   Free Isothermal Jet

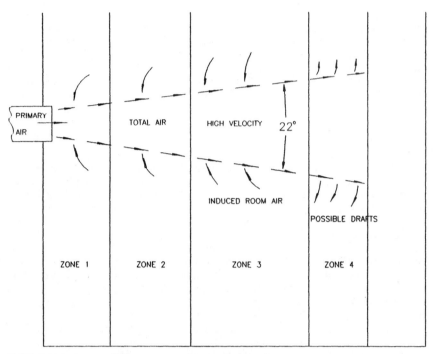

FIGURE 14.3   Zones in an Isothermal Expansion

Zone 2 is a transitional zone where the air velocity at the central axis decreases inversely with the square root of the distance from the discharge. The width of zone 2 depends on the type of outlet, the aspect ratio of the outlet, and the initial air flow turbulence.

Zone 3 is called the main zone and is the zone which the air flow has become fully turbulent. The maximum velocity decreases inversely with the distance from the outlet. Even if the air is discharged from a rectangular outlet, the cross section of the air stream becomes circular. In zone 3, the jet becomes a mixture of the supply air from the jet stream and the room air.

Zone 4 is called the terminal zone since the jet air velocity decreases rapidly to a value less than 50 ft/min. The distance to this zone depends on the velocities and turbulence characteristics of the ambient air.

Zone 3 is the most important as far as room air distribution is concerned. In most cases, zone 3 is the zone in which the air stream enters the occupied area of a room or space. In zone 3, the jet is a mixture of the air discharged from the outlet and the room air. The jet expands as it flows, due to the induction of room air into the total air stream. The relationship between the initial velocity of the jet and the centerline velocity of the air stream at some distance is given by Equation 14.1.

$$V_X = \frac{(V_0 K \sqrt{A})}{x} = \frac{Q_0 K}{x \sqrt{A}} \tag{14.1}$$

where

V_X = the centerline velocity at distance x (terminal velocity) in ft/min

V_0 = initial velocity of the jet in ft/min

x = the distance from the outlet to the point of consideration in ft

K = a proportionality constant

Q_0 = the jet discharge rate in ft³/min

The value for the constant K varies with the configuration of the outlet. For free openings, K has a value of 6; for grilles a value of about 5; and for round ceiling diffusers, a value of about 1. Table 14.1 lists the value of K for various types of outlets.

Equation 14.1 may be rearranged and solved for the distance. Equation 14.2 gives the distance, or throw, that a free jet stream will travel before it reaches a given terminal velocity, V_x.

$$x = \frac{QK}{X_x \sqrt{A}}$$ (14.2)

In addition to the throw of an isothermal jet, the total air flow of the jet is an important consideration. In order to determine the total air flow for a jet, it is necessary to calculate the amount of air that will be entrained by the jet as it flows through the still room air. Equation 14.3 may be used to determine the entrainment ratio of an isothermal jet.

$$\frac{Q_x}{Q_0} = \frac{\sqrt{2}V_0}{V_x}$$ (14.3)

where

Q_x = the total air flow rate at distance x from the outlet in ft³/min

Q_0 = the air discharge rate from the outlet in ft³/min

x = the distance from the face of the outlet in ft

V_0 = the air discharge velocity at the outlet in ft/min

V_x = the air velocity at distance x from the outlet in ft/min

The entrainment ratio is also referred to as the induction ratio.

Example 14.1 An 8 in × 6 in high sidewall grille with a free area of 0.26 ft² is discharging a free isothermal jet of air into a room. The rate of air flow is 180 ft³/min. At what distance from the outlet will the jet reach a velocity of 100 ft/min and what will the total air flow rate be at that distance?

TABLE 14.1 Recommended Values of Centerline Velocity Constant, K

Outlet Type	Discharge Pattern	K
High sidewall grilles	0° deflection	5.0
	Wide deflection	3.7
High sidewall linear	Core less than 4 in high	3.9
	Core more than 4 in high	4.4
Low sidewall	Up and on wall, no spread	4.4
	Wide spread	2.6
Baseboard	Up and on wall, no spread	3.9
	Wide spread	1.8
Floor	No spread	4.1
	Wide spread	1.4
Ceiling circular directional	360° horizontal	1.0
	4-way, little spread	3.3
Ceiling linear	1-way, horizontal along ceiling	4.8

Assuming a constant air density, Equation 2.16 may be used to calculate the discharge velocity.

$$V_O = \frac{180 \text{ ft}^3/\text{min}}{0.26 \text{ ft}^2} = 692 \text{ ft/min}$$

From Table 14.1, the velocity constant, K, has a value of 5.0. For a terminal velocity of 100 ft/min, the throw is

$$x = \frac{(180 \text{ ft}^3/\text{min}(5))}{(100 \text{ ft/min})}(\sqrt{0.26}) = 17.6 \text{ ft}$$

The total air flow may be determined by rearranging the entrainment ratio in Equation 14.3.

$$Q_x = \left(\frac{(\sqrt{2})(180 \text{ ft}^3/\text{min})(692 \text{ ft/min})}{100 \text{ ft/min}} \right) = 1763 \text{ ft}^3/\text{min}$$

14.1.3   Non-Isothermal Free Jets

In the previous section, the characteristics of isothermal jets (jets with the same temperature as the ambient space temperature) were discussed. This situation, however, rarely occurs in actual HVAC applications. Jets of air supplied to spaces must be at a different temperature in order to offset any heating or cooling loads in the space. A non-isothermal jet is one that is discharged at a temperature different from the ambient space temperature.

Non-isothermal free jets behave differently than isothermal free jets, due to the buoyancy forces of the air. Buoyancy forces are caused by the difference in air density, resulting from the difference in temperature of two masses of air. For jets that have a temperature higher than the ambient temperature, the tendency is for the jet to rise. In cooling applications, the jet has a tendency to fall, or drop.

The factors affecting the degree to which the buoyancy forces cause a jet to rise, or drop, include the velocity of the jet and the magnitude of the difference in air temperatures. The greater the difference in temperatures, the greater the effect of buoyancy forces. In general, the higher the velocity of the jet, the less of an effect the buoyancy forces will have. Buoyancy forces do not have a significant effect at jet velocities above 150 ft/min.

14.1.4   Confined Jets and Mixing Jets

Air jets in HVAC applications are rarely free jets (a jet that is not confined by obstructions). Jets are usually confined by obstructions, such as room walls and/or room ceilings. These surfaces have a significant effect on the performance characteristics of jets. Other jets of air, from nearby outlets, also affect the characteristic of an air jet. Multiple air jets located in the same surface, discharging in parallel, behave quite differently than jets in opposite facing surfaces, discharging toward each other.

An air jet discharged parallel to and within a few inches of a smooth surface will tend to follow, or "hug," the surface as it flows. This tendency is referred to as the surface effect or **Coanda effect**. The effect of a surface, such as a ceiling, is shown in Figure 14.4.

The surface will tend to draw a jet flowing parallel to the surface toward the surface itself. Normally, a free jet will take the shape of a cone as it flows from the outlet. A confined

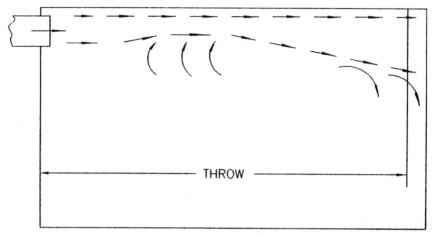

FIGURE 14.4   Isothermal Jet Near Ceiling

jet flowing near a parallel surface will take the shape of a half cone, since one side will be confined by the surface. Essentially, the same amount of air must pass through a cone that would be half as large as for a free jet. This will result in higher air velocities near the surface, which will cause a low pressure area near the surface. The low pressure area is the result of the surface effect on the air stream. The surface effect will occur whenever the angle of discharge between the air stream and the surface is less than 40° and the outlet is within 24 in of the surface.

The surface effect tends to increase the throw of jets from all types of outlets and decrease the drop of horizontally flowing air streams. Figure 14.5 shows a typical cooling jet near a ceiling. For isothermal jets, the values of K used in Equations 14.1 and 14.2 should be multiplied by $\sqrt{2}$.

In non-isothermal cases, the throw and drop of a jet are determined by a balance between the thermal buoyancy and the surface effect, which depends on the jet velocity and the distance between the outlet and the solid surface.

When designing room air distribution systems, it is advisable to take advantage of the surface effect when selecting and locating air outlets. The surface effect tends to keep the velocities in the total air stream out of the occupied zone and allow longer throws for the air devices. Types of air devices and their selection are discussed later in this chapter.

Two jets that are being discharged from opposing surfaces will behave just as if each were discharging against an opposing solid surface. The two jets will act independently until they collide. After colliding, the two air streams will project downward in a vertical plane. The downward projection of the air streams would be acceptable, as long as the high velocity region does not extend down into the occupied zone. In order to avoid this situation, the distance between two opposing outlets should be greater than twice the throw distance of the outlets. On the other hand, the maximum distance between two opposing outlets should be no more than twice the throw distance for a terminal velocity of 50 ft/min.

Two jets can also be confined if they are in the same surface and have parallel air flows. Twin parallel jets will act independently until the spreading air streams interfere. The point of interference, and the distance from the outlets, varies with the distance between the outlets. The maximum velocity for each jet, prior to the point of interference, is along the centerline of each jet.

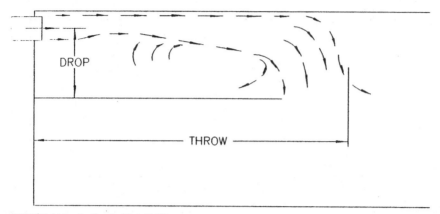

FIGURE 14.5   Cooling Jet Near Ceiling

Past the point of interference, the two air streams combine and the maximum velocity occurs on a line midway between the two outlets. At the point of interference, the air streams combine to form a secondary jet that is conical in shape. The secondary jet then behaves as a single jet of twice the area of the original outlets, emanating from the point of intersection.

The following general statements may be made concerning the characteristics of air jets:

1. Surface effect increases the throw and decreases the drop, compared to free space conditions.

2. Increased surface effect may be obtained by moving the outlet away from the surface, somewhat, so that the jet spreads over the surface after impact.

3. Increased surface effect may be obtained by spreading the jet when it is discharged.

4. Spreading the air stream reduces the throw and drop.

5. Drop primarily depends on the quantity of air and only partially on the outlet size or velocity. Thus, the use of more outlets with less air per outlet reduces drop.

6. Surface effect is essential to good air distribution because it generally prevents high velocity air from entering the occupied zone of the space.

14.1.5   Return/Exhaust Air Inlets

A stagnant area is a zone, or volume, with natural convective air currents. Stagnant areas occur in areas where forced air currents are less than 20 ft/min. In stagnant areas, natural convective air currents rise from warm surfaces and fall from cool or cold surfaces. The air volume stratifies in layers with, sometimes, large temperature gradients from the top to the bottom of the stagnant zone.

A return or exhaust intake has very little effect on room air motion. It only affects the room air in the immediate vicinity of the intake.

Since the air intake location has little effect on room air motion, the best location for return air intakes is in areas that may be stagnant areas. The openings should be located so there is a minimum of short-circuiting of the supply air. Short-circuiting occurs when the supply air flows directly into the return inlet, without being distributed throughout the room. Locating the return/exhaust intakes in stagnant areas will tend to return the warm air during cooling and the coolest air during heating to the HVAC system. This will tend to reduce areas of stagnant air in a room.

14.2   ROOM AIR OUTLETS

Room air outlets can be classified according to their discharge pattern and location within the room or space. Each type and location of outlets produces a different pattern of air motion within a room. Whether the discharge air is cooling or heating the space, it has an effect on the pattern of room air motion. The basic types and locations of supply air outlets are:

- High sidewall discharge (Group A)
- Floor outlet adjacent to an interior wall (Group B)
- Floor outlet adjacent to an exterior wall (Group C)
- Baseboard and low sidewall outlets (Group D)
- Ceiling diffusers (Group E)

14.2.1   High Sidewall Outlets (Group A)

These outlets are mounted in or near the ceiling and have a horizontal discharge pattern. As the air jet is discharged from the outlet, the surface effect tends to keep the air in contact with the ceiling as it flows into the room.

These types of outlets tend to perform particularly well in cooling applications and can be used with high air flow rates and large supply/room temperature differentials. If the cool air jet is only confined by a ceiling, the surface effect attaches the air stream to the ceiling until the jet velocity slows enough for the buoyancy forces to cause the jet to drop. As the jet flows, it entrains room air above the occupied zone.

The performance of these outlets is a function of various factors, including outlet vane deflection settings, the supply/room air temperature difference discharge velocity, and the extent to which the jet is confined within the space. Vane deflection settings reduce the jet flow and drop by changing the air stream from a single straight jet to a wide-spreading fanned out jet. Wide deflection settings also tend to increase the surface effect of the ceiling. Multiple small outlets tend to reduce the drop of the jet. In cooling applications with Group A outlets, the distance to any opposing wall must be taken into consideration. During cooling, the air jet entrains room air until it strikes the opposing wall, where it deflects downward, as shown in Figure 14.6.

If the throw of a high sidewall outlet is longer than the distance to the opposing wall, as shown in Figure 14.6, the air jet may enter the occupied zone with an excessive velocity. This situation is referred to as "overblow." In some cases, overblow may be acceptable since the occupied zone of a room begins one foot out from the wall.

A Group A outlet used in a heating situation with an opposed wall must be used with some precautions. During heating, the warm air of the jet tends to rise and a shorter throw results. As the total air stream rises toward the ceiling, a stagnant zone may form between the floor and the total air stream, as shown in Figure 14.7. When the airstream strikes the opposing wall, the air envelope does not descend down the wall, as in the case for a cooling jet. The buoyant warm air forces the air envelope to curl back toward the ceiling, causing a stagnant area in the occupied zone.

If the air is discharged from an interior wall toward an opposing exterior wall, the air stream is converted into a cold air stream when it comes in contact with a cold surface

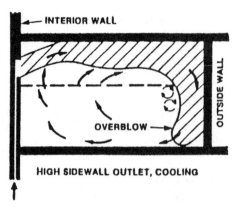

FIGURE 14.6   High Sidewall Outlet Cooling Mode

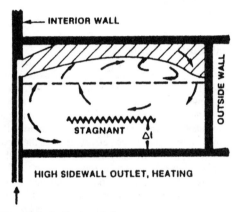

FIGURE 14.7   High Sidewall Outlet Heating Mode

(especially a window). The cold air stream then drops and causes a cold draft as it enters the occupied zone.

Group A outlets also include ceiling outlets with horizontal discharge patterns. The performance characteristics of these types of ceiling diffusers are similar to the characteristics of the high sidewall outlets. The characteristics of Group A ceiling diffusers are shown in Figures 14.8 and 14.9.

In general, Group A outlets are best suited for cooling only applications, interior spaces, or areas where perimeter heating loads in winter are small (mild climates). Supply/room air temperature differentials in the heating mode should be less than 25°F. If they are used for heating applications, some means of projecting the warm airstream down to the floor is required. If that is not possible, additional heat should be supplied at the perimeter exterior walls, especially under glass.

14.2.2.   Floor Outlets Adjacent to Interior Walls (Group B)

Group B outlets are floor outlets mounted next to an interior wall in the room. The air discharge from the outlet is vertical and parallel to the wall. Typically, these outlets do not

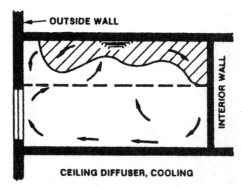

FIGURE 14.8 Ceiling Diffuser Cooling

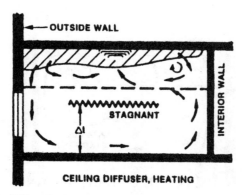

FIGURE 14.9 Ceiling Diffuser Heating

have vanes to deflect the air as it is discharged. As the air is discharged vertically along the wall, air is drawn into the air stream toward the wall. This results in the air stream attaching to the wall surface in a manner similar to the surface effect for a ceiling.

When Group B outlets are in the cooling mode, the air flows up the wall until it strikes the ceiling, at which point it deflects and follows the ceiling for a short distance, as shown in Figure 14.10.

As the cool air begins to flow parallel to the ceiling, the buoyancy forces of the air begin to take effect. The result is that the air stream begins to drop into the occupied zone. This causes a stagnant area to form near the ceiling, away from the air stream, and above the occupied zone. The occupied space, directly below the total air envelope, will have adequate air motion and will receive adequate cooling. Areas farther away from the air envelope will tend to be more stagnant and have greater temperature stratification.

The distance that the total air envelope travels before dropping into the occupied zone depends on the air discharge rate from the outlet and the supply/room air temperature difference. Typically, the distance is around 10 ft. It is recommended that the opposite inside wall should not be more than 20 ft beyond the point where the total air envelope drops. This limits the use of Group B outlets for cooling applications to rooms not more than 30 ft in width.

When Group B outlets are used in heating applications, the air flows up the wall and across the ceiling, as in the case for cooling. In heating applications, however, the warm air

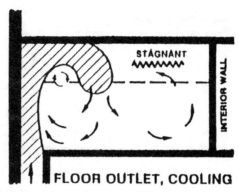

FIGURE 14.10   Floor Outlet, Cooling

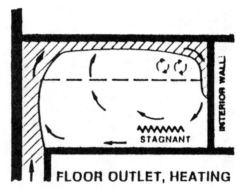

FIGURE 14.11   Floor Outlet, Heating

jet does not fall after striking the ceiling. The buoyancy forces cause the air stream to remain attached to the ceiling, as shown in Figure 14.11.

The stagnant air zone for Group B outlets is much smaller than in the cooling mode. Group B outlets in the heating mode will tend to have a smaller stagnant area than Group A outlets in the heating mode.

In general, Group B outlets perform better than Group A outlets in heating applications. However, Group B outlets perform poorer in cooling applications than do Group A outlets.

14.2.3   Floor Outlets Adjacent to Exterior Walls (Group C)

Group C outlets are similar to Group B; however, they are located in a floor adjacent to an outside wall. They are also characterized by wide-spreading jets, high supply air velocities, and higher entrainment ratios. Since the outlets are near the wall, the primary air is drawn toward the wall in a manner similar to the surface effect for a ceiling. This results in a "scrubbing" effect on the wall, which increases the overall heat transfer coefficient for that wall. The increased U value of the wall increases the wall's heat gain or loss. Installing the outlets some distance from the exterior wall will reduce the scrubbing effect of the air

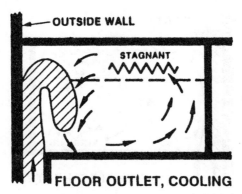

FIGURE 14.12 Floor Outlet Cooling–Exterior Wall

stream. A distance of 6 in and a deflection angle of 15° away from the wall will reduce the scrubbing effect.

In the cooling mode, the wide diffusion of the primary air jet along the wall reduces the velocity of the jet. The result is that the jet of cool air tends to fold back due to buoyancy forces of the air, as shown in Figure 14.12.

The folding of the cool air jet back to the floor, before it reaches the ceiling, results in a larger stagnant zone than for Group B outlets. The size of the stagnant zone may be reduced by limiting the supply/room air temperature difference to a maximum of 15°F.

Group C outlets perform much better in heating applications than in cooling applications. They are especially effective for exterior walls with large glass areas. The primary advantage to Group C outlets in heating applications is that they tend to counteract the cold draft that flows down the interior of a cold surface, such as a window, when the outdoor air temperature is below 30°F.

As the air is discharged vertically, it attaches to the inside of the exterior wall. When it reaches the ceiling, it attaches to that surface and flows across the room, as shown in Figure 14.13.

Since the air is warm, the buoyancy forces, as well as the surface effect, attach the air stream to the ceiling. This results in a stagnant air area in the occupied zone away from the exterior wall. Closer to the exterior wall, room air is inducted into the air jet, which reduces air stagnation near the wall. The induction of room air also tends to reduce the problem of cold air near the floor.

14.2.4 Baseboard and Low Sidewall Outlets (Group D)

Group D outlets discharge air horizontally near the floor of a room. Typically, these types of outlets are baseboard type or sidewall outlets mounted in a wall near a floor. Since they are mounted low and discharge the air horizontally, the air jet is discharged directly into the occupied zone at high velocities. Since they discharge air directly into the occupied zone, they are not recommended for comfort heating or cooling applications.

In the cooling mode, the horizontal air jet remains near the floor, as shown in Figure 14.14. Both the surface effect and the buoyancy forces tend to keep the jet near the floor. This results in a stagnant area just above the occupied zone.

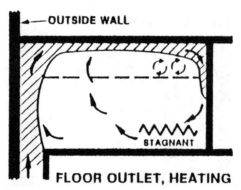

FIGURE 14.13   Floor Outlet Heating–Exterior Wall

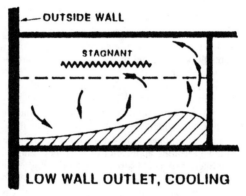

FIGURE 14.14   Low Sidewall Outlet, Cooling

In the heating mode, the horizontal air jet tends to rise as it flows across the floor, due to the buoyancy forces of the warm air. There is very little air stagnation for Group D outlets when they are used for heating, as shown in Figure 14.15.

If a low sidewall outlet is used for heating, the outlet velocities should be less than 300 ft/min. Velocities greater than 300 ft/min will result in excessive air velocities in the occupied zone.

14.2.5   Vertical Discharge Ceiling Diffusers (Group E)

Group E outlets include outlets such as ceiling diffusers, nozzles, grilles, and similar devices. The discharge from these types of outlets is directed in a downward direction toward the floor. They direct a high velocity stream of air directly into the occupied zone.

During cooling, the air stream flows directly downward, due to the discharge pattern and due to the buoyancy forces of the air temperature difference. This situation is frequently referred to as **"dumping."** The air comes out the device and falls directly to the floor, as shown in Figure 14.16.

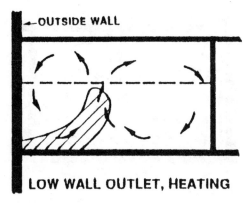

FIGURE 14.15   Low Sidewall Outlet, Heating

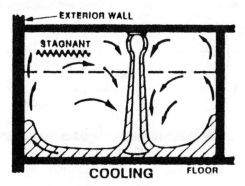

FIGURE 14.16   Vertical Discharge Ceiling Diffuser, Cooling

After the air stream reaches the floor, it is deflected horizontally and flows across the surface of the floor. This results in a stagnant air volume near the top of the occupied zone and above it. These types of outlets are not recommended for cooling applications.

In the heating mode, the warm air jet is again directed toward the floor as it is discharged. However, the buoyancy forces of the warm air slow the descent of the air stream until it eventually stops and begins to rise, as shown in Figure 14.17.

As the air stream begins to rise, it tends to spread. This tendency to spread results in a more even air distribution pattern in the space. If this type of outlet is used in a room with an exterior wall, cold downdrafts can result along the inside surface of the exterior wall.

14.2.6   Horizontal Discharge Ceiling Diffusers (Group F)

Horizontal discharge ceiling diffusers are the most common air outlets used for comfort HVAC applications. They are similar to Group E outlets, except they produce a horizontal discharge pattern in multiple directions. The horizontal discharge produces a surface effect along the ceiling which results in a relatively short throw, an even distribution of air

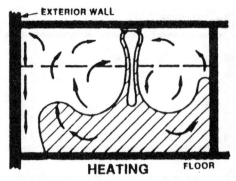

FIGURE 14.17 Vertical Discharge Ceiling Diffusers, Heating

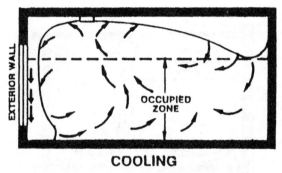

FIGURE 14.18 Group F Outlet, Cooling

velocity, and an even temperature distribution in the occupied zone. Group F outlets also have a high induction ratio which results in rapid diffusion of the primary air jet into the room air. This allows Group F outlets to discharge larger quantities of air at higher velocities than other types of outlets.

In the cooling mode, the primary air jet maintains contact with the ceiling due to the surface effect, as shown in Figure 14.18. The jet tends to entrain large amounts of room air, reducing the jet velocity and mixing with the jet to increase the temperature of the total air jet. Buoyancy forces bring the jet down into the occupied zone as the jet velocity decreases. When the jet strikes a vertical wall, it flows down the wall, entraining room air until it reaches the floor.

In the heating mode, Group F outlets do not perform as well as for cooling applications. Since the surface effect and the buoyancy forces tend to keep the warm air near the ceiling, a stagnant area forms in the occupied zone. If Group F outlets are used in heating applications, the jet discharge should be directed toward an exterior wall or surface, as shown in Figure 14.19.

Group F outlets may be used in heating applications where the climate is moderate and the perimeter heat loss is no greater than 400 Btu/hr•linear ft of wall. For greater heating loads, some form of supplemental perimeter heating is suggested.

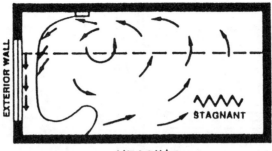

HEATING

FIGURE 14.19   Group F Outlets, Heating

14.3   COMMERCIAL AIR DISTRIBUTION DEVICES

The size, quantity, location, and type of air device used to supply air to a room or space has an important effect on the air distribution, air movement, temperature, and comfort conditions in a space. The basic types and locations of air outlets were discussed in the previous section. This section discusses the types of commercially available air distribution devices for HVAC systems and some of their characteristics. Guidelines for selecting and locating air devices within a space are discussed later in this chapter.

The following basic types of supply air outlets are commercially available for comfort HVAC systems:

1. grilles and registers
2. louvered and perforated panel ceiling diffusers
3. linear diffusers
4. nozzles

Each type of device has its own performance characteristics, such as air flow patterns, surface effect, and sound level. Each characteristic must be considered in the selection and application of air devices.

Another important characteristic of an air device is its tendency for **smudging**. Smudging is a problem that occurs with ceiling mounted devices. Smudging occurs when dirt particles, held in suspension in the room air, are subjected to air turbulence at the outlet face of the air device. As a result of the air turbulence, the dirt particles become attached to the ceiling. Over time, the particles build up on the ceiling resulting in a noticeable dark smudge on the ceiling surface. Precautions should be taken in the selection of supply air devices to reduce the problem of smudging in high traffic areas, such as lobbies, corridors, etc.

14.3.1   Grilles and Registers

An air grille typically consists of a square or rectangular frame with vertical and/or horizontal vanes. A typical air grille is shown in Figure 14.20. A grille is usually mounted on the end of ductwork or on the side of ductwork.

A grille can have a single or double set of vanes. The vanes are used to direct the air as it is discharged and control the air discharge pattern. Vertical vanes control the horizontal

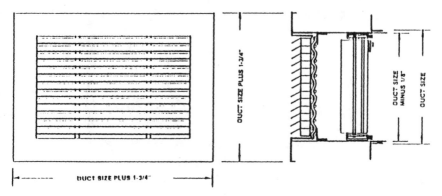

FIGURE 14.20   Supply Air Grille/Register

air discharge pattern, while horizontal vanes control the vertical pattern. A grille can have two sets of vanes to control the pattern vertically and horizontally. A grille that has a volume control damper with it is called a register.

Grilles and registers are typically used in high sidewall applications. They tend to have a relatively low entrainment ratio, a longer drop, longer throws, and high air velocities. Ceiling mounted grilles tend to discharge air in a downward direction into the occupied zone. Ceiling mounted grilles and registers are generally not recommended for comfort air conditioning applications, since they may cause a draft in the occupied zone.

Manufacturers' catalogs for grilles and registers usually include performance data such as supply air flow rates, pressure losses, deflection angles, and noise criteria (NC) for various sizes of grilles and/or registers. A typical manufacturer's catalog performance table for registers is shown in Table 14.2.

In addition to the supply performance for various sizes of grilles and registers, manufacturers also list **core area** (A_C), which is the total unobstructed, or free, area of the device. This is the total open area of the device through which air can flow. Another outlet characteristic frequently included with the outlet performance is the **effective area** (A_K) of an outlet. Whenever a fluid flows through an opening with sharp corners (such as air flowing through the openings of an outlet), the flow separates from the sharp corners due to inertia in the fluid. Consequently, the area through which the fluid is flowing is less than the actual open area. The area through which the fluid actually flows is called the **vena contracta** or A_K. The discharge velocity for an outlet may be calculated by rearranging Equation 2.16, and using the A_K of the outlet, for the cross-sectional area of the conduit.

14.3.2   Louvered and Perforated Panel Ceiling Diffusers

Square and rectangular ceiling diffusers are commonly available in two types, louvered and perforated panel, or perforated face. These types of diffusers are among the most commonly used air devices in commercial HVAC systems. A major feature of these types of outlets is that they can discharge air across a ceiling in all four directions. Diffusers are available in one-, two-, three-, and four-way discharge patterns. Some diffusers have adjustable louvers that allow the discharge pattern to be adjusted to suit the needs of a particular application. Ceiling diffusers are considered Group F type outlets.

The louvered diffusers have multiple passages, or louvers, through which air may be discharged. They are available in round, square, and rectangular shapes. Figure 14.21 shows a typical square louvered diffuser.

TABLE 14.2 Supply Register Performance

SIZE		300			400			500			600			700			800			900			1,000			1,200		
V-Duct Vel. Blade Set / P_1		0 / .010	22½ / .012	45 / .027	0 / .014	22½ / .021	45 / .050	0 / .023	22½ / .034	45 / .062	0 / .034	22½ / .051	45 / .120	0 / .047	22½ / .071	45 / .165	0 / .083	22½ / .096	45 / .220	0 / .081	22½ / .125	45 / .280	0 / .100	22½ / .155	45 / .340	0 / .150	22½ / .225	45 / .480
6×6, 8×4 — CFM		75			100			125			150			175			200			225			250			300		
Throw		8			10			13			16			18			21			24			26			30		
NC		L			L			L			L			L			L			21			24			29		
8×5, 10×4 — CFM		85			110			140			165			195			220			250			275			335		
Throw		8			11			14			17			19			22			25			28			32		
NC		L			L			L			L			L			20			22			25			31		
8×6, 10×5, 12×4 — CFM		100			135			170			200			235			270			305			335			405		
Throw		9			12			15			18			21			24			27			30			35		
NC		L			L			L			L			L			21			24			27			33		
14×4 — CFM		115			155			195			235			270			310			350			390			465		
Throw		9			12			16			19			23			26			29			31			38		
NC		L			L			L			L			L			23			26			29			35		
8×8, 16×4, 10×6, 12×5 — CFM		130			180			220			260			310			350			400			440			530		
Throw		9			13			17			20			24			28			30			33			40		
NC		L			L			L			L			20			24			27			30			36		

(*Courtesy of Carnes*)

14.19

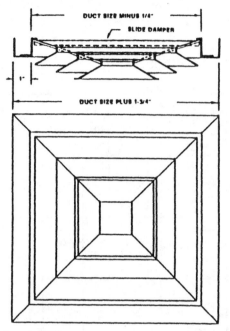

FIGURE 14.21   Square Louvered Diffuser

Louvered diffusers are comprised of concentric rings, or cones, made of vanes arranged in fixed directions and an outer shell, or frame. A flange, or neck, is used to connect the diffuser to the ductwork. The neck can be square or round. These types of diffusers can be flush with the ceiling or "step-down," which project down beyond the frame.

Ceiling diffusers are also available with perforated panels or faces. The diffuser has multiple small louvers that are concealed behind a perforated panel, or face, as shown in Figure 14.22.

The performance of perforated face diffusers is very similar to the performance of the louvered diffuser. The appearance of the diffuser is the main difference between the two styles of diffusers. Most perforated face diffusers have multiple vanes to allow for multiple discharge patterns. Table 14.3 shows typical manufacturers' performance data for ceiling diffusers.

Ceiling diffusers are typically mounted near the center of the space they serve, or are evenly distributed if there are multiple diffusers. Since diffusers can discharge air in multiple directions, they are able to supply large quantities of air. A high induction ratio makes it possible for diffusers to supply air with greater supply/room air temperature differences, 25°F and greater.

14.3.3   Linear Diffusers

Linear diffusers, also called slot diffusers, are long diffusers with a large aspect ratio (ratio of length to width). Aspect ratios for these types of air devices are as high as 25 to 1 or greater. A typical linear diffuser may be 4 in wide × 48 in long. The appearance of a linear diffuser is similar to that of a grille or register. A typical linear diffuser is shown in Figure 14.23.

Linear diffusers can be mounted in a high sidewall, in a floor, in a sill, or in a ceiling. Linear diffusers mounted in a high sidewall have performance characteristics similar to Group A outlets.

TABLE 14.3   Typical Diffuser Performance Data

12x12 Face, One-Way Pattern Adjustment

Neck Size (Inches)		Duct Velocity — FPM								
		200	300	400	500	600	700	800	900	1000
6x6	CFM	50	75	100	125	150	175	200	225	250
	PS	X	.01	.02	.03	.04	.06	.08	.10	.13
	PT	.01	.02	.03	.05	.07	.09	.12	.15	.19
	NC	L	L	L	L	L	22	26	30	34
	Throw - 150 FPM	—	—	1	1	2	3	4	6	6
	Throw - 100 FPM	—	1	2	4	6	7	8	9	10
	Throw - 50 FPM	2	6	8	10	12	14	16	18	19
8x8	CFM	88	133	177	222	266	311	355	400	444
	PS	.01	.02	.03	.05	.07	.09	.12	.16	.19
	PT	.01	.02	.04	.06	.09	.12	.16	.21	.26
	NC	L	L	L	L	24	29	33	37	40
	Throw - 150 FPM	1	3	5	8	9	11	13	14	16
	Throw - 100 FPM	3	6	9	12	14	17	19	21	22
	Throw - 50 FPM	9	14	19	22	24	26	27	29	31

(Courtesy of Carnes Corporation)

14.21

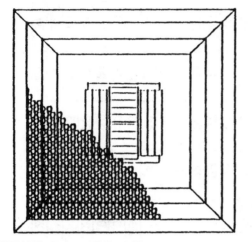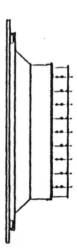

FIGURE 14.22   Perforated Ceiling Diffuser

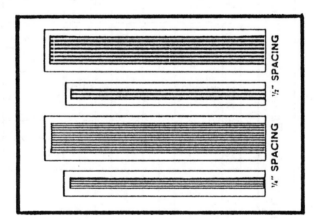

FIGURE 14.23   Typical Linear Diffusers

The air discharge from a high sidewall linear diffuser will not drop in cooling applications if the diffuser is located within 6 to 12 in of the ceiling and is long enough to establish the surface effect. If the diffuser is more than 2 ft below the ceiling, the air stream will drop cold air into the occupied zone.

Floor and sill mounted linear diffusers have performance characteristics similar to that of Group B and C outlets. When the diffuser is located within 8 in of a perimeter wall, the air stream may either be directed up toward the ceiling or deflected slightly toward the wall. If the diffuser is more than 8 in from the wall, air should be deflected toward the wall at an angle of about 15°.

Linear diffusers are also frequently mounted in ceilings. Ceiling mounted diffusers are also called slot diffusers and have performance characteristics similar to those of Group F outlets. Slot diffusers usually include a plenum box, to facilitate connection to the supply

FIGURE 14.24 Slot Diffuser and Plenum (*Courtesy of Carrier Corporation*)

air ductwork and to provide even air distribution along the entire length of the diffuser. A typical slot diffuser and plenum are shown in Figure 14.24.

Slot diffusers typically have one or more air distribution slots of various widths that run the entire length of the diffuser. The number and width of each slot is determined by the required air flow and the desired performance of the diffuser, such as the throw.

The slots in the diffuser also usually have adjustable turning vanes to direct the air flow in a particular direction. Since the diffusers have a large aspect ratio, they can only have one- or two-way air distribution patterns. A slot diffuser with two-way air distribution pattern is shown in Figure 14.25.

Slot diffusers are a good application for ceiling installations because they discharge across the ceiling. Slot diffusers should not be used in downward discharge applications because this would direct a cold jet of air directly into the occupied zone. The diffusers should be located near the center of the space they serve. For large spaces, multiple slot diffusers should be located to provide even air distribution throughout the space. In spaces with exterior walls, slot diffusers are frequently located in the ceiling, near the exterior surface, in order to provide a "washing" effect with the supply air along the perimeter of the space.

14.3.4 Nozzles

Nozzles are round or rectangular supply air outlets that are contracted just before the outlet opening. The contraction, just before the discharge opening, results in a high discharge velocity, a long throw, and a small spread for the air stream. A drum louver, which is a typical nozzle type outlet, is shown in Figure 14.26.

These types of outlets are not used frequently in comfort HVAC applications. They are most widely used in industrial applications, where spot cooling is required and air devices must be mounted in high locations.

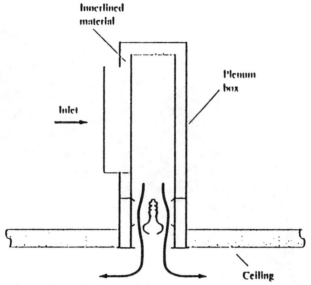

FIGURE 14.25   Two-Way Slot Diffuser

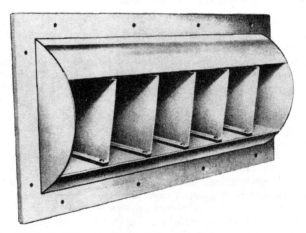

FIGURE 14.26   Drum Louver (*Courtesy of Krueger Manufacturing*)

14.3.5   Air Distribution Accessories

There are a number of accessories commonly used with air distribution devices to help improve the performance of the device. Figure 14.27 shows several of the devices used in air distribution systems.

Adjustable **volume dampers** are frequently installed in air ducts connected to air devices (runout ducts). The purpose of the volume damper is to adjust the air flow to the desired rate.

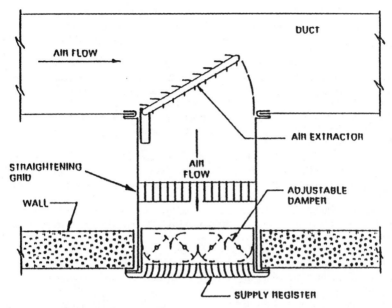

FIGURE 14.27   Air Distribution Accessories

These dampers may be opened or closed to the extent required to provide the required air flow through the air distribution device. Straightening grids are used to straighten the flow of the air as it approaches the air device. Frequently, when the air flow makes an abrupt turn, turbulence, or swirling, occurs in the air flow which results in an uneven discharge from the device. The straightening grid removes the turbulence so that an even flow of air passes through the air device.

An **air extractor** may be mounted in the collar, or the branch of runout ducts, to air devices. These extractors pivot, to remain parallel to the air flow in the duct, regardless of the setting. Air extractors help to extract some of the air flowing through the main duct and turn it into the branch duct to the air device. They should be used only when the main duct is wide enough to allow the extractor to open to its maximum position without causing undue restriction of air flow in the duct.

These accessories may be used to help improve the performance of supply air devices. However, it must be kept in mind that these devices cause some restriction to the air flow and result in considerable loss in duct air pressure. Pressure losses in ductwork and fittings are discussed in Chapter 15.

14.3.6   Selection and Application of Air Distribution Devices

When sizing, selecting, and locating air devices, there are a number of considerations that must be taken into account. Among the considerations are such issues as the location and magnitude of heating and/or cooling loads, the type of HVAC system, the conditions required in the space, and architectural features, such as size, shape, and ceiling height of the space.

The location and magnitude of local sources of heat gain or loss promote convective air currents or cause stratification. Sources of heat gain or loss have a significant effect on both the type and size of supply air outlets and return air inlets. Supply air outlets and return air inlets should be located to neutralize any undesirable convective currents caused by concentrated heating or cooling loads. If a concentrated heat source is located in the occupied zone, the heating effect can be counteracted by either directing cool air at the source or by locating an exhaust or return air inlet adjacent to the heat source. Returning or exhausting air from a heat source is advisable for cooling applications, since heat is removed at its source rather than dissipated into the conditioned space. It is recommended that the air be exhausted from a heat source whenever the wet bulb temperature of the return/exhaust air is above that of the outdoor air.

One criteria used in selecting air devices is the velocity of the air passing through the device. The space air flow, the number of devices, the type of device, and the size of the device will affect the velocity of the air passing through the device. Table 14.4 gives the recommended maximum discharge air velocities for various types of spaces.

When sizing and selecting return air devices, the intake velocity should also be taken into consideration. Table 14.5 shows recommended face velocities for return air intakes.

The air velocity through the devices in a space is directly related to the required air flow for the space and the number of devices serving the space. Table 14.6 provides some rule of thumb guidelines for selecting the type of supply air outlet to use in a particular application, based on the required air flow and the difference between the supply and room air temperature.

Closely related to the air velocity for a device are the acoustical characteristics of the device. Generally speaking, the higher the air velocity through the device, the greater the sound generated at the device. Chapter 12 discussed the basics of the sound levels in various types of spaces. Usually, manufacturers of air devices will list the noise criteria for their devices, with the performance data for the device, as shown in Table 14.2. Noise generated by air devices is transmitted directly to the space and cannot be attenuated. In order to ensure that the sound criteria for the space is not exceeded, air devices should be selected at least 3 dB lower than the desired noise criteria level for the space.

When locating air outlets for a space, they should be located to provide an even air distribution pattern throughout the occupied zone. The throw to the devices must be analyzed at the design air flows. Obstructions such as walls and beams must be considered. The effect of the other outlets must also be considered. Outlets should be selected and located so

TABLE 14.4   Maximum Air Jet Discharge Velocities

Application	Recommended Maximum Jet Velocity (ft/min)
Broadcast Studios	500
Residences	750
Apartments	750
Churches	750
Hotel Rooms	750
Legitimate Theaters	1000
Private Offices	1000
Movie Theaters	1250
General Offices	1500
Retail Stores	1500
Industrial Buildings	2000

TABLE 14.5 Recommended Return Air Inlet Face Velocity

Inlet Location	Velocity Over Gross Inlet Area (ft/min)
Above occupied zone	800 & up
Within occupied zone (not near seats)	600–800
Within occupied zone (near seats)	400–600
Door or wall louvers	200–300
Undercut doors	200–300

TABLE 14.6 General Guide for Type of Supply Air Outlet

Outlet Type	ft^3/min•ft^2 Floor Space	Max. Air Changes/hr	Temperature Difference (°F)
Grilles & Registers	0.6 to 1.2	7	15–25
Slot Diffusers	0.8 to 2.0	12	20–25
Perforated Panel	0.9 to 3.0	18	20–25
Ceiling Diffuser	0.9 to 5.0	30	20–35

the throw will enter the occupied zone without interference from air jets from nearby air outlets. The performance of the outlet, if it is near an exterior wall, must also be taken into account. When analyzing the throw of the air outlet, the design, as well as reduced air flow rates, must also be considered for variable air volume (VAV) systems. In VAV systems, the reduced air flow rates, at reduced cooling loads, result in reduced discharge velocities and reduced throw distances. Low flow rates may cause the outlet to "dump" the air directly down into the occupied space. Therefore, different operating modes must be considered when sizing, selecting, and locating air outlets.

When sizing and selecting air distribution devices for a space, the following procedure is recommended:

1. Determine the air flow requirements for the space based on the space heating/cooling loads, the space temperature, and the supply air temperature. Heating and cooling loads may be determined by the methods presented in Chapters 4 through 9. Typically, the space set point temperature will be 75°F and the supply air temperature will be around 55°F for cooling (a 20°F temperature difference). Higher temperature differences will result in reduced supply air flow necessary to satisfy the space load.

2. Based on the required air flow rate and the guidelines previously discussed, tentatively select an air outlet type, air flow rate for each device, and location within the space. Evaluate the air flow rate for each outlet, throw and discharge pattern, building/room constraints, and esthetic/architectural requirements.

3. Locate outlets to provide a uniform air distribution pattern and temperature distribution throughout the space. Consider the projection pattern of the outlet, room constraints, and the outlet heating and cooling requirements.

4. Select the outlet that most closely meets the above requirements from the air device manufacturers' cataloged performance data. In addition to evaluating the throw and discharge pattern for the outlet, evaluate the pressure loss and the noise criteria at the design air flow rate.

5. Size, select, and locate return/exhaust air inlets for the space. Inlets should be located in areas most likely to be areas of stagnant air. Avoid locating inlets too close to air outlets, which could result in "short circuiting" of the air flow.

14.3.7 Duct Arrangement

When designing air distribution systems for spaces, it is also important to consider the duct connection to the air device. The manner in which the air stream approaches an outlet has a significant effect on the performance of that outlet. To achieve the desired air diffusion from the outlet, the velocity of the air stream must be as uniform as possible as it approaches the air device. The air should approach the device perpendicularly. The effect of an improper duct approach cannot be corrected by the air device.

Supply air devices should be installed at the end of a straight section of connecting duct. Frequently, though, this is not possible. If it is not possible to provide a section of straight duct before the air device, turning vanes or air extractors should be provided to straighten the air flow before it approaches the air outlet.

Frequently, flexible ductwork is used to supply air to a ceiling mounted diffuser. Whenever flexible ductwork is used, its length should be kept to a minimum. It is important that there be no kinks or pinches in the ductwork. Again, the ductwork should be installed so the air flow is perpendicular to the diffuser as it approaches.

14.4 AIR TERMINAL UNITS

Air systems employ a number of different methods to provide air to the conditioned space at the required rate and supply air temperature. Frequently, it is necessary to control the air volume flow rate and/or air temperature for each space or group of spaces. This requires an individual air flow control and/or air temperature control device for each space or zone (group of spaces with similar thermal load characteristics).

Air terminal units are frequently used to provide control of comfort conditions within the space. These terminals are located in the supply air duct to the space to regulate the volume of air flow and/or the temperature of the air supplied to the space. Control of the terminals is provided by thermostats located within the space. The thermostat causes the air terminal to modulate the supply air volume and/or temperature. Terminal units are also used to reduce the pressure of high pressure systems before the air is supplied to the conditioned space.

The two main categories of terminal units are constant flow rate and variable flow rate units. They may also be categorized as being **pressure dependent** or **pressure independent**. Pressure dependent units will vary the rate of air flow in accordance with the pressure changes in the supply air pressure. Pressure independent units will not vary the air flow in response to changes in system pressure. Pressure independent units are also referred to as pressure compensating units.

The flow rate may be controlled by one of several methods. The most common methods of flow control are modulating mechanical pneumatic or electric volume control actuators. These types of terminals are referred to as **externally powered**. The actuators modulate a damper, or air valve, in the terminal unit to control the air flow rate. Flow control can also be provided by the air pressure in the ductwork itself. These terminals are referred to as **system powered**. The flow control assembly derives all of the energy necessary for its operation from the supply air distribution system.

The types of terminal units (or boxes) commonly found in HVAC systems include single duct terminals, dual duct terminals, bypass terminals, induction terminals, and fan powered terminals. Each of these types of terminals may be used to control comfort conditions within a space.

14.4.1 Single Duct Terminals

In this type of system, there is generally a single supply air duct serving a large number of spaces, such as an entire floor of a building. Since each room, or group of rooms, may have different heating and cooling load characteristics, these devices may be used to control the conditions within the space.

The simplest single duct systems with individual space control are constant volume reheat systems, as shown in Figure 14.28.

This system employs a terminal with a heating coil. The reheat coil heats the supply air as it enters the conditioned space. The heating medium may be hot water, steam, or electricity. Cooled air is supplied to the terminal at a constant temperature. As the air flows through the reheat coil in the terminal, it is heated in response to the room air temperature sensed by the room thermostat. These systems are not widely used in HVAC systems today because of their inherent energy inefficiency.

Another type of single duct system that is widely used in HVAC systems today is the **variable air volume (VAV) terminal**. This type of terminal modulates the volume of air flow entering the space in response to the cooling load sensed by the space thermostat. A typical VAV system schematic is shown in Figure 14.29. In a VAV system, air is supplied by an air handling unit at a variable rate and usually at a constant temperature, typically around 55°F. When the cooling load in the space increases, a damper, or valve, in the terminal opens to allow increased air flow to the space. As the cooling load decreases, the cooling air flow is decreased.

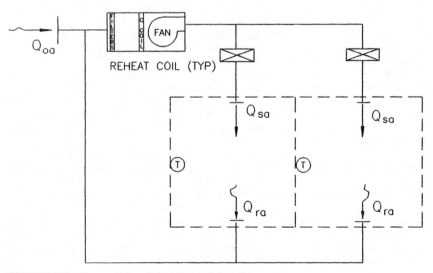

FIGURE 14.28   Constant Volume Reheat System Schematic

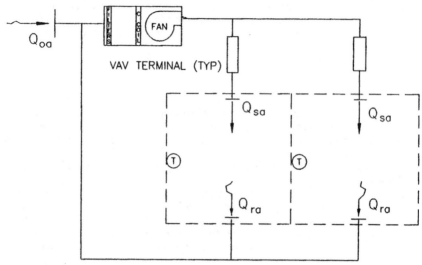

FIGURE 14.29   VAV System Schematic

When using VAV type air terminals, it is important that the reduced air flow at light loads not be less than that required for good indoor air quality. Some provision should be included on the VAV terminal to ensure that the air flow rate does not drop below that required by code and that recommended in Chapters 11 and 13. Occasionally, the minimum air flow rates will provide cooling air in excess of that required for the cooling load. In such a situation, the HVAC system may overcool the space. Since the VAV terminal controls only the cooling air flow, it is usually necessary to provide some means of heating in perimeter spaces. Heating may be supplied by a separate heating system or by providing a reheat coil in the VAV terminal.

In order to supply air at a temperature that more closely matches that of the space requirements, a reheat coil may be provided in the VAV terminal. A typical VAV terminal with hot water reheat is shown in Figure 14.30.

An air valve, or damper, is provided on the inlet to the terminal to vary the air flow rate in response to the cooling load. A reheat coil is provided on the terminal discharge to provide heating for space heating loads and to prevent overcooling of the space at minimum air flow rates. In order to provide an energy-efficient system, the reheat coil should not provide heat when the system is in the cooling mode, unless the air flow is at the minimum setting for the terminal.

A variation of the single duct VAV terminal is the **bypass terminal**. It functions in a manner similar to the VAV terminal, except in how it controls the volume of cool air flow to the space. Instead of throttling the air flow like a VAV terminal, the bypass terminal varies the air flow rate by either supplying the air to the space or by bypassing the air. Air is usually bypassed directly to a ceiling return air plenum, or in some cases ducted directly to the return air ductwork. A room thermostat controls the relative proportions of supply and bypassed air. These types of terminals are frequently used on small- to medium-sized air systems where VAV control is desired; however, the expense of variable air volume control for the air handling unit is not justified.

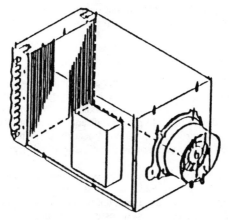

FIGURE 14.30 VAV Terminal with Hot Water Reheat (*Courtesy of the Trane Company*)

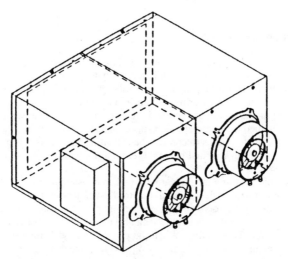

FIGURE 14.31 Dual Duct Terminal Unit (*Courtesy of the Trane Company*)

14.4.2 Dual Duct Terminals

Dual duct systems are similar to single duct constant volume systems in that they supply a constant volume of air to the conditioned space. However, as their name implies, there are two ducts that supply air to the unit, one with cool air and one with warm air. A typical dual duct terminal is shown in Figure 14.31.

A damper, or air valve, is provided in the inlet for each duct. The dampers modulate the flow of cool air and warm air to satisfy the load for the space. The mixing dampers are controlled by a room thermostat.

Dual duct units are typically constant volume devices; however, they are also available as variable flow units. Flow control can be either pressure independent or pressure dependent.

14.4.3   Induction Terminals

Induction terminals use a jet of high velocity air from the HVAC system to induce room air into the terminal unit. The cold air supplied by the system is referred to as primary air, and the warm room air is sometimes called secondary air. As the warm room air is drawn into the unit, it is mixed with the cold primary air and supplied to the space. The amount of primary air supplied to the space is varied in accordance with the cooling load in the space, and is controlled by a room thermostat. Space heating is usually provided by a different system.

The primary advantage to using an induction system over a standard variable air volume system is that it allows the use of colder air. Most VAV systems supply air to the space around 55°F. This produces a total air stream temperature that is not objectionably cold as it enters the occupied zone. Induction systems can have supply air temperatures that are lower, as low as 45°F. The lower supply air temperature requires a lower primary air flow rate to satisfy the room cooling load. The lower supply air flow rate, in turn, permits smaller ductwork and smaller air handling units.

14.4.4   Fan Powered Terminals

Fan powered terminals, like induction terminals, mix cold primary air with warm room air to provide a total air flow at some intermediate temperature that is more acceptable than the temperature of the primary air. Unlike the induction terminal, however, the fan terminal uses a fan to draw warm room air into the unit, rather than the jet of cold air. The advantage of the fan terminal over the induction unit is that a constant air circulation rate may be maintained in the conditioned space. Fan terminal units may also include reheat coils to allow the terminals to control space heating as well as space cooling. There are two basic types of fan terminal powered terminal units: constant volume and bypass type.

Constant volume fan terminal units are used when constant air flow in the space is desired. There are two inlets on the terminal: one for the primary air from the central system and the other to draw warm room air from the ceiling plenum (space). All the air supplied to the space passes through the fan in the terminal. The fan operates continuously whenever the primary air system is operating. When the primary air system is off (such as during night setback), the fan can cycle on and off to provide space heating from the reheat coil. During normal cooling, the space thermostat modulates the primary air damper in response to the space cooling load. During normal heating operation, the space thermostat moves the primary air damper to the minimum position and modulates the heating medium to the reheat coil.

In bypass type fan terminal units, the cold primary air bypasses the fan in the terminal and is delivered directly to the space. The amount of primary air is modulated by a room thermostat in response to the room cooling load. The fan, which only draws in warm room air, is mounted parallel to the primary air damper. The fan is not energized until the primary air damper is closed to a minimum position. During normal cooling operation, the fan is off. The fan only operates in the heating mode when the primary air damper is closed, or in the minimum position. Some heat is picked up from the warm plenum air; additional heat can be provided by a reheat coil.

14.5   SIZING AND SELECTING AIR TERMINAL UNITS

When selecting air terminals for HVAC systems, there are several considerations. Some of the more basic considerations include the type of HVAC system serving the space, the pressure of the system (high or low), the temperature of the air supplied by the system, and the requirements of the space. Other considerations include the performance of the terminal, the terminal acoustical characteristics, available ceiling space, and costs.

14.5.1   Single Duct Terminals

The first consideration for selecting single duct terminals is the type of controller that will be used to control the flow of the air through the terminal. Some terminals use butterfly type dampers, while others use a valve to control the primary air flow. Flow control valves are typically used where a more precise control of the air flow rate is required. The method by which the terminal is controlled (**direct acting or reverse acting**) and the method of actuation (pneumatic, electronic, direct digital control) should be considered. Direct acting and reverse acting controls are discussed in Chapter 20.

After the type of control has been determined, the maximum and minimum air flow rates for the terminal must be determined. Based on the air flow requirements, a preliminary selection should be made from the unit manufacturers' catalog data. Typical performance data for a single duct terminal is shown in Table 14.7.

If sound level is an important consideration, the terminal should be selected so the design air flow is near the low end of the cataloged performance data. If cost is an important issue, then terminals should be selected with the design air flow near the upper end of the performance data. In most applications, however, it is best to select the terminal with the design air flow near the midrange of the performance data.

In all cases, the terminals must be selected so that the system pressure at the terminal inlet is not less than that required in the performance data. Or conversely, the fan and duct system must be designed to provide adequate pressure at the terminal inlet. The design of duct and fan systems is discussed in Chapters 15 and 16.

In all cases, there must be adequate space in the ceiling cavity for the terminal unit. To reduce the possibility of problems resulting from sound generated by the terminal, it may be possible to locate the terminal above the ceiling of a non-sensitive area, such as a corridor.

14.5.2   Fan Powered Terminals

Many of the same considerations for the single duct terminals are applicable to the fan powered terminal units. In addition to the primary air flow through the unit, the designer must also consider the flow of the fan. In the case of constant volume units, the fan should be selected on the basis of the total air flow (primary plus secondary). For bypass type fan terminal units, the fan is selected on the required secondary air flow rate. The secondary air flow is based on the minimum air flow requirements for the space heating load.

In order to select a fan, calculate the static pressure of the ductwork, downstream of the fan, at the design air flow rate. Static pressure calculations for ductwork are discussed in Chapter 15. After the air flow and static pressure requirements are determined, a manufacturer's fan curves may be used to select an appropriate fan terminal unit. The capacity of a reheat coil, if any, should also be considered when selecting a unit from catalog data.

TABLE 14.7   Single Duct Terminal Performance Data

SIZE	CFM	MIN. ΔPs	DISCHARGE NC @ ΔPs				NUMBER OF OUTLETS	RADIATED NC @ ΔPs				CFM	SIZE
			MIN.	1.5"	2.5"	3.0"		MIN.	1.5"	2.5"	3.0"		
5	100	.75	20	23	32	34	1	—	—	—	—	100	5
	200	.75	22	26	34	37	1	—	—	20	20	200	
	300	.75	21	25	33	35	2	—	—	21	21	300	
	400	.75	24	28	37	39	2	—	20	24	24	400	
6	300	.75	21	25	33	35	2	—	—	21	21	300	6
	400	.75	24	28	37	39	2	—	20	23	24	400	
	500	.75	27	32	39	41	2	—	23	26	27	500	
	600	.75	30	33	40	42	3	20	25	27	29	600	

14.5.3 Common Errors in Selecting Air Terminals

In addition to the above guidelines, the designer should avoid some common errors that are made in the selection of air terminals and the design of air systems using air terminals. These errors include:

- Oversizing terminals. The result of oversizing terminals is low air velocity. At low air velocities, the air flow control damper or valve must operate near the closed position most of the time. This results in poor control at the design air flow rate.

- Capacity concentrated in too few terminals. If a single large terminal serves a space that should be served by several smaller units, comfort problems can occur due to different loads throughout the space.

- Improper discharge ductwork. The duct connections at the terminal discharge can have a significant effect on pressure drop in the downstream ductwork. Tees and sharp elbows in the discharge should be avoided. Flexible ductwork should also be used sparingly because it has a higher pressure loss per unit length than rigid metal ductwork.

- Improper inlet connections. Changes in direction of ductwork, immediately upstream of a terminal inlet, should be avoided to reduce air turbulence as it enters the control damper, or valve, and to reduce pressure losses.

- Excessive air temperature rise through the reheat coil. Reheat coils should be sized and selected for no greater temperature rise than that required by the space heating load. This will reduce the possibility of excessive supply air temperatures entering the conditioned space, which can result in distribution problems. The effects of air buoyancy, due to temperature differences, were discussed earlier in this chapter.

Chapter 15

DUCT SYSTEM DESIGN

In the previous chapter, air distribution in conditioned spaces was discussed. The method by which air is supplied to the room air devices was only briefly mentioned. This chapter deals with the air distribution system through which air is delivered to the air devices serving the conditioned spaces at the required quantities and at a suitable pressure. A discussion of fans and other air moving equipment is included in Chapter 16.

The ducts in an HVAC system are conduits through which air is delivered to the various conditioned spaces throughout a building. The volume and temperature of the air are a function of the heating and cooling loads in the space, as well as the requirements of ventilation codes. The design of the duct system will have a significant effect on the power requirements of the system fan, the operating costs of the system, and the noise generated by the system. A good duct design must consider the fundamentals of fluid flow in closed conduits, as well as the economics and esthetics of the system.

During the design of a system, there are several important factors that should be taken into consideration. These include:

Air flow and pressure requirements. The required rate of air flow is a direct function of the heating and cooling loads for the spaces served by the system and the temperature at which the air is supplied. Ventilation codes and standards, such as ASHRAE Standard 15, also set minimum air flow requirements for spaces.

Space air distribution. The manner in which air is supplied or returned from a space will have a significant impact on the design of the system. The types of air distribution devices and pressure requirements will also affect the duct system design.

Space available for ductwork. The cost of space in most buildings is at a premium. Although ductwork is frequently routed through interstitial spaces, such as above ceilings and below floors, this space is also at a premium cost as it adds to the height of a building. In an attempt to keep building costs at a minimum, the space available for ductwork is limited. When routing and sizing ductwork, particular attention must be paid to building structural members, such as beams and joists, in ceiling spaces.

Acoustics. As air flows through ductwork and air devices, it generates sound as it flows. Generally speaking, the higher the air velocity, the greater the sound generated. When routing ductwork through various types of spaces, the designer must be aware of the maximum allowable sound levels in the space, and design the duct system accordingly.

The reader should refer to Chapter 12 for additional discussion of HVAC system acoustics.

Friction loss and operating costs. As a fluid flows through a closed conduit, such as air flowing through ductwork, it loses energy through pressure reduction. This pressure loss is a result of friction from the air flow in the duct. The system fan must supply enough energy to the system in the form of pressure to overcome these losses. As a result, the greater the system friction losses, the greater the energy consumed by the fan. Friction losses in ductwork are discussed in more detail later in this chapter.

First cost. The cost of air distribution systems is a significant part of the total cost of an HVAC system. Larger ducts mean lower friction losses and lower operating costs. Larger ducts also result in higher installed or first costs. Larger ducts require more material and more fabrication time. As discussed earlier, larger ducts require more space inside a building.

Heat losses/gains and air leakage. Whenever ductwork passes through unconditioned spaces or is routed outdoors, heat is lost or gained to/from the surrounding air. The rate of heat loss or gain is a direct function of the difference in temperature between the air inside the duct and the air outside the duct. The rate of heat transfer is also a function of the thermal resistance of the ductwork (how well it is insulated) and also the surface area of the ductwork. Air leakage is another important consideration. The quality of construction obviously has a significant effect on air leakage. The pressure of the air inside the ductwork also directly affects the rate of air leakage.

System balance. In order for an HVAC system to perform the function for which it was designed, it must provide the design air quantities to each space. Since the outlets nearest the supply fan would tend to be subject to a higher air pressure, there is a tendency for the outlets closest to the fan discharge to flow more air than desired, and the farthest outlets to flow less, due to the differences in air pressure. If a system were not balanced, a situation of uneven air flow would result. For systems to be properly balanced, the flow rate from each outlet or flow into each inlet should be (1) equal to the design value for constant volume systems and (2) equal to predetermined values at maximum and minimum flow for variable air volume (VAV) systems. System balancing is discussed in greater detail later in this chapter.

Fire and smoke control. Duct systems must be designed so they will not spread fire and/or smoke throughout a building or from one space to another in the event of a fire. Some of the considerations for fire and smoke control were discussed in Chapter 11. Occasionally, duct systems will be designed so smoke may be evacuated from some areas of a building, while pressurization to prevent the entrance of smoke is provided in other areas of a building.

When sizing and designing duct systems, there are a number of methods available to the designer. The most common methods are the equal friction method, the static regain method, and the T-method. Other methods include the velocity reduction method, the constant velocity method, and the balanced capacity method. Each method has its advantages and disadvantages, which will be discussed later in this chapter. The three most popular methods will be discussed in some detail.

15.1 AIR FLOW IN DUCTWORK AND CLOSED CONDUITS

The physical laws governing the flow of fluids in conduits apply to the flow of air in ductwork. The Continuity Equation, Bernoulli's Equation, Darcy's Equation, the laws of pressure, and the laws governing the flow of viscous fluids are all applicable to the flow of air in ductwork. You may want to review Section 2.4, Concepts of Fluid Mechanics, in order to have a better understanding of air flow in ductwork.

15.1.1 Air Pressure in Ductwork

Section 2.4.1.1 introduced you to the concept of pressure, absolute pressure, and gauge pressure. In supply air systems, almost all pressures are measured in gauge pressure above atmospheric pressure. Only occasionally are pressures below atmospheric encountered in HVAC systems, such as at the inlet to fans or in return/exhaust air ductwork.

Pressures in HVAC systems are usually only slightly greater than atmospheric pressure. Since such pressures are relatively small with respect to atmospheric pressure, pressures in HVAC systems are usually measured in inches of water gauge (in. wg). To convert from pressure in psi (lb/ft^2) to in. wg, it is necessary to multiply by 27.68 in. wg/psi.

Section 2.4 also introduced the concept of static pressure and velocity pressure. Static air pressure in a duct is the pressure which exists in the duct, independent of any motion or velocity. Static pressure is exerted in all directions equally. In other words, it is uniform in all directions. For static pressure to exist, the medium must be confined in some sort of container, such as ductwork. Static pressure exerts the same pressure on all sides of the ductwork, equally in all directions.

Velocity pressure, or head, is pressure that is exerted as a result of the motion of a fluid, such as air. The pressure resulting from the motion of the fluid is due to the mass and momentum of the fluid as it is flowing. The greater the velocity of the air, the greater the velocity pressure. Velocity pressure is also referred to as dynamic pressure. Velocity pressure is a measure of the kinetic energy of the air in the ductwork and is defined by Equation 2.19. Equation 2.19 may be modified such that the velocity pressure defines the kinetic or dynamic energy of the air flow.

$$P_v = \rho\left(\frac{V^2}{2g_c}\right)$$
(15.1)

where

P_v = the velocity pressure in lb/ft^2

ρ = the density in lb/ft^3

V = the velocity in ft/sec

g_c = the acceleration due to gravity in ft/sec^2

The standard density for air at sea level and standard conditions is 0.075 lb/ft^3 and can be assumed to be constant for most HVAC applications. Equation 15.1 may be modified, assuming a standard air density and using standard units, to yield Equation 15.2.

$$P_v = \left(\frac{V}{4005}\right)^2$$
(15.2)

where

P_v = the air velocity pressure in in. wg

V = the velocity in ft/min

4005 = a constant to make the units consistent

As air flows through ductwork, it possesses both static and velocity pressures. The total pressure of an air stream is defined as the sum of the static pressure and velocity pressure, and is given in Equation 15.3.

$$P_T = P_s + P_v \qquad\qquad (15.3)$$

where

P_T = the total pressure in in. wg

P_s = the static pressure in in. wg

P_v = the velocity pressure in in. wg

Static pressure can be measured directly, while velocity pressure must be derived by subtracting the static pressure from the total pressure. Static pressure can easily be measured with a pressure gauge. Since air pressures in HVAC systems are usually less than 6 in. wg, the pressure measurements are usually made with a manometer which measures pressures less than 1 psi. Velocity pressure is measured with an instrument called a pitot (pitot-static) tube and a manometer. A typical pitot tube is shown in Figure 15.1.

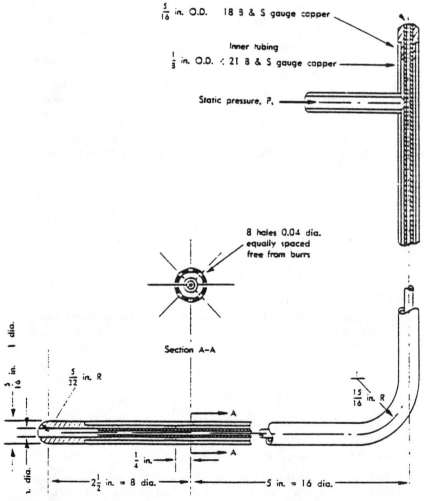

FIGURE 15.1   Pitot-Static Tube

The pitot tube is actually a tube within another tube. The outer tube has eight holes in it to measure the static pressure. Pressure measured from these holes is unaffected by the velocity of the fluid as it flows past the tube. The inner tube is used to measure total pressure (velocity pressure + static pressure). Velocity pressure readings are taken by inserting the pitot tube in a duct with flowing air, and pointing the open end of the tube in the upstream direction. As the air flows into the inner tube, pressure is created in the inner tube, which is a function of the air velocity and the static pressure in the ductwork. The holes in the side of the outer tube measure the static pressure of the air and are not affected by the air velocity. Measurements are made, on a manometer, of the total pressure less the static pressure, which results in the velocity pressure in the duct. The velocity pressure may then be used to calculate the air velocity in the duct. Determining the air velocity from the velocity pressure is discussed later in this chapter.

15.1.2   Basic Fluid Flow for Air Systems

In Chapter 2, the basic concepts and equations for fluid flow were discussed. The Continuity Equation (2.16), Bernoulli's Equation (2.22), and Darcy's Equation (2.24) may be applied to air flow in ductwork. The Darcy Equation will be discussed in the next section.

The Continuity Equation, when applied to air flow in ductwork, states that the air flow past one point in a closed conduit is equal to the flow past another point in the same conduit. This assumes that the density remains constant, which is a reasonable assumption for air flow in HVAC systems. Equation 2.16 may be used to determine the air flow rate through ductwork.

$$Q = AV \qquad (2.16)$$

where

Q = the air flow rate in ft^3/min

A = the cross-sectional area of the duct in ft^2

V = the air velocity in ft/min

Bernoulli's Equation is also used frequently in the design of duct systems. Bernoulli's Equation states that energy will be conserved (neglecting any losses) as it flows and that the energy will take one of three forms: potential energy (PE), kinetic energy (KE), and flow energy (FE) or flow work. In most HVAC systems, the change in elevation is negligible; therefore, the potential energy, due to elevation changes, may be neglected. In a short straight section of ductwork, the flow work may also be negligible. Therefore, the energy of the air flow at any point is the sum of the potential energy and the kinetic energy of the air. The potential energy is a function of the static pressure of the air, and the kinetic energy is a function of the air velocity. With these assumptions in mind, Equation 15.3 and Bernoulli's Equation (Equation 2.22) may be combined to yield Equation 15.4.

$$P_T = P_s + P_v = P_s + \frac{\rho V^2}{2g_c} \qquad (15.4)$$

where

$\rho V^2/2g_c$ = the velocity pressure in lb/ft^2

P_s = the static pressure in lb/ft^2

V = the velocity in ft/sec

g_c = the acceleration due to gravity = 32.2 ft/sec^2

ρ = the density in lb/ft^3

The velocity pressure term in Equation 15.4 may be rewritten so the units are consistent with those more commonly used in HVAC system design.

$$P_v = \rho\left(\frac{V}{1097}\right)^2 \tag{15.5}$$

where

P_v = the velocity pressure in in. wg

V = the air velocity in ft/min

As indicated earlier, the air density in HVAC systems is usually assumed to be constant, and equal to standard density, i.e., $\rho = 0.075$ lb/ft³. Substituting the value of standard air density into Equation 15.5 yields Equations 15.2 and 15.6.

$$P_v = \left(\frac{V}{4005}\right)^2 \tag{15.2}$$

or

$$V = 4005\,(P_v)^{\frac{1}{2}} \tag{15.6}$$

Equation 15.2 may be substituted into Equation 15.4 to yield the following equation for total air pressure, for air flowing in a duct, assuming standard density for air.

$$P_T = P_s + \left(\frac{V}{4005}\right)^2 \tag{15.7}$$

Example 15.1   Air is flowing through a 36 in $\times$ 12 in duct. The duct is traversed (pressure measurements taken at various cross-sectional locations) and an average total pressure of 2.0 in. wg and a static pressure of 1.75 in. wg are measured. What is the air velocity and the flow rate in the duct?

Using Equation 15.3,

$$P_v = P_T - P_s = 2.0 - 1.75 = 0.25 \text{ in. wg}$$

From Equation 15.6, the air velocity is

$$V = (4005)\sqrt{0.25} = 2002 \text{ ft/min}$$

Using the Continuity Equation, the air flow rate may be determined:

$$Q = AV = (36 \times 12)\left(\frac{1}{144}\right)(2002) = 6006 \text{ ft}^3/\text{min}$$

15.1.3   Friction Losses in Straight Ductwork

In the previous section, Bernoulli's Equation was used to determine the air velocity in a duct, and the flow work was assumed to be negligible for short sections of ductwork. This is a reasonable assumption for very short sections of ductwork, but is not reasonable for entire systems or even sections of finite length.

For straight sections of ductwork, there are losses in energy due to the friction flow resistance in the ductwork. These friction losses result in a reduction in the total pressure of the air in the duct. These losses are an irreversible transformation of mechanical energy into heat.

In Chapter 2, a simplified version of Darcy's Equation (Equation 2.24) was presented that provided a relationship between fluid flow friction losses: the fluid velocity and a loss coefficient, or friction factor. Friction losses in straight ductwork are a function of the turbulence and the velocity of the air as it flows through the ductwork. The greater the turbulence, the greater the friction loss.

As a fluid flows through a closed conduit, the fluid turbulence is a function of the fluid velocity, the fluid viscosity, and the relative smoothness or roughness of the conduit. A measure of the turbulence, due to the fluid velocity and viscosity, is the **Reynolds Number (Re)**. The Reynolds Number is a dimensionless number that relates the fluid velocity, viscosity, and density.

$$Re = \frac{\rho V D}{\mu} \tag{15.8}$$

where

D = the diameter of the conduit

V = the velocity of the fluid

ρ = the density of the fluid

μ = the dynamic viscosity of the fluid

If Re < 2000, the flow is laminar. If Re > 2000, the flow is turbulent.

Friction losses for turbulent flow are considerably higher than for laminar flow. In laminar flow, the fluid flows in parallel flow streams, as shown in Figure 2.11. For laminar flow, the friction losses are mainly due to the viscosity of the fluid, and result from friction between the fluid molecules themselves as they flow. In turbulent flow, the fluid molecules do not flow in smooth flow streams; rather, their motion is random in all directions as the bulk of the fluid flows. In turbulent flow, friction losses are caused by the friction between the fluid and the sides of the conduit, as well as from the random motion of the fluid molecules. Air flow in ductwork is almost always turbulent flow.

As air flows through ductwork, the total pressure is constantly decreasing due to friction losses. Referring to the Continuity Equation, it may be observed that the air velocity may increase or decrease for a constant air flow rate, depending on the cross-sectional area of the duct. As the air velocity increases, the static pressure of the air decreases and the velocity pressure increases by an equivalent amount (neglecting any friction losses). Conversely, as the velocity decreases, the static pressure increases and the velocity pressure decreases.

Referring to Figure 15.2, it can be observed that the total pressure is decreasing, as the air flows through ductwork, due to friction losses.

However, the static pressure and velocity pressure may increase or decrease as the air velocity changes. The only method by which the total pressure may be increased is by adding flow work with a fan or similar device.

In Chapter 2, a simplified version of Darcy's Equation was presented:

$$h_L = \frac{C_L V^2}{2g} \tag{2.24}$$

where

h_L = the energy loss due to friction

C_L = a loss coefficient or friction factor

V = the velocity of the fluid

g = the acceleration due to gravity

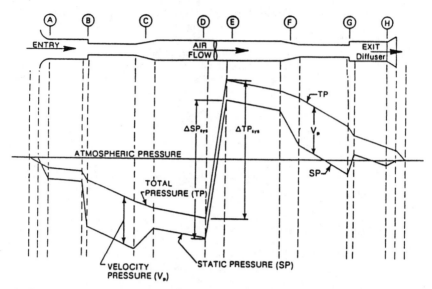

FIGURE 15.2   Pressure Changes During Flow in Ductwork

This equation relates the friction loss, due the fluid velocity, and a **loss coefficient or friction factor (C_L).** The friction factor is a dimensionless number that is directly proportional to the friction loss. The friction factor is a function of the roughness in the conduit and the Reynolds Number of the flowing fluid.

For flow in ductwork, a friction chart is used to determine the friction factor for various sizes of round ductwork and various flow rates. The chart gives the friction loss per 100 ft of straight ductwork. A friction chart is shown in Figure 15.3. The friction chart plots friction loss vs. duct air flow rate for various sizes of duct. In accordance with the Continuity Equation, the air velocity in a duct is a function of the air flow rate and the cross-sectional area of the duct. For calculation purposes, the mean velocity is used. The friction chart is applicable for the following situations:

- the ductwork is round
- the air temperature is between 41°F and 95°F
- elevations up to 2000 ft
- duct materials of medium roughness

Figure 15.3 gives the friction loss per 100 ft of straight duct. Equation 15.9 may be used to calculate the friction loss for a duct of a given length and friction factor.

$$h_L = LC_{L-100} \tag{15.9}$$

where

L = length of duct in ft

C_{L-100} = friction factor per 100 ft of duct

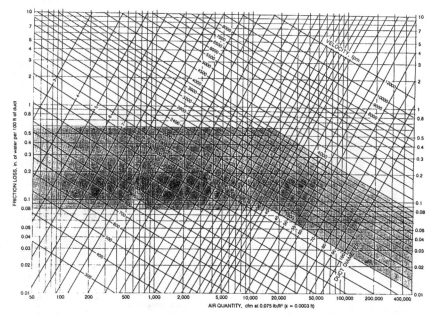

FIGURE 15.3 Duct Friction Chart (Copyright 1985 by the American Society of Heating, Refrigerating, and Air Conditioning Engineers, Inc., from ASHRAE Handbook-Fundamentals. Used by permission.)

Example 15.2   A 125 ft straight section of 14 in diameter galvanized steel duct has 1000 ft³/min of air flowing through it. Assuming an air temperature between 41°F and 95°F and an elevation under 2000 ft, what is the loss for the entire length of ductwork?

Locate the air flow rate at the bottom of the friction chart, as shown in Figure 15.4, and follow the line up until it intersects with the line for a 14 in diameter duct (where a point is located on the chart). From the location of the point, a friction factor of 0.09 in. wg/100 ft is read from the left side of the chart. The friction loss for the entire length of ductwork may then be calculated with Equation 15.9 by multiplying the friction factor by the length of the duct as follows:

$$h_L = C_{L-100}L = \left(\frac{0.09 \text{ in. wg}}{100 \text{ ft}}\right)(125 \text{ ft}) = 0.11 \text{ in. wg}$$

As indicated earlier, the friction chart is for round ductwork only. However, square and rectangular ductwork is frequently used in HVAC systems. Equation 15.10 may be used to convert square and rectangular ductwork to equivalent round ductwork.

$$D_e = \frac{[1.30(ab)^{0.625}]}{(a+b)^{0.25}} \qquad (15.10)$$

where

D_e = equivalent round ductwork in in

a and b = the dimensions of the two sides in in

Table 15.1 may also be used to make the conversion from square or rectangular to round and visa versa. Flat oval ductwork may also be converted to round with similar conversion

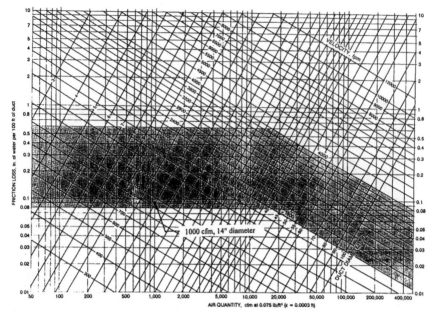

FIGURE 15.4   Duct Friction Chart for Example 15.2

tables. Since duct design is part of practically every HVAC design project, several slide rule type duct calculators are available for sizing ductwork and determining friction factors. These calculators automatically make the conversion from square or rectangular to equivalent round. They may also be used to determine air velocities at various flow rates for various sized ducts.

Example 15.3   A 75 ft long section of 24 in wide × 12 in high duct has 2500 ft³/min of air at standard conditions flowing through it. What is the equivalent diameter of the rectangular duct and what is the pressure loss?

From Equation 15.10, the equivalent diameter is calculated as

$$D_e = \frac{(1.3)(24 \text{ in} \times 12 \text{ in})^{0.625}}{(24 \text{ in} + 12 \text{ in})^{0.25}} = 18.2 \text{ in or } 18 \text{ in } \phi$$

Plotting the intersection of 2500 ft³/min and 18 in ϕ duct on a friction chart, a pressure loss of 0.15 in. wg/100 ft is obtained. Using Equation 15.9,

$$h_L = \left(\frac{0.15 \text{ in}}{100 \text{ ft}}\right)(75 \text{ ft}) = 0.11 \text{ in. wg}$$

As indicated earlier, the friction chart may only be used for applications where the elevation (altitude) is 2000 ft or less. For elevations greater than 2000 ft, Figure 15.3 may still be used; however, it must be adjusted for the change in density, due to altitude. Equation 15.11 and Table 15.2 may be used to adjust the values, obtained from the duct friction chart, for altitudes other than sea level.

$$h_{L2} = h_{L1}\left(\frac{\rho_1}{\rho_2}\right) \tag{15.11}$$

TABLE 15.1 Circular Equivalents of Rectangular Ducts

DIMENSIONS IN INCHES

SIDE RECTANGULAR DUCT	6	7	8	9	10	11	12	13	14	15	16	17	18	19	20	22	24	26	28	30	SIDE RECTANGULAR DUCT
6	6.6																				6
7	7.1	7.7																			7
8	7.5	8.2	8.8																		8
9	8.0	8.6	9.3	9.9																	9
10	8.4	9.1	9.8	10.4	10.9																10
11	8.8	9.5	10.2	10.8	11.4	12.0															11
12	9.1	9.9	10.7	11.3	11.9	12.5	13.1														12
13	9.5	10.3	11.1	11.8	12.4	13.0	13.6	14.2													13
14	9.8	10.7	11.5	12.2	12.9	13.5	14.2	14.7	15.3												14
15	10.1	11.0	11.8	12.6	13.3	14.0	14.6	15.3	15.8	16.4											15
16	10.4	11.4	12.2	13.0	13.7	14.4	15.1	15.7	16.3	16.9	17.5										16
17	10.7	11.7	12.5	13.4	14.1	14.9	15.5	16.1	16.8	17.4	18.0	18.6									17
18	11.0	11.9	12.9	13.7	14.5	15.3	16.0	16.6	17.3	17.9	18.5	19.1	19.7								18
19	11.2	12.2	13.2	14.1	14.9	15.6	16.4	17.1	17.8	18.4	19.0	19.6	20.2	20.8							19
20	11.5	12.5	13.5	14.4	15.2	15.9	16.6	17.5	18.2	18.8	19.5	20.1	20.7	21.3	21.9						20
22	12.0	13.1	14.1	15.0	15.9	16.7	17.6	18.3	19.1	19.7	20.4	21.0	21.7	22.3	22.9	24.1					22
24	12.4	13.6	14.6	15.6	16.6	17.5	18.3	19.1	19.8	20.6	21.3	21.9	22.6	23.2	23.9	25.1	26.2				24
26	12.8	14.1	15.2	16.2	17.2	18.1	19.0	19.8	20.6	21.4	22.1	22.8	23.5	24.1	24.8	26.1	27.2	28.4			26
28	13.2	14.5	15.6	16.7	17.7	18.7	19.6	20.5	21.3	22.1	22.9	23.6	24.4	25.0	25.7	27.1	28.2	29.5	30.6		28
30	13.6	14.9	16.1	17.2	18.3	19.3	20.2	21.1	22.0	22.9	23.7	24.4	25.2	25.9	26.7	28.0	29.3	30.5	31.6	32.8	30
32	14.0	15.3	16.5	17.7	18.8	19.8	20.6	21.8	22.7	23.6	24.4	25.2	26.0	26.7	27.5	28.9	30.1	31.4	32.6	33.8	32
34	14.4	15.7	17.0	18.2	19.3	20.4	21.4	22.4	23.3	24.2	25.1	25.9	26.7	27.5	28.3	29.7	31.0	32.3	33.6	34.8	34
36	14.7	16.1	17.4	18.6	19.8	20.9	21.9	23.0	23.9	24.8	25.8	26.6	27.4	28.3	29.0	30.5	32.0	33.0	34.6	35.8	36
38	15.0	16.4	17.8	19.0	20.3	21.4	22.5	23.5	24.5	25.4	26.4	27.3	28.1	29.0	29.8	31.4	32.8	34.2	35.5	36.7	38
40	15.3	16.8	18.2	19.4	20.7	21.9	23.0	24.0	25.1	26.0	27.0	27.9	28.8	29.7	30.5	32.1	33.6	35.1	36.4	37.6	40
42	15.6	17.1	18.5	19.8	21.1	22.3	23.4	24.5	25.6	26.6	27.6	28.5	29.4	30.4	31.2	32.8	34.4	35.9	37.3	38.6	42
44	15.9	17.5	18.9	20.2	21.5	22.7	23.9	25.0	26.1	27.2	28.2	29.1	30.0	31.0	31.9	33.5	35.2	36.7	38.1	39.5	44
46	16.2	17.8	19.2	20.6	21.9	23.2	24.3	25.5	26.7	27.7	28.7	29.7	30.6	31.6	32.5	34.2	35.9	37.4	38.9	40.3	46
48	16.5	18.1	19.6	20.9	22.3	23.6	24.8	26.0	27.2	28.2	29.2	30.2	31.2	32.2	33.1	34.9	36.6	38.2	39.7	41.2	48
50	16.8	18.4	19.9	21.3	22.7	24.0	25.2	26.4	27.6	28.7	29.8	30.8	31.8	32.8	33.7	35.5	37.3	38.9	40.4	42.0	50
52	17.0	18.7	20.2	21.6	23.1	24.4	25.6	26.8	28.1	29.2	30.3	31.4	32.4	33.4	34.3	36.2	38.0	39.6	41.2	42.8	52
54	17.3	19.0	20.5	22.0	23.4	24.8	26.1	27.3	28.5	29.7	30.8	31.9	32.9	33.9	34.9	36.8	38.7	40.3	42.0	43.6	54
56	17.6	19.3	20.9	22.4	23.8	25.2	26.5	27.7	28.9	30.1	31.2	32.4	33.4	34.5	35.5	37.4	39.3	41.0	42.7	44.3	56
58	17.8	19.5	21.1	22.7	24.2	25.5	26.9	28.2	29.3	30.5	31.7	32.8	33.9	35.0	36.0	38.0	39.8	41.7	43.4	45.0	58
60	18.1	19.8	21.4	23.0	24.5	25.8	27.3	28.7	29.8	31.0	32.2	33.4	34.5	35.6	36.5	38.6	40.4	42.3	44.0	45.6	60
62	18.3	20.1	21.7	23.3	24.8	26.2	27.6	29.0	30.2	31.4	32.6	33.8	35.0	36.0	37.1	39.2	41.0	42.9	44.7	46.5	62
64	18.6	20.3	22.0	23.6	25.2	26.5	27.9	29.3	30.6	31.8	33.1	34.2	35.5	36.5	37.6	39.7	41.6	43.5	45.4	47.2	64
66	18.8	20.6	22.3	23.9	25.5	26.9	28.3	29.7	31.0	32.2	33.5	34.7	35.9	37.0	38.1	40.2	42.2	44.1	46.0	47.8	66
68	19.0	20.8	22.5	24.2	25.8	27.3	28.7	30.1	31.4	32.6	33.9	35.1	36.3	37.5	38.6	40.7	42.8	44.7	46.6	48.4	68
70	19.2	21.0	22.8	24.5	26.1	27.6	29.1	30.4	31.8	33.1	34.3	35.6	36.8	37.9	39.1	41.3	43.3	45.3	47.2	49.0	70
72															39.6	41.8	43.8	45.9	47.8	49.7	72
74															40.0	42.3	44.4	46.4	48.4	50.3	74
76															40.5	42.8	44.9	47.0	49.0	50.8	76
78															40.9	43.3	45.5	47.5	49.5	51.5	78
80															41.3	43.8	46.0	48.0	50.1	52.0	80
82															41.8	44.2	46.4	48.6	50.6	52.6	82
84															42.2	44.6	46.9	49.2	51.1	53.2	84
86															42.6	45.0	47.4	49.6	51.6	53.7	86
88															43.0	45.4	47.9	50.1	52.2	54.3	88
90															43.4	45.9	48.3	50.6	52.6	54.8	90
92															43.8	46.3	48.7	51.1	53.4	55.4	92
96															44.6	47.2	49.5	52.0	54.4	56.3	96

Equation for Circular Equivalent of a Rectangular Duct:

$$d_c = 1.30 \frac{(ab)^{0.625}}{(a+b)^{0.250}} = 1.30 \sqrt[8]{\frac{(ab)^5}{(a+b)^2}}$$

where

a = length of one side of rectangular duct, inches.
b = length of adjacent side of rectangular duct, inches.
d_c = circular equivalent of a rectangular duct for equal friction and capacity, inches.

(Courtesy of the Trane Company)

where h_{L1} = the duct friction loss at sea level

h_{L2} = the duct friction loss at the desired altitude (elevation)

ρ_1/ρ_2 = the density ratio for the desired altitude from Table 15.2

15.1.4 Dynamic Losses in Ductwork

In the previous section, friction losses for straight sections of ductwork were discussed. Friction losses for straight ductwork make up a significant portion of the losses for a complete duct system; however, they are not the only source of pressure losses in duct systems. Dynamic losses also are significant in duct systems. Dynamic losses include losses due to changes in direction of the ductwork, changes in area, abrupt expansions and contractions,

TABLE 15.2   Density Ratios at Various Altitudes

Altitude (ft)	ρ_1/ρ_2
10,000	0.69
8,000	0.74
7,000	0.77
6,000	0.80
5,000	0.83
4,000	0.86
3,000	0.90
2,000	0.93
1,000	0.96
sea level	1.00

and similar situations. Dynamic losses occur in duct fittings which are used to accomplish changes in direction, changes in area, etc. Basically, dynamic losses can be divided into two main categories:

1. those that result from changes in direction of the duct

2. those that result from changes in the cross-sectional area of the duct

Dynamic losses result from disturbances in the air flow caused by the changes in direction or flow area. Changes in flow direction or area result in flow separation and the formation of flow disturbances and eddies. These flow separations and disturbances occur in the immediate vicinity of the change in direction or area.

Dynamic losses are calculated from the air velocity and a loss coefficient for the type of duct fitting. Dynamic losses cannot be separated from the friction losses at a fitting; therefore, fitting loss coefficients include both friction and dynamic losses. For calculation purposes, the loss is assumed to occur entirely at the fitting. Friction loss through the fitting itself is assumed to be negligible. However, for relatively long fittings, friction losses must be added to the normal dynamic loss.

Pressure losses, through duct fittings, vary as the square of the mean air velocity in the ductwork and according to the type of fitting used. Therefore, Equation 15.2 may be modified to produce Equation 15.12, which may be used to calculate dynamic losses for duct fittings.

$$h_L = C\left(\frac{V}{4005}\right)^2 \tag{15.12}$$

Dynamic loss coefficients have been determined experimentally for various types of fittings commonly used in HVAC duct systems. ASHRAE has developed a duct fitting database which includes over 220 round and rectangular fittings.

An elbow is a duct fitting in which the flow of the air changes direction. The angle of the change in direction is typically 45° or 90°; however, elbows can be fabricated for any angle. As the air flows through an elbow, a centrifugal force is created that acts toward the outer wall of the fitting. When the air flows from the straight part of the duct to the curved or angled part, an increase in static pressure and a decrease in air velocity occurs at the outer wall. At the same time, a decrease in static pressure and an increase in air velocity occur at the inner wall. The magnitude of the loss for an elbow is a function of:

• the turning angle of the elbow

• the relative radius of curvature

- whether turning vanes or splitter vanes are installed
- the cross-sectional area of the duct

Typical fitting loss coefficients for 90° round duct elbows are shown in Figure 15.5 and loss coefficients for radius rectangular elbows are shown in Figure 15.6.

As the relative radius of curvature becomes larger, the flow loss coefficient becomes smaller.

Dynamic loss coefficients are affected by the roughness of the duct walls, as well as the air velocity for any particular fitting. The installation of splitter dampers, turning vanes, and duct runs in rectangular ducts can significantly reduce the loss for an elbow. Figure 15.7 shows the loss coefficients for rectangular mitered elbows with turning vanes.

Turning vanes provide a more uniform air velocity through the bend, thereby reducing losses. Splitters may also be used to reduce dynamic losses in elbows. Splitters are non-adjustable sheet metal dividers installed in elbows. The splitters ensure that any point on one blade is equidistant from a similar point on an adjacent blade. The precise blade shape maintains a constant duct area and ensures a constant air velocity through the turn. This reduces losses due to changes in air velocity.

Tees and wyes are duct fittings that combine or diverge air streams. A converging or diverging fitting that has an angle of 90° is called a tee. If the converging or diverging flow angles are between 15° and 75°, the fitting is called a wye. A converging tee or wye combines two air streams, while fittings that divide air flows are called diverging tees or wyes. Diverging fittings are also called branches.

A tee can be used to change the direction of one or more air streams. A tee that changes the air flow direction of both air streams is sometimes called a double deflection tee. A double deflection tee and loss coefficients for both converging and diverging flows are shown in Figure 15.8.

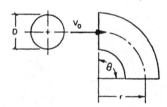

$$C_o = K_\theta C_o'$$

Coefficients for 90° Elbows

r/D	0.5	0.75	1.0	1.5	2.0	2.5
C_o'	0.71	0.33	0.22	0.15	0.13	0.12

Angle Correction Factors K_θ

θ	0	20	30	45	60	75	90	110	130	150	180
K_θ	0	0.31	0.45	0.60	0.78	0.90	1.00	1.13	1.20	1.28	1.40

FIGURE 15.5　Loss Coefficients for Round Radius 90° Elbows (Copyright 1989 by the American Society of Heating, Refrigerating, and Air Conditioning Engineers, Inc., from the ASHRAE Handbook-Fundamentals. Used by permission.)

$$C_o' = R_\theta R_{Re} C_o'$$

Coefficients for 90° Elbows:

					C_o'						
					11/W						
r/W	0.25	0.5	0.75	1.0	1.5	2.0	3.0	4.0	5.0	6.0	8.0
0.5	1.3	1.3	1.2	1.2	1.1	1.1	0.98	0.92	0.89	0.85	0.83
0.75	0.57	0.52	0.48	0.44	0.40	0.39	0.39	0.40	0.42	0.43	0.44
1.0	0.27	0.25	0.23	0.21	0.19	0.18	0.18	0.19	0.20	0.27	0.21
1.5	0.22	0.20	0.19	0.17	0.15	0.14	0.14	0.15	0.16	0.17	0.17
2.0	0.20	0.18	0.16	0.15	0.14	0.13	0.13	0.14	0.14	0.15	0.15

				R_{Re}				
				Re•10^{-4}				
r/W	1	2	3	4	6	8	10	14	≥20
0.5	1.40	1.26	1.19	1.14	1.09	1.06	1.04	1.0	1.0

FIGURE 15.6   Loss Coefficients for Rectangular Radius Elbows (Copyright 1985 by the American Society of Heating, Refrigerating, and Air Conditioning Engineers, Inc., from the ASHRAE Handbook-Fundamentals. Used by permission.)

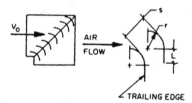

Design No.	Dimensions, in.			
	r	s	L	C_o
1[a]	2.0	1.5	0.75	0.12
2	4.5	2.25	0	0.15
3	4.5	3.25	1.60	0.18

[a] When extension of trailing edge is not provided for this vane, losses are approximately unchanged for single elbows, but increase considerably for elbows in series.

FIGURE 15.7   Loss Coefficients for Rectangular Mitered 90° Elbows with Turning Vanes (Copyright 1989 by the American Society of Heating, Refrigerating, and Air Conditioning Engineers, Inc., from the ASHRAE Handbook-Fundamentals. Used by permission.)

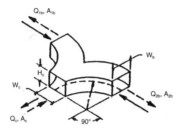

$$r/W_c = 1.5$$

$$Q_{1b}/Q_c = Q_{2b}/Q_c = 0.5$$

Converging		
A_{1b}/A_c or A_{2b}/A_c	0.50	1.0
$C_{c,1b}$ or $C_{c,2b}$	0.23	0.07
Diverging		
A_{1b}/A_c or A_{2b}/A_c	0.50	1.0
$C_{c,1b}$ or $C_{c,2b}$	0.30	0.25

FIGURE 15.8 Loss Coefficients for a Double Deflection Tee (Copyright 1985 by the American Society of Heating, Refrigerating, and Air Conditioning Engineers, Inc., from the ASHRAE Handbook-Fundamentals. Used by permission.)

Single deflection tees (those in which only one air stream changes direction) are also frequently referred to as branching tees or just branches. For branching tees, there is a reduction in the air velocity in the straight through section as well as in the branch. The velocity reduction occurs immediately downstream of the branch takeoff, which results in an increase in static pressure after the branch takeoff. A branching tee and loss coefficients for the main and branch sections are shown in Figure 15.9.

The junction of two air streams, moving at different velocities in a converging wye, is characterized by turbulent mixing of the air streams, which results in pressure losses. The air stream with the higher velocity loses part of its kinetic energy by transporting it to the slower jet. The slower jet increases in kinetic energy as a result of mixing. A 45° wye and loss coefficients are shown in Figure 15.10.

Entrances and exits are end openings at the end of a duct or connection to equipment, such as coils. At entrances or intakes, the change in direction of the air streams causes eddies, and large scale turbulence develops along the duct wall as the air passes the entrance. Losses for abrupt contractions, however, are relatively small; therefore, long tapered reducers are usually not justified. Angles for contractions can be up to 30° per side, in many cases, before the loss becomes large compared to a long taper. Loss coefficients for contracting and expanding transitions are shown in Figure 15.11.

Dynamic losses that result from expanding areas are always more severe than losses for contracting areas. When air flows out an abrupt exit, eddies and turbulence occur near the edges of the duct, as the air velocity is suddenly expanded and the air velocity suddenly decreases (the velocity pressure is almost entirely converted to static pressure). When making an enlargement, the angle should be kept as small as possible, allowing the air to spread out evenly. If possible, angles should be kept to a maximum of 15° per side.

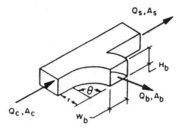

$\theta = 90^\circ$

$r/W_b = 1.0$

Branch $C_{c,b}$

$\dfrac{A_b}{A_s}$	$\dfrac{A_b}{A_c}$	\multicolumn{9}{c}{Q_b/Q_c}								
		0.1	**0.2**	**0.3**	**0.4**	**0.5**	**0.6**	**0.7**	**0.8**	**0.9**
0.25	0.25	0.55	0.50	0.60	0.85	1.2	1.8	3.1	4.4	6.0
0.33	0.25	0.35	0.35	0.50	0.80	1.3	2.0	2.8	3.8	5.0
0.5	0.5	0.62	0.48	0.40	0.40	0.48	0.60	0.78	1.1	1.5
0.67	0.5	0.52	0.40	0.32	0.30	0.34	0.44	0.62	0.92	1.4
1.0	0.5	0.44	0.38	0.38	0.41	0.52	0.68	0.92	1.2	1.6
1.0	1.0	0.67	0.55	0.46	0.37	0.32	0.29	0.29	0.30	0.37
1.33	1.0	0.70	0.60	0.51	0.42	0.34	0.28	0.26	0.26	0.29
2.0	1.0	0.60	0.52	0.43	0.33	0.24	0.17	0.15	0.17	0.21

Main, $C_{c,s}$

$\dfrac{A_b}{A_s}$	$\dfrac{A_b}{A_c}$	\multicolumn{9}{c}{Q_b/Q_c}								
		0.1	**0.2**	**0.3**	**0.4**	**0.5**	**0.6**	**0.7**	**0.8**	**0.9**
0.25	0.25	– .01	– .03	– .01	0.05	0.13	0.21	0.29	0.38	0.46
0.33	0.25	0.08	0	– .02	– .01	0.02	0.08	0.16	0.24	0.34
0.5	0.5	– .03	– .06	– .05	0	0.06	0.12	0.19	0.27	0.35
0.67	0.5	0.04	– .02	– .04	– .03	– .01	0.04	0.12	0.23	0.37
1.0	0.5	0.72	0.48	0.28	0.13	0.05	0.04	0.09	0.18	0.30
1.0	1.0	– .02	– .04	– .04	– .01	0.06	0.13	0.22	0.30	0.38
1.33	1.0	0.10	0	0.01	– .03	– .01	0.03	0.10	0.20	0.30
2.0	1.0	0.62	0.38	0.23	0.23	0.08	0.05	0.06	0.10	0.20

FIGURE 15.9   Loss Coefficients for a Branching Tee (Copyright 1989 by the American Society of Heating, Refrigerating, and Air Conditioning Engineers, Inc., from the ASHRAE Handbook-Fundamentals. Used by permission.)

Only the more commonly used fittings and loss coefficients are presented here for illustration purposes. You should consult References 1 and 27 for additional information on loss coefficients for various fittings.

Example 15.4   A 75 ft section of 24 in diameter duct has 4000 ft³/min of air, at standard conditions, flowing through it. The duct makes two 90° turns before it transitions to 18 in diameter and runs for another 25 ft, as shown in Figure 15.12.

The elbows are 90° and have a radius of 18 in. The total angle of convergence for the transition is 30° (15° per side). What is the pressure loss for the entire section of duct and fittings?

The duct pressure loss chart may be used for the straight section of duct. A loss coefficient for the two radius elbows may be obtained from Figure 15.5 and a loss coefficient

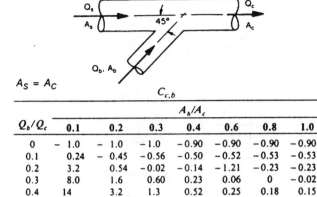

$A_S = A_C$

$C_{c,b}$

Q_b/Q_c	A_b/A_c						
	0.1	0.2	0.3	0.4	0.6	0.8	1.0
0	− 1.0	− 1.0	− 1.0	− 0.90	− 0.90	− 0.90	− 0.90
0.1	0.24	− 0.45	− 0.56	− 0.50	− 0.52	− 0.53	− 0.53
0.2	3.2	0.54	− 0.02	− 0.14	− 1.21	− 0.23	− 0.23
0.3	8.0	1.6	0.60	0.23	0.06	0	− 0.02
0.4	14	3.2	1.3	0.52	0.25	0.18	0.15
0.5	22	5.0	2.1	0.65	0.33	0.25	0.22
0.6	32	7.0	3.0	0.91	0.81	0.61	0.51
0.7	43	9.2	3.9	1.2	0.56	0.39	0.33
0.8	56	12	4.9	1.5	0.66	0.39	0.36
0.9	71	15	6.2	1.8	0.72	0.44	0.35
1.0	87	19	7.4	2.0	0.78	0.44	0.32

$C_{c,s}$

Q_b/Q_c	A_b/A_c						
	0.1	0.2	0.3	0.4	0.6	0.8	1.0
0	0	0	0	0	0	0	0
0.1	0.05	0.12	0.14	0.16	0.17	0.17	0.17
0.2	− 0.20	0.17	0.22	0.27	0.27	0.29	0.31
0.3	− 0.76	− 0.13	0.08	0.20	0.28	0.32	0.40
0.4	− 1.7	− 0.50	− 0.12	0.08	0.26	0.36	0.41
0.5	− 2.8	− 1.0	− 0.49	− 0.13	0.16	0.30	0.40
0.6	− 4.3	− 1.7	− 0.87	− 0.45	− 0.04	0.20	0.33
0.7	− 6.1	− 2.6	− 1.4	− 0.85	− 0.25	0.08	0.25
0.8	− 8.1	− 3.6	− 2.1	− 1.3	− 0.55	− 0.17	0.06
0.9	− 10	− 4.8	− 2.8	− 1.9	− 0.88	− 0.40	− 0.18
1.0	− 13	− 6.1	− 3.7	− 2.6	− 1.4	− 0.77	− 0.42

FIGURE 15.10 Loss Coefficients for a 45° Converging Wye (Copyright 1989 by the American Society of Heating, Refrigerating, and Air Conditioning Engineers, Inc., from the ASHRAE Handbook-Fundamentals. Used by permission.)

for the transition from Figure 15.11. Equation 15.12 may be used to determine the losses for the fittings.

The areas for the 18 in and 24 in diameter ducts are

$$A_{18} = \frac{\pi(18)^2}{4} = 254 \text{ in}^2 = 1.76 \text{ ft}^2$$

$$A_{24} = \frac{\pi(24)^2}{4} = 453 \text{ in}^2 = 3.14 \text{ ft}^2$$

The velocities for 4000 ft³/min are

$$V_{18} = \frac{4000 \text{ ft}^3/\text{min}}{1.76 \text{ ft}^2} = 2272 \text{ ft/min}$$

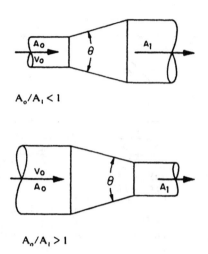

					C_a					
					θ, degrees					
A_o/A_1	10	15	20	30	45	60	90	120	150	180
0.06	0.21	0.29	0.38	0.60	0.84	0.88	0.88	0.88	0.88	0.88
0.1	0.21	0.28	0.38	0.59	0.76	0.80	0.83	0.84	0.83	0.83
0.25	0.16	0.22	0.30	0.46	0.61	0.68	0.64	0.63	0.62	0.62
0.5	0.11	0.13	0.19	0.32	0.33	0.33	0.32	0.31	0.30	0.30
1	0	0	0	0	0	0	0	0	0	0
2	0.20	0.20	0.20	0.20	0.22	0.24	0.48	0.72	0.96	1.0
4	0.80	0.64	0.64	0.64	0.88	1.1	2.7	4.3	5.6	6.6
6	1.8	1.4	1.4	1.4	2.0	2.5	6.5	10	13	15
10	5.0	5.0	5.0	5.0	6.5	8.0	19	29	37	43

FIGURE 15.11   Loss Coefficients for Transitions (Copyright 1989 by the American Society of Heating, Refrigerating, and Air Conditioning Engineers, Inc., from the ASHRAE Handbook-Fundamentals. Used by permission.)

$$V_{24} = \frac{4000 \text{ ft}^3/\text{min}}{3.14 \text{ ft}^2} = 1274 \text{ ft/min}$$

For the elbows, the radius is 18 in. Therefore, referring to Figure 15.5,

$$\frac{r}{D} = \frac{18 \text{ in}}{24 \text{ in}} = 0.75; \text{ from Figure 15.4, } C = 0.33$$

For the transition, refer to Figure 15.11; $\theta = 30°$:

$$\frac{A_0}{A_1} = \frac{3.14}{1.76} = 1.78$$

The closest value in Figure 15.11 for A_0/A_1 is 2; therefore, use $C = 0.2$.

The losses for the straight section of ductwork may be obtained from a duct friction chart, as described in the previous section. The losses for the entire section of ductwork are:

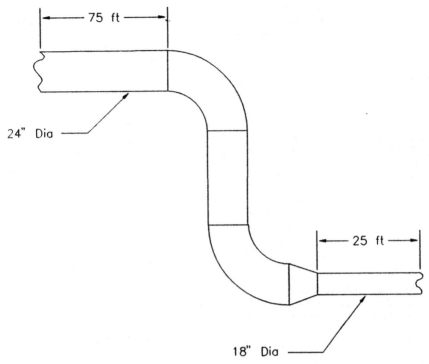

FIGURE 15.12   Ductwork for Example 15.4

24 in diam. duct:	75(0.09/100) =	0.07 in. wg
90° elbow:	0.33(1274/4005)² =	0.03
90° elbow	0.33(1274/4005)² =	0.03
transition	0.2(1274/4005)² =	0.02
18 in diam. duct:	25(0.39/100) =	0.10
loss for entire section		= 0.25 in. wg

15.2   DUCT DESIGN METHODS

The objective of the duct system in an HVAC system is to deliver the design air quantity to each air device in the conditioned space. The design air quantity is usually based on the space heating and/or cooling loads. The purposes of various duct sizing procedures are to size the ductwork in an air distribution system and to determine the pressure requirements for the air system. Once the air flow and pressure requirements for the system have been determined, a suitable fan or air handling unit may be selected and the power requirements for the fan determined. After the ductwork system has been sized, the total pressure requirements may then be determined. The total pressure for the fan is the sum of the total pressure losses throughout the system (including the supply duct, return duct, air devices, duct accessories, and losses internal to the air handling unit itself).

The most widely used duct sizing methods are the equal friction method, the static regain method, and the T-method. Other methods include the velocity reduction method, constant velocity method, and the total pressure method. Each method has advantages and disadvantages. The methods vary in calculation complexity and accuracy. The equal friction method, static regain method, and T-Method will be discussed here.

15.2.1 Duct Design Criteria and Considerations

The design of duct systems frequently involves an number of issues, or considerations, that must be taken into account for the system design to be successful. Below are some of the criteria and considerations that are important in the design of duct systems:

- **Air velocity and pressure loss.** Earlier in this chapter, methods for determining air velocity and pressure requirements of systems were discussed. Duct systems may be classified according to their velocity and air pressure. Systems that have air velocities greater than 2000 ft/min may be considered high velocity systems. Systems with lower velocities are referred to as low pressure systems. Systems with pressures over 2 in. wg are considered high pressure systems and those below 2 in. wg are low pressure systems. The velocity and pressure in a duct system determine the construction requirements for the ductwork.

- **Space available for ductwork.** Frequently, architectural and other factors limit the space available for ductwork. Since ductwork is often installed above ceilings, the space available for ductwork is usually limited by ceiling height, building structure, light fixtures, etc.

- **Noise.** Chapter 12 discussed acoustics and noise generated by HVAC equipment. In addition to noise generated by the equipment itself, noise can be generated by the duct system, as the air flows through the ductwork. Generally speaking, the greater the air velocity, the greater will be the noise generated by the air flow in the duct. Some measures, such as duct lining and duct silencers, can be used to mitigate duct borne noise. Some types of ductwork, such as round, tend to naturally attenuate duct borne noise in some frequencies.

- **Air balance.** The duct design should attempt to provide the design air flow rates to all spaces. Frequently, duct branches near the air handling unit or fan are at a higher pressure and may result in air flow rates greater than desired. In order to reduce this problem, air balancing dampers are required in some duct sections to provide an artificial pressure loss to control the air flow rate. A proper duct design should attempt to reduce the need for balancing dampers by providing a naturally balanced system.

- **Critical path.** The critical path, sometimes called the longest run, is the path that the air will follow which will result in the greatest pressure drop. The fan and motor must be selected based on the critical path, in order for the fan to develop sufficient pressure and air flow to meet the needs of the entire air distribution system.

- **Material costs and operating costs.** Generally speaking, for any given air flow rate the smallest size duct will result in the lowest installed cost for the system. However, the smallest duct will usually result in higher fan power consumption and higher energy costs. The design of duct systems is frequently an economic trade-off between first cost (installed cost) and operating costs.

- **Duct construction.** The type of duct construction has an effect on duct system costs. Some ductwork, such as round, may be easier to fabricate than others. The size and aspect ratio (ratio of width to height) of rectangular ducts has an important effect on the duct cost. Large ducts, in general, and rectangular ducts with large aspect ratios require heavier gauge material to provide sufficient rigidity.

15.2.2 Equal Friction Method

The equal friction method is the most widely used method to size ductwork. It is used frequently for low pressure, low velocity, constant volume systems. It is used almost exclusively for return and exhaust air ductwork. The equal friction method is a straightforward, easy-to-use method. Ductwork may be sized by manual calculation methods. It does not, however, result in an optimum duct system in terms of first cost or operating costs.

Duct systems designed by the equal friction method have a constant pressure loss per ft of length of straight duct. For low velocity systems, the loss factor is between 0.08 and 0.5 in. wg/100 ft. A factor of 0.1 in/100 ft is most commonly used.

After all of the ductwork in a system has been sized, based on the required air flow and the desired friction factor, the critical air flow path must be determined in order to select a suitable fan and motor. The air flow is based on the sum of the air flow rates in each branch, which in turn are based on the space requirements. The pressure loss for the critical path may then be calculated to determine the **external pressure** requirements for the fan or air handling unit. (The external pressure losses are all pressure losses for the system that are not internal to the air handling unit.) When sizing the ductwork, the actual duct sizes should be rounded to the nearest standard size. Standard sizes are typically in 1 in increments for ducts, with the greatest dimension under 20 in, and 2 in increments over 20 in.

The equal friction method usually requires the use of balancing dampers to achieve proper air flow rates and system balance. The equal friction method is best suited for constant volume systems. Since balancing dampers are required, air flows are only balanced for a constant air flow rate (the design air flow). A system is no longer balanced at part load air flow rates in variable air volume systems.

The main advantages to the equal friction method are the ease of calculation and that the method automatically reduces the air velocity in the direction of air flow. The main disadvantages include designs with higher total pressure drops, larger duct sizes, and poor system balance, since only the critical path is analyzed.

Example 15.5 A constant volume air handling unit supplies air to a low velocity, low pressure air distribution system, as shown in Figure 15.13. The required air pressure at each air device is 0.10 in. wg. Using the equal friction method, determine the rectangular duct

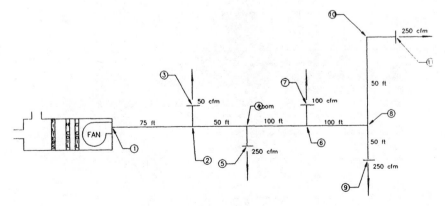

FIGURE 15.13 Air Distribution System for Example 15.5

sizes for each duct section and determine the pressure needed at the fan discharge. All elbows and tees are radius fittings with a radius to width ratio of 1.5. The ceiling height limits the vertical dimension of the ductwork to 12 in.

Since this is a low velocity, low pressure system, the friction factor should be between 0.08 and 0.5 in. wg/100 ft. A value of 0.10 in. wg is chosen since it is most commonly used.

First, each section of ductwork is sized, based on the friction factor and the required air flow rate. From Figure 15.13, it can be seen that the air flow rate for sections 2–3 and 6–7 is 100 ft³/min. The air flow rate of 100 ft³/min and a friction factor of 0.10 in. wg/100 ft are then plotted on the friction chart, as shown on Figure 15.14. An equivalent duct diameter of 5³/₄ in and a velocity of about 550 ft/min are read from the chart. From Table 15.1, the closest rectangular (or square) duct size is 6 in × 6 in.

Referring to Figure 15.13, the air flow for sections 4–5, 8–9, and 8–11 is 250 ft³/min. For a friction factor of 0.10 in. wg/100 ft, the equivalent round duct diameter, from Figure 15.14, is about 8 in and the velocity is equal to 650 ft/min. From Table 15.1, an equivalent rectangular duct size of 8 in × 7 in is chosen as the closest.

Using a friction factor of 0.10 in. wg/100 ft throughout the duct distribution system, and limiting the vertical height of the ductwork to 12 in, the rest of the system is sized as follows:

Section	Air Flow(ft³/min)	Equivalent Diam. (in)	Rectangular Size (in)	Velocity (ft/min)
6–8	500	10¹/₄	11×8	825
4–6	600	11	10×10	875
2–4	850	13	14×10	950
1–2	950	13¹/₂	14×11	975

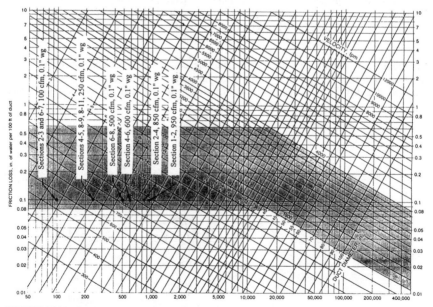

FIGURE 15.14   Stratic Regain Chart *(Courtesy of Carrier Corporation)*

Note: There is actually an infinite number of equivalent rectangular duct sizes for any given diameter. The actual dimensions selected will vary depending on the space requirements. As a rule of thumb, rectangular ducts should be sized as close to square as possible, because square ducts are typically easier and less costly to fabricate than large rectangular ducts.

The total air flow requirement for the fan is the sum of the air flow rates from all outlets, 950 ft³/min. In order to determine the external pressure (pressure not internal to the air handling unit) requirements, it is first necessary to determine the **critical path (longest run)**. The critical path is usually the path to the most remote outlet, although it may also be the path with the greatest air flow. Occasionally, it may be necessary to calculate the loss for several paths, in order to determine which has the greatest drop from the beginning to end.

In order to determine the pressure losses for the distribution system, it is necessary to start at the AHU discharge and calculate the losses for each section and each fitting, as the air flows through the critical path. Although the pressure losses for the actual duct will be slightly different than for rectangular, the values for equivalent round will be used for simplicity purposes without any significant loss in accuracy.

Beginning at the fan discharge, the total air flow is 950 ft³/min and the 14 in × 11 in duct has a loss of about 0.10 in. wg/100 ft. For section 1–2, the friction loss is calculated with Equation 15.9.

$$h_{L,1-2} = \left(\frac{0.10 \text{ in}}{100 \text{ ft}}\right)(75 \text{ ft}) = 0.075 \text{ in. wg}$$

At point 2, the air flow divides and the duct reduces to 14 in × 10 in, in a branching tee before section 2–4, as previously determined and shown in the table above. The branch from 2–3 has 100 ft³/min flowing through it, so the remaining flow in section 2–4 is 850 ft³/min. A dynamic fitting loss must now be calculated for the branching tee. Since the loss for the critical path is being calculated, the loss must be determined for the air as it flows through the tee, instead of the loss for the air as it branches.

Referring to Figure 15.9, the upstream, downstream, and branch air flow rates and areas are used to determine a loss coefficient for the fitting.

The upstream area = A_c = 14 in × 11 in = 154 in²

The downstream area = A_s = 14 in × 10 in = 140 in²

The branch area = A_b = 6 in × 6 in = 36 in²

$$\frac{A_b}{A_s} = \frac{36}{140} = 0.25$$

$$Q_c = 950 \text{ ft}^3/\text{min}, \; Q_s = 850 \text{ ft}^3/\text{min}, \; Q_b = 100 \text{ ft}^3/\text{min}$$

$$\frac{Q_b}{Q_c} = \frac{100}{950} = 0.10$$

Referring to Figure 15.9, the table labeled "Main" is used since the loss is to be calculated for the main, or straight through, section of the duct and not the branch. Using the value for A_b/A_s in the table of 0.33, a value of $Q_b/Q_c = 0.1$, a loss coefficient of $C_{c,s} = -0.01$ is read from Figure 15.9, indicating a gain in static pressure.

The air velocity entering the fitting is equal to the velocity in the duct upstream of the fitting. The velocity for section 1–2 was previously determined from Figure 15.14 as 950 ft/min. Equation 15.12 is used to calculate the dynamic loss (gain) for the branch tee at point 2.

$$h_L = C\left(\frac{V}{4005}\right)^2 = -0.01\left(\frac{950}{4005}\right)^2 = -0.006 \text{ in. wg}$$

Since this is such a small number compared to other losses in the duct system, it may be assumed to be negligible or 0.

For section 2–4, the duct is 14 in $\times$ 10 in, the air flow rate is 850 ft^3/min, and the friction factor is again 0.10 in/100 ft (hence equal friction method). The length is 50 ft, as before when using Equation 15.9.

$$h_{L,2-4} = 50 \text{ ft}\left(\frac{0.10 \text{ in. wg}}{100 \text{ ft}}\right) = 0.05 \text{ in. wg}$$

For the tee at point 4,

$$A_b = 8 \text{ in} \times 7 \text{ in} = 56 \text{ in}^2$$
$$A_c = 14 \text{ in} \times 10 \text{ in} = 140 \text{ in}^2$$
$$A_s = 10 \text{ in} \times 10 \text{ in} = 100 \text{ in}^2$$
$$Q_b = 250 \text{ ft}^3/\text{min}, \ Q_c = 850 \text{ ft}^3/\text{min}, \ Q_s = 600 \text{ ft}^3/\text{min}$$
$$\frac{A_b}{A_s} = \frac{56}{100} = 0.56, \ \frac{Q_b}{Q_c} = \frac{250}{850} = 0.29$$

From Figure 15.9, the loss coefficient for the main is

$$C_{c,s} = -0.05$$

From the table above, the velocity of the air as it enters the fitting is $V = 850$ ft/min. Using Equation 15.12, the loss for the branching tee is

$$h_L = -0.05\left(\frac{850}{4005}\right)^2 = -0.002 \text{ in. wg, again negligible}$$

The losses for the various sections of duct and fittings can be calculated in a similar fashion and a table can be developed showing these pressure losses. The methods for calculating for the tee (symmetrical wye) at point 8 and the radius elbow at point 10 are similar to those for the branching tees. For the symmetrical wye, from Figure 15.8,

$$A_c = 11 \text{ in} \times 8 \text{ in} = 88 \text{ in}^2, \ A_{b1} = 8 \text{ in} \times 7 \text{ in} = 56 \text{ in}^2$$

$A_{b1}/A_c = {}^{56}/_{88} = 0.6$, which is between the tabulated values; therefore, it is necessary to interpolate between the tabulated values. The loss coefficient for a diverging wye is calculated by interpolating as

$$C_{c,1b} = 0.30 + \left(\frac{0.6 + 0.5}{1.0 - 0.5}\right)(0.3 - 0.25) = 0.31$$

The air velocity entering the fitting is 750 ft/min. Using Equation 15.12, the loss for the wye is

$$h_L = 0.31\left(\frac{875}{4005}\right)^2 = 0.01 \text{ in. wg}$$

For the 90° elbow at point 10, the width of the duct was set at 7 in and the height at 6 in.

$$\frac{H}{W} = \frac{7}{8} = 0.88$$

The radius to width ratio (r/W) was given as 1.5. Referring to Figure 15.6, it is determined that it will be necessary to interpolate between values of H/W to obtain a loss coefficient. For $H/W = 0.88$ and $r/W = 1.5$.

$$C'_0 = 0.19 - \left(\frac{0.88 - 0.75}{1.0 - 0.75}\right)(0.19 - 0.17) = 0.18$$

The velocity was determined above as 700 ft/min. Using Equation 15.12, to calculate the loss for the elbow:

$$h_L = 0.18\left(\frac{700}{4005}\right)^2 = 0.005, \text{ negligible}$$

The following table has been developed to show all the losses for the critical path:

Element	Element Type	Loss (in. wg)
1–2	duct	0.08
2	branch tee	neg.
2–4	duct	0.05
4	branch tee	neg.
4–6	duct	0.10
6	branch tee	neg.
6–8	duct	0.10
8	wye	0.01
8–10	duct	0.05
10	90° ell	neg.
10–11	duct	0.05
11	air device	0.10
		0.54

Therefore, the fan must supply 950 ft³/min at a discharge static pressure of 0.54 in. wg in order to produce the air flow rate at the air device at point 11 and the other outlets, as shown in Figure 15.13.

Note that pressure at the outlets closer to the fan would have a pressure greater than the 0.10 in. wg required at the most remote outlet. It would probably be necessary to provide volume dampers in all branches, other than the critical path, in order to achieve the desired air flow rates throughout the duct system. The volume dampers would create an artificial pressure loss in order to have an equal friction loss for each branch.

Example 15.6 The fan in Example 15.5 is providing a flow rate of 950 ft³/min at a total pressure of 0.54 in. wg. A 14 in × 10 in duct is connected to the fan outlet. What is the velocity pressure in the duct, just after the fan discharge?

In Example 15.5, it was determined that the air velocity in a 14 in × 10 in duct would be 950 ft/min for an air flow rate of 950 ft³/min. Using Equation 15.2, the velocity pressure is calculated as

$$P_V = \left(\frac{950}{4005}\right)^2 = 0.06 \text{ in. wg}$$

The total pressure P_T in the duct was calculated as 0.54 in wg. Using Equation 15.3 and solving for the static pressure,

$$P_S = P_T - P_V = 0.54 - 0.06 = 0.48 \text{ in. wg}$$

The air in the duct at the fan discharge is flowing at the rate of 950 ft³/min, at a velocity of 950 ft/min, and has a total pressure of 0.54 in, a static pressure of 0.48 in, and a velocity pressure of 0.06 in. wg.

15.2.3   The Static Regain Method

The need for volume dampers in duct systems designed to be the equal friction method was discussed in the preceding section. Since the static pressure in the ductwork nearest the fan is higher than more remote sections for duct systems designed by the equal friction method, dampers are usually needed to reduce the static pressure at the outlets near the fan discharge in order for the system to balance. This results in a need for the system to be adjusted, or balanced, to achieve the desired air flow rates, and also results in some inefficiency due to the artificial pressure loss through the dampers.

The objective of the static regain duct design method is to design a system for which volume dampers are not needed, or at least to minimize the need for such dampers and system balancing and adjusting. In other words, the static regain method attempts to design a self-balancing system at the design air flow rates. A duct system properly designed by the static regain method will result in nearly equal static pressures at the entrances to all branches and outlets. The main distribution ducts are sized to achieve an approximate balance at each branch. The static pressure at an outlet determines the rate of flow through the outlet for any given sized outlet. Therefore, if the static pressure remains constant throughout the system, only the size, or the free area, of the outlet will determine the discharge rate. Since static pressures throughout the system are roughly equal at both the design flow and part load flow, static regain duct designs are suitable for both constant and variable air volume systems.

In order to accomplish such a system, it is necessary to analyze the pressure losses and air velocities that occur in the duct distribution system. The static regain method utilizes the velocity reduction downstream of each branch in the duct, the magnitude of the velocity reduction being just sufficient to provide a loss in velocity pressure equal to the loss in total pressure that will occur in the succeeding section.

The static regain method requires more calculation time and effort than the equal friction method. Systems designed by the static regain method usually result in duct systems that are slightly larger and higher in first cost than others. However, the static regain method also results in the reduced need for balancing, and lower pressure losses, with a result in lower power requirements and lower operating costs.

Earlier in this chapter, the relationship between static pressure and velocity pressure was discussed. It was also noted that static pressure in ductwork can be transformed into velocity pressure and visa versa. The static regain method is based on sizing the ductwork so the static pressure increase (static regain) due to the reduction in velocity at each branch takeoff offsets the pressure loss of the succeeding section. As the velocity (V_1) of the air flow at some point (1) slows to a slower velocity (V_2), and at some point (2) some of the velocity pressure will be converted to static pressure, hence, static regain. As the air flows through ductwork, the total pressure always decreases in the direction of air flow. For systems that were designed with the static regain method, it can be observed that as the air flows through the ductwork from section to section, the velocity pressure decreases, while the static pressure increases. There is a conversion of velocity pressure to static pressure (static regain).

For duct systems designed by the static regain method, the static pressure for an outlet or branch near the fan discharge is determined based on the air flow requirements of the outlet. All other outlets in the system are selected so they provide the design air flow at approximately the same static pressure as the first outlet. The static regain method then attempts to design the duct distribution system so the same static pressure occurs at all other outlets. The loss in static pressure as the air flows is offset by the gain, due to the reduction in velocity and the conversion of velocity pressure to static pressure.

The first section of duct at the fan discharge is usually selected, based on recommended velocity or static pressure drop, and also based on the space available for ductwork. The first section of duct is the duct between the fan outlet and the first air outlet or branch take-off. Since the first section of duct has no changes in size, the change in velocity pressure is zero. The change in total pressure is equal to the change in static pressure only. The flow and pressure requirements for the fan, therefore, are based on the flow and loss in the first section of duct. Recommended velocities for various applications are given in Tables 15.3 and 15.4.

If there were no friction or dynamic losses at the junction of the outlet, or branch, there would be no loss in total pressure, and all the velocity pressure would be converted to static pressure as the air velocity was reduced. For standard air, the theoretical gain in static pressure would be:

TABLE 15.3 Recommended and Maximum Duct Velocities for Low Velocity Systems

Recommended Velocities (ft/min)			
Designation	Residences	Schools, Theaters, Public Bldgs.	Industrial Buildings
Outdoor Air Intakes	500	500	500
Filters	250	300	350
Heating Coils	450	500	600
Cooling Coils	450	500	600
Air Washers	500	500	500
Fan Outlets	1000–1600	1300–2000	1600–2400
Main Ducts	700–900	1000–1300	1200–1800
Branch Ducts	600	600–900	800–1000
Branch Risers	500	600–700	800
Maximum Velocities (ft/min)			
Outdoor Air Intakes	800	900	1200
Filters	300	350	350
Heating Coils	500	600	700
Cooling Coils	450	500	600
Air Washers	500	500	500
Fan Outlets	1700	1500–2200	1700–2800
Main Ducts	800–1200	1100–1600	1300–2200
Branch Ducts	700–1000	800–1300	1000–1800
Branch Risers	650–800	800–1200	1000–1600

TABLE 15.4 Recommended Maximum Duct Velocities for High Velocity Systems

Air Flow in Duct (ft³/min)	Maximum Velocities (ft/min)
60,000 to 40,000	6000
40,000 to 25,000	5000
25,000 to 15,000	4500
15,000 to 10,000	4000
10,000 to 6,000	3500
6,000 to 3,000	3000
3,000 to 1,000	2500

$$P_{sr} = \left(\frac{V_1}{4005}\right)^2 - \left(\frac{V_2}{4005}\right)^2 \tag{15.13}$$

where

P_{sr} = the theoretical pressure regain, in. wg

V_1 = the velocity in the main upstream branch, ft/min

V_2 = the velocity in the main downstream branch, ft/min

4005 = units conversion factor

Equation 15.13 assumes no friction or dynamic losses for the fitting at the outlet or branch. It has been determined that the actual regain for most duct fittings is between 60% and 90%, depending on the design and construction of the fitting. The average recovery, or regain, is about 75%. Therefore, Equation 15.13 may be modified to provide the average static regain for typical duct fittings.

$$P_{sr} = 0.75\left[\left(\frac{V_1}{4005}\right)^2 - \left(\frac{V_2}{4005}\right)^2\right] \tag{15.14}$$

Equation 15.14 calculates the static regain at a given fitting for an outlet or branch. The static regain method attempts to set this regain, equal to the friction loss for the section of ductwork, downstream of the fitting. The loss for the downstream section of ductwork may be calculated by using the appropriate terms in Equation 15.12.

$$h_{L2} = C\left(\frac{V_2}{4005}\right)^2 \tag{15.12}$$

It is necessary to find values for V_1 and V_2 so $P_{sr} = h_{L2}$. Usually, it is necessary to find V_1 and V_2 by an iteration process, which can be quite lengthy and tedious, especially for large and complex duct distribution systems. Since an iteration process would be so tedious and time-consuming, for all but the simplest of duct systems, a method was developed whereby P_{sr} may be solved by known quantities and reasonable assumptions.

$$P_{sr} = 3.9 \times 10^{-9} V_2^{2.43}\left(\frac{L_2}{Q_2^{0.61}}\right) \tag{15.15}$$

In any given design, the values of V_1, Q_2, and L_2 are known, and Equation 15.15 may be solved for V_2. The area and diameter of the duct may then be determined from Equation 2.16 and the Duct Friction Chart, Figure 15.3. If the duct is to be square or rectangular, Table 15.2 may be used to convert from round to equivalent rectangular duct.

Equation 15.15 may also be solved by using the Static Regain Chart, Figure 15.14. The static regain procedure, using the Static Regain Chart, is as follows:

1. Size the first section of duct, after the fan discharge, for a suitable velocity and/or pressure drop, e.g., 0.1 in. wg/100 ft or recommended velocity from Tables 15.3 or 15.4.

2. Select a suitable required static pressure at the design air flow for the first branch, or outlet, based on the outlet requirements and the guidelines presented in Chapter 14. Select all other outlets for the same static pressure at the design air flow.

3. Subtract the air flow discharge for the first outlet from the total air flow to get the air flow (Q_2) for the next (succeeding) section of duct in ft³/min. Determine the length (L_2) of the next section in ft.

4. Calculate a value for $L_2/Q_2^{0.61}$ for the succeeding section of duct.

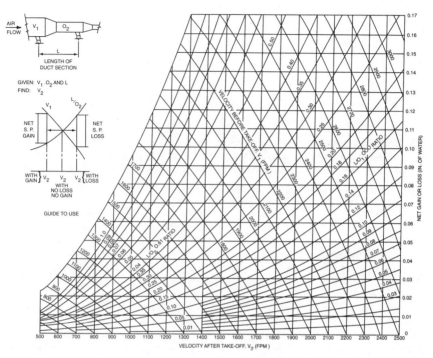

FIGURE 15.15 Static Regain Chart

5. Locate the calculated value for $L/Q^{0.61}$, and the velocity (V_1) for the *previous* section of duct on the Static Regain Chart, Figure 15.15. Read the air velocity (V_2) for the *succeeding* (next) section on the scale at the bottom of the chart.

6. Use the Continuity Equation (Equation 2.16) and V_2 to calculate the required area for the succeeding duct section. Convert to rectangular, with Table 15.1, if rectangular or square duct is required.

7. Locate the air flow and velocity for the succeeding section on a duct friction chart, Figure 15.3. Read the loss coefficient (C_L) from the left side of the chart.

8. Use the value of C_L, the velocity, the length (L), and Equation 15.12 to calculate the static pressure loss for the section.

9. Use the velocities for the preceding and succeeding sections in Equation 15.15 to calculate the static regain.

10. Compare the static regain from Step 9 to the static pressure loss from Step 8. The two values should be within 15% to be acceptable.

11. Repeat Steps 3 through 10 for each section of ductwork.

12. Calculate the static pressure loss for the first section of ductwork using Equation 15.12, and add the static pressure requirement for the first outlet to get the external static pressure for the fan.

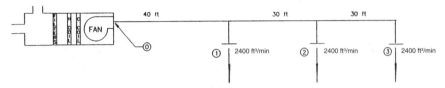

FIGURE 15.16   Duct Distribution System for Example 15.7

13. Using the air velocity for the first section of duct, use Equation 15.2 to calculate the velocity pressure. Use the velocity pressure, the total external static pressure calculated in Step 12, and Equation 15.3 to calculate the total external pressure for the fan.

14. Use the total air flow, the total external pressure, and external static pressure as a basis for selecting a fan. See Chapter 16 for fan selections.

Example 15.7   A centrifugal fan is supplying air to the low velocity duct distribution system for an industrial building, as shown in Figure 15.16.

The fan is supplying 7200 ft³/min of air. Air outlets were selected so each outlet supplies 2400 ft³/min at a static pressure of 0.1 in. wg. Use the static regain method to size the ductwork for round duct. What are the external static and external total pressure requirements for the fan?

Referring to Table 15.3, the recommended air velocity for low velocity main ducts in an industrial application is 1200 to 1800 ft/min with a maximum velocity of 2200 ft/min. Since the air is typically discharged from a centrifugal fan at a high velocity, a relatively high velocity is recommended to reduce static pressure gain at the fan discharge. (See Chapter 16.) A velocity of 1800 ft/min is selected for the first duct section, 0–1. An air flow rate of 7200 ft³/min and a velocity of 1800 ft/min are plotted on the Duct Friction Chart, Figure 15.17, to obtain a duct diameter of 28 in and a loss coefficient of about 0.12 in. wg/100 ft for section 0–1.

For section 1–2,

$$Q_{1-2} = 7200 - 2400 = 4800 \text{ ft}^3/\text{min and, from Figure 15.15, } L = 30 \text{ ft}$$

The L/Q ratio for section 1–2 is

$$\frac{L}{Q^{0.61}} = \frac{(30 \text{ ft})}{(4800 \text{ ft}^3/\text{min})^{0.61}} = 0.17$$

Plotting V = 1800 ft/min, and $L/Q^{0.61}$ ratio of 0.17 on the Static Regain Chart, Figure 15.18, a velocity of 1550 ft/min for the next section is read from the chart. Plotting an air flow rate of 4800 ft³/min and a velocity of 1550 ft/min on Figure 15.17 produces a duct diameter of 24 in and C_L= 0.11 in. wg/100 ft. Using Equation 15.9,

$$h_L = LC_{L-100} = 30 \text{ ft}\left(\frac{0.11 \text{ in. wg}}{100 \text{ ft}}\right) = 0.033 \text{ in. wg}$$

Using Equation 15.14 to calculate the static regain at point 1,

$$P_{sr1} = 0.75\left[\left(\frac{V_1}{4005}\right)^2 - \left(\frac{V_2}{4005}\right)^2\right] = 0.75\left[\left(\frac{1800}{4005}\right)^2 - \left(\frac{1550}{4005}\right)^2\right] = 0.038 \text{ in. wg}$$

Notice that the pressure loss for section 1–2 is approximately equal to the static regain at point 1. Checking the static pressure at point 2 (outlet 2),

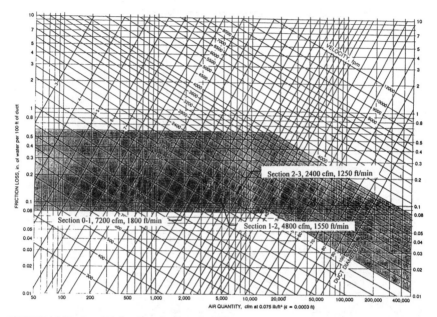

FIGURE 15.17 Duct Friction Chart for Example 15.7

$$P_{s2} = P_{s1} - h_{L,1-2} + P_{sr1} = 0.1 - 0.033 + 0.038 = 0.105 \text{ in. wg}$$

P_{s2} is within the required 15% of P_{s1}.
For section 2–3,

$$Q_{2-3} = 4800 - 2400 = 2400 \text{ ft}^3/\text{min, and } L_{2-3} = 30 \text{ ft}$$

$$\frac{L}{Q^{0.61}} = \frac{30}{(2400)^{0.61}} = 0.26$$

Plotting $V_2 = 1550$ ft/min, and the $L/Q^{0.61}$ ratio of 0.26 on Figure 15.18, yields

$$V_3 = 1250 \text{ ft/min}$$

Plotting $V_3 = 1250$ ft/min, and the air flow rate of 2400 ft³/min on the Friction Chart, Figure 15.17, yields a duct diameter for section 2–3 of 19 in and a $C_{L,2\text{-}3} = 0.15$ in. wg/100 ft. Using Equation 15.9,

$$h_{L,2-3} = L_{2-3}C_{L,2-3} = 30(0.15) = 0.045 \text{ in. wg}$$

$$P_{sr2} = 0.75 \left[\left(\frac{1550}{4005}\right)^2 - \left(\frac{1250}{4005}\right)^2 \right] = 0.04 \text{ in. wg}$$

$$P_{s3} = 0.105 - 0.045 + 0.04 = 0.109 \text{ in. wg}$$

All three outlets essentially have the same static pressure without balancing dampers.

Equation 15.9 may be used to calculate the static pressure loss for section 0–1. From above,

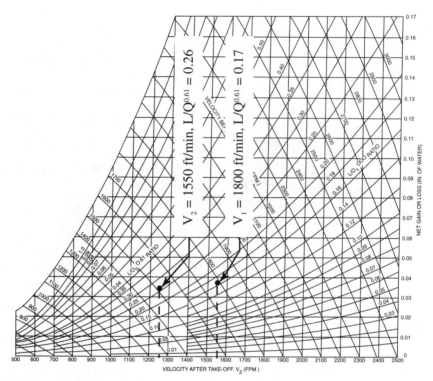

FIGURE 15.18   Static Regain Chart for Example 15.7

$$C_{L,0-1} = 0.12 \text{ in. wg}/100 \text{ ft}$$

$$h_{L,0-1} = LC_{L,0-1} = 40\left(\frac{0.12}{100}\right) = 0.05 \text{ in. wg}$$

Since the first outlet requires 0.10 in. wg, the static pressure at the fan outlet discharge should be

$$P_{s0} = 0.05 + 0.10 = 0.15 \text{ in. wg}$$

The velocity in section 0–1 was determined above as 1800 ft/min. The velocity pressure is calculated using Equation 15.2.

$$P_{v0} = \left(\frac{1800}{4005}\right)^2 = 0.2 \text{ in. wg}$$

Using Equation 15.3,

$$P_T = 0.15 + 0.20 = 0.35 \text{ in. wg}$$

The fan should be selected to provide 7200 ft³/min at an external total pressure of 0.35 in. wg.

The above example is of a very simple system, in order to show the static regain calculation procedure without having too many calculations to obscure the procedure. It can be observed that the calculation process for large, complex systems can become quite long and tedious. Most duct systems designed by the static regain method are designed using computer programs. The advantages of the static regain method have already been discussed. Disadvantages of the static regain method include:

- very low velocities and large duct sizes may result at the end of long duct runs. Main or trunk duct sections, remote from the fan, usually have larger dimensions than a similar system, sized by the equal friction method.
- manual calculations require tedious bookkeeping and trial-by-trial calculations.
- the total pressure requirements of each part of the duct system are not readily apparent.

High velocity/high pressure and variable air volume systems are usually designed using the static regain method. Return and exhaust air systems are not generally suitable for the static regain method. They should be designed using the equal friction method, as described earlier in this chapter.

15.2.4 T-Method

Earlier in this chapter, the relationships among duct size, installed cost, and operating costs were briefly discussed. When the cost of energy is high and/or the installed cost of ductwork is low, a lower initial air velocity is more economical. For low energy costs and/or high installed duct costs, a higher velocity is more economical. Systems designed with large ducts for a given sized air flow will tend to have lower velocities, lower pressure losses, and lower operating costs. Systems with larger ducts will tend to have higher installed costs. Conversely, systems with smaller duct sizes for the same air flow will tend to have a lower first cost and higher operating costs. The initial cost of the duct system, hours of operation per year, energy cost interest rates, energy cost escalation, and the amortization period are all factors in the T-Method for sizing ducts.

The T-Method is an optimizing procedure which sizes ductwork on life cycle costs. The method is based on procedures that attempt to minimize the life cycle cost of the duct system. The life cycle cost of the system is the total cost to install, own, and operate the system, and is a function of the initial cost of the system as well as the present worth of energy costs.

Designing a duct system by the T-Method consists of the following major procedures:

- **System condensing.** This procedure condenses a multi-section duct system into a single imaginary, or representative, duct section. This representative duct section has the same hydraulic characteristics and same owning cost as the actual entire system.
- **Fan selection.** A fan with the optimum total pressure is selected for the representative system.
- **System expansion.** Based on the flow rate and pressure characteristics of the selected fan, the imaginary system is expanded into the original system so the flow is distributed in accordance with the ratio of pressure losses calculated in the system condensing step.

In order to perform the various procedures, an iteration process is necessary to reach optimization, based on the factors to be considered. Since this requires extensive calculations for even the simplest of systems, the T-Method requires that the calculations be made with a computer.

Systems designed by the T-Method are static pressure balanced, just like systems designed by the static regain method. Since systems designed by the T-Method are optimized, and are balanced for both velocity and pressure, systems designed by this method are suitable for both constant volume and VAV systems.

Only an overview of the T-Method is presented here since the procedure is so involved. You should refer to Reference 1 for additional information on the T-Method.

15.3 GENERAL DUCT DESIGN GUIDELINES
AND PROCEDURES

Before a duct system is designed, the system must be zoned and air flow rates determined for each space. Supply and return air devices should be located in the space, based on the methods presented in Chapter 14. The duct design procedure is as follows:

1. Propose a preliminary duct layout that has the required zoning, connects supply air ducting to all supply air outlets, connects return air ductwork to all return air inlets (if not a plenum return), and connects exhaust ductwork to all exhaust inlets. Try to keep systems as direct and compact as reasonably possible.

2. Determine the space available for ductwork. For example, check the space between ceilings and building structural members. Do not forget to allow for lights and other obstructions that may share the same space as the ductwork. Select an allowable maximum height for the ductwork.

3. Divide the duct layout into consecutive sections which converge and diverge at fittings, such as branches and tees.

4. Select a duct sizing method (equal friction, static regain, etc.) based on the type of system (supply, return, constant volume, VAV, etc.).

5. Calculate the air flow rate, static pressure losses, total pressure losses, etc. by the selected duct design method discussed earlier in this chapter.

6. Select a suitable fan or air handling unit, based on the required air flow rate, selected static pressure, total pressure, and operating characteristics. See Chapter 16 for fan selection guidelines.

7. Lay out the duct system in detail. Perform revised calculations if significant changes were made.

8. Resize duct selections as required.

9. Analyze the design for excessive air velocities and objectionable noise. Refer to Chapter 12 for guidelines on HVAC system acoustics. Revise ductwork sizes or provide sound attenuators as required.

General Design Guidelines

1. Get the air where it is required as directly as is reasonably possible. Avoid unnecessarily long duct runs.

2. Keep maximum velocities and unit pressure losses within recommended values. See Tables 15.3 and 15.4.

3. Try to design duct systems that are pressure balanced (self-balancing) as much as possible. Provide volume dampers where this is not possible.

4. Avoid sharp elbows and bends in the ductwork. Use vanes whenever mitered elbows and tees must be used.

5. Avoid sudden enlargements and abrupt contractions in the ductwork. The angle of divergence for enlargements should not exceed 20°. The angle of convergence for contractions should not be greater than 60°.

6. Consider using round duct whenever possible. It has the greatest air-carrying capacity per square foot of sheet metal. It also provides better sound attenuation in some frequencies.

15.4. AIR SYSTEM TESTING AND BALANCING (TAB)

System TAB is the process of checking and adjusting the HVAC system, especially the air distribution system, to produce the design air flow rates, temperatures, and humidity levels throughout a building. The process generally includes:

1. balancing the air distribution for the system to achieve the required air flow rates for each space.
2. adjusting the total system to provide design quantities.
3. establishing quantitative performance of all equipment and systems.
4. verifying that automatic controls are functioning as required.
5. taking sound and vibration measurements.

System testing and balancing is achieved by:

1. checking the installation for conformity to design specifications.
2. measuring and establishing the air and other fluid flow rates of the system.
3. recording and reporting the results.

In order to achieve proper air balance, the air system design should include volume control dampers at the required locations in the system.

The generally accepted method for measuring air flow in ducts is with a pitot tube traverse. A pitot tube is shown in Figure 15.1. The air velocity is then calculated using Equation 15.6. The Continuity Equation (Equation 2.16) may then be used to determine the air flow quantities in the duct. A hood that captures the air as it discharges from an air device may be used to measure air flow from diffusers and registers, etc. Rotating vane anemometers are also used to measure air flow rates from air devices, especially side wall devices.

Generally speaking, it is not possible to achieve perfect air balance in a system. For a system to be considered "in balance," all air flow rates should be within ± 10% of the design values. Supply and exhaust fans are typically adjusted to +10% to account for air leakage in ductwork. For determining if a system is in balance or not, the following tolerances may be used as a guideline:

Supply fans	+5% to 10%
Return/exhaust fans	+5% to 10%
Diffusers/supply air grilles	0% to 10%
Return grilles	0% to − 10%
Exhaust grilles	0% to − 10%

Chapter 16

FANS AND CENTRAL AIR SYSTEMS

A fan is the prime mover of air in an HVAC system. It provides continuous circulation of air throughout the system and conditioned spaces. The mechanical energy, in the form of pressure, must exactly equal the energy lost by the entire system as the air flows.

An air handling unit (AHU) is a complete unit, including a fan, fan motor, fan drive, heating coils, cooling coils, filters, and accessories. The various components may all be housed in a single cabinet or may be individual components assembled to form a complete unit. A central air system typically consists of an air handling unit and an air distribution system of ductwork and air devices.

A fan is a device used to convert rotational mechanical energy (work) to a total pressure increase of a moving gas (usually air). Total pressure consists of both static and velocity pressure. (See Chapter 15.) The conversion of work to pressure is accomplished by changing the momentum of the fluid. Fans produce pressure by altering the velocity of the gas flow. The conversion of mechanical energy to gas velocity is accomplished by means of a fan wheel (or impeller) which imparts a spin to the gas. A fan impeller produces velocity via two related methods:

- centrifugal force created by the rotating air column enclosed between the fan blades.

- kinetic energy imparted to the air, due to its absolute velocity, as it leaves the impeller blade

A distinction between a fan and a compressor should be made. Typically, a fan increases the density of the air by no more than 7% as it travels from inlet to outlet (discharge). This would result in an increase in air pressure of about 30 in. wg, based on air at standard conditions.

16.1 TYPES OF FANS AND THEIR CHARACTERISTICS

There are many types of fans used for HVAC applications. The type of fan used for a particular system depends on the application or required characteristics. For example, some types of fans are best suited for high air flow volumes and low static pressures. Others are better suited for high pressure, low flow situations. Some fan types are better suited for VAV applications than others. Other considerations include space available, noise levels, and the system in which they are installed.

The most common types of fans used for HVAC systems which will be discussed in this chapter include axial flow fans and centrifugal fans. Axial flow fans include propeller, tube-axial flow, and vane-axial flow fans. Centrifugal fans include forward curved, backward inclined, backward curved, and airfoil fans. Special applications of fans include inline centrifugal, plug fans, and power exhaust/ventilator fans. The special fans are usually an adaptation of the two basic types of fans.

Both centrifugal and axial flow fans increase the total pressure of the air by producing velocity pressure, which is converted to static pressure in the air distribution system (ductwork). In an axial flow fan, the air flow direction is parallel to the rotational axis of the fan. In centrifugal fans, air is radially discharged from an impeller. The air turns a net 90° from the fan inlet, to the fan, to the outlet.

One important indicator of the performance of axial flow fans is the hub ratio. The hub ratio is the ratio of the diameter of the hub (or root of the blades) to the tip diameter of the blades. Generally, the higher the hub ratio, the higher the potential pressure capability of the impeller.

A typical propeller fan is shown in Figure 16.1. Propeller fans are used for moving air against a relatively low static pressure, generally less than 1 in. wg. The performance of these fans is very sensitive to changes in air flow resistance. Propeller fans are well suited for moving high volumes of air at little or no static pressure differentials. The propeller fan can be selected over its entire operating range, as long as there are no high noise level restrictions. These fans tend to be noisy. Propeller fans may either be belt driven or direct drive. A propeller fan usually has a hub ratio of less than 0.15.

Tube-axial (duct) fans have propeller type blades in a short cylindrical housing, as shown in Figure 16.2. There are no straightening vanes. These fans are suitable for rela-

FIGURE 16.1 Propeller Fan (*Courtesy of The Trane Company*)

FIGURE 16.2 Tube-axial Fan (*Courtesy of The Trane Company*)

tively moderate static pressures, less than 2 in. wg. Tube-axial fans usually have a hub ratio less than 0.3.

Vane-axial fans are very similar to tube-axial fans. The main difference is the addition of airflow straightening (guide) vanes on the fan discharge, and occasionally on the fan inlet. Vane-axial fans have a propeller type configuration with a hub and airfoil type blades mounted in a cylindrical housing. The housing normally includes straightening vanes on the discharge side of the impeller. Vane-axial fans are more efficient than other axial flow fans, and can develop higher static pressures. A vane-axial fan is typically mounted on the centerline of the duct and produces an axial flow of air. Vane-axial fans generally have hub ratios equal to or greater than 0.3.

Vane-axial fans and tube-axial fans are generally used for handling large volumes of air at relatively low static pressures. The maximum efficiency of these fans is about 65%. The operating range for tube-axial and vane-axial fans is between 65% and 90% of the air flow range. Disadvantages of these fans include relatively high noise levels and lower efficiency.

Figure 16.4 shows a typical centrifugal fan and Figure 16.5 shows the terminology for centrifugal fan components. There are three commonly used types of centrifugal fans based on the design of the impeller. The types are forward curved (FC), backward inclined/backward curved (BI), and radial. Flow within a centrifugal fan is primarily radial through the fan impeller. In a centrifugal fan, air enters the impeller and is accelerated out through the blade passages by centrifugal force, as the impeller rotates. The blast area is the cross-sectional area, just above the cutoff, which prevents the air from recirculating inside the fan,

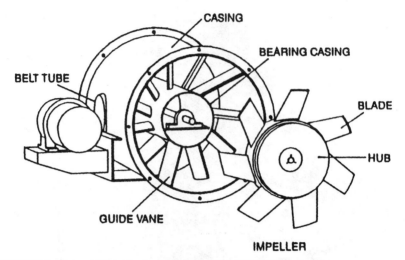

FIGURE 16.3   Vane-axial Fan (*Reprinted with permission of AMCA International*)

FIGURE 16.4   Centrifugal Fan (*Courtesy of Twin City Fan and Blower*)

and discharges the air out of the fan housing. The blast area is always smaller than the outlet area of the centrifugal fan. The blast area and the airflow rate determine the maximum discharge air velocity from the fan. Single width centrifugal fans draw air into the impeller from one side only. Double width fans draw air in from both sides of the impeller.

A double width fan is basically two single width fans, side by side, with two inlets and a single outlet with no partition in the scroll.

Forward curved (FC) centrifugal fans have impeller blades which curve toward the direction of rotation. They are also referred to as "squirrel cage" fans. A forward curved fan

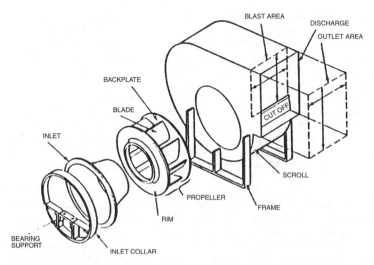

FIGURE 16.5 Terminology for Centrifugal Fan Components (*Reprinted with permission of AMCA International*)

FIGURE 16.6 Forward Curved Centrifugal Fan (*Courtesy of Twin City Fan and Blower*)

impeller is shown in Figure 16.6. These fans have relatively low tip speeds, require relatively little space, and are relatively quiet in operation. They are generally used in low static pressure applications. Although FC fans rotate at relatively slow speeds, they are used for producing high volumes of airflow. The typical operating range of FC fans is from 30%

to 80% of the wide open volume of the fan. The maximum static efficiency of these fans is around 60% to 80%, and generally the peak static efficiency is slightly to the right of the peak static pressure on a fan performance curve. The horsepower curve has an increasing slope, and the curve is said to have an overloading characteristic. Overloading of the fan motor occurs if the system resistance is overestimated and the fan operates at a higher air-flow rate, thus overloading the motor.

Air leaves a forward curved fan at a greater velocity than the tip speed of the impeller. For this reason, the FC fan operates at a lower rotational speed than other types of fans. Since the fan impeller rotates at a relatively low speed, the fan can have a lightweight construction for low pressure applications. Advantages of the FC fan include low cost, low speed, and a wide operating range. Disadvantages include the shape of the fan performance curve, which may allow overloading of the fan motor if the system static pressure decreases.

Backward inclined (BI) centrifugal fans have impeller blades which are inclined opposite to the direction of the impeller rotation, which causes the air to leave the fan at a velocity less than the blade tip speed. A typical BI fan impeller is shown in Figure 16.7. BI fans typically rotate at relatively higher speeds, which result in higher tip speeds and higher efficiency. They have relatively low noise levels. These fans have non-overloading motor horsepower characteristics, i.e., the point of maximum horsepower occurs near the optimum operating point.

BI fans typically rotate faster than an equivalent FC fan. The normal selection range for a BI fan is approximately 40% to 85% of the wide open airflow. Maximum static efficiency is about 80% and generally occurs near the high airflow end of its normal operating range.

Advantages of the BI fan include higher efficiencies, quiet operation, and a non-overloading motor horsepower curve. Since BI fans rotate at relatively high speeds, their

FIGURE 16.7   Backward Inclined Fan Impeller (*Courtesy of Twin City Fan and Blower Co.*)

construction is much stronger than FC fans, which makes them more suitable for higher pressures. Disadvantages include the higher rotational speeds, larger physical sizes, and unstable operation at block tight (shut-off) static pressure.

Airfoil fans (AF) are a refinement of the backward inclined fan. AF fans are similar to BI fans with airfoil shaped blades. The airfoil shaped blades improve the static efficiency of the fan up to about 86% and slightly reduce the noise generated by the fan.

Centrifugal fans with radial impellers have blades which are straight or are mounted in a radial direction from the hub. Radial bladed fans are generally narrower than other types of centrifugal fans. Consequently, they require a larger diameter impeller for a given capacity. This increases the cost of the fan, which is a major reason why they are not commonly used in HVAC applications. Radial fans are frequently used for material handling systems. They are well suited for handling low volumes of air at relatively high static pressures.

Table 16.1 gives a synopsis of the advantages and disadvantages of the most widely used centrifugal fans.

Centrifugal fans are also available in an in-line configuration, as a tubular centrifugal fan, as shown in Figure 16.8. Tubular centrifugal fans generally consist of a single width airfoil impeller in a cylindrical housing. The impeller discharges the air against the inside of the cylinder to produce static pressure at the fan outlet. These fans typically have a static efficiency of about 72% and have a slightly higher noise level than other types of centrifugal fans.

16.1.1 Fan Classifications and Arrangements

Centrifugal fans are frequently categorized according to the width of the impeller wheel and the configuration of the air inlet. Single-width-single-inlet (SWSI) fans have a single inlet cone on one side of the fan. Air is drawn into the fan from one side only. Double-width-double-inlet (DWDI) fans have impellers that are approximately twice the width of single width fans, and have a dividing plate in the middle, as shown in Figure 16.9. DWDI fans have air inlets on both sides of the fan wheel.

Centrifugal fans are also classified by the location or arrangement of the fan drive. The fan drive includes the fan motor, pulleys (if belt driven), bearings, and fan impeller shaft support. Drive arrangements generally describe the bearing location, means of supporting the motor, and how the fan is driven. There are eight standard arrangements for fan width, fan inlet, and drive arrangement. The standard AMCA arrangements are shown in Figures 16.10 and 16.11.

Fans are also classified by external static pressure. The pressure classifications are based on the static pressure that is external to the fan casing. The classifications are:

Class	Static Pressure Range (in. wg)
A	0 - 3
B	3 - 5.5
C	Above 5.5

TABLE 16.1 Comparison of Centrifugal Fan Types

Feature	Forward Curved Blades	Radial Blades	Backward Inclined Blades
Efficiency	Medium	Medium	High
Space Required	Small	Medium	Medium
Speed for Press Rise	Low	Medium	High
Noise	Poor	Fair	Good

FIGURE 16.8   Tubular Centrifugal Fan (*Courtesy of Twin City Fan and Blower*)

FIGURE 16.9   Double Width Centrifugal Fan Impeller (*Courtesy of Twin City Fan Co.*)

AMCA also designates, or describes, centrifugal fans according to rotation and dis-charge location. The method of specifying rotation is to view the fan from the drive side, and to indicate whether the rotation is clockwise or counter-clockwise. There is no official designation of drive sides for axial fans.

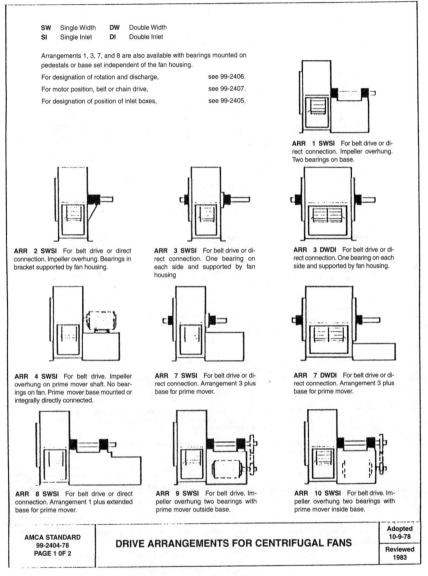

SW Single Width   DW Double Width
SI Single Inlet   DI Double Inlet

Arrangements 1, 3, 7, and 8 are also available with bearings mounted on pedestals or base set independent of the fan housing.

For designation of rotation and discharge,   see 99-2406.

For motor position, belt or chain drive,   see 99-2407.

For designation of position of inlet boxes,   see 99-2405.

ARR 1 SWSI For belt drive or direct connection. Impeller overhung. Two bearings on base.

ARR 2 SWSI For belt drive or direct connection. Impeller overhung. Bearings in bracket supported by fan housing.

ARR 3 SWSI For belt drive or direct connection. One bearing on each side and supported by fan housing

ARR 3 DWSI For belt drive or direct connection. One bearing on each side and supported by fan housing.

ARR 4 SWSI For belt drive. Impeller overhung on prime mover shaft. No bearings on fan. Prime mover base mounted or integrally directly connected.

ARR 7 SWSI For belt drive or direct connection. Arrangement 3 plus base for prime mover.

ARR 7 DWDI For belt drive or direct connection. Arrangement 3 plus base for prime mover.

ARR 8 SWSI For belt drive or direct connection. Arrangement 1 plus extended base for prime mover.

ARR 9 SWSI For belt drive. Impeller overhung two bearings with prime mover outside base.

ARR 10 SWSI For belt drive. Impeller overhung two bearings with prime mover inside base.

AMCA STANDARD 99-2404-78 PAGE 1 OF 2	DRIVE ARRANGEMENTS FOR CENTRIFUGAL FANS	Adopted 10-9-78
		Reviewed 1983

FIGURE 16.10 Drive Arrangements for Centrifugal Fans (*Reprinted with permission of AMCA International*)

Fans are designated as horizontal if the air leaves the fan in a horizontal direction, as up blast if the air is discharged vertically upward, and as down blast if the air is discharged vertically downward from the fan. In addition to horizontal and vertical discharge fans, there are fans that have angular discharges.

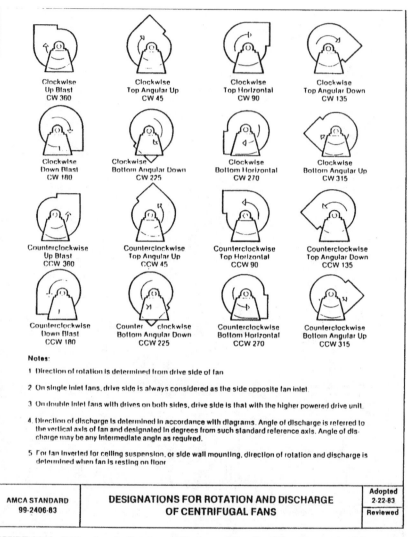

Clockwise Up Blast CW 360	Clockwise Top Angular Up CW 45	Clockwise Top Horizontal CW 90	Clockwise Top Angular Down CW 135
Clockwise Down Blast CW 180	Clockwise Bottom Angular Down CW 225	Clockwise Bottom Horizontal CW 270	Clockwise Bottom Angular Up CW 315
Counterclockwise Up Blast CCW 360	Counterclockwise Top Angular Up CCW 45	Counterclockwise Top Horizontal CCW 90	Counterclockwise Top Angular Down CCW 135
Counterclockwise Down Blast CCW 180	Counterclockwise Bottom Angular Down CCW 225	Counterclockwise Bottom Horizontal CCW 270	Counterclockwise Bottom Angular Up CCW 315

Notes:

1. Direction of rotation is determined from drive side of fan

2. On single inlet fans, drive side is always considered as the side opposite fan inlet.

3. On double inlet fans with drives on both sides, drive side is that with the higher powered drive unit.

4. Direction of discharge is determined in accordance with diagrams. Angle of discharge is referred to the vertical axis of fan and designated in degrees from such standard reference axis. Angle of discharge may be any intermediate angle as required.

5. For fan inverted for ceiling suspension, or side wall mounting, direction of rotation and discharge is determined when fan is resting on floor

AMCA STANDARD 99-2406-83	DESIGNATIONS FOR ROTATION AND DISCHARGE OF CENTRIFUGAL FANS	Adopted 2-22-83
		Reviewed

FIGURE 16.11   Designations for Rotation and Discharge of Centrifugal Fans (*Reprinted with permission of AMCA International*)

16.2 FAN TESTING AND RATING

Fans are tested for aerodynamic performance rating in accordance with ANSI/AMCA Standard 210 (Air Movement and Control Association International, Inc.). Several test configurations may be used with any of four configurations of inlet/outlet ducts, which simulate various types of actual installations. The test configurations most often used are those which employ a test chamber. While others produce equivalent test results, the test chamber is the most efficient with respect to time.

The test method itself calls for operation of the test unit at a constant speed, while the fan pressure and power consumption are measured at several airflow delivery points, ranging from shut-off (no flow) to free-delivery conditions. At shut-off, the inlet (or outlet) of the fan is completely shut off. At free delivery, the resistance at the fan outlet (or inlet) is minimized. Between these two operating points, the airflow through the fan is restricted so eight other delivery points can be determined.

In ANSI/AMCA Standard 210, the mounting of the fan to the chamber, or test duct, is specified. A propeller fan is usually mounted directly into the wall of the test chamber or directly to a section of the test duct. The velocity profile is usually uniform at the fan inlet, but not at the fan outlet. Centrifugal fans and axial fans also have distorted profiles. Test configurations involving ducts require a long section of duct in order to allow the velocity profile to stabilize. An airflow straightener is also used to minimize turbulence and swirl, especially in the case of an axial fan. Testing with a chamber allows the use of a much shorter duct. Figure 16.12 illustrates the velocity profiles of both a centrifugal fan and an axial fan.

It is important to remember that fan aerodynamic performance ratings are based on the assumption of good, smooth, unobstructed airflow path into and out of the fan. If this kind of path does not exist in actual field installation, the fan's performance will be reduced. See Section 16.5 for additional information on the effect of the system on fan performance. AMCA Publication 211 requires that a manufacturer's published certified rating specify

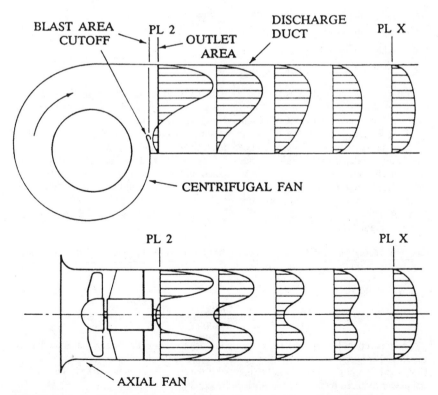

FIGURE 16.12   Fan Velocity Profile in a Straight Duct (*Reprinted with permission of AMCA International*)

which of the four Installation Types apply to the ratings that are published. With this information, along with the fan performance requirements, a fan may then be selected for the exact type of installation proposed. In most cases, it is possible to test a few sizes of fans and to interpolate to determine the ratings for sizes in between. In cases where dimensions of different sizes are not completely proportional, each size must be tested. AMCA 211 does not permit extrapolation of ratings for sizes smaller than the six actually tested.

16.3   FAN PERFORMANCE AND FAN CURVES

At any fixed volume flow rate (ft³/min) through a given system, a corresponding pressure loss, or resistance to the airflow, would exist. (See Chapter 15.) The system resistance is the sum of all pressure losses through filters, coils, dampers, and ductwork. If the flow rate is changed, the resulting pressure loss, or resistance to airflow, will also change. For a fixed system (a system with no changes in damper settings), the system resistance varies as the square of the airflow rate. The relationship governing this change for such a system is

$$\frac{P_{s2}}{P_{s1}} = \left(\frac{Q_2}{Q_1}\right)^2 \tag{16.1}$$

where

P_{s1} = the static pressure at the initial conditions

P_{s2} = the static pressure at the final conditions

Q_1 = the airflow at the initial conditions

Q_2 = the airflow at the final conditions

If the airflow rate is increased from 100% to 120% of the initial design, the system resistance will increase to 144% of the initial design. A decrease to 50% flow would result in a decrease in resistance to 25% of design.

Using Equation 16.1, the system static pressure can be determined for an airflow rate, if the static pressure for the system has been determined at a given flow. If the system resistance was calculated for a number of rates of flow, and plotted on a graph with the airflow on the horizontal axis and the system resistance on the vertical axis, a system resistance curve would result. A system curve is a graphical representation of the relationship between the system airflow and the static pressure loss for the entire system at various airflow rates. A system curve is a graphical representation of Equation 16.1. A typical system curve is shown in Figure 16.13.

Example 16.1   An air distribution system has a static pressure loss of 2 in. wg at an air flow rate of 5000 ft³/min. If the airflow rate was increased to 7000 ft³/min, what static pressure loss for the system would result?

Using Equation 16.1,

$$P_{s2} = P_{s1}\left(\frac{Q_2}{Q_1}\right) = (2 \text{ in. wg})\left(\frac{7000 \text{ ft}^3/\text{min}}{5000 \text{ ft}^3/\text{min}}\right)^2 = 3.92 \text{ in. wg}$$

The theoretical horsepower (hp) required to drive a fan is the power that would be required if there were no losses in the fan itself (such as friction losses, drive losses, etc.). That is, it is the power that would be required if the fan efficiency was 100%. The theoretical power required by a given fan is given by Equation 16.2.

$$W = \frac{QP_T}{6356} \tag{16.2}$$

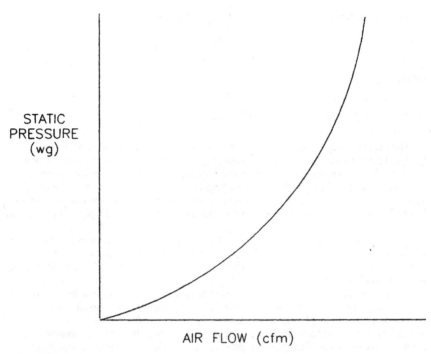

FIGURE 16.13 System Curve

where

W = the power to drive the fan in hp

Q = the airflow in ft³/min

6356 = a conversion factor to make the units consistent

Brake horsepower (Bhp) is the actual power a fan requires for a given set of operating conditions, i.e., airflow, static pressure, and fan efficiency. It is the power to input to the fan shaft. The brake horsepower required to drive a fan is always larger than the theoretical horsepower, due to energy losses inside the fan itself. The brake horsepower required for a fan can only be determined by an actual test of the fan.

After testing a fan to determine the required brake horsepower, both the total efficiency and the static efficiency of the fan can be calculated. The total efficiency of the fan is based on the total pressure produced by the fan, whereas the static efficiency is based on the static pressure. The total fan efficiency may be calculated using Equation 16.3.

$$\eta_{FT} = \left[\frac{(QP_T)}{(6356 \; \text{Bhp})} \right] \times 100\% \qquad (16.3)$$

where

η_{FT} = the total efficiency of the fan in %

Bhp = the brake horsepower required to drive the fan

P_T = the total pressure of the fan in in. wg

The static efficiency may be calculated using Equation 16.4.

$$\eta_{FS} = \left[\frac{(QP_s)}{(6356 \text{ Bhp})} \right] \times 100\% \qquad (16.4)$$

where

η_{FS} = the static efficiency of the fan in %

P_s = the static pressure of the fan in in. wg

The fan efficiency is the flow work output of the fan divided by the work input.

A fan performance curve (fan curve) is a graphical representation of the relationship between the fan airflow and the static pressure produced by the fan. Just as the system curve, a fan has the airflow rate plotted along the horizontal axis and the static pressure along the vertical axis. Fan curves are obtained from a series of laboratory tests on a fan with various levels of restriction at the end of the test duct. Lines connecting the test points form the fan performance curve.

Fan performance curves typically show the static pressure vs. airflow. Some fan curves also include total pressure, static efficiency, and total efficiency. A typical fan curve for a centrifugal forward curved fan is shown in Figure 16.14, and a typical fan curve for a backward inclined and airfoil fan is shown in Figure 16.15.

Usually, fan curves show the entire range of fan performance, from free delivery (no obstruction or resistance to airflow) to no delivery (airtight system with no airflow).

The performance of a fan at various flow and pressure conditions may also be presented in a tabular form called a multi-rating table. In addition to publishing performance curves for their fans, most manufacturers also publish multi-rating tables for their fans. A separate fan curve and multi-rating table is usually provided for each type of fan and each diameter

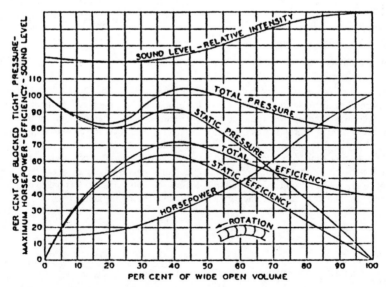

FIGURE 16.14   Typical Fan Curve for a Forward Curved Centrifugal Fan

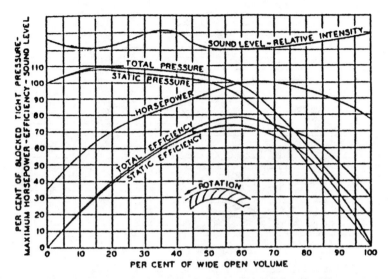

FIGURE 16.15 Typical Fan Curve for a Backward Inclined and Air Foil Centrifugal Fan

impeller manufactured. Multi-rating tables show the fan rotational speed (rpm), airflow rate, static pressure, and horsepower for each size impeller. Interpolation between values is frequently necessary to obtain specific values, such as speed. A typical fan curve and multi-rating table for a BI fan is shown in Figure 16.16.

Most multi-rating tables do not cover the entire range of performance from no delivery to free delivery. They usually cover only the typical operating range.

The operating point at which the fan and system will perform is determined by the intersection of the system curve and the fan curve. A typical fan and system curve is shown in Figure 16.17. Note that a fan will operate along its performance curve. If the actual air system resistance is different from the design resistance, the operating point will be different than the design point, and the volume of air delivered will be different than the design airflow. Figure 16.18 shows the variation from system design that would occur if the actual static pressure for a system was greater than the calculated or design static pressure.

Notice that an increase in system static pressure would cause the system curve to shift to the left and a reduction in airflow would result. The operating point for the fan and air system will shift up or down, along the fan curve, depending on the static pressure for the system.

Example 16.2 An air distribution system is designed to supply 30,000 ft³/min at a static pressure of 4.0 in. wg. Using the fan curves in Figure 16.19, what is the required brake horsepower for the fan, and at what speed (rpm) would the fan operate? What would the static pressure, brake horsepower, and speed be at 20,000 ft³/min?

Plotting 30,000 ft³/min and 4 in. wg on the fan curves of Figure 16.19.

motor Bhp = 30 and fan speed = 950 rpm

If the airflow was reduced to 20,000 ft³/min, the static pressure would be determined using Equation 16.1.

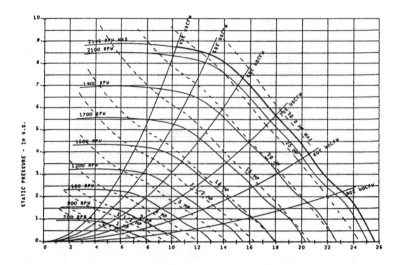

FIGURE 16.16   Typical Fan Curve and Multi-Rating Table (*Courtesy of The Trane Company*)

SIZE 30 22.25 BI HDT

CFM Std Air	Out-Let Vel	\multicolumn{26}{c}{Total Static Pressure}

CFM Std	Out-Let Vel	0.25		0.50		0.75		1.00		1.25		1.50		1.75		2.00		2.25		2.50		2.75		3.00		3.25	
		RPM	BHP	RPM	BHP	RPM	BHP	RPM	BHP	RPM	BHP	RPM	BHP	RPM	BHP	RPM	BHP	RPM	BHP	RPM	BHP	RPM	BHP	RPM	BHP	RPM	BHP
10000	1587	883	2.08	953	2.48	1012	2.92	1058	3.38	1104	3.86	1150	4.36	1194	4.86	1234	5.35	1272	5.84	1308	6.32	1344	6.81	1379	7.31	1415	7.83
11000	1746	962	2.68	1019	3.11	1082	3.58	1129	4.07	1170	4.58	1212	5.10	1254	5.65	1295	6.20	1333	6.75	1369	7.29	1403	7.82	1436	8.36	1469	8.90
12000	1905	1042	3.38	1088	3.86	1151	4.35	1201	4.88	1241	5.42	1278	5.97	1317	6.55	1356	7.15	1394	7.75	1429	8.34	1464	8.93	1496	9.52	1528	10.10
13000	2063	1122	4.21	1162	4.73	1216	5.24	1271	5.80	1313	6.39	1349	6.97	1384	7.57	1419	8.19	1455	8.83	1491	9.49	1525	10.14	1557	10.77	1588	11.41
14000	2222	1203	5.17	1239	5.73	1284	6.28	1339	6.86	1384	7.48	1421	8.10	1454	8.73	1486	9.38	1518	10.04	1552	10.72	1585	11.42	1618	12.13	1650	12.83
15000	2381	1285	6.29	1317	6.87	1355	7.47	1404	8.06	1453	8.70	1492	9.37	1526	10.04	1557	10.72	1587	11.40	1617	12.10	1648	12.83	1679	13.57	1710	14.33
16000	2540	1367	7.54	1396	8.16	1428	8.79	1470	9.42	1519	10.08	1562	10.78	1597	11.50	1629	12.22	1658	12.93	1686	13.66	1714	14.40	1742	15.17	1771	15.95
17000	2698	1448	8.96	1476	9.62	1505	10.29	1541	10.96	1585	11.64	1629	12.34	1668	13.11	1700	13.86	1730	14.63	1757	15.38	1783	16.15	1810	16.94	1836	17.74
18000	2857	1530	10.55	1556	11.24	1583	11.95	1614	12.68	1652	13.39	1695	14.10	1735	14.87	1771	15.68	1801	16.48	1829	17.29	1856	18.11	1880	18.90	1905	19.72
19000	3016	1612	12.33	1637	13.05	1662	13.79	1690	14.56	1722	15.30	1761	16.06	1802	16.82	1839	17.65	1872	18.51	1900	19.35	1927	20.21	1952	21.07	1976	21.93
20000	3175	1695	14.30	1718	15.05	1742	15.83	1766	16.62	1795	17.43	1829	18.22	1867	19.00	1905	19.83	1940	20.71	1971	21.61	1998	22.49	2024	23.40	2048	24.31
21000	3333	1777	16.47	1799	17.26	1822	18.07	1845	18.89	1871	19.75	1899	20.57	1934	21.41	1972	22.26	2008	23.14	2039	24.04	2069	24.99	2095	25.95	2119	26.87

$$P_{s1} = P_{s2}\left(\frac{Q_2}{Q_1}\right)^2 = 4.0\left(\frac{20,000}{30,000}\right)^2 = 1.78 \text{ in. wg}$$

$$\text{motor Bhp} = 10 \text{ and fan speed} = 650 \text{ rpm (approx)}$$

Example 16.3   If the same fan selected in Example 16.2 (30 Bhp and 950 rpm) was operating against a static pressure of 4.5 in. wg instead of 4 in. wg, what (if any) effect would it have on the airflow rate? Sketch the new system/fan curve.

Plotting 4.5 in. wg on the 950 rpm fan curve, as shown in Figure 16.20, the airflow rate is read from the bottom of the curve as $Q = 26,000$ ft³/min.

Using Equation 16.1 to calculate a third point (note that the system curve will always pass through the point of zero airflow and zero static pressure) and any other airflow rate, say 10,000 ft³/min,

$$P_{s2} = P_{s1}\left(\frac{Q_2}{Q_1}\right)^2 = 4.5\left(\frac{10,000}{26,000}\right)^2 = 0.67 \text{ in. wg}$$

Plotting $Q = 10,000$ ft³/min and $P_{s2} = 0.67$ in. wg on the curve produces the new fan curve, as shown in Figure 16.20. Notice that the system curve has shifted upward and to the left, due to the increased static pressure of the system.

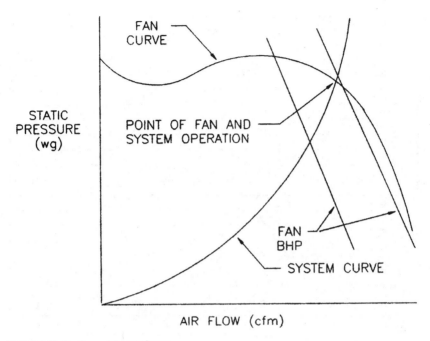

FIGURE 16.17 Fan and System Curves

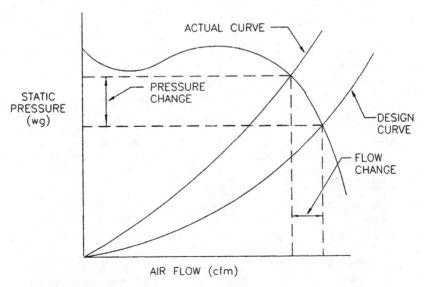

FIGURE 16.18 Variations from Design

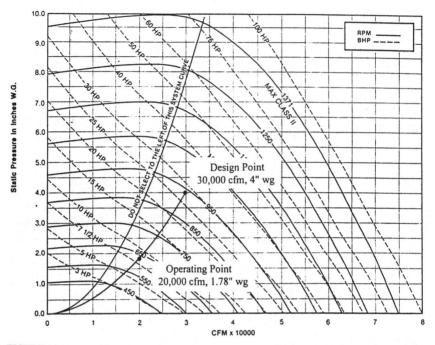

FIGURE 16.19   System and Fan Curves for Example 16.2

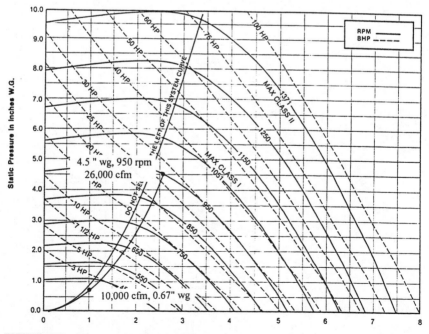

FIGURE 16.20   System and Fan Curve for Example 16.3

TABLE 16.2 Typical Fan Characteristics and Applications

TYPE		IMPELLER DESIGN	HOUSING DESIGN
CENTRIFUGAL FANS	AIRFOIL	• Highest efficiency of all centrifugal fan designs. • Ten to 16 blades of airfoil contour curved away from direction of rotation. Deep blades allow for efficient expansion within blade passages. • Air leaves impeller at velocity less than tip speed. • For given duty, has highest speed of centrifugal fan designs.	• Scroll-type design for efficient conversion of velocity pressure to static pressure. • Maximum efficiency requires close clearance and alignment between wheel and inlet.
	BACKWARD-INCLINED BACKWARD-CURVED	• Efficiency only slightly less than airfoil fan. • Ten to 16 single-thickness blades curved or inclined away from direction of rotation. • Efficient for same reasons as airfoil fan.	• Uses same housing configuration as airfoil design.
	RADIAL	• Higher pressure characteristics than airfoil, backward-curved, and backward-inclined fans. • Curve may have a break to left of peak pressure and fan should not be operated in this area. • Power rises continually to free delivery.	• Scroll. Usually narrowest of all centrifugal designs. • Because wheel design is less efficient, housing dimensions are not as critical as for airfoil and backward-inclined fans.
	FORWARD-CURVED	• Flatter pressure curve and lower efficiency than the airfoil, backward-curved, and backward-inclined. • Do not rate fan in the pressure curve dip to the left of peak pressure. • Power rises continually toward free delivery. Motor selection must take this into account.	• Scroll similar to and often identical to other centrifugal fan designs. • Fit between wheel and inlet not as critical as for airfoil and backward-inclined fans.
AXIAL FANS	PROPELLER	• Low efficiency. • Limited to low-pressure applications. • Usually low cost impellers have two or more blades of single thickness attached to relatively small hub. • Primary energy transfer by velocity pressure.	• Simple circular ring, orifice plate, or venturi. • Optimum design is close to blade tips and forms smooth airfoil into wheel.
	TUBEAXIAL	• Somewhat more efficient and capable of developing more useful static pressure than propeller fan. • Usually has 4 to 8 blades with airfoil or single-thickness cross section. • Hub usually less than transfer by velocity pressure.	• Cylindrical tube with close clearance to blade tips.
	VANEAXIAL	• Good blade design gives medium- to high-pressure capability at good efficiency. • Most efficient of these fans have airfoil blades. • Blades may have fixed, adjustable, or controllable pitch. • Hub is usually greater than half fan tip diameter.	• Cylindrical tube with close clearance to blade tips. • Guide vanes upstream or downstream from impeller increase pressure capability and efficiency.
SPECIAL DESIGNS	TUBULAR CENTRIFUGAL	• Performance similar to backward-curved fan except capacity and pressure are lower. • Lower efficiency than backward-curved fan. • Performance curve may have a dip to the left of peak pressure.	• Cylindrical tube similar to vaneaxial fan, except clearance to wheel is not as close. • Air discharges radially from wheel and turns 90° to flow through guide vanes.
	POWER ROOF VENTILATORS — CENTRIFUGAL	• Low-pressure exhaust systems such as general factory, kitchen, warehouse, and some commercial installations. • Provides positive exhaust ventilation, which is an advantage over gravity-type exhaust units. • Centrifugal units are slightly quieter than axial units.	• Normal housing not used, since air discharges from impeller in full circle. • Usually does not include configuration to recover velocity pressure component.
	POWER ROOF VENTILATORS — AXIAL	• Low-pressure exhaust systems such as general factory, kitchen, warehouse, and some commercial installations. • Provides positive exhaust ventilation, which is an advantage over gravity-type exhaust units.	• Essentially a propeller fan mounted in a supporting structure. • Hood protects fan from weather and acts as safety guard. • Air discharges from annular space at bottom of weather hood.

Table 16.2 gives a synopsis of the characteristics of each type of fan, along with recommended applications for each type of fan.

TABLE 16.2 Typical Fan Characteristics and Applications (Concluded)

PERFORMANCE CURVES[a]	PERFORMANCE CHARACTERISTICS	APPLICATIONS
	• Highest efficiencies occur at 50 to 60% of wide open volume. This volume also has good pressure characteristics. • Power reaches maximum near peak efficiency and becomes lower, or self-limiting, toward free delivery.	• General heating, ventilating, and air-conditioning applications. • Usually only applied to large systems, which may be low-, medium-, or high-pressure applications. • Applied to large, clean-air industrial operations for significant energy savings.
	• Similar to airfoil fan, except peak efficiency slightly lower.	• Same heating, ventilating, and air-conditioning applications as airfoil fan. • Used in some industrial applications where airfoil blade may corrode or erode due to environment.
	• Higher pressure characteristics than airfoil and backward-curved fans. • Pressure may drop suddenly at left of peak pressure, but this usually causes no problems. • Power rises continually to free delivery.	• Primarily for materials handling in industrial plants. Also for some high-pressure industrial requirements. • Rugged wheel is simple to repair in the field. Wheel sometimes coated with special material. • Not common for HVAC applications.
	• Pressure curve less steep than that of backward-curved fans. Curve dips to left of peak pressure. • Highest efficiency to right of peak pressure at 40 to 50% of wide open volume. • Rate fan to right of peak pressure. • Account for power curve, which rises continually toward free delivery, when selecting motor.	• Primarily for low-pressure HVAC applications, such as residential furnaces, central station units, and packaged air conditioners.
	• High flow rate, but very low-pressure capabilities. • Maximum efficiency reached near free delivery. • Discharge pattern circular and airstream swirls.	• For low-pressure, high-volume air moving applications, such as air circulation in a space or ventilation through a wall without ductwork. • Used for makeup air applications.
	• High flow rate, medium-pressure capabilities. • Performance curve dips to left of peak pressure. Avoid operating fan in this region. • Discharge pattern circular and airstream rotates or swirls.	• Low- and medium-pressure ducted HVAC applications where air distribution downstream is not critical. • Used in some industrial applications, such as drying ovens, paint spray booths, and fume exhausts.
	• High-pressure characteristics with medium-volume flow capabilities. • Performance curve dips to left of peak pressure due to aerodynamic stall. Avoid operating fan in this region. • Guide vanes correct circular motion imparted by wheel and improve pressure characteristics and efficiency of fan.	• General HVAC systems in low-, medium-, and high-pressure applications where straight-through flow and compact installation are required. • Has good downstream air distribution. • Used in industrial applications in place of tubeaxial fans. • More compact than centrifugal fans for same duty.
	• Performance similar to backward-curved fan, except capacity and pressure is lower. • Lower efficiency than backward-curved fan because air turns 90°. • Performance curve of some designs is similar to axial flow fan and dips to left of peak pressure.	• Primarily for low-pressure, return air systems in HVAC applications. • Has straight-through flow.
	• Usually operated without ductwork; therefore, operates at very low pressure and high volume. • Only static pressure and static efficiency are shown for this fan.	• Low-pressure exhaust systems, such as general factory, kitchen, warehouse, and some commercial installations. • Low first cost and low operating cost give an advantage over gravity flow exhaust systems. • Centrifugal units are somewhat quieter than axial flow units
	• Usually operated without ductwork; therefore, operates at very low pressure and high volume. • Only static pressure and static efficiency are shown for this fan.	• Low-pressure exhaust systems, such as general factory, kitchen, warehouse, and some commercial installations. • Low first cost and low operating cost give an advantage over gravity flow exhaust systems.

[a] These performance curves reflect general characteristics of various fans as commonly applied. They are not intended to provide complete selection criteria, since other parameters such as diameter and speed, are not defined.

16.4 FAN LAWS

Fan laws relate the performance variables for any dynamically similar series of fans. These laws apply to all fans. The performance variables are the fan size (impeller diameter), D; the speed of rotation, N; the gas (air) density, ρ; the volume flow rate, Q; the pressure, P_T or P_s; the power, W; and the fan efficiency, η. The fan laws are mathematical expressions that relate the above performance variables. Unless otherwise indicated, fan performance is based on dry air at standard conditions, i.e., 14.7 psia, 70°F and 0.075 lb/ft³.

Special care must be exercised when applying the fan laws in the following situations:

1. where any component of the system does not follow Equation 16.1.

2. where the system has been physically altered, or for any reason operates on a different system curve.

For changes in the rotation speed, N, the flow rate varies with the speed of rotation, the pressure varies as the square of the speed of rotation, and the power varies as the cube of the rotation speed. It is assumed that the fan size is constant, the air density is constant, and the system itself is constant.

$$Q_2 = Q_1\left(\frac{N_2}{N_1}\right) \tag{16.5}$$

$$P_2 = P_1\left(\frac{N_2}{N_1}\right)^2 \tag{16.6}$$

$$W_2 = W_1\left(\frac{N_2}{N_1}\right)^3 \tag{16.7}$$

For changes in gas (air) density, ρ, the flow is not affected by a change in gas density. The pressure and power vary directly with the density. It is assumed that the fan size, airflow rate, fan speed, and system are constant.

$$Q_2 = Q_1 \tag{16.8}$$

$$P_2 = P_1(\rho_2/\rho_1) \tag{16.9}$$

$$W_2 = W_1(\rho_2/\rho_1) \tag{16.10}$$

For changes in the fan size, D, the airflow and power vary as the square of the impeller diameter, the rotation speed varies inversely as the impeller diameter, and the static pressure remains constant. This set of laws are used in the design of fan, but are rarely used in the design of HVAC systems. It is assumed that the rotation speed, the gas (air) density, and the system are constant.

$$Q_2 = Q_1\left(\frac{D_2}{D_1}\right)^2 \tag{16.11}$$

$$W_2 = W_1\left(\frac{D_2}{D_1}\right)^2 \tag{16.12}$$

$$P_{s1} = P_{s2} \tag{16.13}$$

$$N_2 = N_1\left(\frac{D_1}{D_2}\right) \tag{16.14}$$

Since the tolerances of fans are not proportional, slightly better performance is generally obtained when projecting from a given fan to a dynamically similar larger fan. The most frequently made error in applying the fan laws is trying to predict the fan performance after changing some physical aspect of the system to which the fan is attached.

Example 16.4   A fan delivers 20,000 ft³/min of air at 1 in. wg static pressure at a rotation speed of 414 rpm, and at a brake horsepower of 4.72. If the heating or cooling load were changed so the system was circulating 22,000 ft³/min without any changes in the distribution system, what will be the resulting change in the fan rotation speed, brake horsepower, and static pressure?

Rearranging Equation 16.5 to calculate rotation speed:

$$N_2 = N_1 \left(\frac{Q_2}{Q_1}\right) = 414 \text{ rpm} \left(\frac{22{,}000 \text{ ft}^3/\text{min}}{20{,}000 \text{ ft}^3/\text{min}}\right) = 455 \text{ rpm}$$

By Equation 16.5, $Q_2/Q_1 = N_2/N_1$, substituting into Equation 16.6,

$$P_2 = P_1 \left(\frac{N_2}{N_1}\right)^2 = (1.0 \text{ in. wg}) \left(\frac{22{,}000}{20{,}000}\right)^2 = 1.21 \text{ in. wg}$$

Substituting $Q_2/Q_1 = N_2/N_1$ into Equation 16.7,

$$W_2 = W_1 \left(\frac{N_2}{N_1}\right)^3 = W_1 \left(\frac{Q_2}{Q_1}\right)^3 = (4.72 \text{ Bhp}) \left(\frac{22{,}000}{20{,}000}\right)^3 = 6.28 \text{ Bhp}$$

Example 16.5   If the temperature of the air circulated by the fan in Example 16.4 was changed from 70°F to 150°F, and the rotational speed of the fan held constant, what would the static pressure and brake horsepower of the fan be at the higher temperature? The density of air at 70°F is 0.075 lb/ft³ and at 150°F, it is 0.065 lb/ft³.

Since the fan rotational speed is constant, the pressure and power vary directly with the change in density. Using Equations 16.9 and 16.10,

$$P_1 = P_2 \left(\frac{\rho_2}{\rho_1}\right) = (1 \text{ in. wg}) \left(\frac{0.065}{0.075}\right) = 0.87 \text{ in. wg}$$

$$W_2 = W_1 \left(\frac{\rho_2}{\rho_1}\right) = (4.72 \text{ Bhp}) \left(\frac{0.065}{0.075}\right) = 4.11 \text{ Bhp}$$

16.5   SYSTEM EFFECT

In Section 16.2, the testing and rating of fans was discussed. Fans are tested and rated with almost ideal inlet and outlet conditions. In general, the duct connections to the fan should be made so the air may enter and leave the fan as uniformly as possible, with no abrupt changes in direction or velocity. Frequently, it is not possible to have an installation that duplicates these ideal conditions in actual applications. Space is often limited for fan installations, and less than optimum connections may have to be used. If good inlet and outlet conditions are not provided in actual applications, the performance of the fan will suffer. In this case, the HVAC system designer must be aware of the penalties (loss in total pressure and efficiency). Note that for a given system capacity, total pressure and efficiency are reduced by a poor installation and the power requirement is increased.

In order to account for less than ideal fan inlet and outlet conditions, AMCA has developed **System Effect Factors (SEF)**, which predict the effect of less than ideal inlet and outlet conditions. The system effect is defined as the estimated loss in fan performance from non-uniform airflow caused by restrictions at the fan inlet and/or outlet. The SEF is a pressure loss which recognizes the effect of fan inlet restrictions, fan outlet restrictions, or other conditions influencing fan performance when installed in a system. The SEF is given in in. wg and must be added to the total system pressure losses. The SEF, for any given configuration, is velocity dependent and will, therefore, vary across the range of fan flow rates.

System Effect Curves are curves that relate the pressure loss for various fan restrictions to the air velocity at either the fan inlet or fan outlet. Figure 16.21 shows a set of System Effect Curves. The air velocity through a restriction is read at the bottom of the curves, the resultant pressure loss is read from the left side, and letter designations for various restrictions are indicated on the right side of the curves. The various restrictions and letter designations are discussed later in this chapter.

The velocity used when entering the System Effect Curves is either the inlet or outlet velocity of the fan, depending upon the location of the restriction in question. If more than one restriction is included in a system, the SEF for each must be determined separately and then the total of these SEFs must be added to the total system pressure losses.

Assuming that a fan is rated and manufactured correctly, the three most common causes of deficient performance of the fan/system combination are:

- improper outlet connections
- non-uniform airflow at the fan inlet
- swirl or spin in the airstream entering the fan inlet

Other major causes of deficient performance are:

- the air performance characteristics of the installed system are significantly different from the original design intent.
- the system design calculations did not include adequate allowances for the effect of accessories and appurtenances.

One bad connection can reduce fan performance far below the published performance. Only when all of the applicable SEFs have been added to the calculated system pressure losses may the performance data in the fan catalog be used without any further adjustments.

16.5.1 System Effect at the Fan Inlet

Non-uniform flow into the fan inlet is the most common cause of reduced fan performance. Such reductions in performance, due to poor inlet conditions, are not equivalent to a simple increase in system resistance; therefore, they cannot be treated as a percentage decrease in flow and pressure from the fan. A poor inlet condition results in an entirely new fan performance. Pressure fluctuations can be up to ten times the magnitude of fluctuations of the same fan with good inlet and outlet conditions. There are three **inlet** conditions which typically can cause a reduction in fan performance.

- spin in the air stream entering the fan
- installation of the fan inlet(s) too close to a wall or bulkhead
- non-uniform air distribution due to a turn in the inlet ductwork

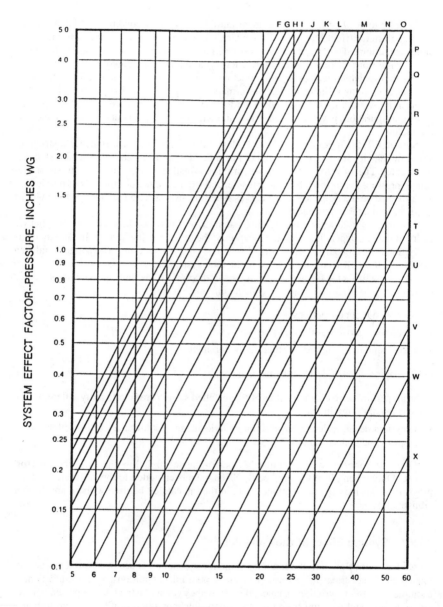

FIGURE 16.21   System Effect Curves (*Reprinted with permission of AMCA International*)

Inlet spin, or vortex, is a frequent cause of reduction in fan performance. Figure 16.22 shows the two spin conditions that can occur as air enters a centrifugal fan.

The ideal condition is one which allows the air to enter the fan axially and uniformly without spin in either direction. A spin in the same direction as the impeller rotation (pre-rotating swirl) reduces the pressure-volume curve by an amount dependent upon the intensity of the vortex. The effect is similar to the change in the pressure-volume curve achieved by the inlet vanes.

If the air spins in a direction counter to the impeller rotation (counter-rotating swirl), the volume and static pressure will be greater than the uniform flow at the inlet, and the brake horsepower input will increase substantially.

In either case, spin always reduces efficiency. Care must be taken in introducing air to a fan inlet to reduce the generation of spin. Figure 16.23 shows how turning vanes in a duct

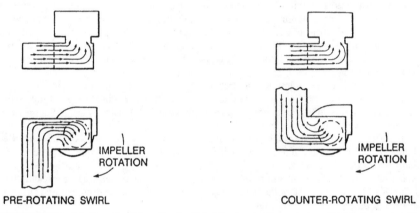

FIGURE 16.22 Inlet Duct Conditions Causing Spin (*Reprinted with permission of AMCA International*)

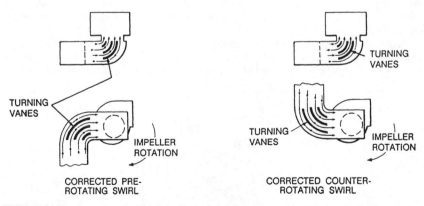

FIGURE 16.23 Corrections for Inlet Spin (*Reprinted with permission of AMCA International*)

connected to the fan inlet can be used to reduce inlet vortices. When turning vanes and/or a suitable length of duct is used between the fan inlet and the elbow (3 to 8 diameters long), the System Effect Factor is not as great.

A reduction in fan performance can also be expected when an obstruction to the airflow is located in the plane of the fan inlet. Inlet obstructions such as walls, bulkheads, structural members, fan supports, and pipes are common examples of inlet obstructions. Fans that do not have smooth entries and are installed without suitable inlet ducts exhibit flow characteristics similar to a sharp edge orifice in that vena contractra is formed at the fan inlet as shown in Figure 16.24.

Large sheaves and belt guards on belt-driven fans can also have the effect of restricting airflow. Obstructions at the fan inlet are classified in terms of unobstructed percentage of inlet area. When an inlet collar is provided, the inlet area is calculated from the inside diameter of this collar. Where no inlet collar is provided, the inlet plane is defined by the points where the radius of the inlet cone becomes tangent with the fan housing. It is preferable to have no obstructions within one impeller diameter of the fan inlet; ³/₄ impeller diameter should be considered as a minimum. Fans within plenums and cabinets or fans next to walls should be located so air may flow unobstructed into the inlet. Fan performance is reduced if the space between the fan inlet and the enclosure is too restrictive. The inlets of multiple double-width centrifugal fans, located in common enclosures, should be at least one impeller diameter apart.

Uneven air distribution into the fan impeller is also caused by the installation of an elbow in the inlet duct, too close to the fan inlet. When using elbows in close proximity to the fan inlet, instability in the fan operation may occur and result in an increase in fan pressure fluctuations and sound power level.

It is strongly recommended that inlet elbows be installed a minimum of three impeller diameters away from the inlet to a centrifugal or axial fan. There should be at least one equivalent duct diameter of straight duct between any inlet elbow and the fan inlet. ANSI/AMCA Standard 210 limits inlet ducts to a cross-sectional area no greater than 112.5%, or less than 92.5% of the fan inlet area. The included angle of transition is limited to 15° converging and 7° diverging.

Where space limitations prevent the use of optimum fan inlet connections, more uniform flow can be achieved by the use of vanes in the inlet elbow. The pressure losses through the elbows with the turning vanes are part of the system pressure losses. If the airflow approaching the elbow is significantly non-uniform due to a disturbance further upstream in the system, the pressure loss through the elbow will be higher than the published or calculated value.

Figures 16.25 through 16.30 show various inlet duct configurations for centrifugal fans, and list System Effect Factor curves for each configuration. The D referred to in Figures 16.25 through 16.30 is the diameter of the fan inlet collar. R is the centerline radius of the

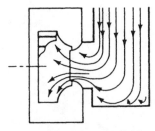

FIGURE 16.24   Non-Uniform Airflow at Fan Inlet (*Reprinted with permission of AMCA International*)

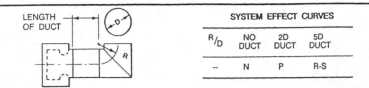

R/D	SYSTEM EFFECT CURVES		
	NO DUCT	2D DUCT	5D DUCT
--	N	P	R-S

FIGURE 16.25 Two Piece Mitered 90° Round Section Elbow—Not Vaned (*Reprinted with permission of AMCA International*)

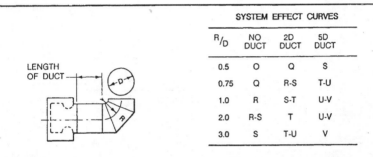

R/D	SYSTEM EFFECT CURVES		
	NO DUCT	2D DUCT	5D DUCT
0.5	O	Q	S
0.75	Q	R-S	T-U
1.0	R	S-T	U-V
2.0	R-S	T	U-V
3.0	S	T-U	V

FIGURE 16.26 Three Piece Mitered 90° Round Section Elbow—Not Vaned (*Reprinted with permission of AMCA International*)

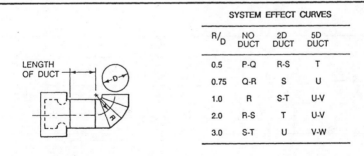

R/D	SYSTEM EFFECT CURVES		
	NO DUCT	2D DUCT	5D DUCT
0.5	P-Q	R-S	T
0.75	Q-R	S	U
1.0	R	S-T	U-V
2.0	R-S	T	U-V
3.0	S-T	U	V-W

FIGURE 16.27 Four or More Piece Mitered 90° Round Section Elbow—Not Vaned (*Reprinted with permission of AMCA International*)

inlet duct elbow. All dimensional units should be consistent, i.e., all in./ft. Figure 16.31 shows inlet configuration for axial fans, along with the corresponding SEF curves. The H/T ratio used in Figure 16.31 is the ratio of fan impeller hub to the fan blade tip diameter and D is the fan inlet diameter.

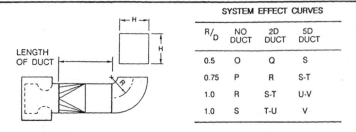

R/D	SYSTEM EFFECT CURVES		
	NO DUCT	2D DUCT	5D DUCT
0.5	O	Q	S
0.75	P	R	S-T
1.0	R	S-T	U-V
1.0	S	T-U	V

FIGURE 16.28   Square Elbow with Inlet Transition—No Turning Vanes (*Reprinted with permission of AMCA International*)

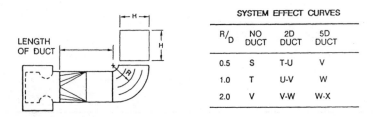

R/D	SYSTEM EFFECT CURVES		
	NO DUCT	2D DUCT	5D DUCT
0.5	S	T-U	V
1.0	T	U-V	W
2.0	V	V-W	W-X

FIGURE 16.29   Square Elbow with Transition—3 Long Turning Vanes (*Reprinted with permission of AMCA International*)

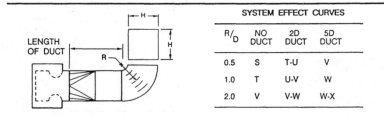

R/D	SYSTEM EFFECT CURVES		
	NO DUCT	2D DUCT	5D DUCT
0.5	S	T-U	V
1.0	T	U-V	W
2.0	V	V-W	W-X

FIGURE 16.30   Square Elbow with Inlet Transition—Short Turning Vanes (*Reprinted with permission of AMCA International*)

Example 16.6   A centrifugal fan is circulating 1500 ft³/min of air and has a 12 in. dia inlet duct. What would the static pressure loss, due to the System Effect, be if the inlet duct had an elbow with a center line radius of 12 in. connected directly to the fan, with no

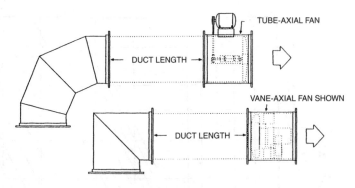

	H/T	90° Elbow	No Duct [1][2]	0.5 D [1][2]	1.0 D [1][2]	3.0 D
Tube-axial Fan	.25	2 piece	U	V	W	---
Tube-axial Fan	.25	4 piece	X	---	---	---
Tube-axial Fan	.35	2 piece	V	W	X	---
Vane-axial Fan	.61	2 piece	Q-R	Q-R	S-T	T-U
Vane-axial Fan	.61	4 piece	W	W-X	---	---

DETERMINE SEF BY USING FIGURE 7-1

[1] Instability in fan operation may occur as evidenced by an increase in pressure fluctuations and sound level. Fan instability, for any reason, may result in serious structural damage to the fan.

[2] The data presented in Figure XX-X is representative of commercial type tube-axial and vane-axial fans, i.e. 60% to 70% fan static efficiency.

FIGURE 16.31   System Effect Curves for Inlet Duct Elbows—Axial Fans (*Reprinted with permission of AMCA International*)

straight section of duct and no turning vanes? What would the loss be if there was a 24 in. long section of straight duct between the elbow and the fan inlet?

Plotting 1500 ft³/min and a 12 in. diameter duct on a duct friction chart (Figure 15.3), the air velocity in the duct is determined to be about 1900 ft/min.

For a 90° elbow of round duct with no turning vanes, Figure 16.28 should be used.

$$\text{the radius, } R = 12 \text{ in.}$$

$$\text{the diameter, } D = 12 \text{ in.}$$

$$\frac{R}{D} = \frac{12 \text{ in.}}{12 \text{ in.}} = 1.0$$

From Figure 16.28, curve R is selected for $R/D = 1.0$ and no inlet duct. An air velocity of 1900 ft/min is plotted on curve R of the System Effect Curves, as shown in Figure 16.32. A pressure loss for the System Effect Factor is read from the left side of the curves,

$$P_s = 0.28 \text{ in. wg}$$

This value is a pressure loss, in addition to the pressure losses for the system, as calculated by the methods in Chapter 15.

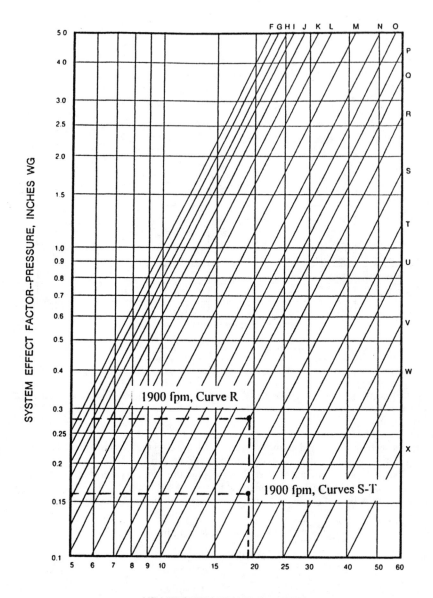

FIGURE 16.32   Systems Effect Curves for Example 16.6 (*Reprinted with permission of AMCA International*)

If the inlet duct has a section of straight duct 2 ft long between the elbow and the fan inlet, Figure 16.28 would again be used. In this case, though,

$$\text{the length } L = 2 \text{ ft} = 24 \text{ in.}$$
$$\text{the duct diameter } D = 12 \text{ in.}$$
$$\frac{L}{D} = \frac{24}{12} = 2.0$$
$$\text{or } L = 2D$$

For $R/D = 1.0$ and $L = 2D$, Figure 16.28 indicates S-T as the System Effect Curve. This indicates that the inlet air velocity should be plotted between curves S and T. Locating 1900 ft/min, between curves S and T on Figure 16.32, the System Effect pressure loss is

$$P_s = 0.16 \text{ in. wg}$$

the addition of the straight duct, on the fan inlet, reduced the System Effect pressure loss by 0.12 in. wg or about 43%.

16.5.2 System Effect at Fan Outlets

Just as in the case of fan inlets, the outlet or discharge configuration can have a significant effect on fan performance. Earlier in this chapter, the testing of fans was discussed along with ideal conditions for a fan discharge. As in the case of test set-ups for fans, it is recommended that a straight section of duct or a gradual transition duct be connected to the fan discharge to convert some of the velocity pressure of the air to static pressure. This process is often referred to as static regain. (See Chapter 15.) The efficiency of the conversion is a function of the length of straight duct or angle of expansion of a transition, the distance to the first change in direction (elbow) or branch, and the direction of the turn with respect to the fan. At a fan discharge, the velocity profile in the duct varies from zero just below the fan cutoff, to very high at the top of the fan housing, as shown in Figure 16.33.

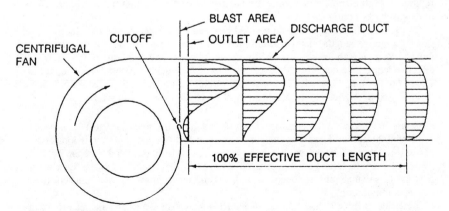

FIGURE 16.33 Fan Discharge Velocity Profile (*Reprinted with permission of AMCA International*)

As the air fills the duct, the velocity profile gradually evens out until it becomes uniform, three to five equivalent duct diameters, downstream of the fan outlet. If the duct is straight and uniform in size, some of the dynamic energy is converted to static pressure.

As a fan impeller rotates, it imparts kinetic energy in the form of velocity pressure to the air. Centrifugal fans produce pressure from the centrifugal force, which results from the rotating impeller blades, while the kinetic energy results from the velocity of the air leaving the impeller. Axial flow fans only produce a change in velocity as the air passes through the impeller. Part of the velocity pressure of the air is converted to static pressure by a diffuser or evase. The efficiency of conversion is a function of the angle of expansion and the length of the diffuser. The straight duct, connected to the fan outlet, also converts some of the velocity pressure to static pressure.

A fan outlet diffuser or duct is also referred to as an evase. AMCA 210 specifies that an outlet duct not be greater than 107.5% or less than 87.5% of the fan outlet area. It also requires that the angle of a transition not be more than 15° for converging transitions and 7° for diverging transitions. For 100% diffusion to a uniform duct velocity, the duct, including the transition, should be extended at least $2\frac{1}{2}$ equivalent duct diameters from the fan outlet.

Frequently, it is not possible to have ideal discharge conditions at a fan outlet. As in the case of less than ideal fan inlet conditions, it is necessary to apply a System Effect Factor to less than ideal fan outlet conditions. If an elbow must be located near a fan discharge, then it should have a minimum turning radius to duct diameter ratio (R/D) of 1.5 and it should be arranged to provide the most uniform airflow possible. When it is necessary to make a turn directly out of a fan discharge, the rotation of the fan should be selected so that the turn direction follows the natural spin of the air from the fan. Turning vanes will usually reduce the pressure loss through an elbow, but where a non-uniform approach velocity profile exists (such as a fan discharge), the vanes may actually cause the non-uniform profile to continue beyond the elbow. If branch takeoffs or splits are too close to the fan outlet, non-uniform flow conditions will exist and the pressure loss may vary widely from the design intent.

If a suitable discharge duct cannot be provided at a fan outlet, a System Effect Factor must be added to the system resistance losses, as calculated by the methods presented in Chapter 15. To select a System Effect Curve, it is necessary to know the blast area of the fan, in addition to the fan outlet area, in order to calculate the blast area to outlet area ratio. Figure 16.5 identifies the blast area and the outlet area for a centrifugal fan. If this data is not available, it is suggested that a ratio of 0.6 be used. Figures 16.34 and 16.35 may then be used to determine the appropriate System Effect Curve for a particular discharge configuration. Figure 16.21 may then be used to determine the resultant static pressure loss due to the discharge configuration.

If a section of straight duct is provided between the fan outlet and the elbow, the percent of effective length must be determined. For a duct to have 100% effective length, it must be long enough for uniform air flow to develop, as shown in Figure 16.33. For 100% effective length, the duct should be a minimum of $2\frac{1}{2}$ duct diameters for air velocities of 2500 ft/min or less. For each additional 100 ft/min of air velocity, the effective length must be increased by 1 equivalent duct diameter. This is demonstrated in Example 16.7.

Example 16.7 An SWSI centrifugal fan is discharging air into an air distribution system at a rate of 4000 ft³/min. The air discharges into a 14 in × 14 in duct that is 36 in long and then passes through a 90° elbow, as shown in Figure 16.36. The blast area for the fan is 12 in × 12 in and the outlet area is the same as the duct, 14 in × 14 in. What is the System Effect pressure loss that should be added to the distribution system losses to account for the discharge configuration?

Referring to Table 15.2, a 14 in × 14 in duct is equivalent to a 15.3 in diameter round duct.

Plotting 4000 ft³/min and a 15 in diameter duct on a duct friction chart (Figure 15.3) results in a discharge air velocity of 3200 ft/min.

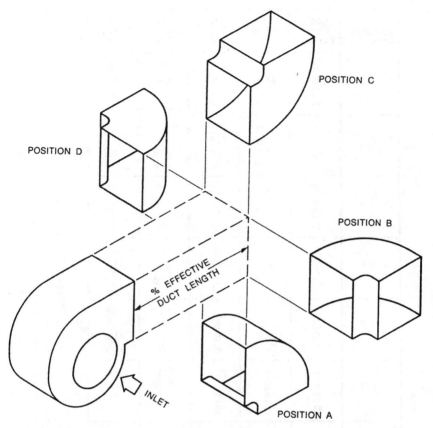

FIGURE 16.34   Outlet Elbows for SWSI Centrifugal Fans (*Reprinted with permission of AMCA International*)

Since the discharge air velocity is greater than 2500 ft/min, the 100% effective length for the 15 in diameter duct is $2\frac{1}{2}$ duct diameters, plus an additional diameter for each additional 100 ft/min greater than 250 ft/min. The 100% effective length is calculated as

$$100\% \text{ effective length} = 2.5D + \left(\frac{3200 - 2500}{100}\right)D = 9.5D$$

$$= 9.5\,(15\text{ in}) = 142.5\text{ in}$$

For a discharge duct length of 36 in, the percent effective length is

$$\% \text{ effective length} = \frac{36}{142.5} \times 100\% = 25\%$$

$$\text{The fan blast area} = 12\text{ in} \times 12\text{ in} = 144\text{ in}^2$$

$$\text{The fan outlet area} = 14\text{ in} \times 14\text{ in} = 196\text{ in}^2$$

Blast Area / Outlet Area	Outlet Elbow Position	No Outlet Duct	12% Effective Duct	25% Effective Duct	50% Effective Duct	100% Effective Duct
0.4	A	N	O	P-Q	S	
	B	M-N	N	O-P	R-S	
	C	L-M	M	N	Q	
	D	L-M	M	N	Q	
0.5	A	O-P	P-Q	R	T	
	B	N-O	O-P	Q	S-T	
	C	M-N	N	O-P	R-S	
	D	M-N	N	O-P	R-S	
0.6	A	Q	Q-R	S	U	
	B	P	Q	R	T	
	C	N-O	O	Q	S	
	D	N-O	O	Q	S	
0.7	A	R-S	S	T	V	
	B	Q-R	R-S	S-T	U-V	
	C	P	Q	R-S	T	
	D	P	Q	R-S	T	
0.8	A	S	S-T	T-U	W	
	B	R-S	S	T	V	
	C	Q-R	R	S	U-V	
	D	Q-R	R	S	U-V	
0.9	A	T	T-U	U-V	W	
	B	S	S-T	T-U	W	
	C	R	S	S-T	V	
	D	R	S	S-T	V	
1.0	A	T	T-U	U-V	W	
	B	S-T	T	U	W	
	C	R-S	S	T	V	
	D	R-S	S	T	V	

NO SYSTEM EFFECT FACTOR

SYSTEM EFFECT CURVES FOR SWSI FANS

For **DWDI** fans determine SEF using the curve for SWSI fans. Then apply the appropriate multiplier from the tabulation below

MULTIPLIERS FOR DWDI FANS

ELBOW POSITION A = ΔP X 1.00
ELBOW POSITION B = ΔP X 1.25
ELBOW POSITION C = ΔP X 1.00
ELBOW POSITION D = ΔP X 0.85

FIGURE 16.35   System Effect Curves for SWSI Fans (*Reprinted with permission of AMCA International*)

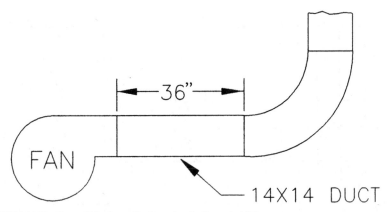

FIGURE 16.36   Fan and Discharge Configuration for Example 16.7

The ratio of the blast area to outlet area is

$$\frac{\text{Blast Area}}{\text{Outlet Area}} = \frac{144 \text{ in}^2}{196 \text{ in}^2} = 0.73$$

Referring to Figure 16.34, the configuration has the elbow in Position C.

Referring to Figure 16.35, for an elbow in Position C, a blast area to outlet area ratio of 0.7 and a 25% effective discharge duct, $R\text{-}S$ are selected as the appropriate System Effect Curves. Plotting an air velocity of 3200 ft/min between System Effect Curves R and S, as shown in Figure 16.37, results in the following System Effect pressure loss for the discharge.

$$P_s = 0.60 \text{ in. wg}$$

This chapter provides an introduction to pressure losses for the System Effect, resulting from various fan/duct configurations. You should consult Reference 26 for more detailed data on fan System Effect.

16.6   AIR HANDLING UNITS AND AIR SYSTEMS

Up to this point, this chapter has dealt with the selection and installation of fans, which are the primary air moving components in HVAC systems. However, in the design of HVAC distribution systems, fans are not usually selected as individual components. Rather, they are selected as part of an air handling unit (AHU). An AHU is a set of modular components consisting, at the minimum, of a fan in a cabinet with a motor. Other components that can make up a complete air handling unit include cooling coils, heating coils, filters, and dampers. Usually, each component is an individual module, such as a coil section, which can be added to the unit as required for a particular application. The cabinets for AHUs are usually constructed of sheet metal panels. The sheet metal is usually either painted or galvanized steel. ARI Standard 430 defines air handling units as "a factory-made, encased assembly, consisting of a fan or fans and other necessary equipment to perform one or more of the functions of circulating, cleaning, heating, cooling, humidifying, dehumidifying, and mixing of air, but which does not include a source of heating or cooling. This device is capable of use with ductwork having a total static resistance of at least $1/4$ in. wg."

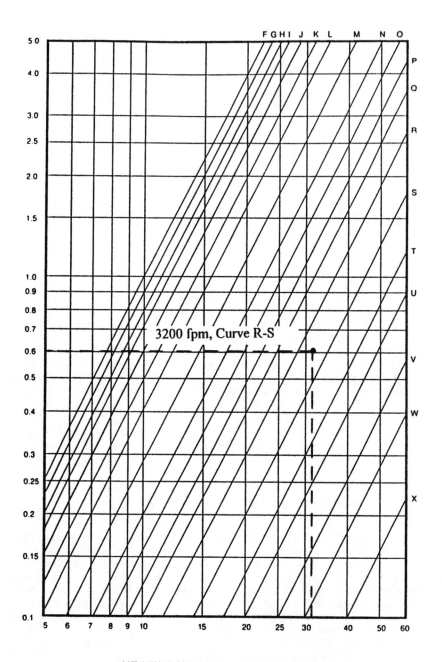

3200 fpm, Curve R-S

AIR VELOCITY, FPM IN HUNDREDS

(Air Density = 0.075 lbm/ft³)

FIGURE 16.37   System Effect Curves for Example 16.7 (*Reprinted with permission of AMCA International*)

Air handling units are selected and sized based on the system heating and/or cooling loads, air circulation rate, and type of air distribution system. Factory-assembled air handling units are available from 900 ft³/min to 63,000 ft³/min. The most commonly used air distribution systems include constant volume, multi-path (multi-zone and dual duct), and variable air volume (VAV). A complete AHU should perform some or all of the following functions:

- provide air movement and circulation
- cool and dehumidify the air
- heat or temper the air
- humidify the air
- filter and clean the air
- provide adequate outdoor ventilation
- provide a means of mixing outdoor and recirculated air

Air handling units can be divided into two basic configurations which describe the location of the fan with respect to the coil section. The two configurations are blow-thru and draw-thru. The terminology for central station AHUs is shown in Figure 16.38, with the various components and their typical location on a modular AHU. A blow-thru unit has the fan located upstream of the coil section. The fan blows or pushes the air through the coils. In the case of the blow-thru unit, the cooling coil removes the heat added to the air stream by the fan. The unit configuration does not affect the total system cooling load; however, the fan heat for a blow-thru AHU does not become part of the space cooling load. Since the air is blown through the coils, it can take more than one parallel path after leaving the fan. This permits what are called multi-zone systems and AHUs. Multi-zone AHUs are units with several separate temperature controlled zones to serve different areas of a building with different heating and/or cooling load characteristics. Blow-thru units can also be used for dual duct units. A more detailed discussion of the different types of air distribution systems can be found later in this chapter.

The majority of single fan applications for central station AHUs have the fan located downstream of the coil section and draw or pull the air through the coils. Since the draw-thru unit has the fan located downstream of the cooling coil, the fan heat enters the cooled air stream and becomes part of the room sensible cooling load by increasing the supply air temperature to the conditioned space. This results in a higher required airflow rate for a given space sensible cooling load.

The unit fan/coil configuration must also be considered when installing a trap in the cooling coil condensate line. The condensate drain line for a blow-thru AHU will be under a positive pressure, while the drain line for the draw-thru unit will be under a negative pressure. The height of the water column in the drain trap must be determined accordingly.

Air handling units can also be classified according to the path of the air as it travels through the unit. In a horizontal unit, the flow of air is completely horizontal. The fan and coils are in a line horizontally. In a vertical AHU, the fan is mounted above the coil section and the air flows vertically to the fan after passing through the coil section. Horizontal and vertical arrangements for AHUs are shown in Figure 16.39.

The AHUs shown in Figure 16.39 are both horizontal discharge units. The discharge for AHUs may also be vertically through the top or bottom (for horizontal units).

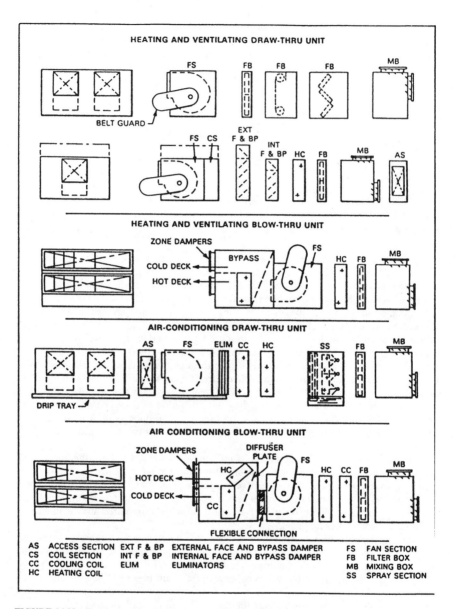

FIGURE 16.38   Terminology for Central Station Air Handling Units

16.6.1 Air Handling Unit Components and Accessories

The fan is the main component in any air handling unit. Air handling units are usually selected on the basis of the fan requirements for the HVAC system. The selection and installation of fan sections for AHUs should be based on the general requirements for fans

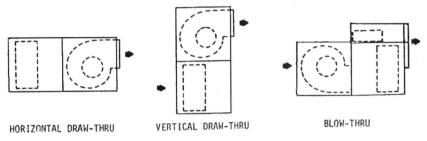

HORIZONTAL DRAW-THRU   VERTICAL DRAW-THRU   BLOW-THRU

FIGURE 16.39   Air Handling Unit Arrangements

presented earlier in this chapter. Once the preliminary selection of a fan section has been made, the other components of the AHUs can be selected and sized. This section presents an overview of the components typically found in complete AHUs.

Most AHUs include some method to cool, dehumidify, and sometimes heat the air as it is circulated through the unit. These cooling, dehumidifying, and heating processes are usually performed by cooling and heating coils.

Mechanical cooling and dehumidification of air is accomplished by one of two types of cooling coils, chilled water and direct expansion (DX). Air is cooled with a chilled water cooling coil by circulating cooled or chilled water through a coil as the heat transfer fluid or heat transfer medium. Air is cooled by DX coils, as the result of circulating a refrigerant through the coil as the heat transfer medium.

Cooling and dehumidification of air is accomplished by passing the circulated air through a coil with a heat transfer medium that is cooler than the air. The coil serves as an extended surface heat exchanger between the heat transfer fluid and the air. Whenever air is cooled below its dew point temperature, moisture will condense out of the air. You should refer to Chapter 9 and Reference 1 for a more detailed discussion of cooling coil psychrometric processes.

Heating of air may also be accomplished by circulating a heat transfer fluid through a coil with air passing through it, although heat may also be produced in a coil electrically. The heating fluids most commonly used in HVAC systems are heated water and steam.

When selecting a cooling or heating coil, important issues to consider include cooling or heating capacity, the quantity of air passing through the coil, the velocity of the air as it passes through the coil, the entering and leaving air temperatures, the heat transfer medium, the static pressure drop as the air passes through the coil, and the cost of the coil. Frequently, there is more than one coil available from a manufacturer that will meet the required performance criteria. The objective is to select a coil which will minimize first cost but not at the expense of performance, operating cost, and system efficiency. A more detailed discussion of heating and cooling coils for HVAC systems is presented in Chapter 17.

Air filters are essential in AHUs because they perform a necessary function that provides two benefits in HVAC systems.

- They remove airborne particulate matter before the circulated air enters the conditional space, thereby reducing the level of contaminants in the space.
- They remove airborne particulate matter before it adheres to the inside of AHU components and air distribution ductwork.

Important considerations for selecting and specifying air filters include filter efficiency, air velocity through the filter, air static pressure drop through the filter (both clean and dirty), the filter media used, the cost of the filters, and the desired cleanliness in the conditioned space. A more detailed discussion of air cleaning and filtration is provided in Chapter 18.

Mixing sections, or mixing boxes as they are commonly called, are used to mix two converging air streams in the AHU, usually system return air and outdoor air. A mixing box is a plenum with two air inlets, each of which has a set of volume control dampers. Mixing boxes provide the proper mixing of return air and outdoor air in the required quantities. The sets of dampers in the mixing box inlets can be permanently adjusted to provide set quantities of return and outdoor air, or can be actuated to modulate the relative quantities of the two air streams. Modulating dampers are usually automatically adjusted by a pneumatic or electric actuator. The modulation range can be from 100% outdoor air (economizer) to the minimum outdoor air required by code. (See Chapters 11 and 13.) Occasionally, filters are included in the mixing box to both mix and filter the air as it enters the AHU. These components are referred to as combination filters/mix boxes.

AHU accessories include such items as dampers, moisture eliminators, and humidifiers. Dampers are used to control the flow of air. In addition to using dampers in mixing boxes, dampers may also be used for temperature control purposes. Dampers may be used to pass air, either through the face of a heating or cooling coil, or to bypass the air around the coil. The air passing through the coil is either heated or cooled, depending on the type of coil. Air passing around the coil is unaffected, i.e., neither heated nor cooled and dehumidified. The two air streams are then mixed inside the AHU, after passing the coil, to achieve the desired temperature. Dampers of this type are called face and bypass dampers.

Moisture eliminators are occasionally required to remove condensed moisture from the air downstream of a cooling coil. If the air velocity through a cooling coil becomes too great, moisture that has been condensed out by the coil may be blown off the coil and become entrained in the circulated air, or be carried beyond the AHU condensate drain pan. A moisture eliminator removes the majority of the condensed moisture that has become entrained in the air stream, reducing moisture carryover.

Humidifiers are occasionally needed to add moisture to the air stream in order to maintain minimum relative humidity conditions in the conditioned space. Humidification is frequently required in winter when low temperature, low moisture air is brought into a building for outdoor ventilation purposes. Low humidity levels in the conditioned space can also be a problem during economizer operation when 100% outdoor air is brought into the building by the air system for cooling purposes during mild weather.

When cold outdoor air is brought into a building through an AHU with a chilled water or heating water coil, freezing of the water inside the coils can easily occur. Even though the outdoor air may be mixed with warmer return air, and the average mixed air temperature is above freezing, stratification caused by incomplete mixing of the air streams can cause water in a coil to freeze. An air blender is a device that promotes more complete mixing of two air streams, reducing the possibility of freezing due to stratification and incomplete mixing. When selecting and specifying air blenders, the static pressure loss through the blender must be considered and added to the system pressure losses (resistances).

16.7 AIR DISTRIBUTION SYSTEMS

Chapter 1 discussed some of the major requirements of air distribution systems for HVAC of buildings. Basic descriptions of the system types were also included in Chapters 1 and 10. This section provides an overview of the most common types or air distribution systems commonly used for building HVAC. Central air distribution systems can basically be divided into three major categories which describe how heating, ventilating, and cooling capacity control is achieved. The major types of HVAC air systems are constant volume/single path, constant volume multi-path, and variable air volume systems.

16.7.1 Constant Volume Single Path

The most basic type of central air HVAC system is the constant volume single path system. As the name implies, the airflow rate is constant. There is only one path for the air to follow as it leaves the AHU and is distributed to the conditioned space. This type of system is best suited for a single zone because the system is either in the heating mode or cooling mode at any given time. The system may be in the "dead band" mode where it is neither heating nor cooling, just circulating air. It is not possible to simultaneously heat one zone and cool another with a single system. Typical applications for the constant volume single path systems are single family houses, large auditoriums with very few exterior surfaces, spaces with a single exterior exposure, and other spaces with fairly uniform heating and/or cooling loads. A schematic diagram of a constant volume single path system is shown in Figure 16.40.

Constant volume single path systems normally have an AHU with heating and/or cooling coils and ductwork in a series flow path. Sometimes, the heating coil may be replaced with a gas or oil-fired furnace, as in the case of a residential system. The heating or cooling is cycled on or off, in response to a load in the conditioned space sensed by a thermostat.

A variation on the constant volume single path system is the constant volume reheat system. This system can provide either heating or cooling to multiple zones simultaneously, with a constant air volume system. A schematic diagram of a constant volume reheat system is shown in Figure 16.41.

In this type of system, air is circulated at a constant rate and leaves the AHU at a set temperature, usually about 55°F. Just before the air enters the conditioned zone, a duct mounted reheat coil increases the temperature of the air entering the zone in accordance with the

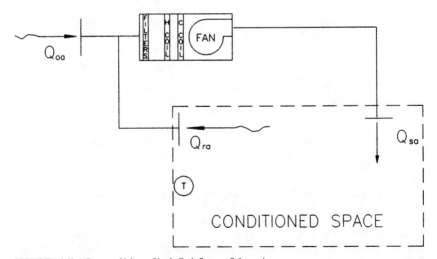

FIGURE 16.40 Constant Volume Single Path System Schematic

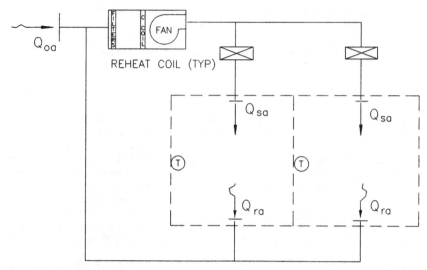

FIGURE 16.41   Constant Volume Reheat System Schematic

zone heating or cooling load sensed by the zone thermostat. Reheat coils can be electric, steam, or heating water. A separate preheat coil may also be provided in the AHU to preheat the incoming outdoor air as necessary.

Constant volume reheat systems are rarely used on new installations because they are very energy inefficient. Air is first cooled to a set supply air temperature and then heated to meet the zone requirements. Some energy codes no longer allow constant volume reheat systems.

16.7.2 Multi-Path Systems

As the name implies, multi-path systems allow the conditioned air to follow multiple parallel paths to each zone. The temperature of the air supplied to each zone is a function of the heating or cooling load in that zone. Typically, the air handling unit has both heating and cooling coils, in order to provide either heated or cooled air to a zone as required. The two most common multi-path air systems are the multi-zone system and the dual duct system.

Multi-zone systems usually have an air handling unit with a cooling coil and multiple heating coils. A typical multi-zone air handling unit is shown in Figure 16.42.

Each zone has its own supply air duct from the AHU, enabling the air to flow through multiple parallel supply air ducts to the zones. Each zone has its own reheat coil inside the AHU. The zone reheat coil is controlled by a zone thermostat. Air is passed through the cooling coil in the AHU and is cooled to a predetermined temperature, usually about 55°F. The airflow is then divided inside the AHU into separate paths for each zone. Still inside the AHU, the air for each zone is then either passed through a reheat coil or bypassed around the reheat coil according to the zone heating or cooling load. The proportion of air passed through or around the reheat coil is controlled by the zone thermostat.

Since the supply air is first cooled, then reheated or bypassed, to provide the desired "mixed" air temperature, these systems are rather energy inefficient and energy wasteful. Very few multi-zone systems are installed in new buildings today. Most multi-zone systems are found in older buildings.

FIGURE 16.42   Multi-Zone Air Handling Unit (*Courtesy of York International*)

The second multi-path system in use today is the dual duct system. As the name implies, there are dual (two) ducts to each space or zone. One duct carries heated air, while the other carries cooled air. There can be single, or dual, air handling units for the system. Either cooled air, heated air, or both are supplied to the space according to the thermal load in the space. In order to provide air to the space at the required temperature, a dual duct terminal is provided for each zone to mix the heated and cooled air in the required proportion. The dual duct terminal mixes the two air streams in accordance with the requirements of the zone thermostat. A typical dual duct terminal is shown in Figure 16.43.

Notice that there are two air valves on the inlet of the terminal. One is for the cooled air and the other is for heated air. The air valves throttle each airflow according to the requirements of the space thermostat.

16.7.3   Variable Air Volume

Other than the constant volume single path system, the variable air volume (VAV) system is the most common type of air system installed in nonresidential buildings today. As the name implies, the rate of airflow in the system is not constant, but varies in accordance with the total sensible cooling load for the building. The supply air is usually cooled and supplied to the distribution ductwork at a constant temperature, although the air temperature may be reset in accordance with the cooling load. A schematic of a typical VAV system is shown in Figure 16.44.

As may be observed from Figure 16.44, the VAV system is also a single path system; however, it is not constant air volume. The amount of cooled air admitted to the conditioned space is a function of the sensible cooling load in the space which is sensed by a thermostat. As the temperature of the space increases, an air valve in a VAV terminal opens, increasing the airflow as the cooling load is satisfied.

Heat for the space is usually not provided by heating the supply air in the AHU. Frequently, heat is provided by baseboard radiators around the exterior perimeter of the

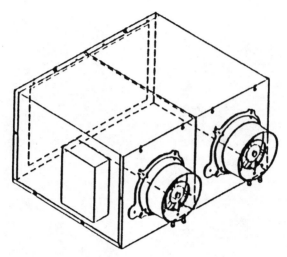

FIGURE 16.43   Dual Duct Terminal Unit (*Courtesy of the Trane Company*)

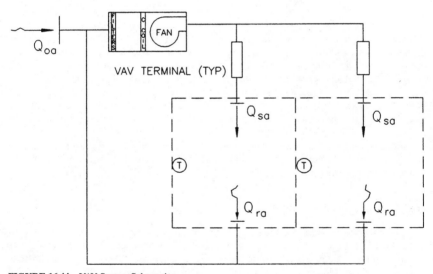

FIGURE 16.44   VAV System Schematic

conditioned space. Heat may also be provided in the VAV terminal with a reheat coil in the VAV terminal itself. A typical VAV terminal with a reheat coil is shown in Figure 16.45.

Notice that an air valve is on the inlet of the VAV terminal. The air valve throttles, or varies, the flow of conditioned air to the space, in response to a space thermostat. For terminals with reheat, the reheat coil is usually located on the discharge of the terminal.

Since the quantity of air supplied to the spaces varies, it is good design practice to provide some method to control the airflow of the air handling unit fan. This is normally accomplished in one of three ways. The three methods are fan discharge dampers, fan inlet

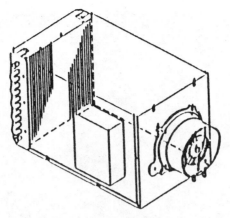

FIGURE 16.45   VAV Terminal with Hot Water Reheat (*Courtesy of the Trane Company*)

vanes, and variable frequency motor drives. Each method varies the flow of the AHU fan in response to changes in static air pressure in the main supply air duct.

The first, and least desirable, are discharge dampers. Discharge dampers are automatically actuated dampers that are placed on the outlet, or discharge, of the AHU. The dampers open and close in response to the pressure in the main supply air duct. As the VAV terminals open and the pressure in the main duct drops, the discharge dampers open to allow more air from the fan into the ductwork. As the VAV terminals close and the duct pressure increases, the discharge dampers close. Discharge dampers are somewhat energy inefficient because the fan is rotating at full speed while the dampers block off the airflow. Discharge dampers may also result in undesirable high air pressures in the air handling unit.

A variation on discharge dampers is the bypass system. Air is simply bypassed around the conditioned space and returned to the AHU. Air is normally "dumped" into a return air plenum or duct by a motorized relief damper. The fan in the AHU is still rotating at full speed, fully loaded. Bypass systems are typically found on smaller commercial air systems.

A second, and more energy efficient, method to control the AHU fan air volume is with inlet vanes on the fan. These vanes are located at the air inlet on the fan, and open or close in response to the pressure in the main supply air duct. Figure 16.46 shows a centrifugal fan with inlet vanes.

Because the inlet vanes restrict the amount of air entering the fan, the fan rotates in an unloaded condition, even though it is rotating at full speed. This reduces the load on the electric motor driving the fan, and reduces the power consumed by the motor. Inlet vanes are more costly than discharge dampers; however, they are also more effective.

The third, and most effective, way to control the airflow of a fan is with a variable frequency motor drive. These drives vary the frequency of the electric power to the fan motor in response to the pressure in the main supply air duct. The varying electrical frequency, in turn, causes the motor and fan to rotate at different speeds. When the air pressure in the duct drops, the fan speed increases to supply more air. When the air pressure increases, the fan speed decreases accordingly. Although variable frequency drives tend to be more expensive than the other methods to control air volume, they are also much more efficient. Frequently, the energy cost savings of the variable frequency drives justify the additional first cost. Variable frequency drives are discussed in more detail in Chapter 19.

FIGURE 16.46   Fan with Inlet Vanes (*Courtesy of York International*)

16.8   FAN AND AHU SELECTION AND INSTALLATION

Proper fan and/or AHU selection and installation play a key role in the performance of an HVAC system. Selection involves not only finding a fan or AHU to match the required airflow, heating/cooling capacities, and pressure requirements, but all aspects of an installation, including the air stream characteristics, operating temperatures, drive arrangements, and mounting. In order to properly select a fan or AHU for a given system, it is necessary to know the capacity and total pressure requirements of the system. The type and arrangement of the prime mover (fan or AHU), the possibility of fans in parallel or series, the nature of the load (variable or steady), and noise constraints must be considered. Steps must also be taken to avoid fan and system performance problems. After the system characteristics have been determined, the main considerations in the actual fan selection are efficiency, reliability, size, weight, speed, noise, and cost. It is important that the fan be efficient and quiet in its operation. Generally, a fan will generate the least noise at peak efficiency. Operation considerably beyond this point will be noisy and may result in performance problems.

Generally, fans and AHUs are selected at the peak efficiency or just to the right of the peak on a fan performance curve. Fans selected to the right will be slightly smaller (lower in cost), but will require more power, run at higher speeds, and may have a higher sound rating. For forward curved (FC) fans, the pressure on the fan curve drops steadily, from a point of maximum efficiency to full open (maximum flow) operation. The power curve rises continuously from low to peak power consumption at peak airflow. Low pressures and high airflow rates result in higher power consumption which can overload the fan motor.

Referring to a fan curve, for both a backward inclined (BI) and airfoil (AF) fans, you will see that these fans have a peak static pressure (other than blocked tight, zero flow) and the pressure quickly decreases to the right of the peak. If a fan has a steep pressure curve, a slight change in pressure from one point at which the fan is operating will not greatly affect the rate of airflow. Since BI and AF fans have steep pressure curves, they have non-overloading characteristics. Steep pressure curves tend to produce nearly constant flow rates under changing pressures. The power consumption of both BI and AF fans is minimum at zero airflow. The brake horsepower increases with increasing air flow, up to a point to the right of peak efficiency. As a result, the fan is non-overloading. The exact performance and operating limitations of a particular fan should be obtained from the fan or AHU manufacturer.

When designing an air distribution system for a fan or AHU, the designer should pay particular attention to the fan inlet and discharge configurations. The designer should attempt to minimize the System Effect caused by the connecting ductwork. Non-uniform airflow into the fan inlet is the most common cause of deficient fan performance. In general, the duct connections should be such that the air may enter and leave the fan or AHU as uniformly as possible, with no abrupt changes in direction or velocity. Space is often limited for the fan or AHU installations, and less than ideal connections may have to be used. Figure 16.47 shows a comparison of various fan inlet conditions, and Figure 16.48 shows a comparison of fan outlet connections. Note that for a given system, the capacity, total pressure, and efficiency will be reduced and the power consumption increased as the result of poor inlet and/or outlet conditions.

As discussed earlier in this chapter, the operating point at which a fan and system will perform, is determined by the intersection of the system resistance curve and the fan performance curve. Selection of a system/fan operating point at other than near fan peak efficiency can result in performance problems and in unstable air flow. The main reasons for unstable airflow in fans and systems are:

System surge

Fan surge

Fan stalling

System surge occurs when the system resistance and fan performance curves do not intersect at a distinct point, but instead intersect over a range of volumes and pressures, as shown in Figure 16.49.

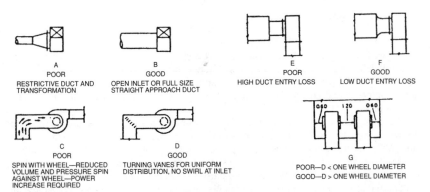

FIGURE 16.47 Fan Inlet Connections (Copyright 1979 by the American Society of Heating, Refrigerating, and Air Conditioning Engineers, Inc., from the ASHRAE Equipment Handbook. Used by permission.)

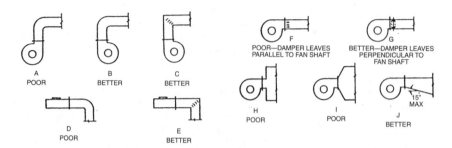

FIGURE 16.48 Fan Outlet Connections (Copyright 1979 by the American Society of Heating, Refrigerating and Air Conditioning Engineers, Inc., from the ASHRAE Equipment Handbook. Used by permission.)

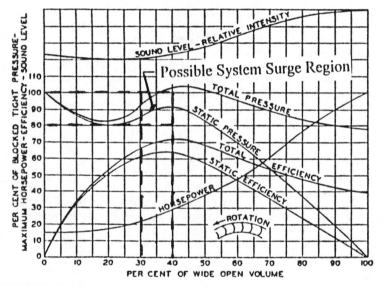

FIGURE 16.49 System Surge Region

In this situation, the system resistance curve and the fan performance curve may almost be parallel. The operating point is actually a range, because multiple airflow rates can occur for any given pressure. The operating range can be over a range of airflow volumes and static pressures. This will result in unstable operation known as system surge, pulsation, or pumping.

Fan surge can occur when there is a high pressure differential between the fan inlet and discharge and when there is little or no airflow. For any fan, the point of minimum air pressure occurs at the center of rotation of the fan impeller. The maximum pressure occurs at the discharge side of the fan impeller. Fan stall is a condition where the airflow through the fan reverses direction, and flows from discharge to the fan inlet at the center of the fan impeller. Stall can occur when there is a high pressure differential across the fan and low airflow. As the fan impeller is rotating, there is insufficient air entering the fan impeller to completely fill the space between the fan blades. This condition will result in a temporary

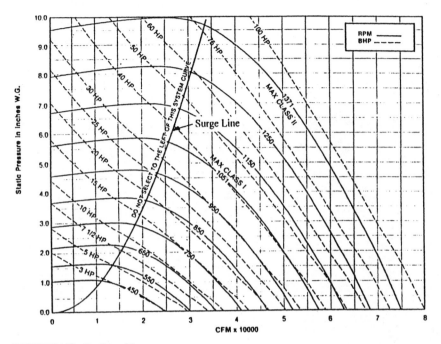

FIGURE 16.50 Fan Surge Line

reversal in airflow as the air attempts to flow backward, from the high pressure of the discharge to the low pressure of the fan inlet. The reversal of airflow occurs briefly until the pressure differential is reduced, at which time the airflow moves forward again. These fluctuations in airflow direction and pressure result in large pulsations or surges in the fan. Figure 16.50 shows the fan surge line on a fan performance curve. Operation at a point to the left of this line could result in fan surge. The fan should be selected so the point of fan/system operation is to the right of the fan surge line.

Fan stall is an operating phenomenon of most axial fans. For airfoil blades in an axial fan, the angle of attack is the angle between the plane of rotation of the fan blades, and the centerline (chord line) of the blade airfoil. The angle at which the air passes over the airfoil blade is a function of the blade angle of attack. If the angle of attack becomes too great, the airflow separates from the blade and fan stall occurs. The smooth flow of air across the blades suddenly breaks, and the pressure difference across the blades decreases. The fan drastically loses its pressure capability. The stall region for an axial fan is generally located at the upper part of the fan performance diagram, beyond the region of highest total efficiency. Operation near the upper part of the curve may result in fan stall.

Chapter 17

AIR SYSTEM HEATING, COOLING, AND HUMIDITY CONTROL

The primary function of a building HVAC system is to provide a comfortable environment in occupied spaces of the building. The principal method by which an HVAC system provides a comfortable indoor environment is to control the indoor air temperature and relative humidity of the conditioned space. Air systems circulate air as the heating and cooling medium for the conditioned spaces. Air is used to transfer the heating and/or cooling loads of the conditioned spaces to the air conditioning system as the air is circulated between the spaces and the air handling unit. As air is circulated through a conditioned space, the space heating or cooling load is transferred to the air. As air is circulated through the AHU, heat is added to and/or removed as required to maintain the desired conditions within the space. Moisture is frequently removed from and occasionally added to the air as it passes through the AHU.

This chapter provides an overview of the methods by which HVAC systems heat, cool, humidify, and dehumidify the air circulated to a conditioned space.

17.1 HEATING, COOLING, AND PSYCHROMETRICS OF AIR

Whenever air is heated or cooled, the dry bulb temperature of the air increases or decreases accordingly. The dry bulb temperature of an air-water vapor mixture is simply the temperature that would be read from an ordinary thermometer. Heat that changes the dry bulb temperature of air is referred to as sensible heat. The sensible heat absorbed by air as it flows and changes dry bulb temperature may be expressed by Equation 2.4, which is a simplified version of the First Law of Thermodynamics. Equation 2.4 is used so frequently in HVAC design that is has been modified to contain units most frequently used in HVAC calculations and design. Equations 17.1 and 17.2 are the First Law of Thermodynamics for air.

$$q = 1.08Q(T_2 - T_1), \text{ for heating} \tag{17.1}$$

$$q = 1.10Q(T_2 - T_1), \text{ for cooling} \tag{17.2}$$

where

q = the sensible heat transferred to or from the air as it flows, Btu/hr

T_1 = initial or entering dry bulb temperature of the air, °F

T_2 = final or leaving dry bulb temperature of the air, °F

Q = the air flow rate, ft³/min

Equation 17.1 is used extensively in sizing and designing air heating systems. As the air is heated in an AHU, all heating is sensible heat.

Example 17.1   Air is flowing at the rate of 4000 ft³/min through an AHU with a heating device. The air enters at 50°F and leaves at 90°F. How much sensible heat is required to heat the air?

Using Equation 17.1,

$$q = 1.08(4000 \text{ ft}^3/\text{min})(90° - 50°) = 172,800 \text{ Btu/hr } (172.8 \text{ MBh})$$

Most HVAC systems serving buildings are required by code to bring in certain minimum amounts of outdoor air. The amount of outdoor air required is a function of the usage of the space and the number of occupants in the space. (See Chapter 11.) The system return air (*ra*) volume is usually less than the supply air (*sa*) volume, due to exhaust air and exfiltration of air out of the building. Air may also be relieved from the building in order to bring in the required outdoor air without overpressurizing the building. The supply air is a combination, or mixture, of re-circulated (return) air and outdoor air, as shown in Equation 17.3.

$$Q_{sa} = Q_{ra} + Q_{oa} \tag{17.3}$$

where

Q_{sa} = supply air flow rate, ft³/min

Q_{ra} = return air flow rate, ft³/min

Q_{oa} = outside air flow rate, ft³/min

The minimum supply air flow rate is usually determined by the total cooling loads for the conditioned space. (See Chapter 9.) It may also be determined by a ventilation code.

Since the air entering the cooling and/or heating section of an AHU is usually a mixture of return air and outdoor air, it is usually necessary to calculate an entering, or mixed air, temperature of the air as it enters the cooling and/or heating section. The mixed air dry bulb temperature is a function of the quantities of return air and outdoor air, and the temperatures of each air stream. Equation 9.8 is used to calculate the mixed air dry bulb temperature.

$$T_{mdb} = \frac{Q_{oa}T_{oa} + Q_{ra}T_{ra}}{Q_{ra} + Q_{oa}} \tag{9.8}$$

where

T_{mdb} = the mixed air temperature, °F

Q_{oa} = outside air flow rate, ft³/min

T_{oa} = outside air temperature, °F

Q_{ra} = return air flow rate, ft³/min

T_{ra} = return air temperature, °F

Example 17.2   An AHU is supplying heated air to a conditioned space at the rate of 8000 ft³/min and at a leaving air temperature of 90°F. The return air temperature is 70°F. Outdoor air is brought into the system at a rate of 2000 ft³/min and at a temperature of 10°F. At what rate must the AHU heating coil provide heat under these conditions?

Using Equation 17.3 to solve for the return air flow rate,

$$Q_{ra} = Q_{sa} - Q_{oa} = 8000 \text{ ft}^3/\text{min} - 2000 \text{ ft}^3/\text{min} = 6000 \text{ ft}^3/\text{min}$$

Equation 9.8 is then used to calculate the mixed air or entering air dry bulb temperature.

$$T_{mdb} = \frac{(2000 \text{ ft}^3/\text{min})(10°F) + (6000 \text{ ft}^3/\text{min})(70°F)}{2000 \text{ ft}^3/\text{min} + 6000 \text{ ft}^3/\text{min}} = 55°F$$

Therefore, the entering (initial) air temperature is 55°F and the final (leaving) air temperature is 90°F. Using Equation 17.1 to calculate the rate at which heat must be supplied to heat the air,

$$q = (1.08)(8000 \text{ ft}^3/\text{min})(90° - 55°) = 302,400 \text{ Btu/hr } (302.4 \text{ MBh})$$

Standard atmospheric air is considered to be at a temperature of 60°F, has a density of 0.075 lb/ft³ of dry air, and a pressure of 14.7 psia. Standard atmospheric air is actually a mixture of a number of gases and vapors, including water vapor (moisture).

Since air is a mixture of dry air and water vapor, it is necessary to specify two independent properties in order to define a given state point for the air at a given pressure. The two properties most commonly used to define a state point in HVAC system design are the dry bulb and the wet bulb temperature of the air. Dry bulb temperatures for air were discussed above. The wet bulb temperature of an air-vapor mixture is a measure of the humidity level, or moisture content of the mixture, at any given dry bulb temperature. The higher the wet bulb temperature at any given dry bulb temperature, the greater the moisture content of the air-vapor mixture. Specifying the dry bulb and wet bulb temperature of air at a given pressure completely defines the state point and the internal energy of the air.

The dew point of air is the dry bulb and wet bulb temperature at which the air-vapor mixture becomes saturated. At the saturation temperature, the air can no longer hold any additional moisture, and the moisture begins to condense out. Whenever air is cooled to a dry bulb temperature below its dew point, moisture will condense out of the mixture. Both sensible and latent heat are removed from the mixture, as the moisture goes from the vapor phase to the liquid phase.

Enthalpy (h) is a measure of the internal energy or total heat of an air and water-vapor mixture. Since the enthalpy of perfect (ideal) gases, or an air-vapor mixture, is the sum of the enthalpies of each constituent gas or vapor, the enthalpy of an air-vapor mixture is the total heat of the mixture. The customary units for the enthalpy of an air-vapor mixture are Btu/lb of dry air. Since the enthalpy of atmospheric air is a measure of the total heat content of the air-vapor mixture, it is used extensively in the design of HVAC systems that add and/or remove sensible and latent heat to/from air-vapor mixtures.

As an air-vapor mixture is cooled and the moisture undergoes a change in phase from the vapor phase to the liquid phase, the easiest way to calculate the total heat removed in the cooling process is to use the initial and final enthalpies. The easiest way to determine the enthalpy of an air-vapor mixture is to plot the dry bulb and wet bulb temperatures on a psychrometric chart and read the enthalpy from the chart. Figure 9.2 shows a typical psychrometric chart and Figure 9.3 shows the location of the dry bulb, wet bulb, and enthalpy scales on a psychrometric chart. Equation 17.4 is used to calculate the total heat removed from an air-vapor mixture.

$$q_T = 4.5Q(h_1 - h_2) \tag{17.4}$$

where

q_T = the total heat transferred, Btu/hr

Q = air flow rate, ft³/min

h_1 = the initial or entering enthalpy of the air-vapor mixture, Btu/lb

h_2 = the final or leaving enthalpy of the air-vapor mixture, Btu/lb

4.5 = a constant to make the units consistent

You should refer to Chapter 9 for a more detailed discussion of psychrometric processes.

Example 17.3   Air is flowing through an AHU at a rate of 5000 ft³/min. It is entering a cooling coil at 80°Fdb/70°Fwb. The air is cooled and dehumidified to a point where it leaves at 55°Fdb/55°Fwb. At what rate is the total sensible and latent heat being removed by the cooling coil?

When plotting the entering conditions on a psychrometric chart, as shown in Figure 17.1, the entering enthalpy is read as 34.0 Btu/lb of dry air. When plotting the leaving conditions on the chart, the leaving enthalpy is read as 23.2 Btu/lb. Using Equation 17.4, the total heat removed by the cooling coil is calculated as

$$q_T = 4.5(5000 \text{ ft}^3/\text{min})(34.0 \text{ Btu/lb} - 23.2 \text{ Btu/lb}) = 243{,}000 \text{ Btu/hr}$$

Equation 17.2 is used to calculate the sensible heat:

$$q_S = 1.1(5000 \text{ ft}^3/\text{min})(80° - 55°) = 137{,}500 \text{ Btu/hr}$$

The total heat of the air-vapor mixture is the sum of the sensible and latent heat of the mixture. Therefore, the latent heat may be calculated by subtracting the sensible heat from the total heat.

$$q_L = q_T - q_S = 243{,}000 \text{ Btu/hr} - 137{,}700 \text{ Btu/hr} = 105{,}000 \text{ Btu/hr}$$

17.2   AIR HEATING COILS

Air may be heated in an AHU at two different locations within the unit. If the air is heated as it comes into the AHU before passing through the cooling coil, the heating coil is referred to as a preheat coil. A preheat coil may heat only the incoming outside air or it may heat the mixed return and outdoor air. If the heating coil is located downstream of the cooling coil, it is referred to as a reheat coil. As the name implies, air is reheated after it has been cooled by the cooling coil. Reheat coils are frequently used in multi-zone AHUs for zone temperature and humidity control. Reheat coils may also be located in supply air ducts outside the AHU and in terminal units serving individual spaces. Chapter 9 discusses the application of reheat coils for humidity control. The heating medium used for reheat coils is usually heating hot water, steam, or electricity.

Heating water coils are the most commonly used type of heating coil for HVAC systems. Water coils consist of at least one set, or row, of tubes that are mounted perpendicularly to the flow of air passing through the coil. The tubes are usually mounted horizontally, although they may be vertical. A single row water coil is shown in Figure 17.2.

A coil may contain additional rows of tubes. Each set of tubes, in the direction of air flow, is considered a row. U-bends are provided at the end of each row to provide a complete circuit. A coil with one set of tubes is called a one row coil. A coil with four sets of tubes is called a four row coil. A four row water coil is shown in Figure 17.3.

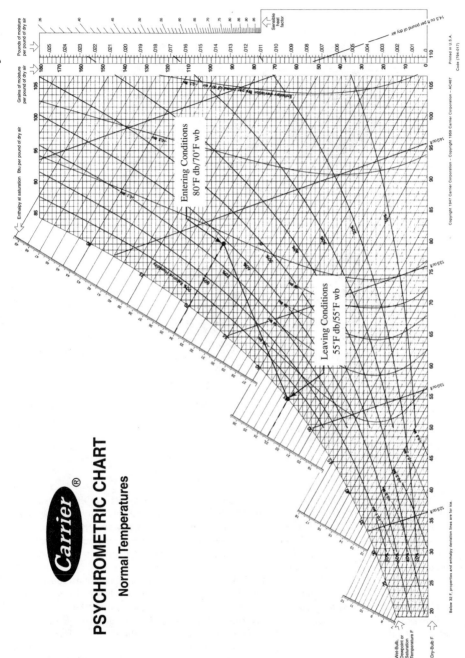

FIGURE 17.1 Psychrometric Chart for Example 17.1

17.5

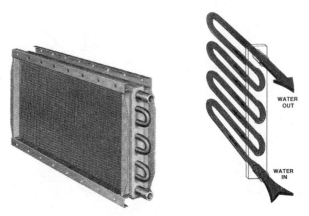

FIGURE 17.2   Single Row Water Coil *(Courtesy of the Trane Company)*

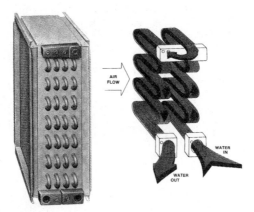

FIGURE 17.3   Four Row Water Coil *(Courtesy of The Trane Company)*

The primary heat transfer surfaces in coils are the tubes themselves. The tubes are usually made of copper, steel, or stainless steel. The heat transfer surface is usually increased by the addition of fins bonded to the tubes. The tubes remain the primary heat transfer surface, although the fins increase the rate of heat transfer from the tubes to the air. Coils are usually specified by the number of rows and fin spacing. Fin spacing is given in fins/in of coil face.

The face area of a coil is the area occupied by the tubes, fins, and the spaces between the tubes and fins. It does not include mounting flanges or pipe headers. Coils are also specified by their face velocity and static pressure drop. The face velocity of a coil is the air flow rate divided by the coil face area. The static pressure drop for a coil is a function of the number of rows, the fin spacing, and the coil face velocity. The greater the rows and face velocity and the lower the fin spacing, the greater the air static pressure loss.

Heating hot water, or just hot water, coils are used frequently in HVAC applications. Water is heated by a device such as a boiler or hot water generator to a temperature of around 200°F. The heated water is then circulated through a coil which serves as a heat exchanger to transfer heat from the hot water to the air as the water and air pass through the coil. The performance of water coils depends primarily on the size of the coil, the number of rows, the air and water velocities, entering air and water temperatures, and the mass flow rates.

Supply water temperatures can be as high as 250°F, although most systems operate at a maximum temperature of 200°F. Since pure water boils at 212°F (under standard pressure of 14.7 psia), pressurization is required for systems that operate with temperatures at or greater than 212°F. Water typically enters the coil around 200°F, and drops in temperature as it transfers heat to the air. Typically, design temperature drops for hydronic heating systems range from 20°F to 40°F.

Multi-row water coils are also classified as a parallel flow or counter flow arrangement. In a parallel flow coil, the air and water both enter the front of the coil. The water flows parallel to the air as it flows from the front row, or tube, to the back row. In counter flow coils, the flow of the air and water are opposite. The water enters the back row of the coil and flows forward, as it goes from row to row, in an opposite direction. The flow of the water is counter to the flow of air. The four row coil, shown in Figure 17.3, is a counter flow coil. Generally, heating water coils have one, two, three, or four rows.

The water velocity inside the tubes usually ranges from 1 ft/sec to 8 ft/sec, with 3 ft/sec to 5 ft/sec as optimum. As a rule of thumb, the water velocity should be about 4 ft/sec, and the water pressure drop for the coil should be limited to 10 ft wg. On multi-row coils, turbulators are occasionally installed within the water tubes to produce a turbulent flow, which increases the rate of heat transfer.

Water coils usually require some sort of air vent in the connecting piping to the coil to allow any air entrained in the system to be removed. A buildup of air in a water coil can cause a stoppage in water flow. Water coils are also typically provided with a means to drain the coil when it is not in use. Water in coils can freeze quickly when subjected to below freezing temperatures. Freezing of water coils can also be prevented by running a pump to circulate water through the coil to provide a certain minimum velocity inside the coil tubes.

Equation 2.4, which is a simplified version of the First Law of Thermodynamics, is also applicable to liquids such as water. Equation 2.4 may be modified to be suitable for water, and for units commonly used in HVAC design to yield Equation 17.5.

$$q = (8.33 \text{ lb/gal})(60 \text{ min/hr}) \, Q_w(T_2 - T_1) \approx 500 \, Q_w(T_2 - T_1) \qquad (17.5)$$

where

q = the total heat transferred, Btu/hr

Q_w = the water flow rate, gal/min

T_1 = the initial or entering water temperature, °F

T_2 = the leaving or final water temperature, °F

Example 17.4   Heating water is flowing through a coil in an AHU at the rate of 100 gal/min. The water enters the coil at 200°F and leaves at 170°F. At what rate is the water transferring heat to the coil, assuming no heat is lost to the surroundings?

Using Equation 17.5,

$$q = 500(100 \text{ gal/min})(200° - 170°) = 1,500,000 \text{ Btu/hr} = 1500 \text{ MBh}$$

Example 17.5   Air is flowing through an AHU at the rate of 1000 ft³/min. The air enters a heating water coil at 50° and is heated to 90°F. Water enters the coil at 200°F and leaves at 170°F. What is the required water flow rate?

This is a classic application of Equation 2.3 and an energy balance on an HVAC system component (the heating coil). It is used repeatedly in the design of hydronic HVAC systems.

First, Equation 17.1 is used to calculate the heat required to raise the temperature of the air.

$$q = 1.08 \, (1000 \text{ ft}^3/\text{min})(90° - 50°) = 43,200 \text{ Btu/hr} = 43.2 \text{ MBh}$$

Since no heat is stored in the coil during steady state conditions, and assuming there is no heat loss to the surroundings, the same amount of heat transferred to the air must be transferred from the water. Equation 17.5 may be solved for the required water flow rate and the appropriate values substituted.

$$Q_w = \frac{q}{500(T_2 - T_1)} = \frac{43,200 \text{ Btu/hr}}{500(200° - 170°)} = 2.9 \text{ gal/min}$$

Low pressure steam is also used frequently as the heating medium in HVAC system heating coils. Steam pressures in HVAC applications are usually 15 psig or less, although higher pressures are possible. Steam heating coils are somewhat different from hydronic heating coils in that the steam undergoes a phase change inside the coil. It condenses from the water vapor (steam) phase to the liquid (condensate) phase. As the stream converts back to liquid water (condenses), it releases large amounts of heat. As an example, steam at 15 psig and 250°F will release about 945 Btu/lb of steam as it condenses from steam to liquid condensate. The phase change occurs at the nearly constant temperature.

In order for the steam to condense in a steam coil, the steam itself must be kept inside the coil until it condenses. After the steam becomes condensate, the condensate must be removed from the coil. This is usually accomplished with a steam trap. A steam trap is placed on the discharge piping of the coil to hold the steam in, while enabling the liquid condensate to flow out.

Heating coils, using steam as the heating medium, are made in two general types. The first type is used for general heating, where it is not necessary to maintain equal temperatures over the entire length of the coil, or which are not used in air streams which will drop below 32°F. The second type are those referred to as steam distributing coils and are made with perforated inner tubes. The inner tubes are used to distribute steam along the entire length of the inner surface. These coils are used where the steam flow must be throttled in order to obtain proper temperature control, when even temperatures are required along the length of the coil, or where freezing temperatures are frequently encountered through the coil.

Standard steam coils are made with single tubes. Steam enters through one end of the coil. A standard steam coil is shown in Figure 17.4.

Distributing steam coils are constructed with orificed inner steam distributing tubes, centered and supported to provide uniform steam distribution through the coil, and to provide maximum protection against freeze ups. This design prevents freezing of the condensate, provided a sufficient amount of steam is supplied to the coil, and the condensate is removed with proper trapping as fast as it is condensed. A typical distributing steam coil is shown in Figure 17.5.

Steam coils are typically constructed of copper, steel, or stainless steel tubes with aluminum, copper, steel, or stainless steel fins.

Example 17.6   Outdoor air at 0°F is brought into an air handling unit at the rate of 2000 ft³/min and heated to 40°F by a non-freeze type steam preheat coil. Steam is supplied to the coil at a pressure of 15 psig. The latent heat of vaporization for saturated steam is 945 Btu/lb of steam. What is the required steam flow rate?

Using Equation 17.1 to calculate the heat required to raise the air temperature:

$$q = 1.08(2000)(40 - 0) = 86,400 \text{ Btu/hr} \, (86.4 \text{ MBh})$$

Since the steam has a heating value of 945 Btu/lb, the steam flow rate is calculated directly.

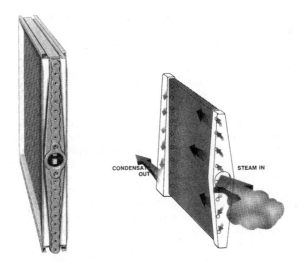

FIGURE 17.4   Standard Steam Coil *(Courtesy of The Trane Company)*

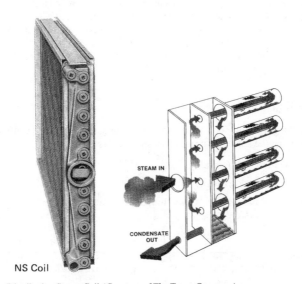

NS Coil

FIGURE 17.5   Distributing Steam Coil *(Courtesy of The Trane Company)*

$$\text{Steam flow } = \frac{86,400 \text{ Btu/hr}}{945 \text{ Btu/lb}} = 91.4 \text{ lb/hr}$$

Electric heating coils are also used frequently in HVAC systems. One of the main advantages of electric heating coils is that they cannot freeze. An electric heating coil consists of a length of resistance wire (usually nickel/chromium) to which a voltage is applied. The resistance wire may either be sheathed (finned tubular) or unsheathed (open). Both types are commonly used.

Open type electric coils consist of a series of electric resistance coils framed in a metal casing. The coils are directly exposed to the airstream. Open coils are frequently used for primary and reheat applications. This design has the advantage of low thermal inertia and quick response to changing conditions. In most cases, open coils tend to be relatively light in weight, making them suitable for mounting in ductwork. Open coils may tend to overheat and have hot spots if the flow of air over the coil is not uniform and steady.

The sheathed, or finned, tubular coil is a resistance wire encased by an electrically insulating layer, such as magnesium oxide, which is enclosed in a finned steel tube. The outer surface temperature of finned tubular electric coils is lower, and the coils are mechanically stronger. Finned tubular heaters were originally used for special applications, such as when the element may have been exposed to human contact, where water from a cooling coil or humidifier may have sprayed directly on it, and where hazardous conditions required an explosion resistant design. The moisture resistant design makes them suitable for use as preheat coils to heat outdoor air, which may be high in moisture content. Finned tube heaters will withstand much greater mechanical abuse. Their greater thermal inertia results in less frequent control cycling.

Both open and finned tube heaters are available in either flanged or slip-in designs. In order to achieve incremental control of the air temperature rise, multiple circuits are usually provided to bring the capacity on in stages. Standard voltages included 208, 240 and 440, in single and three phase. Higher voltages are also available. With electric heating coils, all of the power supplied to the heater is converted to heat. Electric heaters are about 99% efficient. The following conversions are used to convert between electrical energy and thermal energy: 1 W = 3.412 Btu/hr, 1 kW = 3412 Btu/hr.

Air may also be heated by furnaces and with a heat pump cycle. Those heating methods are discussed later in this chapter.

17.2.1 Heating Coil Selection

When selecting a heating coil for a particular application, there are a number of factors to be considered in the selection process. Design factors include:

1. System application for the heating coil:

   a. Preheat for tempering outside air or mixed air

   b. Reheat for providing space temperature and humidity control

2. Type of heating media:

   a. Heating hot water: typically 150°F to 200°F

   b. High temperature hot water: above 212°F, up to 250°F. Additional system pressurization is required for water to remain a liquid above 212°F.

   c. Steam: pressures of 2 to 15 psig are common; however, pressure can be as high as 250 psig.

   d. Electric: common voltages are 208, 240, and 440, in single (208 and 240) and 3 phase.

3. Type of coil:

   a. Hydronic, steam, electric

   b. Tube size, fin spacing, number of rows

4. Standard or non-freeze: for entering air temperatures below 32°F, a non-freeze coil should be used.

17.3   AIR COOLING COILS

Air cooling coils in HVAC systems cool the circulated air by circulating either chilled water or a refrigerant through the coil as a heat transfer medium. Both chilled water coils and refrigerant coils are similar in construction to heating water coils. The coils have tubes through which the heat transfer fluid is circulated. They also have fins to serve as extended heat transfer surfaces. The construction of chilled water coils and refrigerant coils is very similar to heating water coils, with the main difference being that cooling coils usually have more rows than other types of coils. Coil row offerings may vary from two to twelve rows, with four, six, and eight row coils being the most widely used for HVAC applications.

The primary reason that cooling coils have so many rows is that some of the air will pass through the coil and not be cooled or dehumidified. Whenever air is passed through a cooling coil, most of the air is affected by being cooled, and usually, dehumidified. It is inevitable, however, that some of the air passing through the coil will not be affected. The amount of air that passes through the coil unchanged is called bypassed air, and the fraction of the total air flow bypassed is called the bypass factor. The effect of bypass on cooling loads was discussed in Chapter 9. Typical bypass factors for various cooling applications are also discussed in Chapter 9.

Cooling coils that have many rows and close fin spacing have a larger surface area for heat transfer from the flowing air to the cooling coil. These types of coils tend to have low bypass factors. Coils with comparatively wide fin spacing and few rows tend to have higher bypass factors. The velocity of the air passing through the coil also affects the bypass factor. Typical air face velocities through cooling coils range from 300 ft/min to 800 ft/min. Higher face velocities tend to result in higher bypass factors, while lower velocities result in lower bypass factors. The amount of coil surface area has a greater effect on bypass factors than does the face velocity.

The actual surface temperature of a cooling coil with chilled water or with refrigerant is not uniform, but varies with the location on the cooling coil. On a multi-row cooling coil, the surface nearest the entering air is in contact with the warmest air, and therefore is warmer than the average or mean surface temperature. The coil surface near the discharge side of the coil is in contact with air that has already been cooled to near the discharge temperature. The coil surface temperatures near the discharge side of the coil tend to have a lower surface temperature.

In order to analyze the effect of a cooling coil on the air passed through the coil, it is necessary to assume an effective coil surface temperature which would produce the same leaving conditions as the actual nonuniform surface temperatures. This point is commonly called the apparatus dew point (adp) temperature. The adp of a coil is a function of the log mean temperature difference between the fluids (LMTD) and the fluid heat transfer properties of the air and cooling fluid. See Chapter 2 for a discussion of LMTD. The psychrometrics of cooling coil processes were discussed earlier in this chapter and in Chapter 9. A more detailed discussion of cooling coil processes is available in Reference 1.

17.3.1   Chilled Water Cooling Coils

Chilled water coils in HVAC systems use water that has been chilled or cooled by a chiller. The chiller cools the water through a refrigeration process. Chilled water is typically supplied to the cooling coil in an AHU at a temperature between 40°F and 55°F. Lower water temperatures may be used; however, there is a risk of freezing if the water temperature drops to 32°F somewhere in the system and the water does not contain an antifreeze additive such as glycol. Higher water temperatures will result in higher apparatus dew point temperatures, which will result in less dehumidification of the air. Most chilled water systems are designed to operate with a 10°F or 12°F increase in water temperature as the water flows through the coil and absorbs heat.

The main difference between heating water coils and chilled water coils is the number of rows and fin spacing. Chilled water coils require more rows and closer fin spacing since the LMTD for chilled water coils is much less than for heating water coils and because of the dehumidification requirements for cooling coils. In order to provide the required high rate of heat transfer with a smaller LMTD, chilled water coils usually are in a counterflow arrangement. That is, the coldest air meets the coldest water, and the warmest air meets the warmest water. The coil tubes are usually in a staggered arrangement with many rows. This allows lower leaving air temperatures, frequently lower than the leaving water temperatures. Lower leaving air temperatures provide better dehumidification.

Air velocities for cooling coils are usually much slower than heating coils for two reasons: First, the lower LMTD between the air and water requires that the air be in contact with the coil surfaces for a longer period of time to achieve the required heat transfer. Second, since moisture is condensed out of the air as it is cooled, the moisture forms on the coil surfaces. Lower air velocities through the coil are necessary to keep moisture from being blown off the coil and entrained in the air stream as water droplets.

The procedure for calculating the required chilled water flow rate is the same as for the heating water flow rate. Chilled water calculations are based on the total cooling load, the sum of the system sensible and latent cooling loads.

Example 17.7   The total cooling load for the AHU and cooling coil, in Example 17.3, was 243,000 Btu/hr. If the cooling coil was a chilled water coil that received chilled water at 40°F from a system designed for a 10°F water temperature difference, what is the required chilled water flow rate? What would the flow rate be if the system were designed for a 12°F temperature difference? (Neglect any additional load resulting from the coil bypass factor.)

Using Equation 17.5 and a 10°F water temperature difference,

$$Q_w = \frac{q_T}{500(T_2 - T_1)} = \frac{243,000}{500(50 - 40)} = 48.6 \text{ gal/min}$$

Using Equation 17.5 and a 12°F water temperature difference,

$$Q_w = \frac{243,000}{500(52 - 40)} = 40.5 \text{ gal/min}$$

17.3.2 Refrigerant Coils

Although the construction of water coils and refrigerant coils is similar, the operation of refrigerant coils is more complex than that of water coils. Direct expansion (DX) refrigerant systems operate on a thermodynamic cycle that transfers heat from a low temperature area to an area of higher temperature. A typical DX refrigeration system is shown in Figure 2.3.

The refrigerant in a DX coil undergoes a phase change, from a liquid to a vapor, as it absorbs heat from the air passing through the coil. The heat absorbed is referred to as the latent heat of vaporization and is the heat required to change the refrigerant from a liquid to a vapor. For more information on latent heat, refer to Section 2.2.2, Forms of Heat. As the refrigerant changes from the liquid state to the vapor state, it does so at a nearly constant temperature. The temperature at which the refrigerant boils, or vaporizes, in a given coil is dependent upon the pressure of the refrigerant.

While the refrigerant flows into the coil as a liquid and evaporates to a vapor, it absorbs heat from the air. In order to accomplish this, the liquid refrigerant must flow evenly into the coil and be evenly distributed to various tube circuits within the coil. For even coolant distribution and to reduce pressure drop, the flow paths of the refrigerant in a DX coil are always divided into a number of refrigerant circuits. DX coils use a distribution nozzle and multiple feeder tubes to provide a flow of refrigerant to all the circuits in the coil.

In a DX coil, the refrigerant is expanded and evaporated directly inside the tubes of the coil. The DX coil acts as the evaporator in the refrigerating system. In comfort HVAC systems, the evaporating temperature of the refrigerant inside the coil tubes is usually between 37°F and 52°F. At these temperatures, the surface temperature of the coil is usually lower than the dew point temperature of the air entering the coil. The refrigerant flow rate into the coil must be properly metered and distributed to the coil tubes. DX coils used in comfort air conditioning applications have either a capillary tube or a thermostatic expansion valve (TXV) to regulate the flow of refrigerant. The TXV is located at the refrigerant inlet to the evaporator coil. A typical TXV and evaporator coil arrangement is shown in Figure 17.6. The TXV provides smooth evaporator cooling capacity modulation by maintaining a fixed temperature of the refrigerant as it leaves the coil. The majority of the heat transfer from the air to the refrigerant occurs as the refrigerant changes phase from a liquid to a vapor. However, the refrigerant in most DX coils is heated to a temperature higher than that required to completely vaporize the refrigerant. This additional heating of the vapor is called superheating the vapor. The number of degrees that the refrigerant is heated above that required for vaporization is referred to as **"the degrees of superheat."** Superheat represents only a small percent of the coil's heat transfer capacity; however, it is important for refrigeration system control and for compressor protection.

Part load operation and control for DX coils is more critical than for water coils. At part load operation, the evaporating temperature is lowered because less refrigerant is evaporated in the coil. This may result in temperatures below freezing and frost formation on the coil, which blocks the flow of air and severely reduces the rate of heat transfer. A TXV cannot modulate down to zero refrigerant flow. Below a certain point, the clearance between the valve needle and seat alternate between too far open and closed (over capacity with too little superheat and under capacity with too much superheat). The stable

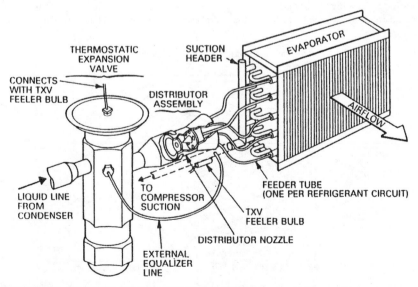

FIGURE 17.6   Thermostatic Expansion Valve and Evaporator Coil

operating range for most TXVs used in HVAC systems is from around 100% to about 50% of design capacity.

DX systems use fluorinated hydrocarbon refrigerants (also called fluorocarbon refrigerants). The most commonly used refrigerants are HCFC-22 (R-22) and HFC-134a (R-134a).

17.4   WARM AIR FURNACES

Warm air furnaces heat air by burning a fossil fuel in a combustion process. The fuels most widely used for furnaces are natural gas, liquefied petroleum (LP) gas, and fuel oil.

A warm air furnace is a combustion and heating device, in which gas or oil is directly fired to heat the air through a heat exchanger, or air is directly heated in order to supply warm air to the conditioned space. A furnace may consist of multiple burners, a heat exchanger, combustion air blower, an ignition device, and controls. The furnaces discussed herein are commercial furnaces used in commercial HVAC systems; however, warm air furnaces are widely used for residential heating applications. The main difference between residential furnaces and commercial furnaces is the size and heating capacity of the equipment. The heating capacity of commercial systems typically ranges from 150 MBh to over 1000 Mbh.

Gas furnaces may be direct fired or indirect fired. Direct fired furnaces have burners located in the conditioned air stream. Figure 17.7 shows a direct fired burner. The fuel is burned directly in the air stream and the combustion products mix with the air passing through the burner. These types of burners are most commonly used on 100% outdoor make-up air units.

Indirect fired furnaces do not allow the products of combustion to mix with the circulated air. A heat exchanger is employed to transfer heat from the burning fuel to the air supplied to the conditioned space. A heat exchanger is shown schematically in Figure 17.8.

Heat exchangers are usually of a clamshell design that completely surrounds the burner. A heat exchanger typically has a two pass arrangement.

Burners may be classified as atmospheric or power burners. Atmospheric burners draw combustion air in at atmospheric pressure. Power burners employ a fan to provide combustion air at a higher pressure. The power burner may be forced draft vented (blower on the upstream side of the burner) or induced draft vented (blower on the downstream side of

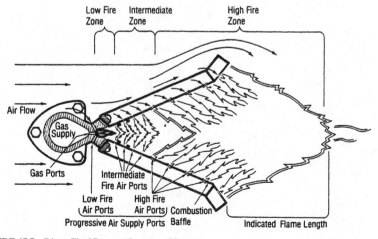

FIGURE 17.7   Direct Fired Burner *(Courtesy of Reznor)*

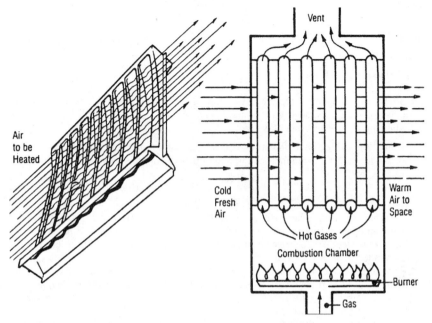

FIGURE 17.8   Heat Exchanger *(Courtesy of Reznor)*

the burner). Power burners are employed to provide better combustion and higher efficiency than atmospheric burners. A power draft blower is used to extract the combustion products and discharge them to a vent pipe or chimney.

The heating capacity of gas furnaces is controlled by a gas valve and an ignition device. The gas supply to a burner is controlled by a pressure regulator and a gas valve, which controls the firing rate. Smaller furnaces are controlled by a room, or duct, thermostat, which opens the gas valve on a call for heat and closes the gas valve when the thermostat is satisfied. Larger furnaces usually have a two stage gas valve, controlled by a two stage thermostat. The two stage gas valve has two open positions (partly open or fully open) to control the flow of gas depending upon the heating load.

There are three methods used to ignite the fuel/air mixture when the gas valve opens. The first is a standing pilot light which is a small continuous flame. Ignition may also be an intermittent pilot flame which is off when the burner is off, and lights prior to the opening of the main gas valve. The third method is by direct spark, or hot surface ignition, which ignites the main burners directly.

Furnace safety controls include a temperature sensitive fan switch which is exposed to the circulating air stream. The fan switch will close the gas valve if excessive temperatures are detected. Other safety controls include a pilot light and gas pressure sensors.

Furnace efficiency is an important factor to consider when selecting a warm air furnace. The performance of warm air furnaces is usually measured by one of the following parameters:

1. Steady State Efficiency is the efficiency of a furnace when operated under steady state, equilibrium conditions. It is calculated by measuring the energy input, subtracting the losses for exhaust gases, and dividing by the full input.

$$\eta = \frac{q_I - q_L}{q_I} \times 100\% \qquad (17.6)$$

where

q_I = fuel input in Btu/hr

q_L = flue losses in Btu/hr

2. **Utilization Efficiency** is obtained by assuming the furnace is 100% efficient and sub-tracting losses for exhaust sensible and latent heat, cyclic effects, infiltration of com-bustion air, and pilot-burner effect.

3. **Annual Fuel Utilization Efficiency (AFUE)** is the same as the Utilization Efficiency, ex-cept that losses from a standing pilot light during non-heating season time are deducted.

$$\eta = \frac{\text{annual energy output (Btu/hr)}}{\text{annual energy input (Btu/hr)}} \times 100\% \qquad (17.7)$$

Gas furnaces presently have steady state efficiencies that vary from 75% to 95%. At-mospheric and power venting furnaces typically have efficiencies from 75% to 85%. Con-densing gas furnaces can have steady state efficiencies as high as 95%. Furnaces are usu-ally rated in terms of fuel, or heat, input and heat output.

LP gas furnaces are very similar to natural gas furnaces. The major difference between the two furnaces is the pressure at which the gas is injected into the burners. For natural gas, the pressure is usually controlled at 3 in. wg to 4 in. wg. For LP furnaces, the pressure is usually 10 in. wg to 11 in. wg.

Fuel oil furnaces are also similar to natural gas furnaces. The major difference between oil and gas furnaces is the combustion systems, heat exchanger, and a barometric draft reg-ulator, which is used in lieu of a draft hood. Heat exchangers for oil furnaces are usually heavy gauge steel formed into a welded assembly.

Furnaces inside buildings must be properly vented and have adequate combustion air. These requirements are usually governed by codes. Some of the code requirements for fuel burning equipment, such as furnaces, are discussed in Chapter 11.

17.5   HEAT PUMPS AND REFRIGERATION SYSTEMS

Another commonly used method of heating air is the heat pump. Basically, a heat pump is a reverse air cooling system that operates on a reverse refrigeration cycle. There are a number of sources available for use as a heat sink in a heat pump cycle, which include am-bient air, water, ground (well) water, and the earth itself. Air is the most commonly used heat source (heat sink) and is the type of heat pump discussed herein. The term heat pump is used to describe a year round heating and air conditioning system that operates on a re-frigeration cycle. Heat is taken from a heat source (or heat sink) and supplied to the condi-tioned space when heat is required, and is removed from the space and discharged to the heat sink when cooling and dehumidification are required. Unitary or packaged heat pump systems are typically available from $1^1/_2$ tons to 25 tons.

Since heat pumps operate on a refrigeration cycle, a basic discussion of refrigeration systems is in order. A basic direct expansion (DX) system is shown in Figure 2.3. The ma-jor components include a compressor, a condenser, an evaporator, and a thermal expansion valve (TXV). Since heat only flows naturally from a high temperature region to a low tem-perature region, it is necessary to apply energy, in the form of work, to a refrigeration sys-tem. In the DX system, a refrigerant is circulated by the compressor. The refrigerant is a substance that will undergo a phase change (change from a gas to a liquid and back to a gas) as it goes through the cycle. The refrigerant enters the compressor as a gas and is compressed. Increasing the pressure of the gas causes its temperature to increase and it becomes a superheated vapor. The superheated refrigerant vapor then passes through a condenser where the refrigerant rejects much of its heat. The heat is usually rejected directly to the atmosphere. When the refrigerant comes out of the condenser, it is a saturated vapor at high pressure. (A saturated substance is one in which vapor, or liquid phases, can be present as a given temperature and pressure.) The saturated vapor refrigerant then passes through a TXV. The TXV reduces the pressure of the refrigerant, which also reduces its temperature. The refrigerant becomes almost 100% saturated liquid at a much lower temperature and pressure. The saturated liquid then passes through the evaporator. As the name implies, the liquid refrigerant evaporates and absorbs heat from the circulated air as it passes through the evaporator. The evaporator in an air conditioning system is the cooling coil. When the refrigerant comes out of the evaporator, it is again a vapor and the cycle repeats.

The compressor is a device used to add work to the system by pressurizing and circulating refrigerant through the system. Most compressors used in DX air conditioning and heat pump systems are the reciprocating type, although they may also be centrifugal type or screw type. Since liquids are incompressible (for the most part), compressors can only compress vapors and gases. All of the refrigerant entering a compressor must be in the vapor phase.

The condenser is a heat exchanger where the heat of vaporization and compression is transferred from the hot refrigerant gas to a heat sink, usually air. The refrigerant in the condenser is changed from a superheated vapor to a compressed liquid state.

The thermal expansion valve (TXV) is a metering device that controls the flow of refrigerant and reduces its pressure. After passing through the TXV, the refrigerant is a low pressure, saturated liquid. Since the refrigerant pressure after the TXV is low, the refrigerant will begin to boil, or evaporate, at a temperature of about 35°F.

The evaporator coil is used to evaporate the saturated liquid refrigerant. As the refrigerant changes from the liquid phase to the vapor phase, it absorbs heat. The heat that is absorbed in the evaporator is from the air circulated through the evaporator coil. The refrigerant is completely evaporated before it leaves the evaporator and is drawn into the compressor to repeat the cycle.

The performance, or efficiency, of a refrigeration system is indicated by the coefficient of performance (COP), or the energy efficiency ratio (EER) of the system.

$$\text{COP} = \frac{q}{W} \tag{17.8}$$

where

q = the refrigerating effect, Btu/hr

W = work of power input to the system, Btu/hr

$$\text{EER} = \frac{q}{W} \tag{17.9}$$

where

q = the refrigerating effect, Btu/hr

W = the work or power input to the system, W

A heat pump is essentially identical to the typical refrigeration system shown in Figure 2.3, with the exception that a reversing valve is added to the system. Figure 17.9 shows a refrigeration system of the most common type of air source heat pump.

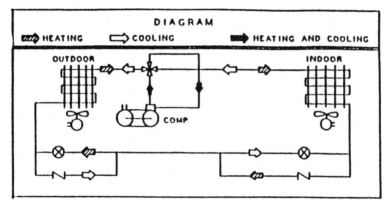

FIGURE 17.9   Air Source Heat Pump Refrigeration System Schematic

In the cooling mode, the refrigerant flow is such that the evaporator coil is located in the indoor air flow and the condenser is located outdoors. In this arrangement, heat is absorbed by the evaporator located indoors and rejected to the outdoors by the condenser. In the heating mode (heat pump mode), the reversing valve is used to switch the refrigerant flow, causing the location of the evaporator coil and condenser coil to be reversed. In the heating mode, the evaporator absorbs heat from the outdoor air and the condenser rejects the heat to the indoor air.

As the outdoor air temperature decreases, the efficiency of the heat pump decreases and the heating capacity also falls off. Selecting equipment for a given outdoor heating design temperature is more critical than for other types of heating systems. Usually, some other means of auxiliary heating is required when outdoor air temperatures are below 30°F. A supplemental heater is often provided with a heat pump system to provide heat when the capacity of the heat pump does not meet the required heating during low outdoor temperatures.

Most air source heat pumps also use the reverse cycle to melt frost that has formed on the outdoor coil when the system operates in the heating mode in cold weather. The reverse operation supplies hot gas to the outdoor coil, which melts the frost that has formed on the coil. After the frost has melted, the system switches back to the heating mode.

For a more detailed discussion of air source heat pumps, as well as other types of heat pumps, consult Reference 3.

17.6   AIR SYSTEM HUMIDITY CONTROL

In addition to controlling the temperature of an occupied space, it is frequently necessary to control the moisture level or relative humidity of the space.

When a cooling coil cools the air passing through it to a temperature that is below the dew point of the air, moisture will condense out of the air. Cooling coils are the primary method for removing moisture from air and for maintaining a maximum relative humidity level in the conditioned space. Air cooling coils were discussed earlier in this chapter.

Whenever cold outdoor air comes into a conditioned space, its moisture level is usually considerably below that of the conditioned space. Low moisture content outdoor air may enter the space by infiltration, or through the air distribution system, in the form of outdoor ventilation air. In order to maintain the relative humidity in the conditioned space at a desired level, it is frequently necessary to add moisture to the air by a humidification process. Moisture is usually provided to a conditioned space via the air circulated by the air system, although it may also be added to the conditioned space directly. The psychrometric processes involved in the humidification process are discussed in Chapter 9.

The additional moisture needed in a conditioned space is usually added by a humidifier, which is controlled by a humidistat located in the conditioned space or in the supply air duct to the space. The rate at which moisture must be added is usually determined by the amount of outdoor air introduced to the space by infiltration, or through the AHU. The humidification load may be calculated with Equation 17.10.

$$H = 4.5Q_{oa}(W_{ia} - W_{oa}) - S + L \qquad (17.10)$$

where

h = humidification load, lb_{H_2O}/hr

Q_{oa} = outside air flow rate, ft³/min

W_{ia} = indoor air humidity ratio, lb_{H_2O}/$lb_{dry\ air}$

W_{oa} = outdoor air humidity ratio, lb_{H_2O}/$lb_{dry\ air}$

S = contribution from internal moisture sources, lb/hr

L = other moisture losses, lb/hr

The most common types of humidifiers used in commercial HVAC systems are the heated pan type, the steam grid type, and the self-contained type. The heated pan type is shown in Figure 17.10.

The heated pan humidifier consists of a pan with water in it, a steam or electric coil to heat and vaporize the water, and a make-up water float valve. The pan is attached to the bottom side of a duct so the air passes over the surface of the water. The heating coil heats the water, causing it to vaporize and be entrained in the air as it passes. A make-up water float valve is used to admit make-up water as it is vaporized and maintain the required water level in the pan.

A more commonly used type of humidifier is the steam grid type, as shown in Figure 17.11.

The steam grid type consists of a steam distribution manifold, mounted in the supply air duct, as well as a steam flow control valve and a steam/condensate separator. Low pressure steam, usually from a central plant, is injected into the air stream by the manifold. The control valve controls the flow of steam into the grid, usually in response to a humidistat in the conditioned space or supply air ductwork.

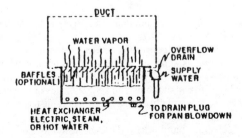

FIGURE 17.10   Heated Pan Type Humidifier

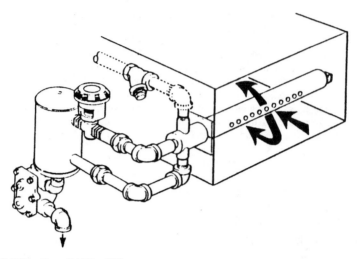

FIGURE 17.11   Steam Grid Humidifier

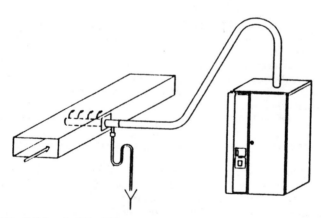

FIGURE 17.12   Self Contained Humidifier

The third type of humidification commonly used in commercial HVAC systems is the self contained type shown in Figure 17.12.

The self contained type usually consists of an electric steam generator and distribution tubing. Domestic cold water is supplied to the steam generator where it is electrically heated to produce low pressure steam. The steam is then injected directly into the air stream in a supply air duct with distribution tubing and spray nozzles or a manifold. The humidity level in the conditioned space is usually controlled by a humidistat, which controls the electrical power to the steam generator.

When sizing and selecting humidifiers, there are a number of issues that should be considered:

- the humidifier should not be located upstream of cooling coils
- if air filters are to be located downstream of the humidifier, care must be taken ensure the moisture from the humidifier is not trapped by the air filters
- a means to shut the humidifier off should be provided in the event of no air flow in the duct
- when using steam from a central plant for humidification, ensure that there are no hazardous chemicals used in the treatment of the steam to prevent corrosion

Chapter 18

AIR CLEANING
AND FILTRATION

The primary objective in the design and selection of HVAC systems is to provide the desired temperature and relative humidity, as well as adequate ventilation, in the occupied space. Although this is the primary consideration for HVAC systems, the cleanliness of the air in the occupied space and in the HVAC system is also very important. Chapter 11 discussed code requirements for HVAC systems, including the requirements for outdoor ventilation air. The main purpose for bringing outdoor air into a building through the HVAC system is to provide clean, uncontaminated air in the occupied space. Outdoor air, as well as recirculated indoor air, contains contaminants which are not desirable in the space or in the air system. In order to provide air that is relatively free of contaminants, the air must be filtered and cleaned by the air handling system.

Airborne atmospheric dust has a broad range of sizes at any particular location. It is a complex mixture of smoke, mists, fumes, dry granular particles, and fibers. The particles in the atmosphere range in size from less than $0.01\mu m$ ($1\mu m$ = 1 micron = 10^{-6} m) to the dimension of lint, leaves, and insects. The extent to which impurities and contaminants are present in air is termed their concentration. Concentration may be expressed in one of two different ways: mass per unit volume of air (g/m^3) or volume per unit of volumes of air (parts per million, ppm).

The effect of dust in HVAC systems primarily depends on the size of the dust particles. Coarse dust will tend to eventually clog the openings between the fins in heating and cooling coils. Fine dust particles will tend to cause discoloration and smudging on surfaces such as walls and ceilings. Indoor air pollutants may be controlled by means of source capture, or dilution with outdoor ventilation air. Pollutants in mixed outdoor and recirculated air may be cleaned through filtration.

The terms air cleaner and air filter may be used interchangeably when applied to HVAC systems. The performance characteristics of air filters that are most important in HVAC systems are the ability of the filter to remove particles from the air stream, its resistance to air flow, and its operating time before cleaning or replacement of the filter media. The major factor influencing filter design and selection is the degree of cleanliness required. In general, the cost of the filter or filtration system will increase as the size of the particles to be removed decreases, as will the static pressure drop through the filter.

18.1 AIR CONTAMINANTS

In Chapter 13, various sources of air contamination in buildings were discussed. The primary method for controlling various contaminants in buildings is the introduction of outdoor ventilation air through the HVAC system. ASHRAE Standard 62 prescribes the required outdoor air rates for various types of building occupancies. The prescribed ventilation rates are based on the assumption that the outdoor ventilation air is relatively clean and free of contaminants.

Outdoor air intake locations should always be away from sources of contamination, such as vehicle loading docks, cooling towers, plumbing vent outlets, kitchen exhaust outlets, exhaust fan discharge, etc. However, even air that appears to be quite clean and free from contamination will also contain some contaminants. Any materials other than oxygen, nitrogen, carbon dioxide, water vapor, and rare gases present in air are considered to be contaminants. The most common contaminant is atmospheric dust, as described above. Atmospheric dust typically contains soot, smoke, silica, clay, decayed animal and plant matter, organic materials, and metallic fragments. It may also contain living organisms such as mold, spores, bacteria, and plant pollens.

Dusts are defined as solid granular particles and fibers of sizes less than 100 m. Living organisms include viruses from 0.003 m to 0.06 m, and bacteria between 0.4 m and 5 m. Fungal spores and pollen vary in size from 10 m to 100 m. Gases and vapors consist of dispersed molecules which are suspended in air. Gaseous contaminants in air can result in odors and, in some cases, irritating or toxic effects. Figure 18.1 shows the relative size of common air contaminants.

The mixture of granular particles, fibers, smoke, fumes, mists, and gases in a mixture with air is referred to as an aerosol. Aerosol particle sizes can be divided into three classifications:

• Coarse particle: mostly soil particles suspended naturally by wind erosion, agricultural activities, construction, or travel on unpaved roads. Particles are produced by mechanical fracture of solids.

• Fine particle: mostly combustion products with some production by atmospheric reactions, agglomeration, and suspension of very fine particles.

• Nuclei: produced by evaporation, recondensation, and reactions of combustion processes such as gasoline and diesel engines.

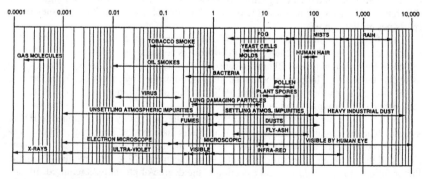

FIGURE 18.1   Relative Size Chart of Common Air Contaminants

18.2   TESTING AND RATING OF AIR FILTERS

In order to make accurate comparisons among air filters, it is necessary to perform standardized tests to determine their relative performance. Air filter testing is complex and no single test adequately describes all air filters. For example, the wide variation in the amount and type of particulate matter in the air being cleaned makes rating difficult. It is also difficult to closely relate measurable performance to specific uses and applications. Recirculated air tends to have a higher proportion of lint than does outdoor air. ASHRAE Standard 52 specifies the test methods used to rate air cleaners.

Caution must be exercised in using published data on filter efficiency, because the performance of two cleaners tested by different methods generally cannot be compared. Complete rating of air cleaners requires data on efficiency, resistance to air flow, dust holding capacity, and the effect of dust loading on efficiency and resistance.

There are three important characteristics that are used to rate and compare air filters. They are efficiency, air flow resistance, and dust holding capacity. Efficiency is a measure of the ability of the air cleaner to remove particulate matter from the airstream. The average efficiency, during the life of the filter, is an important measure for most air filters. The characteristics of aerosols that affect the performance of air filters the most include particle size and shape, specific gravity, concentration, and electrical properties. The removal of particles becomes progressively more difficult as particle size decreases.

Air flow resistance is the resistance to air flow through the filter that results in a pressure drop, from the upstream side of the filter to the downstream side. As the filter efficiency increases, the resistance also increases in general.

The dust holding capacity is a measure of the amount of a particular type of dust that a filter can hold when air is flowing through it at a specified rate before it reaches some specified maximum resistance. The dust holding capacity may also be based on the filter resistance before the filter efficiency is seriously reduced, due to the collected dust. The exact measurement of true dust holding capacity is complicated by the variability of atmospheric dust; therefore, a standardized dust is used. Synthetic dusts are not the same as atmospheric dust.

There are four basic types of tests, plus variations, that are used to rate air cleaners. These are the arrestance, dust spot efficiency, the dust holding capacity, and particle size efficiency.

18.2.1   Arrestance Test

The arrestance test utilizes a standardized synthetic dust (ASHRAE Test Dust), consisting of various particle sizes. During the test, the dust is fed into the airstream, passing through an air cleaner, and the weight fraction of the dust removed by the filter is measured. The test is also called the synthetic dust weight arrestance. A known amount of prepared test dust is fed into the airstream. The concentrations of the dust in the air leaving the filter is determined by passing the entire air flow through a high efficiency air filter, located downstream of the test filter.

The after filter is then weighed to measure its increased weight. The weight percentage, or arrestance of the dust removed by the filter, is then calculated based on the original total weight of the test dust.

The test primarily determines the ability of a filter to remove the largest atmospheric dust particles, and provides little indication of the filter's performance in removing the smallest dust particles. The test is used primarily to test low efficiency filters.

The indicated weight arrestance depends largely on the particle size distribution of the test dust. It also depends on the tendency of the dust to agglomerate. A high degree of standardization of the test is required. The test dust is composed of 72% Standard Air Cleaner Test Dust, Fine, 23% powered carbon, and 5% cotton lint. The dust cloud used in the arrestance test is considerably coarser than typical atmospheric dust.

18.2.2   Dust Spot Efficiency

Atmospheric dust is fed into an airstream which is passed through an air cleaner being tested. The discoloration effect of the cleaned air on filter paper targets is compared to that of the incoming air. By measuring the change in light transmitted by the filters, the efficiency of the filter in reducing the soiling of surfaces may be compared. This type of measurement is called the atmospheric dust spot efficiency.

Since these effects depend primarily on fine particles, the test is mostly used for medium or high efficiency filters that remove very small particles. The dust spot test measures the ability of a filter to reduce the soiling and smudging of interior building surfaces.

18.2.3   Particle Size Efficiency

Atmospheric dust is fed into the airstream entering the air cleaner, and samples are taken upstream and downstream of the filter. The samples are drawn through a particle counter to obtain the filter efficiency for each size of particle. Particles are counted both upstream and downstream of the filter to obtain an average efficiency count. Both laser and white light optical particle counters are used.

The test is highly dependent upon the type of aerosol used for testing and, to a lesser extent, the particle spectrometer used to determine the particle count.

18.2.4   Dust Holding Capacity Test

Synthetic ASHRAE Test Dust is fed into the airstream, and passes through the filter being tested. The resistance (pressure drop) of the filter to the flow of air is measured as the dust is fed into the airstream. The test is stopped when the resistance reaches the maximum allowable resistance set by the filter manufacturer. The ASHRAE Dust Holding Capacity, therefore, is the integrated amount of dust held by the filter up to the time the resistance reached the maximum value and the test was terminated.

18.3   TYPES OF AIR FILTERS

The methods used by air filters to collect and arrest dust particles are based on five main principles, or mechanisms. The mechanisms are straining, direct interception, inertial deposition, diffusion, and electrostatic effects.

Straining involves the collection of the largest and coarsest particles. As the air passes through a membrane opening that is smaller than the particles being removed, the air passes through the media while the particle, which is too large, is arrested or strained. The fibers of the filter media are so close together that a large particle cannot pass. This method is most applicable to the collection of large particles and lint.

Direct interception involves passing the air streamline close enough to the filter media fibers that the passing particle contacts the fibers of the media and remains there. The process is almost independent of air velocity.

Inertial deposition uses inertial forces of the particles to collect dust. This method is also known as viscous impingement or inertial impaction. The mass of the particles is large enough that their inertia prevents them from following the air streamline as it changes direction. The particles cross the streamlines, contact the filter media, and remain there. At high velocities, the particles may not adhere to the media because the drag forces are high.

Diffusion results when very small particles (less than 0.4 μm) have random motion about their basic streamlines (Brownian Motion). This results in deposition on the filter media. This deposition causes a concentration on the media which further enhances filtration by diffusion. The very small particles are bombarded by the random motion of the air molecules that are driven into the filter fibers where they are arrested.

Electrostatic effects result from the particles or filter media being charged by an electrical field. Electrostatic effects are used in electronic air cleaners which are discussed in the next section.

18.3.1 Panel Filters

Panel filters, also known as flat filters, are the most commonly used type of air filters for comfort HVAC applications. They are typically flat panels, mounted perpendicular to the direction of air flow. Panel filters are typically low efficiency (less than 30%), although they are also available as medium efficiency filters. These filters are usually low pressure drop, low cost, good on lint, but low efficiency on atmospheric dust. Panel filters are occasionally used as prefilters; that is, they are installed upstream of higher efficiency filters. They are usually installed on the intake or upstream side of coils. Panel filters typically come in thicknesses ranging from $1/2$ in to 4 in.

Most panel filters use the viscous impingement (inertial deposition) to collect dust particles. The filters may be constructed of inexpensive materials and discarded after use, or they may be a permanent type that may be cleaned and re-used. Most panel filters are throwaway types that are disposed of after they have been used, although they are available with washable media and frames. Panel filters are comprised of coarse fibers with a high porosity as the filter media. The media is usually coarse (15 μm to 60 μm diameter) glass fibers, vegetable fibers, synthetic fibers, metallic wools, expanded metals and foils, crimped screens, random-matted wire, and synthetic open-cell foams. Panel filters are composed of coarse fibers with a high porosity as the filter media. The filter media is coated with a viscous material, such as oil. The viscous material is used as an adhesive, which causes the dust particles to stick to the media when they come in contact with it. The filter media is usually held in place by wire frames. Figure 18.2 shows typical panel filters.

Typical face velocities for panel filters range from 200 ft/min to about 800 ft/min. The limiting factor, other than increased resistance to air flow, is the tendency for agglomerations of collected dust to be blown away by the high air velocity. Panel filters are available with flat or pleated filter media. The pleats are V-shaped and provide an extended surface allowing higher face velocities. Pleating of the media allows a higher ratio of media area to filter face area, which allows higher air velocities at reasonable pressure drop. Panel filters may also be mounted in an arrangement of angled filters, or V-banks, to achieve lower face velocities through the filters. The V-bank arrangement provides additional filter face area for the air to pass through, as shown in Figure 18.3.

Panel filters for commercial HVAC systems usually have an initial resistance from 0.1 in. wg to 1.0 in. wg. The filters usually need to be replaced or cleaned when their final resistance reaches 0.5 in. wg to 2 in. wg, depending on the filter. Most panel filters are replaced after 0.5 in. wg resistance is reached, since their efficiency will decline following it. The de-

FIGURE 18.2   Panel Filters (*Courtesy of the Farr Company*)

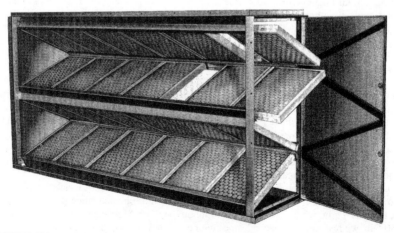

FIGURE 18.3   Angled or V-Filters (*Courtesy of Carrier Corporation*)

FIGURE 18.4   Roll Filter (*Courtesy of Carrier Corporation*)

cline in efficiency results from the absorption of the viscous coating by the dust, rather than the increase in resistance due to the collection of dust. When designing HVAC systems, it is necessary to consider the filter resistance when it is fully loaded with dust.

18.3.2   Renewable Filters

Just as the name implies, renewable filters have a filter media that is renewable. These filters may also be called roll filters. The filter media, which is made of a media that can be woven into a blanket, is mounted on a roll. The filter media is usually disposable material. The renewable filter is usually a viscous impingement type material. The media is typically about 2 in thick and held in place by a wire mesh screen. The efficiency of roll filters is typically between low and medium.

As the filter media becomes loaded with dust, new media is pulled off a roll, while the loaded media is pulled onto another roll. The filter media is usually advanced automatically as it becomes loaded to reduce maintenance. The rolls are usually actuated by electric motors. A typical roll filter is shown in Figure 18.4.

The roll motor is typically actuated automatically by a pressure switch, although time and media light-transmission controls are also used.

Face velocities are usually around 500 ft/min and initial resistances range from 0.40 in. wg to 0.50 in. wg. The resistance remains constant as long as the proper operation of the filter system is maintained.

18.3.3   High Efficiency Air Filters

High efficiency filters have filtration efficiencies of 85% and above. These filters are always extended media surface filters. Since the filtration efficiencies are so high, extended media surfaces are necessary to keep air velocities through the media and resistances to reasonable values. High efficiency filters can be up to, and greater than, 12 in in the direction of air flow. The media surface of high efficiency filters can be 50 times the face area. High efficiency filters come in two styles: pleated media panel type and bag type.

The pleated panel type is similar to the lower efficiency type, except the thickness of the filter is much greater to allow for the necessary pleats. The media has deep space folds to provide extended media surface. The filter media for the panel type high efficiency filters is usually sub-micron glass fiber paper. A typical high efficiency filter is shown in Figure 18.5.

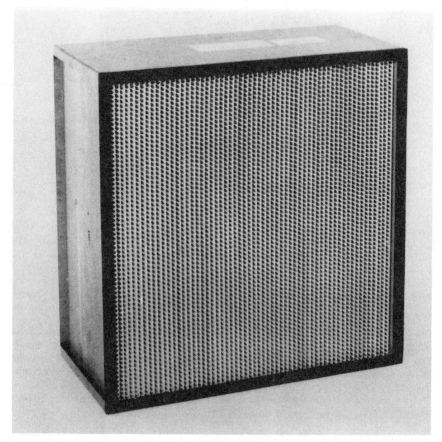

FIGURE 18.5   High Efficiency Panel Filter (*Courtesy of the Farr Company*)

High efficiency filters can also be bag type of extended media surface, as shown in Figure 18.6. The media is arranged as a bag, which provides a significantly larger media surface area.

High efficiency filters can be classified as HEPA (high efficiency particulate air) and ULPA (ultra low penetration air) filters. HEPA filters have efficiencies of 99.97% for dust particles greater than 0.3 μm in diameter. ULPA filters have efficiencies of 99.999% for dust particles greater than 0.12 μm diameter. High efficiency filters operate at duct velocities around 250 ft/min and with resistances from 0.5 in. wg to 2.0 in. wg.

The filter media itself usually has a higher efficiency than the mounted filter assembly. Sealing the filter within its frame and sealing it between the frame and housing are important considerations. Since high efficiency filters remove such small particles, it is advisable to install lower efficiency prefilters upstream of the high efficiency filters. The removal of larger particles by the lower efficiency prefilters reduces the dust loading on the high efficiency (final) filters, which extends the service life of the more expensive high efficiency filter. High efficiency filters are frequently required for hospital operating rooms,

FIGURE 18.6   High Efficiency Bag Filter (*Courtesy of the Farr Company*)

microelectronics manufacturing, pharmaceutical laboratories, and precision manufacturing operations.

18.4 ELECTRONIC AIR CLEANERS

Electronic air cleaners use the principle of electrostatic precipitation to collect dust. These cleaners employ the principle of attraction between bodies with opposite electrical charges. Dust particles are charged either naturally or within the cleaner to attach them to collector plates or media and to agglomerate particles to greater sizes. Due to the numerous points of contact between the collected particles, the bond between the particles holds them together by intermolecular forces. These intermolecular forces are much greater than the attraction forces between the particles and the collectors. As a result, the particles agglomerate and grow in size to a point where they are blown off the collector plates and carried away by the airstream. The agglomerated particles are then typically collected by an afterfilter located downstream of the electronic air cleaner.

Electronic air cleaners are effective for removing dust, smoke, and pollen from the airstream. These cleaners can remove airborne contaminants with average efficiencies of up to 98% at air flow velocities ranging from 150 ft/min to 350 ft/min. The pressure drop of the electronic air cleaner itself is quite low and ranges from 0.15 in. wg to 0.25 in. wg at face air velocities between 300 ft/min and 500 ft/min. The efficiencies of the cleaners usually decrease as the collectors become loaded with particles, with higher air velocities or with nonuniform air flow through the cleaner. Electronic air cleaners must be periodically cleaned to remove collected particulate matter. Detergent and hot water is usually used to clean collector plates. Some electronic air cleaners include automatic washing systems.

There are three types of electronic air cleaners available for commercial HVAC systems: the ionizing plate type, the charged-media nonionizing type, and the charged-media ionizing type.

The **ionizing plate type air cleaner** consists of two parts, or stages. The first stage is the ionizer, which charges the particles in the aerosol, and the second stage is the plate package to collect the charged particles. The principle of operation for an ionization type cleaner is shown in Figure 18.7. Positive ions are generated by high potential ionizer wires as the air flows across them. Any particles that strike the wires become ionized. A direct current, between 6 kVDC and 25 kVDC, is applied to the wires. The high voltage on the wires creates an ionizing field for charging the particles in the airstream. The intense electrical field in the vicinity of the wires ionizes the oxygen molecules nearby. As the aerosol passes through the ionized field, some of the ions attach themselves to the dust particles, giving the dust particles a charge. The ionizing wires are typically made of tungsten steel.

The dust particles then pass through a set of alternating charged and grounded plates. A positive direct current, between 4 kVDC and 10 kVDC, is applied to alternating plates. The other plates are grounded. As the particles pass between the charged plates, the ionized dust particles are driven to the plates by the force exerted by the electric fields on the charges that the plates carry. The positively charged duct particles are attracted by the grounded plates and repelled toward the grounded plates by the positively charged plates.

Collector plates are frequently coated with oil, which serves as an adhesive. Electrical forces drive the particles to the collector plate surface where they are held by intermolecular adhesion forces. Occasionally, air cleaners are used without any adhesive on the collec-

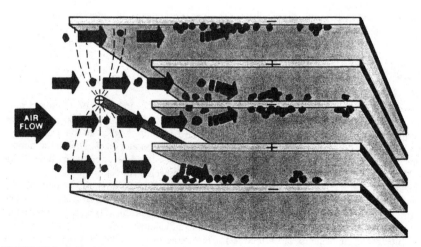

FIGURE 18.7   Ionized Plate Operating Principle

tor plates. In this case, the precipitator may form agglomerates, which are eventually blown off and are collected downstream by an afterfilter.

Ionization cleaners are well suited for removing fine dust and smoke particles from the air stream. The efficiency of these air cleaners is a function of the air velocity through the cleaner. Typically, these filters have little resistance to air flow; however, prefilters are frequently used with the electronic air cleaners. The air flow entering the cleaner should be uniformly distributed across the face of the air cleaner.

Charged media nonionization electronic air cleaners combine the features of both dry and electronic cleaners. They are composed of a di-electric filtering media (usually a pleated panel filter); however, no ionization effects are used. The di-electric pleated filter consists of alternating grounded and positive charged members.

The positive charged members typically have a 12,000 VDC charge. The alternating members create an intense electrostatic field. The airborne particles approaching the field are polarized and drawn into filaments of fibers in the media. These types of electronic air cleaners offer little air flow resistance, typically 0.10 in. wg at air velocities about 250 ft/min, when clean. The resistance of the cleaner increases as the media collects dust or as the air velocity increases. As a result of the dust buildup, the filter tends to equalize the air flow over the face of the filter.

These filters should not be used if the relative humidity of the air exceeds 70%, because the moisture will affect the di-electric properties of the media.

Charged media ionizing electronic air cleaners use a combination of the processes of the ionization-plate type and the charged media nonionization type of electronic air cleaners. Dust particles are charged in a corona-discharge ionizer, then collected by a charged media filter mat. This combination provides higher efficiencies than either of the two types alone.

18.5 ACTIVATED CARBON AIR FILTERS

All the previously discussed air filters are effective in removing particulate contaminants from air to various degrees. None of them, however, are very effective on gaseous contaminants or odors. Filters with activated materials may be used to remove objectionable odors and vapors of gaseous airborne contaminants, such as smoke, through a process called adsorption.

Adsorption is the physical condensation of a gas or vapor on an activated substance. During the adsorption process, gas molecules diffuse into the micropores or macropores of the adsorption material, adhere to these surfaces, and bond with the surfaces. The material used as an adsorber is usually an activated substance. Activated substances are highly porous. As an example, 1 lb of extremely porous carbon contains an internal surface area of more than 50,000 ft². The activation process creates extensive surfaces where adsorption can take place. This results in a high adsorptive capacity for gases and vapors. Adsorbers are made into granules, pellets, and fibers, and are formed into porous beds. The beds provide a large area where air can pass and come in contact with the adsorber surface. In addition to activated carbon, other adsorbent materials include zeolite, silica gel, activated alumina, and mica.

Adsorption capacity is defined as the amount of carbon tetrachloride adsorbed by a given weight of activated carbon. The adsorption capacity is affected by operating temperature and humidity level. In general, higher temperatures and humidity levels decrease adsorption capacity. The maximum operating temperature is about 100°F.

The most commonly used adsorption material used in HVAC applications is activated carbon. 1 lb of activated carbon may adsorb 0.2 lb to 0.5 lb of gases. Activated carbon can

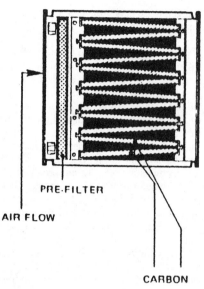

PRE-FILTER

AIR FLOW

CARBON

FIGURE 18.8   Activated Carbon Air Filter

remove particulate matter as small as 0.003 m to 0.006 m in diameter by adsorption. Among the odors and materials that activated carbon is quite effective in neutralizing are alcohol, antiseptics, body odors, cigarette smoke, cooking odors, gasoline, plastics, rubber, sewer odors, and solvents.

Activated carbon is made from coconut shells, coal, or petroleum residues. It is usually provided in the form of granules or pellets. The structure of activated carbon is such that it contains millions of microscopic pores. These pores provide the extremely large surface area which takes up and holds the substances that are adsorbed by the activated carbon.

Activated carbon air filtration units typically consist of a set of prefilters and multiple trays of activated carbon, as shown in Figure 18.8. The activated carbon is usually placed in trays through which the air to be cleaned must pass. Low efficiency panel filters are usually used as prefilters for the carbon.

The air-purifying effectiveness of activated carbon depends, in part, on the length of time the air is in contact with the carbon. Air velocities through activated carbon filters should be kept low. Air face velocities are typically between 375 ft/min and 500 ft/min, with corresponding resistances between 0.2 in. wg and 0.3 in. wg.

The adsorption efficiency of the activated carbon bed remains almost constant for the service life of the filter. When the activated carbon filter has reached a saturation point where it cannot adsorb any more contaminants, the filter must either be reactivated or regenerated. Reactivation is the process of removing the spent carbon and replacing it with new carbon. Regeneration is the process by which the adsorbed contaminants are removed from the spent carbon and converted to fresh. The regeneration process for activated carbon consists of passing air, nitrogen, or steam through the spent carbon. Collected contaminants are disorbed from the carbon by the high temperature.

Airborne gaseous contaminants may also be removed by a process called chemisorption. During chemisorption, gas molecules bond to the surface of the chemisorption media through chemical action instead of physical adsorption.

TABLE 18.1 Typical Filter Applications for HVAC Systems

Application	System Designator[b]	Prefilter		Prefilter 2/Filter		Final Filter	Application Notes
Warehouse, storage, shop and process areas, mechanical equipment rooms, electrical control rooms, protection for heating and cooling coils	A1	None	None	50 to 85% arrestance	Panel-type or automatic roll	None	Reduce larger particle settling. Protect coils from dirt and lint.
	A2	None	None	25 to 30% dust spot	Pleated panel or extended surface	None	
Special process areas, electrical shops, paint shops, average general offices and laboratories	B1	None	None	75 to 90% arrestance 35 to 60% dust spot	Extended surface, cartridge, bag-type, or electronic (manually cleaned or replaceable media)	None	Average housecleaning. Reduces lint in air stream. Reduces ragweed pollen >85% at 35%. Removes all pollens at 60%, somewhat effective on particles causing smudge and stain.
Analytical laboratories, electronics shops, drafting areas, conference rooms, above-average general offices	C1	75 to 85% arrestance 25 to 40% dust spot	Extended surface, cartridge, or bag-type	>98% arrestance 80 to 85% dust spot	Bag-type, cartridge, or electronic (semi-automatic cleaning)	None	Above average housecleaning. No settling particles or dust. Cartridge and bag types very effective on particles causing smudge and stain, partially effective on tobacco smoke. Electronic types quite effective on smoke.
	C2	None	None	>98% arrestance 80 to 85% dust spot	Electronic (agglomerator) with bag or cartridge section	None	
Hospitals, pharmaceutical R&D and manufacturing (nonaseptic areas only), some clean ("gray") rooms	D1	75 to 85% arrestance 25 to 40% dust spot	Extended surface, cartridge, or bag-type	>98% arrestance 80 to 85% dust spot	Bag-type, cartridge, electronic (semi-automatic cleaning)	95% DOP disposable cell	Excellent housecleaning. Very effective on particles causing smudge and stain, smoke and fumes. Highly effective on bacteria.
	D2	None	None	>98% arrestance 80 to 95% dust spot	Electronic (agglomerator) with bag or cartridge section	None	
Aseptic areas in hospital and pharmaceutical R&D and manufacturing. Cleanrooms in film and electronics manufacturing, radioactive areas, etc.[c]	E1	75 to 85% arrestance 25 to 40% dust spot	Extended surface, cartridge, bag-type	>98% arrestance 80 to 85% dust spot	Bag-type, cartridge, electronic (semi-automatic cleaning)	≥99.97% DOP disposable cell	Protects against bacteria, radioactive dusts, toxic dusts, smoke, and fumes.

[a]Adapted from a similar table courtesy of E.I. du Pont de Memours & Company.
[b]System designators have no significance other than their use in this table.
[c]Electronic agglomerators and air cleaners are not usually recommended for clean room applications.

18.6 AIR FILTER SELECTION GUIDELINES

When evaluating filters and cleaners, there are three main factors that initially must be given careful consideration. These are:

- the degree of air cleanliness required
- disposal of the dirt after it has been collected
- the amount and type of contaminants in the air to be cleaned

The above factors directly affect the initial costs, operating costs, and the level of maintenance that will be required. Operating costs, expected service life, and efficiency are as important as first cost.

Filters are normally placed upstream of heating and/or cooling coils, as well as other air conditioning equipment in the system, to protect them from dust. Table 18.1 shows typical filter applications for HVAC systems.

The installation of the filters is as important as the filters themselves. The in-service efficiency of filters is significantly reduced if there are air leaks around the filter, either through leaky bypass dampers or poorly fitting filter holding frames. The higher the efficiency of the filter, the more important the installation becomes. Filters should be installed so that the filter face area is perpendicular to the air flow as much as possible.

The most important considerations for a satisfactory and efficiently operating air cleaning and filtration system include:

- the filter size must be adequate for the air flow and dust load expected
- the filter must be suitable for the operating conditions, such as the required efficiency, amount of dust to be collected, allowable resistance, and operating temperatures (typical efficiencies are shown in Table 18.2)
- the filter should be the most applicable for the particular application
- the installation should provide an even air flow across the filter
- sufficient space should be provided in front and/or behind the filter for inspection and service of the filter
- access doors should be provided in HVAC equipment and ductwork for filter servicing
- access doors on the clean air side of the filter should be gasketed to prevent air leakage
- filters, other than electronic filters, should have indicators to provide a warning when the filter is loaded to maximum capacity or when the renewable media is about to run out
- service life of the air filter
- the addition of a low efficiency filter upstream of HEPA and ULPA filters

TABLE 18.2   Filter Efficiencies

Filter Type (ASHRAE%)	Efficiency at 10 μm	Efficiency at 1 μm	Efficiency at 0.3 μm
20–25%	98%	10%	1%
50–55%	99%	30%	19%
80–85%	99%	90%	49%
90–95%	99%	99%	78%

Chapter 19

INTRODUCTION TO ELECTRICAL SYSTEMS

This chapter provides an overview of electrical systems as they relate to HVAC systems. A complete, operational HVAC system requires a coordinated effort of both an HVAC and electrical system designer. It is important that the HVAC system designer has a working knowledge of electrical systems in order to provide a coordinated system design. Electrical components of most interest to the HVAC system designer include motors, motor controls, overcurrent protection, and transformers, as well as wiring and circuitry.

It is not intended that this chapter provide the reader with complete information for the design of an electrical system or parts of electrical systems. The design of electrical systems should be performed by someone who is qualified to design such systems. Although the design of electrical systems should be performed by a competent electrical designer or engineer, the HVAC system designer should have a basic understanding of electrical systems.

19.1 BASIC ELECTRICITY AND ELECTRICAL CIRCUITS

Electricity is electrons in motion. Electrical potential, or electromotive force (emf), causes free or loosely bound electrons to move along or through a medium. The forces that move electrons are magnetic forces.

The most basic elements of an electrical circuit include a method for producing an electrical potential, a load, and interconnecting wiring to complete the circuit. A basic electrical circuit is shown in Figure 19.1 In order for electricity to flow, there must be a complete circuitous path, or circuit, for the electricity to flow through. The electrical power must return to the source from which it came. If there are any openings or breaks in the circuit, there will be no flow of electricity.

The flow of electrons through a circuit is referred to as electrical current, or just **current**. The current is capable of producing a number of effects including heat and magnetism. Current is defined as the time rate of flow of an electrical charge, and is typically measured in amperes (A). An **ampere** is the rate of flow of electrical charge of one coulomb per second (1C/sec). As the current traverses the circuit, energy is expanded in the form of work. Energy and work are discussed in Chapter 2. Current is measured by an instrument called an ammeter.

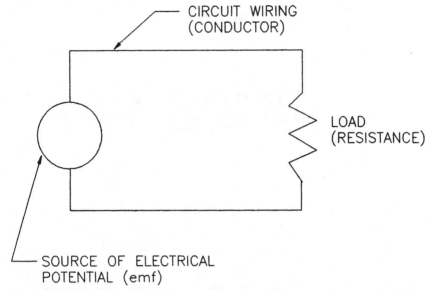

FIGURE 19.1   Basic Electrical Circuit

The source of electrical potential is typically provided by a battery or an electrical generator. It provides the electromotive force (emf) necessary for the electrons to flow. An emf is also called a charge. According to the Law of Conservation of Energy, the energy that a charge receives from an emf source must be equal to that which is expended as the current flows through the complete circuit. Therefore, the emf of the source must equal the sum of the losses around the circuit. Electrical potential (emf) is measured in volts (V). **Voltage** is the electrical pressure that causes current flow in a circuit. 1 V is when 1 coulomb (C) of charge does 1 joule (J) of work in moving from one point to another.

As the current flows through a circuit, it encounters **resistance** to the flow. As the current flows against the resistance, energy is consumed (a voltage reduction occurs). This reduction in voltage (voltage drop) is proportional to the rate of flow and the resistance. The relationship between voltage drop, current flow, and resistance is defined by **Ohm's Law**. Ohm's Law states that the current in a metallic circuit is given by the emf in the circuit, divided by the resistance of the circuit when the temperature is held constant. Ohm's Law is expressed mathematically by Equation 19.1.

$$I = \frac{E}{R}, \; E = IR \tag{19.1}$$

where

  I = the current in A

  E = the electromotive force (voltage) in V

  R = the resistance in ohms (Ω)

An **ohm** (Ω) is a resistance across which there is a voltage drop when the current is 1 A. The reciprocal of resistance is called **conductance**; hence, wires are frequently called conductors since they are low in resistance.

The flow of electrical current through any resistance, no matter how small, produces heat. The heat results from the conversion of electrical energy to thermal energy. The amount of heat generated is exactly equal to the energy supplied electrically, in accordance with the principle of Conservation of Energy. (See Chapter 2.) This is the basis upon which resistance electric heaters operate. The power dissipated in heat is given by Equation 19.2.

$$P = RI^2 = EI \qquad (19.2)$$

where

P = the power in W

19.1.1 Equivalent Resistance

The simple circuit shown in Figure 19.1 contains only one resistance. This simple case rarely occurs in actual practice. Usually, there are many resistances in a circuit. In order to simplify the analysis of circuits, multiple resistances are frequently combined into one single equivalent resistance. Typically, there are two situations for multiple resistances in electrical circuits: resistances connected in series and resistances connected in parallel. Each situation can be converted into a single equivalent resistance.

Figure 19.2 shows two resistances connected in series. The current flowing through one resistor must also be equal to the flow through the other. Therefore, the total equivalent resistance for a circuit with resistances connected in series is equal to the sum of the individual resistances, and is given by Equation 19.3.

$$R_T = R_1 + R_2 + R_3 + \cdots + R_N \qquad (19.3)$$

where

R_T = total equivalent resistance, Ω

$R_1, R_2, \ldots R_N$ = individual resistances, Ω

Since the current flow through each resistance is the same, the voltage drop across each resistance is a direct function of the resistance.

Figure 19.3 shows a circuit with the resistances connected in parallel. In circuits with parallel resistances, the voltage drops across the resistances will always be equal, and the current flow through each resistance will be a function of the resistance. The total equivalent resistance, for resistances connected in parallel, will always be less than the lowest resistance in the parallel combination. The reciprocal of the total equivalent resistance is equal to the sum of the reciprocals of the individual resistances and is given by Equation 19.4.

$$\frac{1}{R_T} = \frac{1}{R_1} + \frac{1}{R_2} + \frac{1}{R_3} + \cdots + \frac{1}{R_N} \qquad (19.4)$$

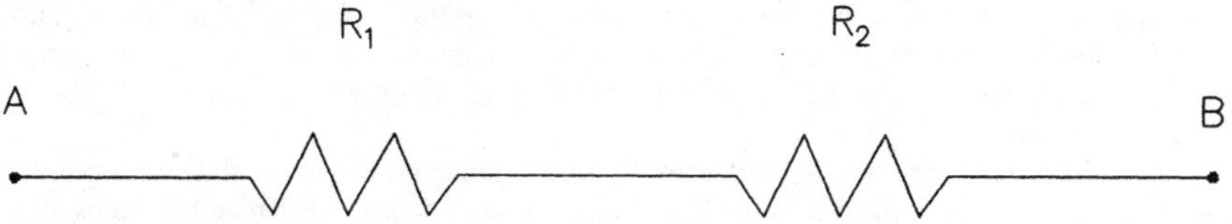

FIGURE 19.2 Resistances in Series

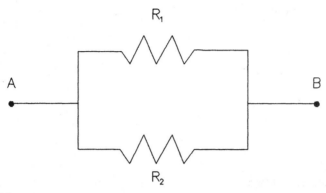

FIGURE 19.3   Resistances in Parallel

19.1.2   Direct and Alternating Current

Up to this point, the flow of current in a circuit has been assumed to be in only one direction. However, this is not the case for most electrical systems that provide power to HVAC systems. Almost all electrical power in buildings is alternating current.

As the name implies, **direct current (DC)** flows in only one direction, from a point of positive potential to a point of negative potential. A very common example of a direct current power source is a battery. The battery has a positive terminal and a negative terminal. The current flow is in one direction, through a circuit connected to the battery, from positive to negative. DC electrical power is not widely used in HVAC applications. The use of DC power in HVAC systems is primarily limited to control systems. Control systems are discussed in Chapter 20.

Whenever a **conductor**, such as a wire, is passed through a magnetic field, a current flow is induced in the conductor by the magnetism. The direction of current flow in the conductor is a function of the location of the field's magnetic poles (north and south) and the direction that the conductor is moving. If the conductor is passed through the magnetic field in one direction, and then reversed and passed through the field in the opposite direction, the induced current will first flow in one direction and then in the opposite direction. The current flow direction alternates, which is referred to as **alternating current (AC)**.

The electrical current supplied by electrical utility companies in the United States to all residential and commercial buildings is alternating current. HVAC systems in buildings use the same electrical power as that supplied to the building by the electrical utility.

Electrical power from utility companies is produced by large generators. The generators have magnets which produce a magnetic field inside the generator. The magnetic field is produced by stators, which are stationary magnets surrounding the inside of the generator. A rotor is located inside the generator, and is comprised of conductors mounted in a cylindrical shape. As the rotor rotates in the magnetic field, each conductor is first passed through the magnetic field in one direction and then the other. The rotation on the magnetic field causes a current flow in one direction and then the other, causing an alternating current. The magnitude of the AC current, produced by a generator, varies with time and results in a sin wave when current flow is plotted against time, as shown in Figure 19.4. This is called a **waveform**.

The voltage (emf) for alternating current follows the same pattern as the current; however, the time when the maximum and minimum voltage peaks occur may not be exactly when the current flow peaks. Common voltages for AC circuits include 120 V, 208 V, 240 V, and 480 V. Voltages up to 600 V are considered low voltages.

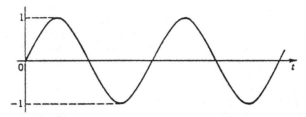

FIGURE 19.4   Waveform for Alternating Current

The frequency of an AC current or voltage is the number of cycles that occur in 1 sec. AC power in the Untied States is 60 cycles per second (60 Hz).

Since AC power follows a waveform as the current and voltage increase and decrease, all calculations for AC power are based on sin waves. There are four values of the AC waveform that are especially important.

1. **Instantaneous value:** The voltage and current in AC power is constantly changing from a maximum value to a minimum value. The value of the voltage or current at any given instant is known as the instantaneous value of that variable.

2. **Maximum value:** The voltage and current each reach a maximum positive and negative value instantaneously. These peak values are known as the maximum value.

3. **Average value:** The equivalent DC value that is equal to the area enclosed by the sin wave. For either the positive or negative halves of the sin wave, the average value is equal to 0.636 times the maximum value.

4. **Effective value:** The **root means squared (rm²)** value of an AC voltage or current is that value which would produce the same heating effect in a resistance as an equivalent direct current. It takes into account that the AC current is varying with time. The rm² value of an AC sin wave is equal to 0.707 times the maximum value.

Generators typically have rotors with multiple conductors that are rotated through the magnetic fields. Each set of conductors produces AC power and a waveform. The waveforms of each set of conductors occur at different times. The time difference between the waveforms is measured in degrees of rotation of the rotor and is referred to as the **phase angle** (ϕ). The number of waveforms produced is referred to as the number of phases. AC power is typically single phase or three phase.

19.1.3   Power for AC Circuits and Power Factor

Power is defined as the time rate of doing work or expending energy. Whenever a force results in motion, work is performed. In electrical systems, the force (emf) is the voltage and the motion is the flow of electrons (current). The rate at which work is performed results in power. Power may be expressed in various types of units, including ft•lb/min and hp. In electrical systems, power is expressed as **watts (W)** or **kilowatts (kW)**.

The calculation of power for DC circuits is straightforward. It is simply the voltage times the amperage and is expressed by Equation 19.2. In AC circuits, the calculation of power is not as straightforward because the resistance in AC circuits is complicated by the alternating current. When there is a flow of electricity through a conductor, a magnetic field is created around the conductor. The direction of the magnetic field depends upon the direction of current flow. When the flow of electrical current stops, the magnetic field collapses. The collapse of the magnetic field, however, is not instantaneous, and as it collapses it induces a current flow in the conductor. Since alternating current is rapidly changing direction (60 Hz), the magnetic field and the induced current are constantly changing. As the current flow changes, however, the induced current flow lags the main current flow by a brief period of time. This results in an additional resistance to the main current flow.

Capacitance is similar to resistance in that it resists the flow of current. However, capacitance offers resistance as the result of changes in voltage across it. Capacitance in an electrical circuit allows electrons to be stored within the circuit. A capacitor has two conducting surfaces separated by a dielectric material (a material with almost infinite resistance). As the voltage increases, the capacitor stores current. As the voltage decreases, the capacitor discharges current in the direction of the original current flow.

Impedance is the total of all resistance to current flow offered by an AC circuit resistance, capacitance, and inductance. Ohm's Law for an AC circuit is expressed by Equation 19.5.

$$I = \frac{E}{Z} \qquad (19.5)$$

where

Z = circuit impedance, Ω

Since capacitance and inductance in AC circuits add impedance to the circuit, the power calculated by Equation 19.2 does not indicate the actual wattage supplied by the source (generator). In circuits where capacitance predominates, the current will lead the voltage. In circuits where inductance predominates, the current will lag the voltage. The measurement of difference in phase between the current and voltage is called **power factor** and is the cosine of the angle between the current and voltage, and is expressed as a decimal. Equation 19.6 expresses the power for AC circuits.

$$P = EI\sqrt{\phi}\,(PF) \qquad (19.6)$$

where

ϕ = phase of the power supplied (i.e., single phase or three phase)

PF = power factor for the circuit

The calculation of the power factor for a circuit is an involved process, depending upon the elements in the circuit. The calculation of the power for AC circuits is beyond the scope of this book.

19.2 MOTORS AND MOTOR CONTROLS

The most common use of electrical power in HVAC systems is motors. Electric motors are devices that convert electrical energy into kinetic energy, usually in the form of a rotating shaft. Rotating shafts from motors are typically used to drive a fan, pump, compressor, etc.

It has been estimated that over 7% of the annual electrical power consumption in the U.S. is motor-driven HVAC equipment. Electrical motors are almost exclusively used to drive equipment such as fans. Other than motors used in control systems, almost all motors used in HVAC systems operate on AC power.

Motors used in HVAC applications can be divided into two classifications, hermetic and non-hermetic. Hermetic motors consist of a stator and a rotor without shaft end seals or bearings. These types are used in sealed refrigeration compressor units. Non-hermetic motors are not in sealed units and are the most common type of motor. An example of a non-hermetic motor would be a fan motor. Since non-hermetic motors are used in air systems, they will be the only motor type discussed here.

Motors are also generally classified according to their enclosures. The type of enclosure is based on the environment in which the motor will operate. The National Electrical Equipment Manufacturers Association (NEMA) has classified motor enclosures by type. The most common are open **drip proof** (suitable for applications where solid or liquid drops may fall on the motor at any angle not greater than 15°), **splash proof** (applications where solid or liquid particles may come at the motor in a straight line), **totally enclosed** (prevents free flow of air between inside and outside of the enclosure, but not airtight), and **explosion proof** (motor is completely enclosed and is suitable for applications where explosive gases and dusts are present).

Motors are also classified by their horsepower. Motors that are less than 1 hp are referred to as **fractional horsepower** motors. Motors that are 1 hp or greater are referred to as **integral** motors.

When selecting motors for a particular application, the insulation of the motor must be selected in accordance with the ambient temperature and humidity. Motors are rated for a maximum temperature rise, and the insulation must be such that it will not deteriorate under the maximum temperature conditions. The maximum allowable temperature rise is an indication on the quality of the motor insulation. Standard motor temperature rise is based on the continuous operation of the motor and is limited for each class of insulation by NEMA standards, as shown in Table 19.1.

Motors used in HVAC systems normally use Class A insulation that allows a 40°C temperature rise for the motor windings.

Under normal operating conditions, the temperature rise results from conversion of electrical energy to mechanical work and the friction of the rotating parts. Motors designed for continuous service can carry the design load for reasonably long periods of time without overheating. Occasionally, a motor may be operated at a higher than rated horsepower load. This is normally acceptable, provided the service factor limit is not exceeded. Standard open type motors have a **service factor (SF)**, which allows a continuous overload above the rated nameplate horsepower, without causing an excessive temperature rise due to overload. Table 19.2 lists service factors for fractional and integral horsepower motors. These service factors are applicable only when the voltage and frequency are at the rated value. If an undervoltage is present, the listed service factors should be reduced by the square of the reduced voltage ratio.

TABLE 19.1   Motor Insulation Class

Class of Insulation	Max. Temp. Rise
A	40–55°C (105–135°F)
B	70–75°C (160–168°F)
F	90–95°C (195–204°F)
H	110–115°C (230–240°F)

TABLE 19.2   Motor Service Factor

Motor Size	Service Factor
$1/20$, $1/12$, $1/8$	1.40
$1/6$, $1/4$, $1/3$	1.35
$1/2$, $3/4$, 1	1.25
$1 1/2$ and up	1.15

Earlier in this chapter, inductance and power factors for AC circuits were discussed. Some motors have permanent magnets to produce the magnetic field wherein the rotor rotates. Many AC motors, however, have coils to produce the magnetic field. Since the coils produce the magnetic field as the result of AC power, the AC power can result in significant inductance for the motor. The inductance of a motor can have a significant effect on the power factor for an AC circuit, particularly when large motors are involved. Typically, the larger the motor, the larger the impedance and the lower the power factor for the circuit. Many large motors have capacitors included with them or installed in the circuit to improve the power factor.

Motor efficiency has become an important criterion in the selection of motors. The Energy Policy Act of 1992 mandates that high efficiency motors be used in many applications, including many HVAC applications. Motor efficiencies can range from a low of 65% to a high of 94%. There are many factors affecting motor efficiency, including sizing of the motor to the load, the type of motor, the motor design, and the type of bearings.

Perhaps the most significant factor that the HVAC designer has control over is the sizing of the motor for equipment such as fans. Oversizing a motor may result in inefficiency. The efficiency of most motors falls off rapidly at loads lighter than the full rated load for the motor. Polyphase motors usually reach maximum efficiency at loads that are just slightly less than full load. Larger output motors typically are more efficient at rated loads than are smaller motors.

19.2.1   Single Phase AC Motors

Single phase AC power is typically available in 120 V 208 V, and 240 V systems. Some 120 V systems are available as 120 V, two conductor systems; however, most 120 V systems in commercial buildings are three conductor.

Most 208 V and 240 V AC circuits are three conductor. A typical 240 V, three conductor, single phase circuit is shown in Figure 19.5. A three conductor single phase circuit has two groups of conductors to produce 120 V between each outside line and the center neutral line which results in 240 V between the two outside lines. This arrangement can supply either 120 V or 240 V motors.

Single phase motors may be used for motors up to 5 hp; however, three phase AC power is preferred for motors larger than $^3/_4$ hp. Single phase induction motors have only one winding in the stator to produce a magnetic field. The AC produces a magnetic field with alternating polarity; however, it does not move or revolve. It remains in a fixed position. Therefore, it is necessary to provide a method of producing torque to accelerate the motor to the full load speed. The most popular single phase motor types are the shaded pole, split

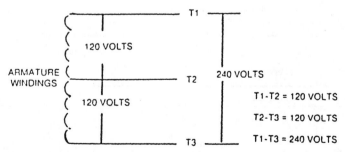

FIGURE 19.5   120/240 V Circuit

TABLE 19.3 Single Phase Motors

Motor Type	Voltage	Horsepower
Shaded Pole	115, 230	$^1/_{100}$ to $^1/_4$
Split Phase	115, 230	$^1/_{20}$ to $^1/_3$
Capacitor Start	115, 115/230	$^1/_{20}$ to 1.5
Split Capacitor	115, 230	$^1/_{20}$ to 5

phase, capacitor start, and split capacitor. Table 19.3 lists the various types of single phase motors, voltages, and sizes.

In **shaded pole** motors, the trailing edge of each pole is wound with a shading coil which produces slip. The current in the main winding of the stator produces the slip. The slip gives the pole the torque necessary for rotation and helps to establish the initial low starting torque which turns the rotor and the load.

Shaded pole motors are typically high in slip, low in starting torque, and low in running torque. They are inexpensive, can operate at variable speeds, and have inherent overload protection. The efficiency of a shaded pole motor is quite low. Shaded pole motors are air cooled. The low starting torque of shaded pole motors limit their application to small direct driven equipment, such as propeller fans and fan coil units. They are always fractional horsepower motors.

Split phase motors have two windings, a main stator winding and an auxiliary start winding. The start winding is mounted so it is 30° electrically from the main winding. This produces the slip necessary for rotation. When the motor reaches a speed of approximately 70% of its full load rotational speed, a centrifugal switch mounted on the rotor, which is in a series with the start winding, removes the start winding from the circuit. The motor then operates as a standard induction motor.

The split phase motor has a low starting torque and requires a large starting current. These motors have inherent overload protection and no slip with constant speed. This type of motor operates with satisfactory noise levels and is relatively inexpensive. The high starting current may result in a flicker of lights served by the same AC system.

Split phase motors are classified as general purpose motors and are frequently used for small propeller type fans. The starting torque allows the motor to be used for direct or belt driven equipment; however, they are limited to loads of $^1/_3$ hp, or less.

The **capacitor start** motor is also similar to the split phase motor in that it has two windings, a start winding and a main stator winding. To improve the performance of the split phase motor, a starting capacitor is added in series with the starter winding and the centrifugal switch. When the motor reaches approximately 70%–75% of the full load rotational speed, the capacitor and starter winding are removed from the circuit by a voltage relay. The motor then operates as a regular induction motor. This arrangement results in higher starting torque and quicker acceleration to operating speed. The increased starting torque is due to the low impedance capacitor.

Capacitor start motors are used on direct or belt driven equipment with loads up to $^1/_2$ hp. They are frequently used for applications such as small direct driven or belt driven centrifugal fans.

The **split capacitor** motor is similar to the capacitor start motor in that it has a capacitor connected in series with a start winding. The capacitor is also in the circuit with the main winding. The capacitor and the starter winding remain in the circuit permanently, with the main winding. The capacitor causes a phase lag in the starter winding which produces the torque necessary for rotation.

These motors have a low starting torque, high slip, variable speed, and inherent overload protection. Since the starting torque is low, the use of these motors is limited to direct connected loads with low starting torque. They are not suitable for belt driven applications.

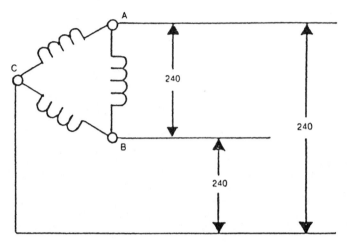

FIGURE 19.6   Three Wire Delta Arrangement

They need no overload protection, are quieter in operation than shaded pole motors, and are quite efficient.

19.2.2   Polyphase AC Motors

The majority of motors that are $^3/_4$ hp and larger in HVAC applications are polyphase, or specifically, three phase motors. Three phase motors are preferred for integral motors because they are self-balancing on three phase power. Three phase AC systems have three current carrying conductors. The phase of these conductors is 120° apart. During one complete cycle, each of the three conductors carries an AC current which peaks at a different time.

The windings of an AC generator can either be connected together to form a delta configuration or a wye configuration. A three wire delta arrangement is shown in Figure 19.6. Three wire delta circuits produce three phase, 240 V AC power.

Three phase AC systems can also have four conductors, three to carry the three phases of AC current and a fourth wire which is a neutral conductor. The **neutral conductor** is grounded and is connected to the center tap of an AC generator. There is a 120 V potential between the neutral conductor and any one of the other three conductors. Three phase, four conductor AC systems can either be a delta configuration, as shown in Figure 19.7, or a wye configuration, as shown in Figure 19.8. Four wire delta circuits can produce 120 V or 240 V, three phase power and four wire wye circuits can produce 120 V or 208 V, three phase power. Four wire systems can also be used to produce 480 V and higher AC power.

The most commonly used three phase motors in HVAC systems are synchronous, wound-rotor induction, and squirrel cage induction.

Synchronous motors are designed to operate at only synchronous speeds. They are exclusively constant speed motors. Their speed is unaffected by changes in the supply voltage or by the load on the motor. Because they operate at synchronous speeds, an external DC exciter is required for the motor starting and running torque. A DC generator is used to provide the required DC power. Synchronous motors have a limited number of applications. They are used only for large loads. When the motor comes up to operating speed, it runs with a slip of only 2%–3%. Large horsepower synchronous motors, operating at low speeds, are relatively low in cost. Some of the characteristics of synchronous motors include:

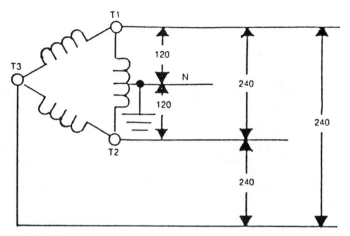

FIGURE 19.7   Four Wire Delta Arrangement

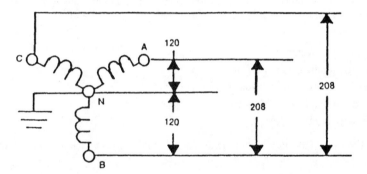

FIGURE 19.8   Four Wire Wye Arrangement

- low starting torque
- constant speed within the motor's load limits
- nearly constant load requirements
- a leading power factor is possible
- high efficiency

  Wound rotor induction motors provide higher torque during starting and a range of adjustable speed operation. The speed of a wound rotor motor is controlled by varying the current supplied to the rotor. The current is varied with external resistors. They may be used where speed control is required. Applications include large fans and centrifugal compressors. Characteristics of wound rotor motors include:

- lagging power factor
- speed control by varying external resistance
- relatively high maintenance

- relatively high initial cost
- high starting torque and low starting current

The most commonly used polyphase motor for HVAC applications is the **squirrel cage** motor. It is the most economical in terms of cost per horsepower.

The two major components are the stator (the stationary magnets and coils) and the rotor. The stator is a laminated iron core within and around the inside of the motor. The rotor is a laminated iron core with bar windings in various shaped slots around the periphery of the core. The rotor looks like a squirrel cage, hence the name squirrel cage motor.

The stator windings of a squirrel cage motor are arranged so the stator will alternate between positive and negative as the current flow alternates. This creates a pulsing magnetic flux. The pulsations of the magnetic flux produce a rotating magnetic field. As the flux from the stator rotates, it cuts the bars or coils of the rotor, which induces a voltage within the rotor circuits. A current is generated within the rotor bars which causes a second magnetic field. The interaction of the two magnetic fields produces the torque that accelerates the rotor.

The speed of a squirrel cage motor is constant and is determined by the number of poles that are active. On a two speed, two wind motor, the two windings determine the number of active poles. The windings are externally switched resulting in speed combinations such as 1800/1200 rpm, 1800/900 rpm, and 1800/600 rpm. These motors are available in a number of speeds; however, they are not variable speed. Characteristics of squirrel cage induction motors are:

- lagging power factor
- multi-speed motors are available
- high efficiency over a wide load range
- rugged and relatively inexpensive
- wide range of starting torque

Squirrel cage induction motors are available in four classes which have been standardized by NEMA, as shown in Table 19.4.

19.2.3   Motor Starters

Motor starters are devices that bring a motor from the off position up to the normal operational rotation at the design load. They are also used to stop the motor. Starting a motor is frequently more involved than closing a switch and applying power to the motor. Part of a starter is a contactor that automatically connects, or disconnects, a motor from a power source. However, starters provide more functions than starting and stopping the motor.

Starters may also provide overload protection for the motor. Operating a motor in an overloaded condition results in overheating in the motor. A starter can include an overload

TABLE 19.4   NEMA Squirrel Cage Induction Motor Classes

Class	Starting Torque	Starting Current	Slip	Full Load Efficiency	Typical Uses
A	Normal	High	Low	Best	Uniform loads such as fans
B	Normal	Medium	Low	Good	Same as for Class A
C	High	Low	Medium	Fair	Uniform high torque loads
D	High	Low	High	Low	Intermittent high torque loads

relay that senses when the motor is operating in an overload condition, and disconnects the motor from the power source when an overload condition occurs. The overload relay contains overload heaters which are made of bi-metallic alloy elements. The temperature of the elements are proportional to the amount of current flow. If the current flow is too high, the elements flex and open the relay contacts.

Starters may also provide protection from damage resulting from single phasing in three phase circuits. Single phasing occurs when one of the three ungrounded conductors, serving a three phase motor, loses its voltage. In this situation, the motor will draw at least 1.7 times the design current flow through the two remaining ungrounded conductors. The current flow can be much higher if the motor stalls. When a single phasing situation occurs, the motor may be destroyed, unless the two remaining ungrounded conductors are immediately disconnected. The overload heaters also serve to disconnect the motor when it is in a single phase condition.

Motor starters may also protect a motor from an undervoltage condition. If the voltage of the power supplied to a motor drops below about 65% of the design voltage, a coil disconnects the motor from the supply voltage.

Starters are frequently used to provide automatic control for motors. A magnetic starter provides the ability to control the starting and stopping of the motor automatically using an electrical interlock device or some other control device.

Motor starters can either be manual or magnetic. Manual starters are normally used only for small fractional horsepower motors. Manual starters may be toggle, selector, or push button switches. They are always across the line type, which means that whenever the starting switch closes, the motor receives whatever line voltage is available. The motor draws whatever current it requires to start and run the motor.

Magnetic starters are used for integral motors. They use an auxiliary circuit within the starter to energize a magnetic coil which closes the starting switch. Control of the magnetic coil may be part of the starter itself or it may be remote. Infrequently started motors may have manual starters; however, motors that are started and stopped frequently or automatically require magnetic starters. A magnetic starter consists of two basic components, a contactor and an overload relay. The contactor consists of stationary contacts and an electromagnetic coil with an armature that controls the position of a moveable contact. When a voltage is applied to the coil, the moveable contacts seat against the stationary contacts. When the voltage, is removed from the coil, a spring or gravity causes the contacts to move away from the stationary contacts. The overload relay holds either overload heaters or bi-metallic sensors. The overload relay senses the overheat that results from the overcurrent condition and breaks the starter's control circuit.

There are two basic classes of starters, **across the line** (full voltage) and **reduced voltage**. Across the line starters connect the motor directly to the full voltage, which results in 100% of the normal inrush current, at locked rotor conditions, and also results in full starting torque. Across the line starters are the lowest cost and simplest method for starting a three phase motor. They are especially well suited to small sized motors where the current draw is low, such as single phase motors up to 5 hp and three phase motors up to 7 hp. On larger motors, they draw full locked rotor current which can cause a momentary voltage drop in the system. Across the line magnetic starters may be used with automatic control systems for automatically starting and stopping motors. They are widely used in HVAC systems.

It is often desirable to start motors with a starter that draws less than full starting current as the across the line starters do. One type of reduced current starter is the **part winding starter** (incremental starter). Part winding starters are used to reduce line disturbances by connecting only part of the motor winding to the line voltage, and connecting the second motor winding after a time interval of 1 sec to 3 sec. A part wind motor has two separate three phase windings, each wound for the full voltage of the system. The application of full voltage to one of the windings results in about 60%–75% of the full voltage, full winding current flow and torque. When the second set of contacts close and the second winding is energized, the current flow and torque reach the normal operating levels.

Another starter used for reduced current starting is the **star-delta starter**. A delta wound motor has six leads and is started with a star delta starter to reduce the inrush current. These types of starters are used for polyphase squirrel induction or synchronous motors. The starter is initially connected to the motor in a star delta configuration, as shown in Figure 19.8. In this configuration, the motor draws about 33% of the locked motor current and provides about 33% of the full torque. After a preset interval, a timer energizes the second set of contacts and the motor windings are then in a delta configuration, as shown in Figure 19.7. After switching to the delta configuration, the full line voltage is applied to the motor. This increases the motor's current draw and torque, bringing the motor up to full speed. The current flow then drops off when the motor reaches full speed conditions. These starters are frequently used on large motor applications. The normal inrush and starting torque are determined by the characteristics of the connection to the motor.

Auto-transformer starters also provide reduced current inrush starting. The auto-transformer typically has three voltage taps, 50%, 65%, or 80% of the full line voltage. This results in an inrush current of 25%–64% of the locked rotor current and full load starting torque. When the motor is started, a starting contactor connects the motor to the line voltage, through the design voltage tap. At the same time, a timer is energized. After a short interval, the auto-transformer is removed from the circuit and the motor is connected to the full line voltage. Auto-transformer starters result in low current draw from the circuit. The starter uses transformers, timers, and extra contacts to control the inrush current. These starters tend to generate heat within the starter itself. They should not be used in applications that require frequent starts.

Primary resistance starters reduce the voltage to the motor windings with a resistance wired in series with the motor stator windings. The inrush current varies directly with the voltage applied. The starting torque also varies as the square of the voltage at full load torque. A number of steps can be provided to reduce the inrush current accordingly. When a motor is started, the motor receives a reduced voltage. After a short interval, the run contacts are closed, bypassing the start contacts and resistors. The motor is then operating on full line voltage. Primary resistance starters are applicable to situations where full locked rotor inrush is undesirable. Applying the voltage to the motor in steps, results in steps of locked rotor current. This allows the system to readjust before the next stage of load is applied.

19.3 TRANSFORMERS

Transformers are devices that are used to change the voltage of AC. They are typically used to reduce the voltage or step the voltage down; however, they can also be used to increase the voltage of AC. Transformers cannot be used on DC circuits. A transformer is typically used to reduce the voltage supplied by a utility company before it enters a building, generally from 4160 V to 480 V. Additional transformers may be located inside a building, to reduce from 480 V to 240 V, 208 V, or 120 V systems. Some transformers have multiple taps (connections) for multiple reduced voltages.

Transformers consist of two (or more) separate coils separated by a dielectric material. There is a primary coil, which has the AC power supplied to it, and a secondary coil(s). The primary and secondary coils are not connected electrically because they are separated by a dielectric material. However, they are located such that the magnetic fields around the coils do "communicate" magnetically. The space between the primary and secondary coils may be either a liquid or a solid material. Transformers with liquid material are referred to as **liquid-filled**. Transformers with a solid material are referred to as **dry-type transformers**.

Earlier in this chapter, the effect of alternating magnetic fields and induced current was discussed. Transformers operate on the principle of alternating magnetic fields. As current flow alternates in the primary coil, it induces an alternating current in the secondary coil. The secondary voltage is a function of the windings in the primary and secondary coils.

Transformers are available for single phase or three phase AC systems. Transformer power capacity is rated in kilovolt-amperes (kVA). Although transformers do not have any moving parts, they do produce a humming noise as they operate. Noise can be an important criterion in the selection of a transformer, depending on the intended location of the transformer.

As transformers operate, they generate heat due to the impedance produced by the magnetic fields in the coils. This impedance causes heat to be generated within the transformer, which is rejected to the surrounding air. Whenever a transformer is located inside a building, the heat rejection rate must be considered in the design loads for the HVAC system.

In summary, transformers are selected based on the primary voltage, secondary voltage, phase, kilovolt-amperes rating, type of insulation, physical size, sound, and heat generation.

19.4 OVERCURRENT PROTECTION

The greatest hazard to electrical systems is short circuits, which result in enormously high current flows. Short circuits are caused by a fault in a power line, motor, or some other electrical device in the circuit. During a short circuit, the current flow can be from tens to hundreds of times the normal flow. In order to protect circuits from dangerous and damaging overload conditions, overcurrent protection must be provided in electrical circuits.

Overcurrent protection serves two purposes: it protects the circuit from a fault in an electrical device or a fault within the circuit itself. The overcurrent device must be capable of successfully interrupting whatever the system can provide.

Overloads occasionally occur in electrical systems as part of the normal operation, such as the starting of motors. Overloads do not require that the circuit be immediately interrupted. However, the circuit should be interrupted if the overload results in temperatures in the circuit that are too high. An overcurrent protective device should have a time-overcurrent characteristic which allows minor overloads in the circuit. A short circuit should be interrupted immediately, however. The two most common types of overcurrent protection devices are fuses and circuit breakers.

A **fuse** is a device with a fusible link or wire that has a low melting temperature. The fusible link is such that, at normal current flow, the link offers little or no resistance to the current flow. At high current flow, however, the resistance of the link is such that it overheats and melts. When the link melts, it breaks the circuit. Fuses are rated not only by voltage and current flow, but also by interrupting capacity. **Interrupting capacity** is the current flow capacity that the fuse can safely interrupt in a short circuit.

There are two basic types of fuses, plug fuses and cartridge fuses. Plug fuses are the screw in type and may be used in circuits up to 30 A. They can only be used in circuits rated at 125 V or less, except in installations having a grounded neutral with a 150 V maximum.

Fuses are also available as cartridge type fuses. Cartridge type fuses have one or more fusible links inside a tube filled with an arc-quenching filler. The purpose of the filler is to reduce the arcing across the gap in the link as it melts. Cartridge fuses can be single or dual element (link) fuses. In a single element fuse, there is only one fusible link to carry the current and to melt if an overcurrent condition occurs. Dual element fuses have a thermal cutout element and a fusible link in series inside the tube. Dual element fuses provide a time delay in the low overload range (such as motor startup) to eliminate needless opening on harmless overload and transient situations. The thermal cutout is designed to open on overcurrents up to approximately 500% of the fuse ampere rating. The link is designed to open on heavier loads such as short circuit conditions.

An alternate overcurrent protection device is the molded case circuit breaker. A **circuit breaker** is an electromechanical device which performs the overcurrent protection function as a fuse; however, it can also function as a switch and can be reset after being tripped. A circuit breaker can be used instead of a switch and fuse to provide both protection and a means to disconnect a circuit. Breakers are available with a time-delay feature which allows for normal overload conditions in a circuit for short periods of time.

Most circuit breakers have both thermal and magnetic trips to open the breaker when an overcurrent condition occurs. The thermal trip is heat sensitive. Heat generated by a excessive current flow causes an element to move and trip (actuate) a latching mechanism that holds the breaker closed during normal operation. When the thermal element trips, the contacts are opened, breaking the circuit. The magnetic trip is comprised of a coil with a movable core. When an overcurrent condition occurs, a magnetic field in the coil actuates the movable core which trips the latch of the circuit breaker.

Molded case circuit breakers are available with voltage ratings up to 600 V and current ratings up to 4000 A. Circuit breakers must be selected so that they have adequate current carrying capacity for the normal circuit loads, but are capable of interrupting the highest fault current possible at the voltage level of the circuit.

19.5   VARIABLE FREQUENCY DRIVES

In Chapter 16, the performance of fans and the fan laws were discussed. Referring to Equations 16.5 and 16.7, it can be observed that the air flow through a fan varies directly with the rotational speed of the fan, and that the power or work required to drive a fan varies with the cube of the rotational speed. The speed of an induction motor is a function of the number of poles in the motor, the motor slip, and the frequency of the applied voltage. The number of poles and the slip are fixed by the motor design and cannot readily be changed. The frequency and the voltage of the power applied to a motor may be readily changed with a variable frequency motor drive.

Variable air volume systems are frequently used in buildings to provide adequate cooling capacity for varying loads in various spaces. There are several methods used with fans to provide for varying air flow rates as discussed in Chapter 16. A commonly used method to vary the rotational speed of the air flow through a fan is to vary the rotational speed of the fan with a variable speed drive. Varying the speed of a fan in accordance with the space load can provide significant energy savings, as indicated by Equations 16.5 and 16.7.

Motor speed is directly proportional to the frequency of the alternating current supplied to the motor. In order to operate at peak efficiency, the voltage applied to the motor must be controlled, as well as the frequency of the AC. A variable frequency motor drive varies the rotational speed of a motor by changing the voltage and frequency of the AC power supplied to a motor. Normal AC line current has a frequency of 60 Hz. The output of a variable frequency drive can range from 0 Hz to 66 Hz, and in some cases, 120 Hz. A constant volts-to-frequency ratio is required to keep a motor operating efficiently.

Almost all variable frequency motor drives consist of a rectifier and an inverter. The rectifier is used to convert the incoming AC to DC. The inverter is then used to convert the DC back to alternating current at the required frequency and voltage.

The rectifier serves two main functions: it converts the incoming AC to DC and it controls the voltage level of the DC. A rectifier is a device that will only allow current flow in one direction, thus creating a pulsating DC when AC is applied to the rectifier. In order to control the voltage of the DC output, the rectifier includes a chopper. A chopper is a switch that regulates the output voltage by turning the current on and off. Rectifiers that create varying DC voltages are called phase-controlled rectifiers.

AC motors cannot operate on DC power; therefore, it is necessary to convert the power supplied to the motor back to AC at the required voltage and frequency. An inverter is used to convert the DC back to AC. There are three types of inverters, the variable-voltage inverter (VVI), pulse width modulation (PWM), and current source inverters (CSI).

Whenever the switches in the variable speed drive switch on, a large momentary current flow will result. The rapid switching by the drive creates current pulses in the incoming power which can result in rapid voltage fluctuations. These voltage fluctuations in the incoming power can create disturbances in other electronic equipment in a building. These disturbances, known as **harmonics**, result from the inductance in the AC distribution system which cannot change instantaneously. As a result, there is a short delay between the time the voltage wave changes and the current wave responds (see discussion of inductance and power factor earlier in this chapter). The resulting irregular current flow rates can be reflected as voltage irregularities on the AC of the distribution system. The problem of harmonics increases with the number of variable frequency drives and other harmonic current producing devices that are installed on an AC distribution system in a building. Poor power quality can affect computers and other microprocessor based equipment adversely.

When selecting variable frequency drives, the following features should be considered:

- the drive should be able to accept and control from the temperature control or building automation system (controls for HVAC systems are discussed in Chapter 20)
- the drive should have a manual override mode that will allow the motor speed to be adjusted manually with a dial or some other device.
- the installation should include manual electrical bypass that would allow the motor to operate at a constant speed in the event of a problem with the drive
- the motor acceleration and deceleration rate
- the drive should have a high power factor, at least 90%
- the drive should have a high efficiency, at least 90%
- the harmonic effects of the drive on the system must be considered

19.6 HVAC SYSTEM/ELECTRICAL SYSTEM INTERFACE

Although the design of electrical systems should always be performed by a fully competent electrical engineer or designer, there are a number of interface points between the two systems, of which the HVAC designer must be aware.

The cooling loads in a building are directly affected by internal electrical loads which become internal cooling loads. Practically all electrical power consuming equipment in buildings, such as lights, generate heat which must be removed by the HVAC system. Rooms that contain electrical equipment should always be directly or indirectly ventilated by the air conditioning system. The HVAC designer should pay particular attention to the following electrical equipment in rooms:

- Transformer Rooms: Assume that 3% to 5% of the active transformer load may be rejected to the room as heat. Consult the transformer manufacturer for more exact heat rejection rates.
- Elevator Equipment Rooms: Most building codes require that elevator equipment rooms either be cooled directly or indirectly, by providing exhaust for the elevator equipment room. Consult the elevator equipment manufacturer for heat rejection rates.

- Motor Control Centers: These centers contain the starters, transformers, etc., necessary to control electric motors, particularly when there are a large number of sizable motors.

  In addition to providing adequate ventilation and cooling to electrical equipment, there are a number of design coordination issues that come under the purview of the HVAC system designer.

- Motor sizes and types for HVAC equipment: It is usually the responsibility of the HVAC system designer for sizing electrical motors for HVAC equipment such as fans. The sizing of the motor for fans should be based on the calculation methods presented in Chapter 15, and the guidelines for fan selection presented in Chapter 16. Motor voltage and phase should be based on the electrical power available to the HVAC system.

- Motor control features: The starters and disconnects for HVAC system motors should be selected based on the requirements of each motor and electrical system requirements. Some packaged HVAC equipment includes starters and disconnects as part of the package. Most equipment, such as air handling units, do not include starters and disconnects, and motor controls must be provided by either the mechanical or electrical contractor. The issue of who will provide these controls must be coordinated.

- Fire detection and alarm: All air systems that circulate air at a flow rate of 2000 ft³/min or more require smoke detectors. The smoke detectors are necessary to shut the air distribution system down, in the event that smoke enters the system. The requirements and type of smoke detectors must be coordinated with the building fire alarm system. Coordination between the HVAC and electrical system designers are required since these smoke detectors can be provided by either the mechanical or electrical contractor.

- Motor efficiency: The efficiency of electrical motors serving HVAC equipment must be considered. Some codes now require high efficiency motors for HVAC equipment, as well as other motor driven equipment.

Chapter 20

BASIC CONTROLS FOR
AIR SYSTEMS

The function of an HVAC system is to control the temperature level, humidity level, ventilation rate, and air cleanliness of the occupied spaces within a building. By necessity, an HVAC system must be designed with sufficient capacity to handle the maximum loads expected for the conditioned space. However, these peak loads may occur for only brief periods of time each year, or even for the entire life of the system. The HVAC system may, in fact, never experience loads which require the system to operate at maximum capacity. As a result, the system must be controlled to operate at capacities required by the actual space load.

The purpose of the control system is to regulate the HVAC system so it operates at the capacities required to handle the space loads at any given time without excessively overshooting or undershooting the actual loads. The control system should automatically adjust the output capacities of the HVAC equipment to match the space load. An automatic control system for an HVAC air system typically includes the control of temperature, humidity level, pressure, and air flow within the conditioned space.

In order to design a properly functioning HVAC system, it is necessary to have a basic understanding of controls for HVAC systems. This chapter provides an overview of control devices and systems for air systems. The actual design of control systems for HVAC systems is a fairly specialized field of engineering. For a more detailed discussion of automatic control, the reader should consult References 2 and 22.

20.1 CONTROL SYSTEM BASICS

Generally speaking, there are two basic methods for providing automatic control: closed loop and open loop. A closed loop (feedback) control system senses changes in the controlled variable (such as air temperature), causing the controlled device to change and take corrective action based on the measurement of the controlled variable. The controlling or modifying action continues to take place until the output (controlled) variable is brought to within an acceptable level. This method is error sensitive and self-correcting. The acceptable level is probably not the exact desired level (or set point) of the variable, but is reasonably close to the desired level of the variable. The difference between the set point and the actual level of the variable is the offset. The closed loop control method is the method most commonly used to control HVAC systems. A closed loop is one in which all parts have an effect on the next step in the loop and are affected by the action of the previous step.

An open loop system is one in which one or more of the steps has no effect or action imposed on the following step or is not affected by other steps in the loop. For open loop (feed forward) control systems, there is no feedback of information from the controlled variable to adjust the controller. Open loop control systems anticipate the effect that a change on the controller will have on the output variable. An example of an open loop control system would be an outdoor air thermostat which resets the control point of a heating system controller based on the outdoor air temperature. The temperature inside the building is not sensed and the heat output is not adjusted accordingly. The only variable that affects the heat output is the outdoor air temperature.

Prior to discussing controls and control systems, it is best to have an understanding of the terminology used to describe control systems. The following are frequently used terms and definitions as they relate to control systems:

- control agent (medium): the medium that is adjusted by the controlled device
- controlled device (actuator): a device that reacts to signals received from a controller
- controller: a device that compares the sensed values of a variable with a set point (the desired level of the controlled variable)
- controlled variable: the temperature, humidity, pressure, or flow being controlled
- damper: a device used to vary the volume of air passing through it
- deviation: the difference between the set point value and the actual value of the controlled variable at any instant
- direct acting: the output signal changes in the same direction as the controlled variable (increasing vs. decreasing, etc.)
- modulating: the controlled device adjusts by increments or decrements in response to the control signal
- modulating control: a mode of automatic control in which the action of the final control element is proportional to the deviation of the controlled medium
- offset: the difference between the set point of the controller and the control point of the controlled variable
- process plant: the apparatus that is being controlled
- reset: the process of automatically adjusting the control point of a controlled variable
- reverse acting: the output signal changing in the direction opposite to the change in the controlled variable
- sensor: a device which measures the controlled variable and conveys representative values to a controller
- sequencer: a mechanical or electrical device that may be used to initiate a series of events in a particular order
- set point: the desired value of the controlled variable
- stability: the ability of a system to maintain a controlled variable at or near the control point without requiring the controller to "hunt" (change continuously from one extreme to another)
- throttling range: the amount of change in the controlled variable required to move the actuator of a controlled device from one end to another

- transducer: a device that converts an input signal from one form to another, e.g., electrical input to a pneumatic output

Control systems can operate in several different modes. A mode of control is the manner in which the control system makes changes and responds to a change in the controlled variable. It relates the operation of the final control element to the measurement signal from the sensor. The type of control mode determines the type of response that the control system has as a result of a change in the controlled variable. The basic modes of control are two position, multi-position, floating, and modulating.

The **two-position (on-off)** control is the simplest mode. The controller output has only two positions, either on or off. When applied to a valve or damper, this means either fully open or fully closed. There are no intermediate positions. When the controlled variable deviates a preset amount from the set point, the controller is actuated to either of its extreme positions, fully on or fully off. This mode allows the controlled variable to vary over a wide range of output. The output is similar to a step function between two settings, as opposed to a near steady condition.

The **multi-position (multistage)** mode is similar to the two-position except there are more intermediate positions or stages. Multistage control is used when the range between fully on and fully off is too great. The multistage control divides the range into multiple stages or steps. Each stage is either on or off; however, the stage is much smaller than the range between fully on and off. An example of multistage control is a heater that can come on in stages of 25%, 50%, 75%, and 100% of full capacity. The greater the number of steps, the smoother the operation.

Floating control is different from two-position or multi-position control in that the final control element may be moved to any position between the extremes (fully on to fully off). The controller has only two positions, with a neutral or dead band in between. When the controlled variable changes enough, the controller moves to one or the other position, depending on the direction of the change in the controlled variable. When the controlled variable returns to its neutral (dead band) range, the controller moves to the off position. This mode of control is called floating because the controller is off when the controlled variable is floating between its two positions.

Modulating (proportional) control is similar to floating control except that the controlled variable may assume any position between its extremes. The controller may also assume any position in response to changes in the controlled variable. There is no neutral zone or dead band. Any changes in the controlled variable will cause the controller to respond. For each movement in the controller, there is a proportional change in the controlled variable. These movements can occur as frequently as changes in the controlled variable occur. Modulating control produces a linear relationship between the change in the controlled variable, and the output of the controller to change the position of the controlled device.

Control systems can also be classified according to the primary source of energy used to power the control devices. The most common types used in HVAC systems are pneumatic, electric, electronic, and self-powered. Some control systems are hybrid combinations of the different classifications of control systems.

Pneumatic control systems use compressed air to actuate control devices. The compressed air used for control systems ranges from 15 psig to 35 psig. Air is supplied to a controller which regulates the pressure supplied to the controlled device. Changes in the output pressure of the controller result in a corresponding change at the controlled device. Pneumatic control systems require an air compressor, with accessories, and a distribution system of pneumatic tubing to the various control devices.

Electric control systems typically operate on 24 V to 120 V AC power. Electric energy is supplied to the controlled device and regulated by a controller. An example of such a controller would be a motor actuating a damper.

Electronic or **direct digital control (DDC)** systems operate on voltages ranging from 0 V to 10 V DC and current flow rates of 4 mA to 50 mA. A DDC controller receives an electronic signal from a sensor and converts the signal to numbers electronically. The system then electronically performs various mathematical operations on these numbers with a microprocessor. The output from the microprocessor is then converted to an electric or pneumatic signal to actuate the controlled device. The current trend in the HVAC industry is to use DDC for new control systems.

Self-powered control systems generate the energy needed to actuate the controller from the controlled medium. These systems incorporate a sensor, controller, and a controlled device in a single package. No external power source is required for the system to provide control. All the energy needed to actuate the controlled device is provided by the reaction of the sensor with the controlled variable. In typical self-powered control systems, temperature changes at the sensor result in pressure or volume changes in an enclosed media that are transmitted directly to the controlled device.

The most basic elements or components that comprise an automatic control system are a sensor, a controller, and a controlled device. These three components can make a complete control system; however, most control systems are composed of many components and accessories.

20.2   SENSING DEVICES

A sensor is a device in the control system that measures the value of the controlled variable (for example, air temperature). If the controlled variable changes, it causes a change in some physical or electrical property of the primary sensing element. The change in the property of the sensing element is then translated or amplified by a mechanical or electrical signal. The signal is then transmitted through the controller by the control system. Typically, sensors are used to measure the temperature, pressure, humidity, or flow of the controlled variable.

20.2.1   Temperature Sensors

Temperature sensors are used extensively in HVAC control systems. Temperature sensors operate on one of three physical characteristics which are:

• a change in physical dimension as the result of thermal expansion

• a change in state of a vapor or a liquid

• a change in electrical properties of a substance due to a change in temperature

Typical temperature sensing elements include a bimetal element, a vapor-filled bellows, a liquid-filled, gas-filled, or refrigerant-filled bulb and capillary, and metals that change in temperature. Two dissimilar metals may also be used to produce a voltage change due to a change in temperature.

Bimetal temperature sensors are the most common type used in HVAC control systems. The element is made up of two strips of different metals. Each metal has a different coefficient of thermal expansion. The two metals are bonded together to form a strip or a spiral element. As the surrounding temperature changes, the temperature of the element changes, causing the two metals to expand or contract differently. In the case of the bimetal strip, the strip bends. The amount of the bend and direction is a function of the direction and amount of temperature change. The spiral element (coil) tends to either coil tighter or straighten as the result of a temperature change. Bimetal temperature sensors can be used for two-position or modulating control.

A rod-and-tube sensor is another type of bimetal temperature sensor. It consists of a rod that is made of a material with a low coefficient of thermal expansion and a tube with a high coefficient. The rod is attached at one end of the tube. As the tube changes length due to temperature changes, the free end of the rod moves. The movement is a function of the temperature change. Rod-and-tube type temperature sensors are frequently used as insertion type sensors.

A sealed bellows type temperature sensor has a bellows type element. The bellows is filled with a vapor, gas, or liquid depending on the expected temperature range for the sensor. Temperature changes cause the fluid inside the bellows to expand or contract with the temperature change, which results in a force or movement of the bellows. Bellows type temperature sensors are frequently used in room thermostats.

A remote bulb and bellows, or diaphragm type sensor, is similar to the bellows type temperature sensor. This type of sensor has a sensor bulb and a capillary tube that are filled with fluid that expands or contracts, in response to temperature changes. The bulb is typically inserted into the material where the temperature is to be sensed. A bellows or diaphragm is connected to the other end of the capillary tube. Temperature changes sensed by the bulb result in a pressure change in the fluid which is transmitted through the capillary tube to the bellows, or diaphragm, where the fluid pressure is converted to a force or movement.

A thermistor is a temperature-sensing device which changes electrical resistance with changes in temperature. The thermistor is made of a semiconductor material that is temperature sensitive. The material typically has a resistance of about 3000 Ω at a temperature of 0°C. Small changes in temperature result in large changes in resistance, which make thermistors quite temperature sensitive. Thermistors are also relatively low in cost and are readily used in DDC control systems.

Resistance temperature detectors (RTD) are made of fine wire, wound into a tight coil. They are similar to thermistors in that they change resistance in response to a temperature change. The RTD-sensing element is available in several different forms. They are suitable for sensing surface temperatures. For immersion sensing, the coil may be encased in a stainless steel bulb for protection. Normally, the RTD is more reliable and more accurate than a thermistor.

A thermocouple is a junction between two dissimilar metals that produces an electrical voltage change in response to changes in temperature. The voltage is a function of the two materials and the temperature of the junction. If the other ends of the conductors are connected, a circuit is formed which causes a flow of current. The other end of the thermocouple must be kept at a constant temperature. The constant temperature is referred to as the cold junction. Thermocouples are frequently used to sense a flame in a pilot light.

20.2.2 Humidity Sensors

Hygroscopic (moisture-absorbing) materials are used for humidity sensors. These materials absorb or reject moisture until a balance is reached with the surrounding air. There are two types of humidity sensors (hygrometers) used in HVAC control systems, mechanical and electronic.

Mechanical hygrometers use materials that expand or contract based on their moisture content. As the material absorbs moisture, it expands (stretches or swells) and as it loses moisture, it shrinks. The change in size is detected by a mechanical linkage which converts the motion to a pneumatic or electrical signal.

Many hygrometers have humidity-sensing elements that are made of organic materials, such as hair, wood fibers, paper, and cotton. Non-organic materials include nylon and other synthetic fabrics.

Electronic hygrometers use materials that change resistance or capacitance in response to changes in moisture level. The resistance type uses a grid that is coated with a hygroscopic substance such as lithium chloride. As moisture is absorbed by the hygroscopic substance, a high-resistance electric circuit is created. The resistance is a function of the level of moisture. The capacitance type is a membrane made of a nonconductive film, stretched between two metal electrodes mounted in a perforated plastic capsule. As the membrane absorbs moisture, the capacitance between the electrodes increases or decreases.

20.2.3   Pressure Sensors

Since pressure is a force exerted over an area, pressure is easily converted to motion or displacement. Pressure sensors convert a change in absolute, gage, or differential pressure into mechanical motion with a bellows, diaphragm, or Bourdon tube mechanism. Pressure sensors are available for measuring pressures from a few in of water to several thousand psi.

Mechanical pressure sensors usually employ a flexible diaphragm, a bellows, or a Bourdon tube. The flexible diaphragm type has a diaphragm that distorts with changes in pressure on one side. The flexible diaphragm is connected to a linkage which converts the motion into a pneumatic or electronic signal. Bellows type pressure sensors have a bellows, made of a corrugated cylinder, which expands or contracts with changes in pressure differential between the inside and outside. The reference pressure is usually atmospheric. One end of the bellows is firmly anchored, while the other end is allowed to move with changes in pressure. The moveable end is connected to a linkage which converts the motion to a pneumatic or electronic signal.

Bourdon tube pressure sensors have a closed semi-circular tube. The end of the tube is connected to a linkage which produces mechanical motion, proportional to the pressure change. The pressure being measured fills the inside of the semi-circular tube. The pressure inside the tube tends to straighten the tube, which results in motion at the end. The greater the pressure inside the tube, the more the tube straightens. Bourdon tubes are frequently used in dial type pressure gages, but rarely in HVAC control systems.

Most pressure sensors are mechanical devices. Pressure is usually not measured directly with electronics. Typically, an electronic signal is produced from a transducer, which converts from a mechanical force or motion to an electrical signal.

20.2.4   Flow Sensors

Flow sensors are frequently used in HVAC control systems. It is often desirable to measure the flow rate of gases and liquids. Examples of applications for flow sensors in HVAC control systems include measuring the rate of air flow in ductwork, measuring water flow in heating and chilled water systems, and measuring the rate of steam flow in steam systems. Flow measuring devices in HVAC control systems include sail switches, paddle wheels, hot wire anemometers, pitot tubes, and orifice plates. Other, more sophisticated, flow sensing devices include venturi meters, turbine meters, magnetic flow meters, vortex shedding meters, and Doppler effect meters. The more commonly used flow measuring devices used in control systems are briefly discussed herein.

Sail switches are typically used to verify that a fluid is flowing; however, they cannot be used to measure the rate of fluid flow. A sail switch has a lightweight plate that is mounted in the fluid, perpendicular to the direction of flow. Since the sail switch offers a relatively large flat surface to the flowing fluid, the fluid moves the sail slightly in the direction of the flow. At the other end of the sail, a switch is mounted which is opened or closed, depending upon whether flow is present or not.

A paddle wheel is similar to a sail switch except that a plate or paddle is used, in lieu of a sail, and the paddles are mounted on a wheel. As the fluid flows, the paddles cause the wheel to rotate. The greater the rate of fluid flow, the faster the paddle wheel rotates. The wheel has a magnetic-measuring mechanism, which measures the rotational speed of the wheel. These devices are used both to measure instantaneous flow rate and to calculate total flow. Total flow is the quantity of fluid that has flowed over a long period of time, such as a month.

Hot wire anemometers are frequently used to measure air flow rates. These consist of a small electric resistance heater and a temperature sensor. As air flows past the heated anemometer, heat is lost to the air. The rate at which the heat is lost to the air is a function of the air velocity. The air velocity is determined by the amount of electric energy required to maintain a reference temperature. Hot wire anemometers require a reference to neutralize the effect of air temperature changes on the output signal.

Pitot tubes are also frequently used to measure air velocities in HVAC systems. A pitot tube is a double tube which may be inserted into a duct to measure the velocity of the air flow. The pitot tube indirectly measures air velocity by measuring the total pressure and static pressure of the flowing air. A typical pitot tube is shown in Figure 15.1. The method by which air velocity is determined, by measuring the total pressure and static pressure of the air, is discussed in Chapter 15.

Orifice plates may be used to measure the flow rate of all types of fluids. The orifice plate consists of a plate with a precise hole or orifice in it, through which the fluid flows. Based on the size of the orifice and the properties of the flowing fluid, the rate of the fluid flow may be determined by measuring the fluid pressure drop, from one side of the orifice to the other. The accuracy of orifice plates drops off as the flow rate decreases below 50% of the design flow rate.

20.3   CONTROLLED DEVICES

Earlier in this chapter, a controlled device was defined as a device that reacts to signals received from a controller. A signal is measured by a sensor which sends a signal to a controller, which in turn signals the controlled device to change as necessary to maintain the desired set point. The controlled device adjusts the flow of the controlled agent in response to a signal from the sensor and controller.

Controlled devices that typically regulate, or vary, the flow of the steam or water are called control valves. Controlled devices that control the flow of air are called dampers. The device that physically causes the control valve or damper to move is called an operator or actuator.

20.3.1   Control Valves

Automatic control valves are frequently used to control the temperature of air leaving a coil in an air handling unit, or some other type of heat exchange component. The leaving air temperature is controlled by regulating the flow of steam, water, gas, or other fluid to a heat exchange component.

A control valve acts as a variable orifice in a fluid flow line such as a pipe. A control valve itself typically consists of a body, a moveable disk or plug, a seat for the moveable

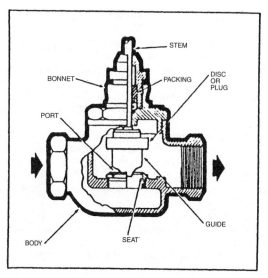

FIGURE 20.1   Components of a Control Valve

disk, a stem to move the disk, and a bonnet with packing to prevent leakage from the valve. A typical control valve is shown in Figure 20.1.

In order for a control valve to move or be actuated, it is necessary that the valve have an actuator. Valve actuators may either be electric motors or pneumatic diaphragm type. A typical control valve with a pneumatic actuator is shown in Figure 20.2.

The control valves shown in Figures 20.1 and 20.2 are two-way control valves. The fluid flows into one port and out another. A schematic for a two-way control valve is shown in Figure 20.3.

A two-way control valve restricts the flow of fluid to the coil, or heat exchange device, by opening or closing in response to the load requirements. As the heating or cooling loads go down, the valve automatically closes, reducing the fluid flow to the heating or cooling coil. As the load increases, the valve automatically opens to increase the rate of flow. Two-way valves are used to control the flow of steam to steam heating coils. Hydronic (heating water or chilled water) coils also frequently use two way control valves, depending upon the pumping/piping arrangement of the hydronic system. If two-way valves are used in hydronic systems, it may be necessary to provide some means to relieve pressure in the system as the valves close and reduce total system flow.

Control valves may also be three-way type. Three-way control valves have three ports. These valves are used on hydronic systems to relieve the pressure built up as flow through coils is reduced. They are not used for steam systems. The valve is piped to the hydronic coil so water may either flow through the coil or be bypassed around the coil, in response to the load requirements. Three-way valves are available in two configurations, depending upon whether they are located in the coil inlet or outlet piping. A three-way valve located in the coil inlet piping is referred to as a three-way diverting valve. A three-way diverting valve is shown in Figure 20.4 and is shown schematically in Figure 20.5.

A three-way diverting valve either allows the water to flow through the coil, or diverts the water around the coil. The flow rate in the supply and return piping remains constant, while the flow through the coil is varied.

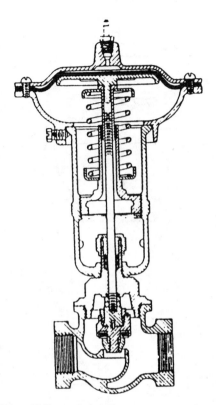

FIGURE 20.2   Control Valve with Pneumatic Actuator

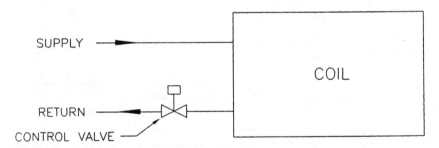

FIGURE 20.3   Schematic Diagram for a Two-Way Control Valve

Three-way control valves may also be mixing valves. Mixing valves, located on the discharge or outlet side of the coil, control the flow through the coil by mixing the flows. Since the flow rate in the piping to the coil remains constant, flow through the coil is varied by mixing water that has flowed through the coil with water bypassed around the coil, in the proportions as required by the load. A typical three-way mixing valve is shown in Figure 20.6 and the schematic for a three-way mixing valve is shown in Figure 20.7.

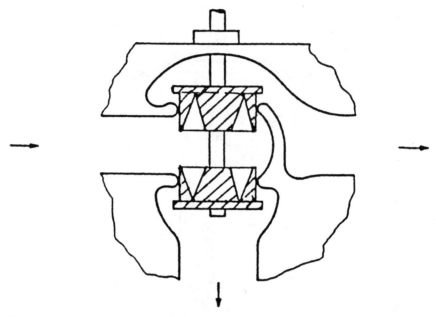

FIGURE 20.4   Three-Way Diverting Valve

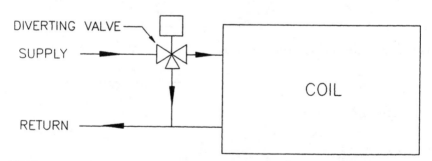

FIGURE 20.5   Schematic Diagram for a Three-Way Diverting Valve

The selection of control valves (two-way or three-way) should be based on the flow characteristics of the valve. The flow characteristics are the relationship between the flow rate and the valve stem travel (stroke) as the stem travel is varied between 0% and 100% of full travel and is based on a constant pressure drop through the valve. The three most common characteristics for control valves are quick opening, linear, and equal percentage. These flow characteristics are shown in Figure 20.8.

The **quick opening** valve approaches maximum flow quickly as the valve is opened. The slope of the percent flow to percent open curve is very steep initially, and then levels off as the valve is opened farther. This type of valve is suitable for two position on-off applications.

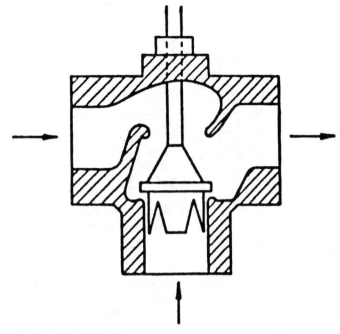

FIGURE 20.6   Three-Way Mixing Valve

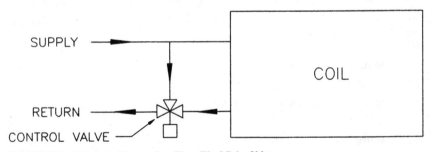

FIGURE 20.7   Schematic Diagram for a Three-Way Mixing Valve

Linear type control valves have a % flow versus % open curve that is a straight line, as the name implies. The opening and flow are directly proportional if the pressure remains constant. Linear type control valves may be suitable for controlling the flow of steam, since the heat output of a steam coil or heat exchanger is directly proportional to the steam flow rate. This is due to the fact that the steam remains at the same temperature as it changes phase from a vapor to a liquid.

Equal percentage type control valves have a flow versus opening curve such that the flow increases exponentially as the valve opens. Each equal increment of opening increases

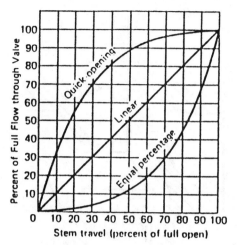

FIGURE 20.8   Control Valve Flow Characteristics

the flow by an equal percentage over the previous value. This type of valve is most suitable for controlling the flow to heating water or chilled water coils because it has a lower ratio of flow increase in the region near full closure. The pressure characteristic of the coil provides a near linear temperature change to the valve control signal change.

Control valves in heating water and chilled water systems are also selected on their **flow coefficient (C_V)**. The flow coefficient is defined as "the flow rate of water at 60°F in gal/min that will pass through a valve that is fully open at a pressure of 1 psig." The flow rate through a valve at pressure drops other than 1 psig may be calculated using Equation 20.1.

$$Q = C_V \sqrt{\Delta P} \qquad (20.1)$$

where

Q = the flow rate, gal/min

C_V = the valve flow coefficient

ΔP = the pressure drop through the valve, psig

20.3.2   Dampers

Dampers are devices used to control the flow of air in an air distribution system. Dampers are made of one or more flat plates mounted in the airstream that rotate to control the air flow. Dampers accomplish air flow control by varying the resistance to the air flow. Automatic or motorized dampers may be used for modulating control or for two-position control. Dampers may have a single large blade (butterfly damper) or may have multiple blades. Multi-blade dampers may be parallel or opposed blade type. Figure 20.9 shows typical multi-blade dampers.

Parallel blade dampers are best suited for two-position control and for applications where two air streams are mixed together. In mixing applications, parallel dampers tend to reduce stratification and improve mixing of two air streams, provided the dampers are properly placed. Better mixing occurs since the two air streams can be directed toward each other and because parallel blade dampers tend to cause more turbulence in the air flow.

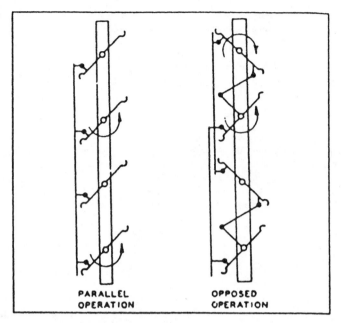

FIGURE 20.9 Typical Multi-Blade Dampers

Changes in blade rotation do not directly result in proportional changes in the rate of air flow. Their flow characteristics are nonlinear.

Opposed blade dampers are constructed so that alternate blades rotate in opposite directions. They provide better control than parallel blade dampers and are used where tight shut-off is required. Opposed blade dampers have flow characteristics that are more linear than parallel blade dampers. The changes in damper rotation are more closely related to changes in air flow rate. This allows a closer control of air flow, especially at low rates.

An important characteristic to consider when selecting dampers is the damper leakage. If dampers do not close tightly, air will leak past the dampers which is frequently undesirable. Damper manufacturers will usually provide leakage information about their dampers. Damper leakage is expressed in one of two ways: (1) percentage of air flow through an open damper, or (2) ft³/min of air/ft² of damper area at a given pressure differential. In both cases, the pressure differential across the damper must also be indicated.

20.3.3 Valve and Damper Actuators

Actuators are devices that are used to move a valve or damper in response to a signal from a controller. The energy used by an actuator to move a control valve or damper is usually dependent upon the type of energy used in the rest of the control system. Almost all actuators are electric or pneumatic. Electric actuators can operate on line voltage (120 V) or low voltage (24 V AC or DC). The most common types of actuators used in HVAC control systems are electric motors and pneumatic actuators.

A solenoid is an electrical device with a magnetic coil and a moveable plunger. Solenoids are most commonly used for two-position control such as open or closed. Solenoids are typically used on small valves, usually less than 2 in.

Electric motors may be used for modulating control as well as two-position. The motor operates the valve or damper through a gear train and linkage. Motor operators are available in three different types:

- unidirectional: used for two-position control
- spring return: used for two-position control in which the valve or damper is actuated to one position, and a spring returns the valve or damper to the original position
- reversible: used for floating and proportional control; the motor can operate in either direction and can stop in any position

Pneumatic actuators for valves consist of a bellows (or diaphragm) which uses compressed air to actuate the valve. A spring is usually used to provide an opposing force to the diaphragm and return it to its original position. A pneumatic actuator for a damper usually consists of a piston and rod connected through linkage to the damper. Pneumatic operators are typically used for proportional control, although they can also be used for two-position control.

20.4 CONTROLLERS

Earlier, a controller was defined as a device that sensed the value of a variable and compared it with a set point. A controller takes the input from a sensor, compares it with the desired set point, then regulates an output signal to cause a control action at the controlled device (control valve, automatic damper, etc.). A controller and sensor can be combined in a single device, such as a room thermostat, or it may be two separate devices. Controllers may be classified by the type of control action and type of energy used for the control signal. As described earlier, the control action may be two-position or modulating. The term **analog control** suggests a continuous modulating signal. The term **digital control** suggests discrete on-off or two-position control. The types of energy used by controllers are pneumatic and electric/electronic. When two separate pneumatic devices are used, the pneumatic controller is usually called a receiver-controller.

20.4.1 Electric/Electronic Controllers

An electronic controller compares a signal from the sensor to a standard reference or set point to produce an electronic error signal. The error signal is then amplified to a high enough level to drive the controlled device. Electric/electronic controllers are available for all modes of control. For two-position control, the controller output may be just an electrical contact that energizes or de-energizes the controlled device. For proportional control, a controller may have a switching circuit with a neutral zone where neither contact is made. Proportional control may also be such that a continuously or incrementally changing output signal is used to position an electrical actuator or controlled device.

A basic electronic controller is composed of a null detector and amplifier with input adjustments and an output switching element to drive the controlled device. The **null detector** compares the difference between the sensor signal and the set point. When the two values are equal, there is a balance condition which results in zero (null) output. The amplifier multiplies the output from the null detector and produces an amplified error signal. The amount of amplification is called the **gain** of the amplifier. Some amplifiers can also differentiate or discriminate between signals of different polarity or phase. The **discriminator** determines if the signal to the amplifier was the result of an increase or decrease in the controlled variable. The signal, therefore, can drive the controlled device in the proper direction to maintain the set point.

20.4.2   Direct Digital Controllers

Direct digital controllers (DDC) are similar to electronic controllers, except that a computer or microprocessor is used to provide control algorithms for the control system. DDCs are different from electronic or pneumatic controllers in that the control algorithm is stored in the computer memory as a set of program instructions. The controller calculates the required output signal. The controlled devices are then actuated to the controlled position through the hardware interface, which converts the digital signal from the computer to an analog signal, which positions the actuator or energizes a relay. Based on the input data, the control algorithms stored in the computer memory make the required control decisions.

20.4.3   Pneumatic Receiver-Controllers

Pneumatic controllers produce a force or position output in order to produce a variable air pressure output signal. Pneumatic receiver-controllers are usually combined with pneumatic sensors. These controllers are usually classified as nonrelay or relay and direct or reverse acting.

A pneumatic nonrelay controller has a restrictor located in the air supply. It also has an air bleed nozzle. The sensor positions an air exhaust flapper which controls the nozzle opening, resulting in a variable air pressure output signal to the controlled device, such as a pneumatic actuator.

A relay type pneumatic controller actuates a relay device that amplifies the air volume available for control. This is accomplished, either directly or indirectly, with a nozzle and a flapper.

20.5   Control System Accessories

In addition to sensing devices, controlled devices, and controllers, there are a number of accessories that are frequently required in order to have a complete control system. Some of the more common accessories found in HVAC control systems include transducers, switches, relays, transformers, step controllers, clocks and timers, air compressors, refrigerated air dryers, pneumatic tubing, and control wiring.

A **transducer** is used to transform one form of energy to another. In control systems, transducers are used to convert electric input signals to pneumatic output or vice versa. Transducers may convert a proportional input signal to either a proportional output signal or a two-position output signal. **Pneumatic/electric (P/E) switches** convert a pneumatic signal to an electric signal. Electro/pneumatic (E/P) switches convert electric signals to a pneumatic signal. P/E and E/P switches are examples of transducers.

Switches are devices typically used to turn another device on or off (two-position), although they may be multi-position. They may be single pole to switch one device, or they may have multiple poles to switch multiple devices. Pneumatic switches are typically manually operated devices which are used to divert air from one circuit to another, or for opening and closing air circuits. Electrical switches may be used to turn electrical devices on or off, and to switch between electrical circuits.

A **relay** is a device used for passing a signal from one device to another, or from one circuit to another. Relays may be pneumatic, electric, or electronic. Pneumatic relays are actuated by the air pressure from a controller to perform several functions. Pneumatic relays are two-position or proportional. Switching relays are pneumatically operated air valves which divert air from one circuit to another or open/close pneumatic circuits. Positive positioning relays are devices which ensure the accurate positioning of a controlled device, such as a valve or damper, in response to changes in pressure from a controller. Relays are also electric or electronic, and may be used to control electric devices.

Transformers were discussed in Chapter 19. They are used in electrical systems to step the voltage in electrical systems up or down. **Amplifiers** in control systems use solid state or vacuum tube components to amplify very small signals to a level where they are usable by the control system. Other types of transformers are used to step a signal down. For example, most electrical control systems require a transformer to step a 120 V AC power supply down to 24 V AC or DC for the control system.

Step controllers are used to operate a number of switches in a certain order or sequence by means of a proportional input signal. Step controllers can be either electric or pneumatic. They are frequently used to bring devices on, in steps or stages, in a predetermined order or sequence.

Timers and time clocks are used to turn components on and off or switch between circuits at predetermined times. An example of such a timer is a time clock on a setback thermostat that switches between day or night operation depending on the time of day. Timers can also be used for other time sequence functions. A time clock is a clockwork mechanism with a rotating dial on which trip tabs can be mounted. A program timer is an electronic timer with several switches which are opened or closed in a preset sequence.

Pneumatic control systems require a source of compressed air to operate, typically around 20 psig. Compressed air for pneumatic control systems is usually provided by a control system air compressor and refrigerated air dryer. Compressors are usually reciprocating type, single stage, air cooled. Air compressors are rated according to the air flow capacity and pressure. Air compressors usually include an air receiver (storage tank) to allow for peak air demand periods. As air is compressed to a higher pressure, the air temperature increases. When the air is cooled, moisture usually condenses out in the form of water. Since water in pneumatic systems is undesirable, the air in a pneumatic system may be mechanically cooled to remove moisture with a refrigerated air dryer. The condensed moisture is then removed from the system by a trap which allows water to pass but not air.

Distribution piping in pneumatic control systems is usually soft copper tubing with soldered fittings. Nylon reinforced plastic tubing is also frequently used; however, codes should be consulted for restrictions regarding where plastic materials may be installed. Some codes restrict the use of plastic in air plenums or in mechanical rooms. Wiring for control components require shielded cable to reduce the interference from stray signals. Cables installed in air plenums should be plenum-rated cable.

20.6   TYPICAL AIR SYSTEM CONTROLS

Up to this point, this chapter has primarily discussed individual control components used in HVAC system controls. In order to have a complete functioning control system, various components must be combined to design a functioning control system. This section introduces the reader to basic HVAC control systems, primarily for air systems. Only elementary control systems are discussed. These simple control systems may be combined to form more complex, complete control systems. A more detailed discussion of controls for HVAC systems may be found in References 2 and 22.

The first step in designing a control system is to write a **sequence of operation**. A sequence of operation, or just sequence, describes verbally how the HVAC system is expected to perform. The sequence may then be used as a basis for selecting and combining various control components that will make up a complete control system, which will result in the desired system performance.

20.6.1 Safety Controls for Air Systems

Most air systems require some sort of controls to shut the system down in the event of a serious problem. The two most commonly used safety controls are fire/smoke detectors and freezestats.

Chapter 11 discussed some of the codes that are applicable to HVAC systems for buildings. Three of the most common codes that are applicable to air distribution systems are the BOCA Mechanical Code, NFPA 90A, and NFPA 90B. All three codes require that **smoke detectors** be provided in the discharge of fans, in air systems circulating air at the rate of 2000 ft³/min or more. The detector must be located downstream of any filters, but before any branch takeoff in the ductwork. If the system circulates 15,000 ft³/min or more, smoke detectors are required in both the supply air and return air of an air handling unit. The purpose of the smoke detectors is to shut the fan motor off in the event that smoke enters the air distribution system. Shutting the fan motor off prevents the air distribution system from circulating smoke throughout a building in the event of a fire. A typical control schematic with a smoke detector, in an air system circulating between 2000 ft³/min and 15,000 ft³/min, is shown in Figure 20.10.

Freezestats, or low temperature sensors, are frequently used in air systems. The freezestat is set to sense the temperature of the air coming into, or passing through, an AHU and shutting off the fan motor if freezing temperatures are sensed. The use of freezestats is especially important in HVAC systems with hydronic heating or cooling coils. If a hydronic coil is subjected to air temperatures below 32°F for a period of time, the water in the coil could freeze and rupture the coil. The purpose of the freezestat is to sense low air temperatures and protect the coils, by shutting the fan motor off when air temperatures that may freeze a coil are present. A typical AHU control schematic with a freezestat is shown in Figure 20.11.

Safety controls can also include high limit temperature controls. **High limit controls** are similar to freezestats except they shut the fan motors down in the event of excessive discharge air temperatures. High limit temperature controls are frequently used with steam heating systems and heating systems with gas or fuel oil furnaces. High temperature limits may also be used with electric heating coils.

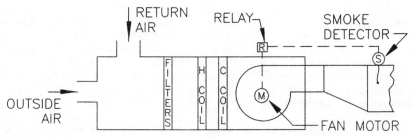

FIGURE 20.10 AHU with Smoke Detector

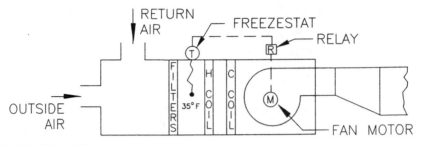

FIGURE 20.11   AHU with Freezestat

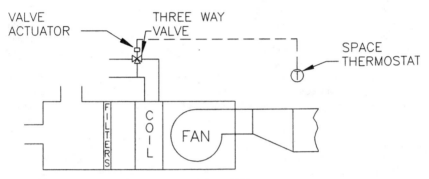

FIGURE 20.12   Space Temperature Control with Control Valve

20.6.2   Air System Temperature and Humidity Controls

Temperature control for air systems with hydronic heating or cooling coils, or with steam heating coils, is usually accomplished by modulating a control valve. Control valves were discussed earlier in this chapter. Frequently, the control valve is controlled directly with signals from a space thermostat, as shown in Figure 20.12.

In this arrangement, the heating or cooling load and the thermostat in the conditioned space determine the coil discharge temperature. The greater the heating load, the higher or lower the discharge air temperature within the capacity limits of the coil. In most HVAC systems, this arrangement is perfectly satisfactory. However, in some cases, it is desirable to limit the discharge air temperature or control the discharge temperature at a predetermined level. To maintain the leaving air temperature within desired limits or at a constant level, a thermostat is mounted in the discharge air duct of the air handling unit. The discharge air thermostat then modulates the control to maintain the desired leaving air temperature. It is also frequently desirable to limit the discharge air temperature to avoid excessively high or low supply air temperatures to the conditioned space. A heating coil may have sufficient capacity to supply air at a temperature over 100°F; however, this probably would not be desirable. To maintain the leaving air temperature to reasonable limits, a discharge air thermostat may be used in series with a space thermostat, as shown in Figure 20.13.

In this arrangement, the space thermostat determines whether heating, cooling, or neither (deadband) is required and sends a signal to the discharge air thermostat. The discharge

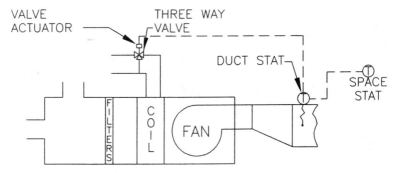

FIGURE 20.13 Discharge Air Temperature Control

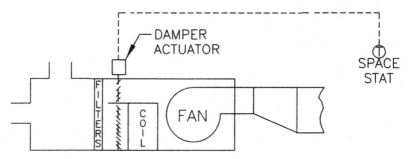

FIGURE 20.14 Face and Bypass Dampers

air thermostat then modulates the control valve to provide discharge air between predetermined limits, such as 55°F for cooling and 90°F for heating.

Temperature control for heating coils and, occasionally, cooling coils may also be provided by **face and bypass dampers**. As the name implies, face and bypass dampers either pass the air over the face of a coil or bypass the air around the coil, as shown in Figure 20.14.

Modulating temperature control is achieved by passing part of the air through the coil in the proportions necessary for the desired leaving air temperature. After the air passes through or around the coil, the two air streams are then recombined (mixed) which results in the desired temperature. Face and bypass dampers are frequently used on steam and heating water preheat coils subjected to freezing air temperatures. Full water flow, through a water coil, or full steam pressure is maintained on a steam coil which reduces the possibility of freezing. Face and bypass dampers on reheat coils are also used for zone temperature control in multizone air handling units. Supply air is either passed through or around the reheat coil by zone face and bypass dampers controlled by a zone thermostat.

Discharge air temperatures may also be controlled with an on-off or staged control. Electric coils are frequently controlled in stages. Stages of the coil are switched on or off depending on the load.

The control of direct expansion (DX) cooling coils is typically on and off. The on and off control is provided by a **solenoid valve** located in the liquid line to the refrigerant coil. The solenoid valve opens and closes, and the compressor starts and stops in response to signals from a thermostat. Control of DX cooling systems can be improved with multiple

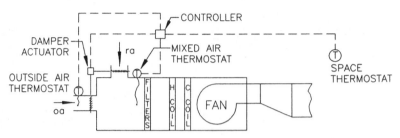

FIGURE 20.15   Economizer Controls

stages of cooling. Multistage cooling is accomplished with multiple refrigerant compressors and multiple refrigerant circuits in the cooling coil. A two-stage system would have two compressors, two solenoid valves, and two circuits in the coil. Each stage could be sized to have approximately 50% of the required cooling capacity. Multistage thermostats are required for multistage cooling systems.

Space temperature control may also be achieved by varying the amount of outdoor air brought in through the air system. Building codes set the minimum amount of outdoor air that must be brought into the building for ventilation purposes. Usually, they do not, however, limit the maximum amount of outdoor air. When a building has internal cooling loads and the outdoor air temperatures are low, it is frequently more economical to bring large amounts of cooler outdoor air in through the HVAC system to satisfy the cooling load. Usually, it is less costly to use outdoor air for cooling than to operate a chilled water or DX cooling system. Such operation of the air system is called **economizer** operation. The air system controls that accomplish economizer operation are called economizer controls, or just an economizer, and are shown in Figure 20.15.

Upon a call for cooling from the space thermostat, the controller senses the outdoor air temperature from an outdoor air thermostat. If the outdoor air is sufficiently cool, the controller modulates the outdoor air and return air dampers in response to a signal from the mixed air thermostat. When the mixed air thermostat senses the mixed air temperature, the damper actuator positions the dampers to provide the proper proportions of outside air and return air, which will result in the desired mixed air temperature. Some economizers operate on only the various dry bulb temperatures. These economizers are called **sensible** or **dry bulb economizers**. Other economizers sense the enthalpy or total energy of the air streams. These admit additional amounts of outdoor air if the enthalpy of the outdoor air is less than that of the conditioned space **(enthalpy economizer)**. It is usually undesirable to bring high humidity, low temperature outside air into a conditioned space.

The maximum relative humidity in a conditioned space is usually controlled by the cooling system. When a cooling coil cools air, it also usually dehumidifies the air. As a result, controlling the maximum temperature in the space indirectly controls the humidity level. In some applications, it is desirable to more directly control the maximum humidity in a space. In that situation a **humidistat** may also be provided in the conditioned space, in addition to the space thermostat. Typically, either the thermostat or the humidistat can control the cooling coil.

The minimum relative humidity in a conditioned space can be controlled with a humidifier and space humidistat. Various types of humidifiers were discussed in Chapter 17. The humidifier is typically cycled on and off, in response to a signal from a space humidistat. A high limit humidistat is usually mounted immediately downstream of the humidifier to prevent excessive amounts of moisture from a malfunctioning humidifier or humidifier control.

20.6.3 Fan and Pressure Control

Many fans in HVAC systems deliver air at a constant volume. The primary method of control is starting and stopping the fan. Start/stop control can be accomplished manually or automatically. Fans that are started and stopped automatically are usually controlled by a signal from a space thermostat. Fans in smaller HVAC systems are frequently off when the temperature is satisfied and come on when the temperature drops below, or rises above, the space temperature set points. Most fans for larger HVAC systems in commercial buildings operate continuously whenever the building is in the occupied mode. The fan may be off during night set back and then be switched on for the occupied mode.

Most HVAC systems are required by code to bring in certain minimum amounts of outdoor air based on the expected number of building occupants. In order to bring outdoor air into a building, an equivalent amount of air must be either exhausted or relieved from the building. Occasionally, toilet and other exhaust systems have capacities that are near the required outdoor air ventilation rate. Frequently though, this is not the case, especially if the air system has an economizer cycle. For an economizer to function properly, a means must be provided to remove as much air from the building as is brought into the building when the HVAC system is in the economizer mode. Occasionally, an air handling unit, such as a packaged rooftop air conditioner, has **relief dampers** in the unit. These dampers are shutter type dampers that open, due to static pressure, and close by gravity when there is no pressure. When gravity relief dampers are not available, an alternate means to relieve building pressure must be provided. This may be accomplished with automatic dampers and static pressure sensors to measure the building pressure. The relief dampers must be located away from the fan, elsewhere in the building. Typically, the dampers are set to open when a static pressure difference of 0.1 in. wg is measured between the indoor and outdoor pressure.

Air circulation systems for large commercial buildings may be variable air volume (VAV). A schematic diagram of a VAV air distribution system is shown in Figure 16.44, and a discussion of VAV system operation is included in Chapter 16. As the VAV terminals in the air distribution system open and close, the static pressure in the supply air ductwork fluctuates. As most of the VAV terminals close to reduce the flow of air to the spaces, the static pressure in the ductwork tends to increase. In large systems, the pressure could increase to a point where the ductwork or AHU casing is damaged. In order to prevent this situation, fans in VAV systems must be provided with air flow capacity control. This is usually accomplished by providing a **static pressure sensor** in the supply air duct to measure the pressure. The pressure sensor is typically located about 75% to 100% of the distance from the first to the last terminal. The signal from the duct pressure sensor is used to modulate the capacity of the fan. Typically, fan capacity control is provided by one of three methods: discharge dampers, fan inlet vanes, or variable frequency drives.

Discharge dampers are simply automatic dampers located in the fan discharge that modulate open or closed in response to duct static pressure changes. They are not very energy efficient. Inlet vanes are automatic dampers placed in the inlet of a centrifugal fan, as shown in Figure 16.46. The inlet vanes are also actuated in response to change in duct static pressure. Fan capacity is varied by restricting the flow of air into the fan, which allows the fan to rotate in an unloaded condition during low air flow rates. Variable frequency drives control the air flow by varying the rotational speed to the fan in response to duct pressure changes. Variable frequency drives are discussed in more detail in Chapter 19.

20.6.4 Individual Space Temperature Control

In large buildings with many spaces that have diverse heating and cooling loads, it is frequently desirable to provide individual temperature control for individual spaces or groups of spaces (zones). In order to individually control temperatures in spaces, it is necessary to provide controls for each space to be controlled in the form of a thermostat. The thermostat usually controls a variable air volume terminal for the space, a reheat coil for the space, or both.

The most widely used type of individual space temperature control for large commercial HVAC systems is the VAV system. A space thermostat is used to control a VAV terminal for the space. A typical VAV terminal is shown in Figure 16.45. Air is usually supplied to the space at a constant temperature, typically around 55°F. Cooling loads in the space are controlled by varying the amount of cooling air supplied to the space. As the load in the space decreases, a damper in the VAV terminal modulates closed, reducing the flow of cooled air. When selecting VAV terminals, care must be taken to not allow the ventilation rate through the terminal to drop below that required by code. Minimum ventilation rates are discussed in Chapter 11. Stops may be placed on the damper to provide a preset minimum damper opening, despite the space cooling load. This can occasionally result in overcooling the space during low cooling loads.

Heat for individual spaces may be provided by reheat coils in the VAV terminal or from a separate source such as baseboard radiators. Reheat coils in VAV terminals are usually either heating water or electric coils. Control for these coils is the same as for coils in air handling units described earlier. Control is normally provided directly from the space thermostat. Controls for VAV reheat coils are normally sequenced so heat is not available unless the cooling air has been modulated to the minimum flow position.

20.7 BUILDING AUTOMATION SYSTEMS

Building automation systems (BAS), also called energy management and control systems (EMCS), are computer-based systems that provide building-wide control for HVAC systems, as well as other systems such as electrical and lighting systems. BAS systems can also be used for security systems, such as fire alarm and intrusion alarm systems. These systems, which are usually DDC systems, allow an operator to monitor and control thousands of elements throughout a building, or building complex, from a single location. These systems can provide central monitoring and recording of local control functions as well as control programs, such as programmed start/stop and reset of local control points. The primary reasons for BAS are energy savings and energy management.

The typical BAS consists of a central computer, software (programs), a display monitor, a printer, and a keyboard, as well as a **distributed network** of control components and wiring. The distributed network typically consists of a communication link (coaxial cable or a twisted pair of wires), the earlier described control elements at the equipment to be controlled, and a **field interface device (FID)**. The FID serves as an interface between the communications link and the local control loop of devices. The FID typically contains a microprocessor which makes many of the control decisions that would otherwise be made by the central computer. The FID and local control loop essentially provide "stand alone control." This reduces signal traffic on the communications link and results in what is called **distributed intelligence**.

A BAS can perform all of the control functions of basic control systems that were described earlier in this chapter. In addition, a BAS can perform more advanced control functions, which more precisely control HVAC systems and keep energy consumption to a minimum. Some of the most common control functions performed by a BAS include the following:

• timed operation: timed operation functions consist of starting and stopping a system based on the time (time of day, day of week, etc.).

- duty cycling: duty cycling consists of shutting a system down for predetermined periods of time during normal operating hours, such as 15 min of each hr of operation.
- temperature-compensated duty cycling: with temperature-compensated duty cycling, each piece of equipment has a space temperature set point assigned to it. Each piece of equipment serving interior zones (cooling only) is cycled off for an assigned maximum period of time as the temperature in the space drops to its assigned maximum value.
- demand control: demand control function stops electrical loads to prevent a predetermined maximum electrical demand from being exceeded.
- outside air temperature cutoff: heating systems have an outdoor air temperature sensor which shuts off the heating system when the outdoor air temperature rises to within 5°F of the indoor air design temperature or a minimum of 40°F.
- warm-up/night cycle: the warm-up/night cycle function controls the outside air dampers when the introduction of outside air would impose a thermal load and the building is not occupied.
- space temperature night setback: the space temperature set point is lowered for heating and raised for cooling when the building is not occupied.
- economizer cycle: outdoor air is used to satisfy all, or part, of a building's cooling load when outdoor air conditions are suitable.
- hot/cold deck temperature reset: with temperature reset, the system selects the individual areas with the greatest heating and/or cooling requirements and adjusts the temperature leaving the heating coil and cooling coil accordingly.
- discharge air temperature reset: discharge air temperature reset function adjusts the cooling coil discharge temperature upward, until the zone with the greatest demand for cooling has closed the control for the terminal reheat coil
- trouble diagnosis: by monitoring certain points in mechanical/electrical systems, detection and diagnosis of problems in mechanical and electrical systems can be performed from a central location
- security and fire detection: the central processing unit can continuously monitor a building's fire detection and alarm systems and provide automatic notification to police and fire departments.

Appendix A

apparatus dew point	adp	heat transfer coefficient	U
approximate	approx	horsepower	hp
area	A	humidity, relative	rh
bypass factor	BF	humidity ratio	W
brake horsepower	bhp	kilowatt	kW
British thermal unit	Btu	kilowatt hour	kWh
conductance	C	leaving air temperture	LAT
conductivity	k	leaving water temperature	LWT
cubic feet	ft³	maximum	max
cubic inch	in³	mercury	Hg
cubic feet per minute	ft³/min	minimum	min
degree	degree or .°	noise criteria	NC
diameter	dia	outside air	oa
dry bulb temperature	db	air flow	Q
effective room latent heat	ERLH	percent	%
effective room sensible heat	ERSH	pounds	lb
effective sensible heat factor	ESHF	pounds per square foot	lb/ft²
entering	ENT	rankine	R
entering air temperature	EAT	relative humidity	rh
enthalpy	h	return air	RA
expansion	exp	room latent heat	RLH
Fahrenheit	F	room sensible heat	RSH
feet per minute	ft/min	room sensible heat factor	RSHF
feet per second	ft/sec	saturation	sat
foot or feet	ft	sensible heat	SH
foot pound	ft•lb	sensible heat ratio	SHR
gallons	gal	shading coefficient	SC
gallons per hour	gal/hr	specific heat	C
grand sensible heat factor	GSHF	square	sq
grand total heat	GTH	static pressure	SP
handling unit	AHU	supply air	sa
heat transfer	q	temperature	T

Abbreviations (*Continued*)

thermal conductivity	k	vertical	vert
thermal resistance	R	volume	vol
time	t	watt	W
U factor	U	wet bulb	wb
variable air volume	VAV		

Letter Symbols

Symbol	Description	Typical Units
A	area	ft^2
c	specific heat	Btu/lb•°F
C	thermal conductance	Btu/hr•ft²•°F
D or d	diameter	ft
h	enthalpy	Btu/lb
k	thermal conductivity	Btu/hr•ft•°F
p or P	pressure	psi
q	time rate of heat transfer	Btu/hr
Q	volumetric flow rate	ft³/min or gal/min
r	radius	ft
R	thermal resistance	ft²•hr•°F/Btu
T	temperature	°F
U	overall heat transfer coefficient	Btu/hr•ft²•°F
W	humidity ratio of moist air	lb (water)/lb (dry air)
Δ	difference between values	
σ	Stefan-Boltzmann constant	Btu/hr•ft²•°R⁴
φ	relative humidity	%
T_{adp}	apparatus dewpoint temperature	
T_{edb}	entering dry bulb temperature	
T_{ewb}	entering web bulb temperature	
T_{ldb}	leaving dry bulb temperature	
T_{lwb}	leaving wet bulb temperature	
T_m	mixture of outdoor and return air	

Conversion Factors

Multiply	By	To Obtain
Atmospheres	29.921	Inches of mercury (at 32°F)
Atmospheres	33.97	Feet of water (at 62°F)
Atmospheres	14.697	Pounds per square inch
Boiler horsepower	33,475	Btu per hour
Btu	778	Foot pounds
Btu	0.000293	Kilowatt hour
Btu per 24 hr	0.347×10^{-5}	Tons of refrigeration
Btu per hour	0.000393	Horsepower
Btu per hour	0.000293	Kilowats
Btu per inch•ft²•hour•°F	0.0833	Btu/ft•ft²•hour•°F
Cubic feet	1,728.0	Cubic inches
Cubic feet	7.48052	Gallons

Conversion Factors (*Continued*)

Multiply	By	To Obtain
Cubic feet of water	62.37	Pounds (at 60°F)
Feet of water	0.881	Inches of mercury (at 32°F)
Feet of water	0.4335	Pounds per square inch
Feet per minute	0.01667	Feet per second
Foot-pounds	0.001286	British thermal units
Gallons (U.S.)	0.1337	Cubic feet
Gallons of water	8.3453	Pounds of water (at 60°F)
Horsepower	33,000.0	Foot pounds per minute
Horsepower	2,546.0	British thermal units per hour
Horsepower	0.7457	Kilowats
Inches of mercurcy (at 62°F)	0.3342	Atmospheres
Inches of mercury (at 62°F)	13.57	Inches of water (at 62°F)
Inches of mercury (at 62°F)	1.131	Feet of water (at 62°F)
Inches of mercury (at 62°F)	70.73	Pounds per square foot
Inches of mercury (at 62°F)	0.4912	Pounds per square inch
Inches of water (at 62°F)	0.07355	Inches of mercury
Inches of water (at 62°F)	0.03613	Pounds per square inch
Inches of water (at 62°F)	5.202	Pounds per square foot
Kilowatts	1.341	Horsepower
Kilowatt hours	3415.0	British thermal units
Pounds	7,000.0	Grains
Pounds of water (at 60°F)	0.01602	Cubic feet
Pounds of water (at 60°F)	27.68	Cubic inches
Pounds of water (at 60°F)	0.1198	Gallons
Pounds per square inch	0.06804	Atmospheres
Pounds per square inch	2.309	Feet of water (at 62°F)
Pounds per square inch	2.0416	Inches of mercury (at 62°F)
Temperature (°C) + 273	1	Absolute temperature (°K)
Temperature (°C) + 17.78	1.8	Temperature (°F)
Temperature (°F) + 460	1	Absolute temperature (°R)
Temperature (°F) − 32	$^5/_9$	Temperature (°C)
Tons of refrigeration	12,000	British thermal units per hour
Watts	3.415	British thermal units per hour
Watt-hours	3.415	British thermal units

Appendix B

Thermal Properties of Building Materials

		Thermal Properties			
Description	Thickness (ft)	Conductivity (Btu/ft²•°F)	Density (lb/ft³)	Specific Heat (Btu/lb•°F)	Resistance (ft²•°F/Btu)
Acoustic Tile					
³/₈ in	0.0313	0.0330	18.0	0.32	0.95
¹/₂ in	0.0417	0.0330	18.0	0.32	1.26
³/₄ in	0.0625	0.0330	18.0	0.32	1.89
Aluminum or Steel Siding	0.005	26.0	480.0	0.1	0
Asbestos–Cement					
¹/₈ in board	0.0104	0.3450	120.0	0.2	0.03
¹/₄ in board	0.0208	0.3450	120.0	0.2	0.06
Shingle					0.21
¹/₄ in Lapped Siding					0.21
Asbestos–Vinyl Tile				0.3	0.05
Asphalt					
Roofing Roll			70.0	0.35	0.15
Shingle and Siding			70.0	0.35	0.44
Tile				0.3	0.05
Brick					
4 in Common	0.3333	0.4167	120.0	0.2	0.80
8 in Common	0.6667	0.4167	120.0	0.2	1.60
12 in Common	1.0000	0.4167	120.0	0.2	2.40
3 in Face	0.2500	0.7576	130.0	0.22	0.33
4 in Face	0.3333	0.7576	130.0	0.22	0.44
Builtup Roofing, ³/₈ in	0.03	0.09	70.0	0.35	0.33
Building Paper					
Permeable Felt					0.06
Two-Layer Seal					0.12
Plastic Film Seal					0.01
Carpet					
With Fibrous Pad				0.34	2.08
With Rubber Pad				0.34	1.23

Thermal Properties of Building Materials (*Continued*)

Description	Thickness (ft)	Thermal Properties			
		Conductivity (Btu/ft²•°F)	Density (lb/ft³)	Specific Heat (Btu/lb•°F)	Resistance (ft²•°F/Btu)
Cement					
1 in Mortar	0.0833	0.4167	116.0	0.2	0.20
1³/₄ in Mortar	0.1458	0.4167	116.0	0.2	0.35
1 in Plaster with Sand	0.0833	0.4167	116.0	0.2	0.20
Aggregate					
Clay Tile, Hollow					
3 in 1 cell	0.2500	0.3125	70.0	0.2	0.80
4 in 1 cell	0.3333	0.2999	70.0	0.2	1.11
6 in 2 cells	0.5000	0.3300	70.0	0.2	1.52
8 in 2 cells	0.6667	0.3600	70.0	0.2	1.85
10 in 2 cells	0.8333	0.3749	70.0	0.2	2.22
12 in 3 cells	1.0000	0.4000	70.0	0.2	2.50
Clay Tile, Paver					
³/₈ in	0.0313	1.0416	120.0	0.2	0.03
Concrete, Heavy Weight Dried Aggregate,					
140 lb					
1¹/₄ in	0.1042	0.7576	140.0	0.2	0.14
2 in	0.1667	0.7576	140.0	0.2	0.22
4 in	0.3333	0.7576	140.0	0.2	0.44
6 in	0.5000	0.7576	140.0	0.2	0.66
8 in	0.6667	0.7576	140.0	0.2	0.88
10 in	0.8333	0.7576	140.0	0.2	0.10
12 in	1.0000	0.7576	140.0	0.2	1.32
Concrete, Heavy Weight Undried					
Aggregate, 140 lb					
³/₄ in	0.0625	1.0417	140.0	0.2	0.06
1³/₈ in	0.1146	1.0417	140.0	0.2	0.11
3¹/₄ in	0.2708	1.0417	140.0	0.2	0.26
4 in	0.3333	1.0417	140.0	0.2	0.32
6 in	0.5000	1.0417	140.0	0.2	0.48
8 in	0.6667	1.0417	140.0	0.2	0.64
Concrete Block, 12 in Heavy Weight					
Hollow	1.000	0.7813	76.0	0.2	1.28
Concrete Filled	1.000	0.7575	140.0	0.2	1.32
Partially Filled Concrete	1.000	0.7773	98.0	0.2	1.29
Concrete Block 4 in Medium Weight					
Hollow	0.3333	0.3003	76.0	0.2	1.11
Concrete Filled	0.3333	0.4456	115.0	0.2	0.75
Perlite Filled	0.3333	0.1512	78.0	0.2	2.20
Partially Filled Concrete	0.3333	0.3306	89.0	0.2	1.01
Concrete and Perlite	0.3333	0.2493	90.0	0.2	1.34
Concrete Block, 6 in Medium Weight					
Hollow	0.5000	0.3571	65.0	0.2	1.40
Concrete Filled	0.5000	0.4443	119.0	0.2	1.13
Perlite Filled	0.5000	0.1166	67.0	0.2	4.29
Partially Filled Concrete	0.5000	0.3686	83.0	0.2	1.36
Concrete and Perlite	0.5000	0.2259	84.0	0.2	2.21
Concrete Block, 8 in Medium Weight					
Hollow	0.6667	0.3876	53.0	0.2	1.72
Concrete Filled	0.6667	0.4957	123.0	0.2	1.34
Perlite Filled	0.6667	0.1141	56.0	0.2	5.84

Thermal Properties of Building Materials (*Continued*)

Description	Thickness (ft)	Conductivity (Btu/ft²•°F)	Density (lb/ft³)	Specific Heat (Btu/lb•°F)	Resistance (ft²•°F/Btu)
Concrete Block, 8 in Medium Weight					
Partially Filled Concrete	0.6667	0.4648	76.0	0.2	1.53
Concrete and Perlite	0.6667	0.2413	77.0	0.2	2.76
Concrete Block, 12 in Medium Weight					
Hollow	1.000	0.4959	58.0	0.2	2.02
Concrete Filled	1.000	0.4814	121.0	0.2	2.08
Partially Filled Concrete	1.000	0.4919	79.0	0.2	2.03
Concrete, Lightweight, 80 lb					
³/₄ in	0.0625	0.2083	80.0	0.2	0.30
1¹/₄ in	0.1042	0.2083	80.0	0.2	0.50
2 in	0.1667	0.2083	80.0	0.2	0.80
4 in	0.3333	0.2083	80.0	0.2	1.60
6 in	0.5000	0.2083	80.0	0.2	2.40
8 in	0.6667	0.2083	80.0	0.2	3.20
Concrete Lightweight, 30 lb					
³/₄ in	0.0625	0.0751	30.0	0.2	0.83
1¹/₄ in	0.1042	0.0751	30.0	0.2	1.39
2 in	0.1667	0.0751	30.0	0.2	2.22
4 in	0.3333	0.0751	30.0	0.2	4.44
6 in	0.5000	0.0751	30.0	0.2	6.66
8 in	0.6667	0.0751	30.0	0.2	8.88
Concrete Block, 4 in Heavy Weight					
Hollow	0.3333	0.4694	101.0	0.2	0.71
Concrete	0.3333	0.7575	140.0	0.2	0.44
Perlite Filled	0.3333	0.3001	103.0	0.2	1.11
Partially Filled Concrete	0.3333	0.5844	114.0	0.2	0.57
Concrete and Perlite	0.3333	0.4772	115.0	0.2	0.70
Concrete Block, 6 in Heavy Weight					
Hollow	0.5000	0.5555	85.0	0.2	0.90
Concrete Filled	0.5000	0.7575	140.0	0.2	0.66
Perlite Filled	0.5000	0.2222	88.0	0.2	2.25
Partially Filled Concrete	0.5000	0.6119	104.0	0.2	0.82
Concrete and Perlite	0.5000	0.4238	104.0	0.2	1.18
Concrete Block, 8 in Heavy Weight					
Hollow	0.6667	0.6060	69.0	0.2	1.10
Concrete Filled	0.6667	0.7575	140.0	0.2	0.88
Perlite Filled Concrete	0.6667	0.2272	70.0	0.2	2.93
Partially Filled Concrete	0.6667	0.6746	93.0	0.2	0.99
Concrete and Perlite	0.6667	0.4160	93.0	0.2	1.60
Concrete Block, 4 in Lightweight					
Hollow	0.3333	0.2222	65.0	0.2	1.50
Concrete Filled	0.3333	0.3695	104.0	0.2	0.90
Perlite Filled Concrete	0.3333	0.1271	67.0	0.2	2.62
Partially Filled Concrete	0.3333	0.2808	78.0	0.2	1.19
Concrete and Perlite	0.3333	0.2079	79.0	0.2	1.60
Concrete Block, 6 in Lightweight					
Hollow	0.5000	0.2777	55.0	0.2	1.80
Concrete Filled	0.5000	0.3819	110.0	0.2	1.31
Perlite Filled	0.5000	0.0985	57.0	0.2	5.08
Partially Filled Concrete	0.5000	0.3189	73.0	0.2	1.57
Concrete and Perlite	0.5000	0.1929	74.0	0.2	2.59

Thermal Properties of Building Materials (*Continued*)

Description	Thickness (ft)	Conductivity (Btu/ft²•°F)	Density (lb/ft³)	Specific Heat (Btu/lb•°F)	Resistance (ft²•°F/Btu)
Concrete Block, 8 in Lightweight					
Hollow	0.6667	0.3333	45.0	0.2	2.00
Concrete Filled	0.6667	0.4359	115.0	0.2	1.53
Perlite Filled	0.6667	0.0963	48.0	0.2	6.92
Partially Filled Concrete	0.6667	0.3846	68.0	0.2	1.73
Concrete and Perlite	0.6667	0.2095	69.0	0.2	3.18
Concrete Block, 12 in Lightweight					
Hollow	1.0000	0.4405	49.0	0.2	2.27
Concrete Filled	1.0000	0.4194	113.0	0.2	2.38
Partially Filled	1.0000	0.4274	70.0	0.2	2.34
Gypsum or Plaster Board					
¹/₂ in	0.0417	0.0926	50.0	0.2	0.45
⁵/₈ in	0.0521	0.0926	50.0	0.2	0.56
³/₄ in	0.0625	0.0926	50.0	0.2	0.67
Gypsum Plaster					
³/₄ in Lightweight Aggregate	0.0625	0.1330	45.0	0.2	0.47
1 in Lightweight Aggregate	0.0833	0.1330	45.0	0.2	0.63
³/₄ in Sand Aggregate	0.0625	0.4736	105.0	0.2	0.13
1 in Sand Aggregate	0.0833	0.4736	105.0	0.2	0.18
Hard Board, ³/₄ in					
Medium Density Siding	0.0625	0.0544	40.0	0.28	1.15
Medium Density Others	0.0625	0.0608	50.0	0.31	1.03
High Density Standard Tempered	0.0625	0.0687	55.0	0.33	0.92
High Density Service Tempered	0.0625	0.0833	63.0	0.33	0.75
Linoleum Tile				0.3	0.05
Particle Board					
Low Density, ³/₄ in	0.0625	0.0450	75.0	0.31	1.39
Medium Density, ³/₄ in	0.0625	0.7833	75.0	0.31	0.08
High Density, ³/₄ in	0.0625	0.9833	75.0	0.31	0.06
Underlayment, ⁵/₈ in	0.0521	0.1796	75.0	0.29	0.29
Plywood					
¹/₄ in	0.0209	0.0667	34.0	0.29	0.31
³/₈ in	0.0313	0.0667	34.0	0.29	0.47
¹/₂ in	0.0417	0.0667	34.0	0.29	0.63
⁵/₈ in	0.0521	0.0667	34.0	0.29	0.78
³/₄ in	0.0625	0.0667	34.0	0.29	0.94
1 in	0.0833	0.0667	34.0	0.29	1.25
Roof Gravel or Slag					
¹/₂ in	0.0417	0.8340	55.0	0.4	0.05
1 in	0.0833	0.8340	55.0	0.4	0.10
Rubber Tile					0.05
Slate, ¹/₂ in	0.0417	0.834	100.0	0.35	0.05
Stone, 1 in	0.0833	1.0416	140.0	0.2	0.08
Stucco, 1 in	0.0833	0.4167	166.0	0.02	0.2
Terrazzo, 1 in	0.0833	1.0416	140.0	0.2	0.08
Wood, Soft					
³/₄ in	0.0625	0.0667	32.0	0.33	0.94
1¹/₂ in	0.1250	0.0667	32.0	0.33	1.87
2¹/₂ in	0.2083	0.0667	32.0	0.33	3.12
3¹/₂ in	0.2917	0.0667	32.0	0.33	4.37
4 in	0.3333	0.0667	32.0	0.33	5.00

Thermal Properties of Building Materials (*Continued*)

Description	Thickness (ft)	Conductivity (Btu/ft²•°F)	Density (lb/ft³)	Specific Heat (Btu/lb•°F)	Resistance (ft²•°F/Btu)
Wood, Hard					
3/4 in	0.0625	0.0916	45.0	0.30	0.68
1 in	0.0833	0.0916	45.0	0.30	0.91
Wood, Shingle					
For Wall	0.0583	0.0667	32.0	0.3	0.87
For Roof					0.94

Thermal Properties of Insulating Materials

Description	Thickness (ft)	Conductivity (Btu/ft²•°F)	Density (lb/ft³)	Specific Heat (Btu/lb•°F)	Resistance (ft²•°F/Btu)
Mineral Wool/Fiberglass					
Batt, 2 1/2 in, R-7	0.1882	0.0250	6.0	0.2	7.53
Batt, 3 1/2 in, R-11	0.2957	0.0250	6.0	0.2	11.83
Batt, 6 in, R-19	0.5108	0.0250	6.0	0.2	20.43
Batt, 8 1/2 in, R-24	0.6969	0.0250	6.0	0.2	27.88
Batt, 9 1/2 in, R-30	0.8065	0.0250	6.0	0.2	32.26
Fill, 3 1/2 in, R-11	0.2917	0.0270	6.0	0.2	10.80
Fill, 5 1/2 in, R-19	0.4583	0.0270	6.3	0.2	16.97
Cellulose					
Fill, 3 1/2 in, R-13	0.2917	0.0225	3.0	0.33	12.96
Fill, 5 1/2 in, R-20	0.4583	0.0225	3.0	0.33	20.37
Preformed Mineral Board					
7/8 in, R-3	0.0729	0.0240	15.0	0.17	3.04
1 in, R-3.5	0.0833	0.0240	15.0	0.17	3.47
2 in, R-6.9	0.1667	0.0240	15.0	0.17	6.95
3 in, R-10.3	0.2500	0.0240	15.0	0.17	10.42
Polystyrene, Expanded					
1/2 in	0.0417	0.0200	1.8	0.29	2.08
3/4 in	0.0625	0.0200	1.8	0.29	3.12
1 in	0.0833	0.0200	1.8	0.29	4.16
1 1/4 in	0.1042	0.0200	1.8	0.29	5.21
2 in	0.1667	0.0200	1.8	0.29	8.33
3 in	0.2500	0.0200	1.8	0.29	12.50
4 in	0.3333	0.0200	1.8	0.29	16.66
Polyurethane, Expanded					
1/2 in	0.0417	0.0133	1.5	0.38	3.14
3/4 in	0.0625	0.0133	1.5	0.38	4.67
1 in	0.0833	0.0133	1.5	0.38	6.26
1 1/4 in	0.1042	0.0133	1.5	0.38	7.83
2 in	0.1667	0.0133	1.5	0.38	12.53
3 in	0.2500	0.0133	1.5	0.38	18.80
4 in	0.3333	0.0133	1.5	0.38	25.06
Urea Formaldehyde					
3 1/2 in, R-19	0.2910	0.0200	0.7	0.3	14.55
5 1/2 in, R-30	0.4580	0.0200	0.7	0.3	22.90

Thermal Properties of Insulating Materials (*Continued*)

Description	Thickness (ft)	Thermal Properties			
		Conductivity (Btu/ft²•°F)	Density (lb/ft³)	Specific Heat (Btu/lb•°F)	Resistance (ft²•°F/Btu)
Insulation Board					
Sheathing, ¹/₂ in	0.0417	0.0316	18.0	0.31	1.32
Sheathing, ³/₄ in	0.0625	0.0316	18.0	0.31	1.98
Shingle Dacker, ³/₈ in	0.0313	0.0331	18.0	0.31	0.95
Nail Base Sheathing, ¹/₂ in	0.0417	0.0366	25.0	0.31	1.14
Roof Insulation, Preformed					
¹/₂ in	0.0417	0.0300	16.0	0.2	1.39
1 in	0.0833	0.0300	16.0	0.2	2.78
1¹/₂ in	0.1250	0.0300	16.0	0.2	4.17
2 in	0.1667	0.0300	16.0	0.2	5.56
2¹/₂ in	0.2083	0.0300	16.0	0.2	6.94
3 in	0.2500	0.0300	16.0	0.2	8.33

Thermal Properties of Air Films and Air Spaces

Description	Thickness (ft)	Thermal Properties			
		Conductivity (Btu/ft²•°F)	Density (lb/ft³)	Specific Heat (Btu/lb•°F)	Resistance (ft²•°F/Btu)
Air Layer, ³/₄ in or less					
Vertical Walls					0.90
Slope 45°					0.84
Horizontal Roofs					0.82
Air Layer, ³/₄ in to 4 in					
Vertical Walls					0.89
Slope 45°					0.87
Horizontal Roofs					0.87
Air Layer, 4 in or more					
Vertical Walls					0.92
Slope 45°					0.89
Horizontal Roofs					0.92

Appendix C

TABLE C.1 Properties of Saturated Steam Above Atmospheric Pressure

Gage Pressure (psig)	Saturated Temperature (°F)	Specific Volume (ft³/lb)	Heat Content (Btu/lb) Sat. Liq.	Heat Content (Btu/lb) Sat. Vap.	Latent Heat of Vaporization (Btu/lb)
0	212	26.8	180	1150	970
1	215	24.3	183	1151	967
2	218	23.0	186	1153	965
3	222	21.8	190	1154	963
4	224	20.7	193	1155	961
5	227	19.8	195	1156	959
6	230	18.9	198	1157	958
7	232	18.1	200	1158	956
8	235	17.4	203	1158	955
9	237	16.7	205	1159	953
10	239	16.1	208	1160	952
12	244	15.0	212	1161	949
15	250	13.6	218	1164	945
18	255	12.5	224	1165	941
20	259	11.1	227	1166	939
25	267	10.4	236	1169	933
30	274	9.4	243	1171	926
40	287	7.7	256	1175	919
50	298	6.6	267	1178	911
60	307	5.8	277	1181	903
75	320	4.9	290	1184	893
100	338	3.9	308	1190	882
125	353	3.2	324	1193	867
150	366	2.8	337	1196	858
175	378	2.4	350	1198	848
200	388	2.1	361	1199	838

TABLE C.2   Approximate Cooling Load Requirements (ft²/ton)

Application	High	Low	Application	High	Low
Apartment			Hotels, motels	250	300
High rise	350	450	Library, museum	200	340
Efficiency	350	450	Office buildings		
Auditoriums, churches	6	15	Large (interior)	300	350
Banks	200	250	Large (exterior)	190	275
Cocktail lounge, bar	90	200	Small	300	375
Computer rooms	50	150	Residences		
Educational			Large	380	600
Classrooms	225	275	Medium, small	400	700
Colleges, universities	150	240	Resturants		
Dormitories	220	350	Large	80	135
Factories			Medium, small	100	150
Assembly areas	90	300	Retail		
Light manufacturing	100	200	Department store	200	400
Heavy manufacturing	60	100	Specialty shops	175	225
Hospitals			Malls	160	375
Patient rooms	165	275			
Public areas	110	175			
Medical centers	250	300			

TABLE C.3   General HVAC Design Criteria for Buildings

Catagory	Inside Design Conditions[1]		Air Movement (ft/min)	Circulation (Air chg/hr)	Noise (NC)	Filter Efficiency
	Winter	Summer				
Dining/Entertain						
Cafeterias	72°/30%	78°/50%	50	12–15	40–50	35%
Resturants	72°/30%	76°/50%	25–30	8–12	35–40	35%
Bar/Nightclub	72°/30%	76°/60%	25–30	20–30	35–45	35%[2]
Kitchens	72°	85°–88°	30–50	12–15	40–50	15%[2]
Office Buildings	72°/30%	76°/50%	25–45	4–10	30–45	35%–60%
Libraries	70°/50%	72°/50%	25	8–12	35–40	35%–60%
Comunication Cntrs						
Tele Termnl Rm	75°/50%	75°/50%	25–30	8–20	up to 60	85%
Radio/TV Studio	76°/35%	76°/50%	25	15–40	15–25	35%
Transport Cntrs						
Airport Term	72°/30%	76°/55%	25–30	8–12	35–50	35%[3]
Bus Terminals	72°/25%	76°/55%	25–30	8–12	35–50	35%
Ship Docks	72°/25%	76°/55%	25–30	8–12	35–50	15%
Garages[4]	40°–55°	80°–100°	30–75	4–6	35–50	15%
Warehouses	varies[5]		varies	1–4	up to 75	10%–35%

Notes:

[1]°F and % relative humidity

[2]Use charcoal for odor control with manual purge control for 100% outside air

[3]With charcoal filters

[4]Also includes service stations

[5]Usually depends upon the material being stored

TABLE C.4   Approximate Heat Gains from Lights (Btu/hr)

Application	Fluorescent			Incandescent		
	Low	Ave.	High	Low	Ave.	High
Armories	5.3	6.0	9.0	12.0	14.0	19.0
Banks	5.0	7.0	10.0	12.0	15.0	20.0
Barber/Beauty Shops	12.0	16.5	24.5	20.0	24.0	30.0
Court Rooms	4.8	6.6	9.8	10.0	14.0	22.0
Dance Halls	1.7	3.7	8.0	3.5	7.0	12.0
Art Studios/Drafting Rooms	10.0	15.0	20.0	25.0	40.0	50.0
Hospitals/Clinics						
General	8.0	12.0	18.0	25.0	35.0	55.0
Medical Offices/Exam Rm.	6.8	11.0	18.0	14.0	20.0	35.0
Operating Rooms						
Major Operations	50.0	100.0	150.0	100.0	200.0	300.0
Minor Operations	20.0	40.0	60.0	40.0	80.0	150.0
Hotels	1.7	2.5	4.0	3.5	5.0	8.0
Library	9.0	15.0	25.0	15.0	35.0	50.0
Offices	12.0	17.0	24.0	25.0	35.0	50.0
Resturants/Dining Areas	6.0	6.0	10.0	7.0	10.0	20.0
Schools						
Auditorium/Gym	3.5	5.0	7.0	7.0	13.0	20.0
Classrooms	9.0	12.0	18.0	15.0	21.0	33.0
Retail/Store Interior	10.0	14.0	18.0	20.0	25.0	35.0
Industrial/Manufacturing	5.0	7.0	9.0	12.0	15.0	20.0

TABLE C.5   Acoustical Design Guidelines for HVAC Systems

Space	RC(N) Level[a,b]
Private residences, apartments, condominiums	25–35
Hotels/motels	
Individual rooms or suites	25–35
Meeting/banquet rooms	25–35
Halls, corridors, lobbies	35–45
Service/support areas	35–45
Office buildings	
Executive and private offices	25–35
Conference rooms	25–35
Teleconference rooms	25 (max)
Open office areas	30–40
Lobbies, corridors, public areas	40–45
Hospitals and clinics	
Private rooms	25–35
Wards	30–40
Operating rooms	25–35
Corridors	30–40
Public areas	30–40
Performing arts spaces	
Drama theaters	25 (max)
Concert and recital halls	c
Music teaching studios	25 (max)
Music practice rooms	35 (max)

TABLE C.5 Acoustical Design Guidelines for HVAC Systems (*Continued*)

Space	RC(N) Level[a,b]
Laboratories (with fume hoods)	
Testing/research (minimal speech communication)	45–55
Research (extensive telephone and speech communication)	40–50
Group teaching	35–45
Churches, mosques, synagogues	25–35
With critical music programs	c
Schools	
Classrooms up to 750 ft^2	40 (max)
Classrooms over 750 ft^2	35 (max)
Lecture rooms, more than 50 (unamplified speech)	35 (max)
Libraries	30–40
Courtrooms	
Unamplified speech	25–35
Amplified speech	30–40
Indoor stadiums and gymnasiums	
School and college gymnasiums and natoriums	40–50[d]
Large seating capacity spaces (with amplified speech)	45–55[d]

[a]The values and ranges are based on judgment and experience, not on quantitative evaluations of human reactions. They represent general limits of acceptability for typical building occupancies. Higher or lower values may be appropriate and should be based on careful analysis of economic, space usage and user needs. They are not intended to serve by themselves as a basis for contractual requirements.

[b]When the quality of sound in the space is important, specify criteria in terms of RC(N). If the quality of the sound in the space is of secondary concern, the criteria may be specified in terms of NC levels.

[c]An experienced acoustical consultant should be retained for guidance on acoustically critical spaces (below RC 30) and for all performing arts spaces.

[d]Spectrum levels and sound quality are of lesser importance in these spaces than overall sound levels.

TABLE C.6 Estimates of Service Lives of Various HVAC System Components

Equipment Item	Median Years	Equipment Item	Median Years
Air conditioners		Coil	
Window unit	10	DX, water or steam	20
Residential, single or split syst.	15	Electric	15
Commercial through-the-wall	15	Heat exchangers	
Water-cooled package	15	Shell-and-tube	24
Heat Pumps		Reciprocating Compressors	20
Residential air-to-air	15	Package chillers	
Commercial air-to-air	15	Reciprocating	20
Commercial water-to-air	19	Centrifugal	23
Roof-top air conditioners		Absorption	23
Single-zone	15	Cooling towers	
Multizone	15	Galvanized	20
Boilers, hot water and steam		Wooden	20
Steel water-tube	30	Ceramic	34
Steel fire-tube	25	Air-cooled condensers	20
Cast iron	30	Evaporative condensers	20
Electric	15		

TABLE C.6 Estimates of Service Lives of Various HVAC System Components (*Continued*)

Equipment Item	Median Years	Equipment Item	Median Years
Burners	21	Insulation	
Furnaces		Molded	20
Gas or oil fired	18	Blanket	24
Unit heaters		Pumps	
Gas or electric	13	Base-mounted	20
Hot water or steam	20	Inline	10
Radient heaters		Sump and well	10
Electric	10	Condensate	15
Hot water or steam	25	Reciprocating Engines	20
Air terminals		Steam turbines	30
Diffusers, grilles, registers	27	Electric Motors	18
Induction and fan coil units	20	Motor starters	17
VAV and double duct units	20	Electric transformers	30
Air Washers	17	Controls	
Ductwork	30	Pneumatic	20
Dampers	20	Electric	16
Fans		Electronic	15
Centrifugal	25	Valve Actuators	
Axial	20	Hydraulic	15
Propeller	15	Pneumatic	20
Ventilating, roof-mounted	20	Self-contained	10

TABLE C.7 Typical Design Velocities for HVAC Components

Element	Face Velocity (ft/min)
Louvers	
Intake	400
Exhaust	500
Filters	
Panel Filters	
Viscous impingement	200–800
Dry, low efficiency flat	Duct Velocity
Dry, pleated medium efficiency	Up to 750
HEPA and ULPA	250
Renewable media filters	
Moving type, viscous impingement	500
Moving type, dry media	200
Electronic air cleaners	
Ionizing type	150–350
Heating Coils	
Steam and heating water	500–1000
Electric	
Open wire	Per mfgs recom
Finned tubular	Per mfgs recom
Cooling and dehumidifying coils	400–500

TABLE C.8 Average Degree Days for Cities in the United States

State and Station	Ave. Winter Temp	Yearly Total	State and Station	Ave. Winter Temp	Yearly Total
Alabama			District of Columbia		
Birmingham	52.4	2,551	Washington	45.7	4,930
Huntsville	51.3	3,070	Florida		
Mobile	59.9	1,560	Jacksonville	61.9	1,239
Montgomery	55.4	2,291	Key West	73.1	108
Alaska			Miami	71.1	214
Anchorage	23.0	10,864	Orlando	60.4	766
Fairbanks	6.7	14,279	Georgia		
Juneau	32.1	9,075	Atlanta	51.7	2,961
Nome	13.1	14,171	Macon	56.2	2,136
Arizona			Savanah	57.8	1,819
Flagstaff	35.6	7,152	Hawaii		
Phoenix	58.5	1,765	Honolulu	74.2	0
Tucson	58.1	1,800	Hilo	71.9	0
Winslow	43.0	4,782	Idaho		
Yuma	64.2	974	Boise	39.7	5,809
Arkansas			Lewiston	41.0	5,542
Fort Smith	50.3	3,292	Pocatello	34.8	7,033
Little Rock	50.5	3,219	Illinois		
Texarkana	54.2	2,533	Chicago	38.9	5,882
California			Moline	36.4	6,025
Long Beach	57.8	1,803	Peoria	38.1	6,025
Los Angeles	57.4	2,061	Springfield	40.6	5,429
Oakland	53.5	2,870	Indiana		
Sacramento	53.9	2,502	Evansville	45.0	4,435
San Diego	59.5	1,458	Fort Wayne	37.3	6,205
San Francisco	53.4	3,015	Indianapolis	39.6	5,699
Colorado			South Bend	36.6	6,439
Colorado Springs	37.3	6,423	Iowa		
Denver	37.6	6,283	Burlington	37.6	6,114
Grand Junction	39.3	5,462	Des Moines	35.5	6,588
Pueblo	40.4	5,462	Dubuque	32.7	7,376
Connecticut			Sioux City	34.0	6,951
Bridgeport	39.9	5,617	Kansas		
Hartford	37.3	6,335	Dodge City	42.5	4,986
New Haven	39.0	5,897	Goodland	37.8	6,141
Delaware			Topeka	41.7	5,182
Wilmington	42.5	4,930	Wichita	44.2	4,620

TABLE C.8   Average Degree Days for Cities in the United States (*Continued*)

State and Station	Ave. Winter Temp	Yearly Total	State and Station	Ave. Winter Temp	Yearly Total
Kentucky			Nebraska		
Lexington	43.8	4,683	Lincoln	38.8	5,864
Louisville	44.0	4,660	Norfolk	34.0	6,979
Louisiana			North Platte	35.5	6,684
Baton Rouge	59.8	1,560	Omaha	35.6	6,612
New Orleans	61.0	1,385	Nevada		
Shreveport	56.2	2,184	Ely	33.1	7,733
Maine			Las Vegas	53.5	2,709
Caribou	24.4	9,767	Reno	39.3	6,332
Portland	33.0	7,511	New Hampshire		
Maryland			Concord	33.0	7,383
Baltimore	43.7	4,654	New Jersey		
Frederich	42.0	5,087	Atlantic	43.2	4,812
Massachusetts			Newark	42.8	4,589
Boston	40.0	5,634	Trenton	42.4	4,980
Worcester	34.7	6,969	New Mexico		
Michigan			Albuquerque	45.0	4,348
Detroit	37.2	6,232	Roswell	47.5	3,793
Flint	33.1	7,377	Silver City	48.0	3,705
Grand Rapids	34.9	6,894	New York		
Lansing	34.8	6,909	Albany	34.6	6,875
Minnesota			Buffalo	34.5	7,062
Duluth	23.4	10,000	New York	42.8	4,871
Minneapolis	28.3	8,382	Rochester	35.4	6,748
Rochester	28.8	8,295	Syracuse	35.2	6,756
Mississippi			North Carolina		
Jackson	55.7	2,239	Asheville	46.7	4,042
Meridian	55.4	2,289	Charlotte	50.4	3,191
Vicksburg	56.9	2,014	Raleigh	49.4	3,393
Missouri			Winston-Salem	48.4	3,595
Columbia	42.3	5,046	North Dakota		
Kansas City	43.9	4,711	Bismark	26.6	8,851
St. Louis	43.1	4,900	Devils Lake	22.4	9,901
Springfield	44.5	4,900	Fargo	24.8	9,226
Montana			Ohio		
Billings	34.5	7,049	Cincinnati	45.1	4,410
Glasgow	26.4	8,996	Cleveland	37.2	6,351
Great Falls	32.8	7,750	Dayton	39.8	5,622
Helena	31.1	8,129	Toledo	36.4	6,494

TABLE C.8   Average Degree Days for Cities in the United States (*Continued*)

State and Station	Ave. Winter Temp	Yearly Total	State and Station	Ave. Winter Temp	Yearly Total
Oklahoma			Texas		
Oklahoma City	48.3	3,725	Amarillo	47.0	3,985
Tulsa	47.7	3,860	Dallas	55.3	2,363
Oregon			Galveston	62.2	1,274
Eugene	45.6	4,726	Houston	61.0	1,396
Portland	45.6	4,635	Utah		
Salem	45.4	4,754	Milford	36.5	6,497
Pennsylvania			Salt Lake City	38.4	6,052
Erie	36.8	6,451	Wendover	39.1	5,778
Harrisburg	41.2	5,251	Vermont		
Philadelphia	41.8	5,144	Burlington	29.4	8,269
Pittsburgh	38.4	5,987	Virginia		
Scranton	37.2	6,254	Lynchburg	46.0	4,166
Rhode Island			Norfolk	49.2	3,421
Rock Island	40.1	5,804	Richmond	47.3	3,865
Providence	38.8	5,954	Washington		
South Carolina			Olympia	44.2	5,236
Chaleston	56.4	2,033	Seattle-Tacoma	44.2	5,145
Columbia	54.0	2,484	Spokane	36.5	6,655
Greenville-			West Virginia		
Spartanberg	51.6	2,980	Charleston	44.8	4,476
South Dakota			Hunington	45.0	4,446
Huron	28.8	8,223	Wisconsin		
Rapid City	33.4	7,345	Green Bay	30.3	8,029
Sioux Falls	30.6	7,839	La Crosse	31.5	7,589
Tennessee			Madison	30.9	7,863
Chattanooga	50.3	3,254	Milwaukee	32.6	7,635
Knoxville	49.2	3,494	Wyoming		
Memphis	50.5	3,232	Casper	33.4	7,410
Nashville	48.9	3,578	Cheyenne	34.2	7,381
			Sheridan	32.5	7,680

TABLE C.9   Average Degree Days for Cities in Canada

Province and Station	Ave. Winter Temp	Yearly Total	Province and Station	Ave. Winter Temp	Yearly Total
Alberta			Nova Scotia		
Banff	—	10,511	Halifax	—	7,361
Calgary	—	9,703	Sydney	—	8,049
Edmonton	—	10,268	Yarmouth	—	7,340
British Columbia			Ontario		
Kamloops	—	6,799	London	—	7,349
Vancouver	—	7,029	North Bay	—	9,219
Victoria	—	5,699	Ottawa	—	8,735
Manitoba			Toronto	—	6,827
Brandon	—	11,036	Prince Edward Ise.		
Curchill	—	16,728	Charlottetown	—	8,164
The Pas	—	12,281	Summerside	—	8,488
Winnipeg	—	10,679	Quebec		
New Brunswick			Montreal	—	8,203
Fredricton	—	8,671	Quebec	—	9,372
Moncton	—	8,727	Saskatchewan		
St. John	—	8,219	Prince Albert	—	11,630
Newfoundland			Regina	—	10,806
Gander	—	9,254	Saskatoon	—	10,870
St. John's	—	8,991	Yukon Territory		
Northwest Terr.			Dawson	—	15,067
Fort Norman	—	16,109	Mayo Landing	—	14,454
Resolution Island	—	16,021			

Bibliography

1. *"ASHRAE Handbook, Fundamentals Volume,"* American Society of Heating, Refrigerating, and Air Conditioning Engineers, Inc., Atlanta, GA, 1989 and 1993.

2. *"ASHRAE Handbook, HVAC Applications Volume,"* American Society of Heating, Refrigerating, and Air Conditioning Engineers, Inc., Atlanta GA, 1991.

3. *"ASHRAE Handbook, HVAC Systems and Equipment Volume,"* American Society of Heating, Refrigerating, and Air Conditioning Engineers, Inc., Atlanta, GA, 1992.

4. *"ASHRAE GRP-158, Cooling and Heating Load Calculation Manual,"* American Society of Heating, Refrigerating, and Air Conditioning Engineers, Inc., Atlanta, GA, 1979

5. Strock, C., Koral, R.L., *"Handbook of Air Conditioning, Heating, and Ventilating,"* Industrial Press, NY, 1965.

6. *"Trane Air Conditioning Manual,"* The Trane Company, La Crosse, WI, 1974.

7. *"Carrier System Design Manual, Part 1, Load Estimating,"* McGraw-Hill Inc., NY, 1972.

8. Sontag, R.E., Van Wylen, G. J., *"Introduction to Thermodynamics, Classical and Statistical,"* John Wiley and Sons, Inc., NY, 1971.

9. Albertson, M. L., Barton, J. R., Simons, D. B. *"Fluid Mechanics for Engineers,"* Prentice-Hall, Inc., Englewood Cliffs, NJ, 1960.

10. Holman, J.P., *"Heat Transfer,"* McGraw-Hill, Inc., NY, 1972.

11. Chapman, A.J., *"Heat Transfer,"* Macmillan Publishing Co., Inc., NY, 1974.

12. Griffin, C.W. Jr., *"Energy Conservation in Buildings: Techniques for Economical,"* The Constructions Specification Institute, Washington, DC, 1974.

13. McQuiston, F.C., *"Heating, Ventilating, and Air Conditioning: Analysis and Design,"* John Wiley and Sons, Inc., NY, 1977.

14. *"The BOCA National Mechanical Code,"* Building Officials & Code Administrators International, Inc., Country Club Hills, IL, 1993.

15. Mull, T.E., *"Design Procedure for a Hybrid Passive Energy Residence"* (Masters thesis, University of Missouri, Rolla, MO, 1982).

16. Mull, T.E., *"Introduction to Heating, Ventilating, and Air Conditioning; How to Calculate Heating and Cooling Loads,"* Business News Publishing, Troy, MI, 1995.

17. *National Fire Codes*, National Fire Protection Association, Quincy, MA, 1993.

18. "Uniform Mechanical Code," International Conference of Building Officials, Whitterier, CA, 1991.

19. ASHRAE Standard 62-1989, "Ventilation for Indoor Acceptable Air Quality," Society of Heating, Refrigerating, and Air Conditioning Engineers, Inc., Atlanta, GA, 1989.

20. ASHRAE/IES Standard 90.1-1989, *"Energy Efficient Design of New Buildings, Except Low Rise Residential Buildings,"* American Society of Heating, Refrigerating, and Air Conditioning Engineers, Inc., Atlanta, GA, 1989.

21. *"HVAC Duct Construction Standards, Metal and Flexible,"* Sheet Metal and Air Conditioning Contractors National Association, Inc., Vennia, VA, 1985.

22. Schaffer, Mark E., *"A Practical Guide to Noise and Vibration Control for HVAC Systems,"* American Society of Heating, Refrigerating, and Air Conditioning Engineers, Inc., Atlanta, GA, 1991.

23. Naeim, Farzad, "The Seismic Design Handbook," Van Nostrand Reinhold, NY, 1989.

24. ASHRAE Standard 55-1981, *"Thermal Environmental Conditions for Human Occupancy,"* American Society of Heating, Refrigerating, and Air Conditioning Engineers, Inc., Atlanta, GA, 1981.

25. *"Industrial Ventilation, A Manual of Recommended Practice,"* 20th Edition, American Conference of Governmental Industrial Hygienists, Cincinnati, OH, 1988.

26. *"AMCA Fan Application Manual,"* Air Movement and Control Association, Inc., Arlington Heights, IL, 1990.

27. *"Engineering Design Reference Manual for Supply Air Handling Systems,"* United McGill Corporation, 1990.

28. *"Handbook of Air Conditioning System Design,"* Carrier Air Conditioning Co., McGraw-Hill, Inc., NY, 1965.

29. *"Industrial and Commercial Power Distribution,"* 7th Edition, The Electrification Council, Washington, DC, 1990.

30. Haines, Roger, W., *"Control Systems for Heating, Ventilating and Air Conditioning,"* 3rd Edition, Van Nostrand Reinhold Company, NY, 1983.

Index

ABOUT THE AUTHOR

Thomas E. Mull has over 22 years experience as a mechanical engineer. His extensive experience includes energy management, heating/air conditioning, utilities, plumbing, and fire protection. His work involved a wide range of facilities including medical, institutional, commercial, industrial, educational, retail, and governmental. His areas of specialized expertise are energy management, indoor environmental control, thermodynamics, heat transfer, and other thermal sciences. During his work, Tom has traveled internationally and has been involved with projects in Europe and the Pacific Rim. He is a mechanical engineer with a St. Louis, MO based consulting engineering firm.

Tom holds Bachelors and Masters degrees in Mechanical Engineering from the University of Missouri-Rolla. He has performed engineering research into passive solar energy and the interaction of building lighting and heating/air conditioning systems. Tom authored and presented a technical paper entitled "Design Procedure for a Hybrid Passive Energy Residence" at the Gateway Energy Conference. He has also taught engineering related courses through the Continuing Education Department at St. Louis Community College.

Tom is a Registered Mechanical Engineer in several states including Missouri and Illinois. He is also a certified energy analyst for the State of Missouri. Tom is a member of the American Society of Mechanical Engineers and the American Society of Heating, Refrigerating, and Air Conditioning Engineers.

Tom's Internet E-mail address is mullt@asme.org.